Atlas Obscura

AN
EXPLORER'S
GUIDE TO
THE WORLD'S
HIDDEN
WONDERS

JOSHUA FOER, DYLAN THURAS
& ELLA MORTON

WORKMAN PUBLISHING · NEW YORK

An Important Note to Readers

Though the publisher and authors have taken reasonable steps to ensure the accuracy and timeliness of the information contained in the book, readers are strongly encouraged to confirm details before making any travel plans. Location and direction information may change; GPS coordinates are approximations and should be treated as such. If you discover any out-of-date or incorrect information in the book, please let us know via book@atlasobscura.com.

Atlas Obscura was written in the spirit of adventure, and readers are cautioned to travel at their own risk and to obey all local laws. Some of the places described in this book are not open to the public and are not meant to be visited without appropriate permissions. Neither the authors nor the publisher shall be liable or responsible for any loss, injury, or damage allegedly arising from any information or suggestions contained in this book.

....

Art direction and design by Janet Vicario
Photo research by Bobby Walsh, Melissa Lucier, Sophia Rieth, and Aaron Clendening
Additional illustrations by Jen Keenan

Workman books are available at special discounts when purchased in bulk for
premiums and sales promotions as well as for fund-raising or educational use.
Special editions or book excerpts can also be created to specification.
For details, contact the Special Sales Director at the address below, or
send an email to specialmarkets@workman.com.

Workman Publishing Co., Inc.
225 Varick Street
New York, NY 10014-4381
workman.com

WORKMAN is a registered trademark of Workman Publishing Co., Inc.

Printed in South Korea
First printing September 2019

10 9 8 7 6 5 4 3 2 1

The beginning of our happiness lies in the understanding that life without wonder is not worth living.

—Abraham Joshua Heschel

CONTENTS

More than 350 feet below the surface, rowboats bob on a lake inside Romania's Turda Salt Mine. Now an underground amusement park, the mine produced table salt continuously from the 11th century until the 1930s. Page 88.

INTRODUCTION

When we launched Atlas Obscura in 2009, our goal was to create a catalog of all the places, people, and things that inspire our sense of wonder. One of us had recently spent two months driving all over the United States searching out tiny museums and eccentric outsider art projects. The other was about to set off for a year of travels in Eastern Europe. We wanted a way of finding the curious, out-of-the-way places that don't often make it into traditional guidebooks—the kinds of destinations that expand our sense of what is possible, but which we would never be able to find without a tip from someone in the know. Over the years, thousands of people from all over the world have joined us in this collaborative project by contributing entries to the Atlas. This book represents just a tiny fraction of what our community has unearthed. Every one of you out there who added a place to the Atlas, made an edit, or sent in a photo: You are all our coauthors. Thank you.

This revised second edition has been updated to include over 100 incredible new places that members of our community have shared with us since the first edition of this book was published in 2016. We've also added a foldout map depicting our idea of the world's most amazing (and longest) road trip. Though Atlas Obscura may have the trappings of a travel guide, it is in truth something else. The site, and this book, are a kind of wunderkammer of places, a cabinet of curiosities that is meant to inspire wonderlust as much as wanderlust. In fact, many of the places in this book are in no way "tourist sites" and should not be treated as such. Others are so out of the way, so treacherously situated, or (in at least one case) so deep beneath the surface, that few readers will ever be able to visit them. But here they are, sharing this marvelously strange planet with us.

This book would never exist without the incomparable and indefatigable Ella Morton, or our ace project manager, Marc Haeringer, who guided this book from beginning to end. Though we have tried to check the accuracy of every fact in these pages, please don't book any plane tickets without first doing your own independent research. Or do! Just be ready for an adventure.

We often ask ourselves just how large a truly comprehensive compendium of the world's wonders and curiosities could ultimately be. The economics of printing and the dimensions of the page set limits on what could be included in this book. But even our website, which faces no such constraints, can never be complete. There is an Atlas Obscura yet to be written that is as comprehensive as the world itself, for wonder can be found wherever we are open to searching for it.

Joshua Foer and Dylan Thuras
cofounders of Atlas Obscura

Europe

Great Britain and Ireland
ENGLAND · IRELAND · NORTHERN IRELAND
SCOTLAND

Western Europe
AUSTRIA · BELGIUM · FRANCE · GERMANY · GREECE
CYPRUS · ITALY · NETHERLANDS · PORTUGAL
SPAIN · SWITZERLAND

Eastern Europe
BULGARIA · CROATIA · CZECH REPUBLIC
ESTONIA · HUNGARY · LATVIA · LITHUANIA · MACEDONIA
POLAND · ROMANIA · RUSSIA · SERBIA · SLOVAKIA · UKRAINE

Scandinavia
DENMARK · FINLAND · ICELAND
NORWAY · SWEDEN

ATLANTIC
OCEAN

SCOTLAND

NORTH
SEA

Fingal's Cave ●

Giant's
Causeway ●
● Vanishing Lake

The Poison
Garden ●

Garden of Cosmic
Speculation ●

NORTHERN
IRELAND

The Silver Swan ●

IRISH
SEA

The Crypts at
Christ Church
DUBLIN ★

MV *Plassey* ●
Leviathan of
Parsonstown ●
● Victor's Way

IRELAND

ENGLAND

Skellig Michael ●
Irish Sky Garden ●

The Chained Books of
Hereford Cathedral ●

WALES

LONDON ★

River Thames
Witley Underwater
Ballroom ●

Maunsell
Sea Forts ●

GREAT BRITAIN AND IRELAND

Mechanical Clock
at Salisbury
Cathedral ●

Sound
Mirrors ●

The Tempest
Prognosticator ●

No Man's Land
Luxury Sea Fort

CELTIC SEA

ENGLISH CHANNEL

ENGLAND

THE SILVER SWAN

NEWGATE, DURHAM

This uncannily lifelike musical automaton mimics a full-size swan floating on a pond of spun-glass rods. Created in the 1770s, it uses three clockwork mechanisms to perform a 40-second routine set to calming bell-like music. When wound, the swan moves its neck from side to side to preen its feathers before dipping its beak into the pond and snatching up a tiny fish.

First displayed in British jeweler James Cox's Mechanical Museum, the swan was purchased by collector John Bowes in 1872 and is now housed in the Bowes Museum—a French chateau in the north of England.

Bowes Museum, Newgate. A museum curator demonstrates the swan at 2 p.m. daily. The Bowes Museum is 17 miles (27.4 km) from Darlington railway station, which is a 2.5-hour train trip from London. Buses run from the station to the museum.
Ⓝ 54.542142 Ⓦ 1.915462 ➥

⇥ Walking, Eating, Moving Machines

Automatons—mechanical figures that move in an eerily lifelike manner—have existed for centuries, but their heyday was during the 18th and 19th centuries.

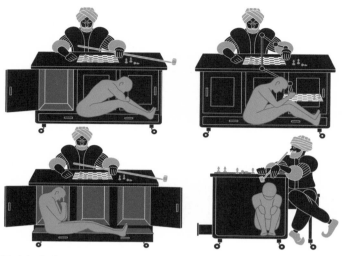

The Turk had a human microcontroller.

The Turk, built in 1770, was one of the most impressive: It consisted of a mechanical man in a turban who played chess against anyone willing to take him on. The machine toured the world, battling opponents like Napoleon Bonaparte and Benjamin Franklin. During the early 19th century, the Turk's apparent intelligence and skill dazzled audiences and frustrated skeptics, who suspected a trick.

In the end, eagle-eyed observers, including Edgar Allan Poe, who encountered the Turk in Virginia in 1835, discovered the secret: a hidden human. The cabinet beneath the chess board held a squashed chess master who made every move by candlelight, pulling levers to operate the Turk's arm and keeping track of the moves on his or her own board. The Turk was nothing but an elaborate hoax.

While the Turk was frustrating its opponents, genuine automatons delighted onlookers with their realistic movements. The Digesting Duck, the 1739 creation of Jacques de Vaucanson, flapped its wings, moved its head, ate grains, and shortly afterward defecated. The digestion process was not authentic— the duck's backside housed a reservoir of droppings that would fall in response to the amount of grains being "eaten"—but it was the first step toward what de Vaucanson hoped would eventually be a genuine eating machine.

Pierre Jaquet-Droz and his two sons spent six years starting in 1768 crafting The Musician, The Draftsman, and The Writer, a trio of dolls now housed in Switzerland's Museum of Art and History. The female musician plays an organ, her chest rising and falling to mimic breathing and her body moving in the manner of an impassioned pianist. The draftsman and writer are dressed identically in lacy shirts, gold satin breeches, and red velvet robes, and each sits at a desk. While the draftsman draws one of four programmed images, including portraits of Louis XV and a dog, the writer dips a goose feather in ink and can write custom text of up to 40 characters.

Steam Men were all the rage in the late 1800s, beginning with 22-year-old New Jersey resident Zadoc Dederick's 1868 model: a 7-foot-9-inch (236 cm) man in a top hat who pulled a carriage. His bulky torso housed a boiler that generated enough power to propel him forward, one footstep at a time.

Canadian George Moore's 1893 version, unattached to a carriage, measured 6 feet (2 m) tall and resembled a medieval knight. An exhaust pipe emerged from his nostril, making him appear to have steamy breath whenever he walked. His movement was limited by one crucial factor: Since he was attached to a horizontal stabilizing arm, he could only walk in circles.

Tipu's Tiger, a tidy representation of the enmity between the residents of India and their 18th-century British colonizers, is an Indian-made, crank-operated toy located in the Victoria & Albert Museum. It depicts a tiger mauling a helpless British officer. Turning the handle makes the man's left hand rise weakly in an attempt to shield his face from the attack. As the hand moves up and down, air rushes through two pairs of bellows. The resulting sounds—beastly growls and the cries of a man in his death throes— leave no ambiguity as to who wins the tussle.

Tipu's Tiger, immortalized mid-meal.

THE POISON GARDEN

ALNWICK, NORTHUMBERLAND

To enter the poison garden of Alnwick, you must first fetch a guide to unlock the black iron gates, which are decorated with a white skull and crossbones and a worrying message: *"These plants can kill."*

Inspired by the poison gardens in 16th-century Padua where the Medicis plotted the frothing ends of their royal enemies, the Duchess of Northumberland created this garden in 2005, dedicating it entirely to poisonous or narcotic flora.

The duchess, Jane Percy, is an unlikely patron. In 1995, her husband unexpectedly became the twelfth Duke of Northumberland following his brother's death, and Alnwick Castle fell into their family's care. Roaming the elaborate gardens, the newly minted duchess decided to transform an overgrown, neglected section into something that was at once both traditional and dangerous. The poison garden now sits nestled among 14 acres of greenery dotted with water sculptures, a cherry orchard, a bamboo labyrinth, and an enormous tree house.

This carefully curated garden contains about 100 plants that have the power to stimulate, intoxicate, sicken, or kill. Guides detail their dangerous properties while enforcing the strict "No touching; no smelling" rules. Poppies, cannabis, hallucinogenic mushrooms, and deadly strychnine are among the innocent-looking greenery. Because of the danger posed by the flora (some can kill or sicken just through touch), some plants are caged, and the garden is secured under a 24-hour security watch.

Denwick Lane, Alnwick. The garden is open from March to October.
Ⓝ 55.414098 Ⓦ 1.700515

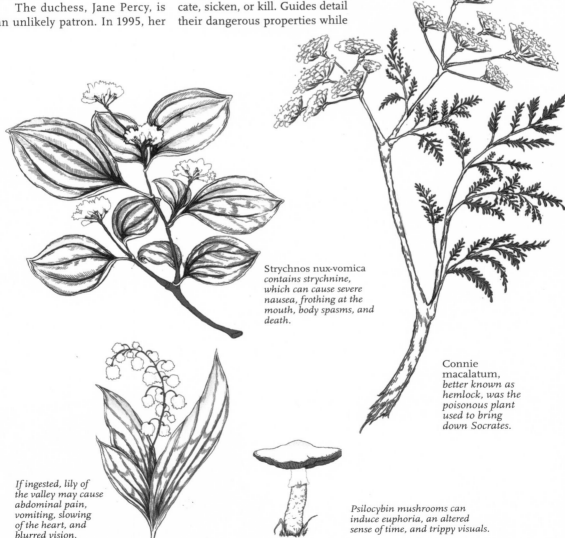

Strychnos nux-vomica *contains strychnine, which can cause severe nausea, frothing at the mouth, body spasms, and death.*

Connie macalatum, *better known as hemlock, was the poisonous plant used to bring down Socrates.*

If ingested, lily of *the valley may cause abdominal pain, vomiting, slowing of the heart, and blurred vision.*

Psilocybin mushrooms can *induce euphoria, an altered sense of time, and trippy visuals.*

THE CHAINED BOOKS OF HEREFORD CATHEDRAL

HEREFORD, HEREFORDSHIRE

This cathedral contains two medieval marvels: a chained library of rare books and one of the earliest maps of the world.

In the Middle Ages, before the availability of the printing press, volumes on law and religion were quite rare and valuable. To protect against theft, the books at Hereford Cathedral were chained to desks, pulpits, and study tables.

The chained library was created in 1611 when a collection of hand-transcribed, hand-bound books was moved into the Lady Chapel. Most of the volumes in the collection are acquisitions dating back to the 1100s, although the oldest book in the collection, the *Hereford Gospels*, dates to about the year 800.

The medieval world map stored at Hereford Cathedral depicts three continents: Europe, Asia, and Africa. On the as-yet-unexplored periphery of these lands roam fire-breathing dragons, dog-faced men, people who survive on only the scent of apples, and the Monocoli, a race of mythical beings who take shade under their giant feet when the sun becomes too bright.

The 5 x 4.5-foot map (1.5 x 1.4 m), created around 1300, is part geography, part history, and part religious teaching aid. A lack of confirmed information on Asian and African geography presented no obstacle for the mapmaker, who used hearsay, mythology, and imagination to fill in the gaps—which explains the four-eyed Ethiopians.

5 College Cloisters, Cathedral Close, Hereford. The cathedral is a 3.5-hour train trip from London and a 15-minute walk from Hereford railway station.
Ⓝ 52.053613 Ⓦ 2.714945

ALSO IN NORTHERN ENGLAND

Steetley Magnesite

Hartlepool · This derelict chemical plant on the North Sea is a photogenic industrial ruin.

Beverley Sanctuary Stones

Beverley · A haven for criminals of all stripes, these stones mark a sacred area where the medieval church provided asylum to thieves and brigands.

MECHANICAL CLOCK AT SALISBURY CATHEDRAL

SALISBURY, WILTSHIRE

The mechanical clock at Salisbury Cathedral is old, but just *how* old is the subject of ongoing debate. The exact date is a matter of importance, for if it was built in 1386, as many horologists believe, it is the oldest working clock in the world.

The faceless clock introduced Salisbury to the new concept of standardized hours, which would replace the season-based increments of the sundial era. It chimed hourly, reminding townspeople to attend church services, and provided a reliable structure for each day.

In 1928, following its rediscovery in the cathedral tower, the clock was disassembled and restored. Although it no longer chimes, today the clock functions in much the same way as it did more than 600 years ago, striking away the hours in the north aisle of the nave.

Salisbury Cathedral, 33 The Close, Salisbury. Trains from London (Waterloo) take 90 minutes. The cathedral is a 10-minute walk from Salisbury station.
Ⓝ 51.064933 Ⓦ 1.797677

Salisbury's 600-year-old timepiece may be the oldest working clock in the world.

George Merryweather's carousel of weather-predicting leeches was more charming than accurate.

THE TEMPEST PROGNOSTICATOR

OKEHAMPTON, DEVON

Surgeon George Merryweather had a passion for leeches. According to Merryweather, the creepy worms possessed humanlike instincts, experienced the hollow ache of loneliness, and were capable of forecasting weather. All this gave him an idea for a machine that he believed could transform meteorology.

In 1851, Merryweather unveiled his "tempest prognosticator" at the Great Exhibition in London. Having witnessed the agitation of freshwater leeches during the lead-up to a heavy storm, the doctor concluded he could build a leech-powered weather forecasting device. The contraption resembled a miniature merry-go-round, but in place of the usual ponies were a dozen glass bottles, each containing a single leech. Should a storm approach, the creatures would make their way to the top of the glass, triggering a wire connected to a central bell.

Though certainly novel, Merryweather's invention did not catch on. His vision of the British government deploying tempest prognosticators nationwide remained mere fantasy, but his invention lives on in the form of a reconstructed version prominently displayed in the Barometer World Exhibition museum in Devon. (Another can be found in the Whitby Museum in North Yorkshire.)

Quicksilver Barn, Merton, Okehampton. The museum is open by appointment.
Ⓝ 50.891854 Ⓦ 4.095316

ALSO IN SOUTHWEST ENGLAND

The World's Largest Greenhouse

St Austell · Huge inflated domes at the Eden Project contain artificial biomes with over one million types of plants.

Lost Gardens of Heligan

St Austell · A 400-year-old garden with fantastical sculptures that has been restored to beauty after years of neglect.

The Museum of Witchcraft and Magic

Boscastle · The world's largest collection of occult- and witchcraft-related artifacts includes dried cats.

The House That Moved

Exeter · A 21-ton Tudor home built in the 1500s and moved 230 feet (70 m) down the street on thick iron rails in 1961 to make way for a new road.

The Cheddar Man and Cannibals Museum

Cheddar · A museum about life, death, and cannibalism in prehistoric Britain.

WITLEY UNDERWATER BALLROOM

GODALMING

The 9,000-acre Victorian estate of James Whitaker Wright was extravagant in every way. With the fortune he made from years of financial fraud, Wright built a 32-room mansion set on lavishly landscaped grounds that included three artificial lakes. Hidden underneath one of the lakes is his most spectacular creation of all: the underwater ballroom.

Built just below the surface of the water, this glorious submerged room has a domed, glass-paneled aquarium roof crowned with an epic statue that rises above the lake as if it's floating on the water. Called a "ballroom" due to its round shape and overall grandeur, the room was actually a smoking room where Wright would entertain his lucky guests. It was splendid, and like everything else on Wright's estate, it was ultimately doomed.

Wright built his palatial home with the money he made from swindling investors out of millions of dollars. In 1904, he was caught, convicted of fraud, and sentenced to seven years in prison. But Wright would never serve time. In a court anteroom, he took his own life by swallowing a cyanide pill immediately after the sentencing. After his death, the estate was purchased by Irish shipbuilder William Pirrie (of SS *Titanic* fame). The mansion was destroyed in a fire in 1952, but the now ancient-looking ballroom is still there, algae-covered and rusting away beneath the lake.

Witley Park is a private property, and while permission is occasionally granted to view the ballroom, it isn't open to the public. Ⓝ 51.147834 Ⓦ 0.683197

A statue of Neptune floats atop this subterranean smoking room.

NO MAN'S LAND LUXURY SEA FORT

GOSPORT, HAMPSHIRE

On maps, it registers as a tiny, nameless speck in the Solent strait between mainland England and the Isle of Wight, but No Man's Land Fort has a dramatic history that belies its cartographic insignificance.

Built in the late 1800s to protect the English coast against a French invasion that never happened, No Man's Land could accommodate 80 soldiers and 49 cannons.

The 200-foot-wide (61 m) fort sat idle for decades, and the Ministry of Defense decommissioned No Man's Land in the 1950s. When the government tried to sell it in 1963, no buyers came forward. In the 1990s, the abandoned fort was transformed into a luxury hotel, complete with two helipads, 21 bedrooms, a roof garden, and restaurants. The submerged center was glassed in as an atrium for the heated pool. Despite its creature comforts and promise of privacy, the hotel never took off.

In 2004, developer Harmesh Pooni bought No Man's Land for £6 million (about $9 million at the time) with the intention of renting it out for special occasions. Unfortunately, contaminated water in the hotel pool caused an outbreak of Legionnaires' disease. Faced with financial ruin and the possibility of losing the island, Pooni took an extreme approach: He covered the helipads with upturned tables, grabbed his keys, and barricaded himself inside the fortress. After a protracted standoff, he was finally evicted in early 2009.

No Man's Land sold for the bargain price of £910,000 (around $1.36 million) in March 2009 to Gibraltar-based Swanmore Estates Ltd. The fort has since been transformed into a venue for weddings and corporate retreats. Standout features include a sauna, a cabaret club, and a laser tag arena located in the former gunpowder storage area.

The Solent is 1.4 miles (2.3 km) north of the Isle of Wight. Ⓝ 50.739546 Ⓦ 1.094995

Now available for laser-themed birthdays.

Before radar, huge concrete "ears" were built to listen for incoming enemy aircraft.

GREATSTONE SOUND MIRRORS

GREATSTONE, KENT

As part of its national defense strategy after World War I, Britain built three massive concrete acoustic mirrors on the southeast coast of England to detect the sound of distant airplane engines in the sky. Working much like giant ears, the trio of reflectors could provide a 15-minute warning of an air invasion by magnifying the sound waves over the English Channel and directing them at microphones. An operator sitting in a nearby booth listened to the transmitted signal through headphones that resembled a stethoscope.

The Greatstone site features three different reflectors, including a 200-foot-long (61 m) curved wall, a 30-foot-tall (9 m) parabolic dish, and a 20-foot-tall (6 m) shallow dish.

Dungeness National Nature Reserve, off Dungeness Road, Romney Marsh. Ⓝ 50.956111 Ⓔ 0.953889

ALSO IN SOUTHEAST ENGLAND

The Little Chapel

Guernsey · One of the smallest chapels in the world, intricately decorated with stones, pebbles, broken china, and glass.

The Margate Shell Grotto

Margate · Discovered in 1835, this mysterious subterranean passageway of unknown age is covered with mystical designs made entirely from seashells.

MAUNSELL ARMY SEA FORTS

THAMES ESTUARY, OFF THE EAST COAST OF ENGLAND

Rising from the water like robotic sentinels on stilts, the Maunsell Army Sea Forts in the Thames Estuary east of London are rusting reminders of World War II's darkest days. Part of the Thames Estuary defense network, the anti-aircraft tower-forts were constructed in 1942 to deter German air raids. Each of the three original forts consisted of a cluster of seven buildings on stilts surrounding a central command tower. Two remain: the Red Sands Fort and the Shivering Sands Fort.

After their wartime career the forts were decommissioned. In the 1960s, pirate-radio broadcasters moved in and established unauthorized stations in the remaining forts. In 1966, Reginald Calvert, manager of the Radio City pirate station, died in a fight with rival Radio Caroline station owner Oliver Smedley. The next year, the British government passed legislation making offshore broadcasting illegal, driving out the pirates and leaving the forts abandoned.

Attempting to enter the decaying forts is not advised. They can be seen by boat or, on a clear day, from Shoeburyness East Beach. Ⓝ 51.361047 Ⓔ 1.024256

An overhead plan of the forts, which were once connected by bridges.

Attention, H. G. Wells:
Your rusty invaders have arrived.

A millennium-long piece of music composed for Tibetan singing bowls will be playing at Trinity Buoy Wharf through 2999.

LONGPLAYER

LONDON

If you miss hearing Longplayer on your next trip to London, you'll get the chance to catch it again—the musical composition will be playing in the old lighthouse at Trinity Buoy Wharf for the next 1,000 years. Longplayer consists of six short recorded pieces written for Tibetan singing bowls that are transposed and combined in such a way that the variations will never repeat during the song's millennium-long run.

It began playing on December 31, 1999, and is scheduled to end in the dying seconds of 2999.

Custodians of the project have established the Longplayer Trust to devise ways of keeping the music alive in the face of the inevitable technological and social changes that will occur over the next ten centuries.

64 Orchard Place, London. Open on weekends. The nearest Tube stop is Canning Town. You can also listen to a livestream of the composition on longplayer.org. Ⓝ 51.508514 Ⓔ 0.008079

ALSO IN LONDON

Clapham North Deep-Level Air Raid Shelter

London · An abandoned World War II bomb shelter, it is the only one of eight deep-level air raid shelters that sits unused.

The Lost River Fleet

London · The largest of London's subterranean rivers now flows through its sewers. The Fleet can be heard flowing through a grate in front of the Coach and Horse pub on Ray Street, Clerkenwell.

The Churchill War Rooms

Westminster · These perfectly preserved underground rooms are where Winston Churchill and his cabinet toiled as they waged war against Hitler's Germany.

Lilliputian Police Station

Trafalgar Square · Now a glorified broom closet, in its heyday London's smallest police station was staffed with eagle-eyed bobbies to keep rowdy Trafalgar Square protesters under control.

The Old Operating Theatre Museum and Herb Garret

Southwark · Sharing the attic of St. Thomas Church with some dusty cobwebs and specimens that seem to be molting, Europe's oldest operating theater looks a lot like it probably did in 1822, minus the screaming, bloody patients.

The Hardy Tree

Camden Town • Inside the walls of Saint Pancras Churchyard is an ash tree with scores of gravestones encircling the trunk. The scalloped rows were formed by novelist Thomas Hardy when he was just a moody young architecture student enlisted to care for the churchyard.

Masonic Lodge of the Andaz Hotel

East London • Forgotten behind a wall for decades, this sumptuous Masonic lodge was rediscovered during renovations, and looks like something out of an Agatha Christie novel.

The Last Tuesday Society's Viktor Wynd Museum of Curiosities, Fine Art & Natural History

Dalston • This reinterpretation of a 17th-century *Wunderkabinett* is a museum, cafe, and art gallery—one part retro-Victorian curiosity shop, two parts dusty horror show, with a jigger or two waiting for you at the bar upstairs.

Wellcome Collection & Library

Bloomsbury • The medical curios of an American pharmacist, collector, and philanthropist run the gamut from Napoleon's toothbrush to a re-creation of a 16th-century barber-surgeon workshop.

Grant Museum of Zoology and Comparative Anatomy

Bloomsbury • With more than 60,000 creatures, including a few that are long extinct and a micrarium of teeny-tiny ones, this is the last of London's zoological museums still open to the public.

Lullaby Factory

Bloomsbury • Wedged between two buildings at the Great Ormond Street Hospital is a secret lullaby-music machine, a network of instruments made of pipes, horns, and pieces of an old hospital boiler.

Novelty Automation

Holborn • These arcade games and satiric automata cover a range of experiences such as "How to practice money laundering," "Buy a house before you get too old," and "Operate a nuclear reactor."

Sir John Soane's Museum

Holborn • A treasure trove of a museum, Soane's is topped with a glass dome and well stacked with tens of thousands of sculptures, paintings, antiquities, and artifacts.

Hunterian Museum

Holborn • John Hunter may have been a bit of a mad scientist, but this anatomical museum befits the unconventional doctor, scientist, and collector of all bodily things, including half of mathematician Charles Babbage's brain and the seven-and-a-half-foot skeleton of Charles Byrne, the 18th-century "Irish Giant" (Hunter may have bribed the undertaker to get that one).

Polly at Ye Olde Cheshire Cheese

Holborn • The taxidermied remains of this beloved parrot still swing in a cage at this 17th-century pub, a former hangout for literary types like Dickens, Tennyson, Twain, and Sir Arthur Conan Doyle.

The First Public Drinking Fountain

Holborn • We can thank the Metropolitan Drinking Fountain and Cattle Trough Association for this 1859 public water fountain, London's very first and an instant hit.

Memorial to Self-Sacrifice

City of London • Slipped into a quiet jewel of a park is a poignant wall of tiled memorials to the bravery of policemen, firemen, and

Headstones arranged by Thomas Hardy.

ordinary Londoners who gave their lives to save others.

Whispering Gallery at St. Paul's Cathedral

City of London • The acoustics of St. Paul's Cathedral Dome, designed by the great Christopher Wren, have a hushed and hidden secret: Speak across the span, all 137 feet (42 m) of glorious light and air, and every word is crystal clear, like a secret whispered in your ear.

Temple of Mithras

City of London • Just blocks from the financial heart of the City of London is a reminder of the city of Londinium: the reassembled remnants of a temple to the Roman god Mithras, whose mystery cult has fueled conspiracy theories for centuries.

London's Original and All-Inspiring Coffee House

City of London • The site of London's first coffee house is still serving drinks of one kind or another, as it has for more than 360 years.

Tomb of the Unknown London Girl

City of London • The 1,600-year-old remains of a Roman girl are reinterred and memorialized at the base of the "Gherkin," a modern architectural landmark.

Shackleton's Crow's Nest

City of London • The lookout from Sir Ernest Shackleton's last ship is inside the crypt of All Hallows-by-the-Tower, one of London's oldest churches.

Houseboats of Regent's Canal

Limehouse • A group of artists and entrepreneurs has built a bohemian community of colorful houseboats along the 8 miles from Little Venice to Limehouse.

Hyde Park Pet Cemetery

Hyde Park • Here lie Dolly, Rex, and Pupsey, in a cemetery devoted to more than 300 Victorian-era furry companions, inside one of the city's largest parks.

Seven Noses of Soho

Soho • Originally numbering 35, it's a challenge to find all the surviving schnoz sculptures (no one really has an exact count) scattered around Soho by artist Rick Buckley—his protest against the proliferation of too many nosy "Big Brother" CCTV cameras. On Bateman Street, Meard Street, D'Arblay Street, Great Windmill Street, Shaftesbury Avenue, Endell Street, and Floral Street.

JEREMY BENTHAM'S AUTO ICON

LONDON

Jeremy Bentham has been sitting in a corridor at University College London since 1850.

The moral philosopher, whose advocacy of animal welfare, prison reform, universal suffrage, and gay rights was far ahead of his time, left a will with specific instructions on the treatment of his corpse. He decreed that his mummified head and skeleton be clad in a black suit, seated upright on a chair in a wooden cabinet, under a placard reading "Auto Icon." He also suggested that his corpse could preside over regular meetings of followers of his utilitarian philosophy.

Bentham's plans for his remains became something of an obsession. For 10 years prior to his death, he reportedly carried a pair of glass eyes in his pocket so that embalmers could easily implant them after his death. Unfortunately, when the time came, something went wrong in the preservation process. Bentham's head took on a mottled, hollow-cheeked look, its leathery skin sagging under a pair of intensely blue glass eyes. In order to create a less grotesque display, preservers created a wax bust of Bentham and screwed it onto the skeleton. They placed the real head between Bentham's feet.

There it sat, undisturbed, until 1975, when a group of mischievous students kidnapped it and demanded a £100 ransom be donated to charity. The university made a counteroffer of £10, and the students caved, returning Bentham's head to its rightful place between his legs. After a few more pranks, including one in which the skull was apparently used as a football, university administrators decided to remove the head from public display. It now sits in the Conservation Safe in the Institute of Archaeology and is removed only for special occasions.

On Gower Street, between Grafton Way and University Street, enter the university grounds at Porter's Lodge. Find the ramp entrance to the South Cloisters, Wilkins Building. Jeremy Bentham is just inside. **Ⓝ 51.524686 Ⓦ 0.134025**

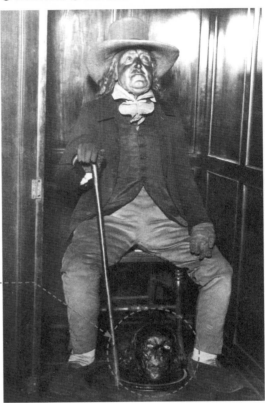

Seated in a hallway at University College London, the long-dead utilitarian philosopher guards his own head from those who might souvenir it.

ARCHIE THE GIANT SQUID

LONDON

The giant squid is often depicted as a sea monster. In Jules Verne's *20,000 Leagues Under the Sea*, a giant squid attacks a boat and devours one of its crew. The kraken of Norse mythology—an enormous creature whose tentacles could supposedly wrap around the tallest of ship masts and rip whole vessels asunder—likely arose from a combination of giant squid sightings, imagination, and exaggeration.

The deep-sea-dwelling giant squid's notorious elusiveness only fueled tall tales. Though records show they have been sighted sporadically since the 16th century, it wasn't until 2002 that photographers were able to capture an image of a live giant squid in its natural habitat—which makes the Natural History Museum's 28-foot (8.5 m) specimen a rare delight. Caught off the coast of the Falkland Islands in 2004 and named Archie in recognition of its species name, *Architeuthis dux*, the giant squid is preserved in a custom-made acrylic tank. **Natural History Museum, Cromwell Road, London. Archie is in the Darwin Spirit Collection of the museum, accessible on special guided tours. Ⓝ 51.495983 Ⓦ 0.176372**

HIGHGATE CEMETERY

LONDON

Opened in 1839, Highgate is one of London's most famous cemeteries. Its residents include Karl Marx (his memorial recognizable by the glowering bearded bust), sci-fi author Douglas Adams, and Adam Worth, the possible inspiration for Sherlock Holmes's nemesis, Professor Moriarty. In the Victorian era, anyone who was anyone wanted to be buried in London's fashionable Highgate Cemetery.

But fashion is fickle. By the 1940s, the Victorian cemetery was in a state of neglect, its once-coveted burial plots covered in vines. In 1970, members of a group interested in the occult claimed to have seen supernatural creatures lurking in the graveyard. Initial reports of ghosts gave way to talk that a vampire—a Transylvanian prince brought to the cemetery in the 1800s—was hiding out in Highgate.

Seán Manchester and David Farrant, self-described magicians and rival monster hunters, vowed to track down and kill the beast. Each proclaimed the other to be a charlatan incapable of finding the vampire. They took their feud to the media—capitalizing on an interest in the occult fueled by the recent release of *The Exorcist*—and announced an official vampire hunt would take place on Friday the 13th of March, 1970. That night, a mob overpowered police and broke into Highgate wielding stakes, garlic, crosses, and holy water. Chaos ensued. No vampires were sighted.

Manchester and Farrant continued to visit the cemetery over the next few years, determined to drive a stake into the heart of the Highgate Vampire. Though neither magician ever found the supposed vampire, real graves were ransacked and real corpses staked and beheaded during the search. In 1974, Farrant received a jail sentence for vandalizing memorials and interfering with human remains in the cemetery.

Debate between Farrant and Manchester continues to this day, while the cemetery remains a popular location for occult, paranormal, and vampire enthusiasts. **Swain's Lane. The cemetery is a 20-minute walk up Highgate Hill and through Waterlow Park.**
Ⓝ 51.566927 Ⓦ 0.147071

Could there be a vampire lurking behind the trees?

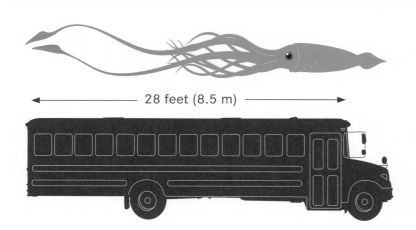

28 feet (8.5 m)

Caught in 2004 and now kept at the Natural History Museum, Archie the Giant Squid is as long as a school bus.

DIFFERENCE ENGINE #2

LONDON

There was almost a Victorian computer. Charles Babbage came achingly close with his 1822 "Difference Engine," a design for a hulking gadget with cranks and gears capable of generating mathematical tables.

The machine offered a well-thought-out solution to the problem of human error in complex calculations, but it proved too large, complicated, and expensive to construct. Government grants allowed Babbage to hire a machinist, Joseph Clement, but after 10 years and much quarreling over prices, Clement had built just a small portion of the prototype.

Undeterred by practical constraints and workplace unpleasantness, Babbage moved on to plans for Difference Engine #2.

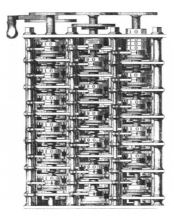

Now functional: Charles Babbage's Victorian computer.

This would be a more streamlined version: 5 tons, 5,000 parts, and 11 feet long (3.4 m). It, too, never progressed to a working model. Babbage died in 1871, leaving reams of notes and sketches for machines that were beyond the era's construction capabilities.

In 1985, more than a century after Babbage's death, London's Science Museum announced plans to investigate the feasibility of his designs by building a difference engine based on the blueprints—and 19th-century materials—devised for Engine #2. The construction team finished the calculating section in 1991, just in time to commemorate the 200th anniversary of Babbage's birth. It functioned flawlessly, confirming Babbage's rightful place in the annals of computing history.

The engine, which features a printing apparatus added in 2002, is now on display at the museum, along with half of Babbage's brain. (The other half is housed at the Hunterian Museum, also in London.)

Science Museum, Exhibition Road, London. Ⓝ 51.498190 Ⓦ 0.173972

IRELAND

VICTOR'S WAY INDIAN SCULPTURE PARK

ROUNDWOOD, WICKLOW

Covering 22 acres, this park includes sculptures of an emaciated Buddha, an enormous disembodied finger, and *The Split Man*, a figure ripping itself in two, representing "the mental state of the dysfunctional human." Victor Langheld established the park in 1989 after traveling to India in search of spiritual enlightenment. The series of sculptures, carved in stone by craftsmen in the south Indian state of Tamil Nadu, represent spiritual progression, from *Awakening* (a child emerging from a decaying fist) to *The Ferryman's End* (a cadaverous old man in a sinking boat, half-submerged in a lake).

Roundwood. The park is a 45-minute drive from Dublin. It's open from May through September. Dress for the damp. Ⓝ 53.085765 Ⓦ 6.219654

The emaciated ferryman invites visitors to contemplate mortality at Victor's Way Sculpture Park.

RUINS OF THE MV *PLASSEY*

INISHEER, GALWAY

More rust than metal at this point, the creaky shell of the steam trawler MV *Plassey* has been sitting on a bed of rocks on the shore of Inisheer Island for over half a century.

Early in the morning of March 8, 1960, the cargo ship was transporting yarn, stained glass, and whiskey across the Atlantic when it got caught in a terrible storm. Fierce winds blew the vessel toward Inisheer, tearing a hole in the bottom and causing water to rush into the engine room.

Using a breeches buoy—a rocket-fired rescue device similar to a zip line—islanders managed to save all 11 members of the crew from the icy sea. No sooner had they warmed up and calmed down with a few shots of local whiskey than another storm hit, delivering the MV *Plassey* to the rocky coast of Inisheer. Locals salvaged wool, lumber, and doors for construction, and made off with a stash of Black & White scotch hidden in the hold.

Today, the bronze-colored wreck, riddled with holes and stripped of all its assets, looks oddly beautiful against the gray rocks, green grass, and blue sky. **The wreck is on the east coast of Inisheer, just south of Killagoola. Ferries from the mainland leave from Doolin. Ⓝ 53.055816 Ⓦ 9.503730**

With its flaking hull full of holes, the MV *Plassey* is slightly less than seaworthy.

LEVIATHAN OF PARSONSTOWN

BIRR, OFFALY

William Parsons, the third earl of Rosse, built this 58-foot (17.7 m) telescope in the 1840s to investigate the space phenomena he knew as "nebulae." At the time, telescopes were not powerful enough to show that these so-called nebulae were actually an assortment of different objects ranging from star clusters and galaxies to clouds of gas and dust. Parsons's 6-foot-wide (2 m) telescope lens exposed the solar system in greater detail than ever before—though it was not until the 1920s that Edwin Hubble discovered that some of the fuzzy objects were in fact galaxies.

Dubbed "the Leviathan of Parsonstown," Parsons's reflecting telescope remained the largest in the world for over 75 years. However, following the death of the earl of Rosse—and that of his son, who tended to the instrument in his absence—the telescope fell into disuse in 1878 and was dismantled in 1908. Thanks to the seventh and current earl, the telescope was reconstructed in the late 1990s with a new mirror and motors. You can now see the restored Leviathan and learn about its workings in the attached science center.

Birr Castle, Birr. Two hours from Dublin by car and about one hour from Shannon Airport and Galway city. Ⓝ 53.097123 Ⓦ 7.913780

The Leviathan may look like a cannon, but it was made for stargazing.

IRISH SKY GARDEN

SKIBBEREEN, IRELAND

Designed by American James Turrell in 1992, this magical knoll is an unparalleled piece of public art. The enormous grassy crater features a central stone plinth reminiscent of ancient Celtic and Egyptian altars. The plinth rises up at an angle, and a stone footrest on each end allows the viewer to lie down and see up and out toward the edge of the crater above. All that is visible from this vantage point is the green grass and sky, creating a uniquely serene experience that invokes past rituals and ancient rites.

The Sky Garden is set in the natural landscape of the Liss Ard ("High Fort" in Gaelic) estate, which is situated on an ancient Celtic ring fort. So even though the sky garden is a contemporary piece of art with a strikingly postmodern shape and aesthetic, it is the perfect complement to a place that is already a bridge between two worlds.

Liss Ard Estate, Castletownshend Road Skibbereen, County Cork. The garden is open on certain days during May and June. You'll need to get a key from reception and pay a small entrance fee.
Ⓝ 51.528405 Ⓦ 9.254247

SKELLIG MICHAEL

THE SKELLIGS, KERRY

The dozen monks who sequestered themselves on this rocky island in the 7th century were a hardy lot.

Skellig Michael—*skellig* is derived from the Irish word *sceillic*, meaning "steep rock"—lies 8 miles (13 km) from the coast of County Kerry. It is beset by wind and rain, which make the ascent to its 714-foot-high (217.6 m) peak extra treacherous.

Despite these harsh conditions, a group of determined Irish Christians established a monastic outpost on the island that remains largely intact 1,400 years later. Using stones, the monks built hundreds of stairs leading up to Skellig Michael's summit, where they erected six beehive-shaped stone huts and a small chapel. They survived on a diet of fish, seabirds, and vegetables grown in the monastery garden. Withstanding multiple Viking raids during the 9th century, the

monks occupied Skellig Michael until the late 12th century, when frequent storms sent them back to the mainland.

Climbing the 670 uneven, steep steps to the top presents both a physical and mental challenge, but at the summit, you'll be able to enter the monastic huts and imagine the grueling life of a 7th-century ascetic.

Boats leave from Portmagee (a 90-minute trip) from April to September, weather permitting.
Ⓝ 51.772080 Ⓦ 10.538858

Beehive-shaped monastic cells, built by medieval monks, top the rocky island of Skellig Michael.

The largest crypt in Ireland is known for its mummified cat and rat, known locally as "Tom & Jerry."

THE CRYPTS AT CHRIST CHURCH CATHEDRAL

DUBLIN

In 1030, the Viking rulers of Dublin built the original Christ Church Cathedral out of wood. When the Normans invaded in 1171, they tore the cathedral down and built one of stone, adding an enormous crypt.

Laurence O'Toole, the archbishop of Dublin, oversaw the rebuilding process. After O'Toole was canonized as the patron saint of Dublin in 1225, his heart was preserved in a heart-shaped reliquary and kept in the cathedral inside an iron cage.

The saintly heart remained in an alcove until March 2012, when two men pried open the cage and stole the icon. Six years later the heart was recovered by police and restored to its proper place.

Other fascinating relics worth seeing are a set of stocks made in 1670 that were once used outdoors to punish criminals publicly and a marble monument depicting Nathaniel Sneyd, an Irish politician who died in 1833. Text beside the sculpture states that he "perished by the indiscriminating violence of an unhappy maniac"—in other words, he was shot.

The most unusual objects in the crypt are the mummified cat and rat, whose poses suggest they died mid-chase. According to church lore, the cat pursued the rat into a pipe of the church organ during the 1850s, and both became stuck. James Joyce used both cat and rat as a simile in *Finnegan's Wake* when he described someone as being "as stuck as that cat to that mouse in that tube of that christchurch organ."

Christchurch Place, Dublin. Ⓝ 53.343517 Ⓦ 6.271057

ALSO IN IRELAND

The Calendar Sundial

Galway · A modern sundial uses ancient methods to tell time and date perfectly.

St. Michan's Mummies

Dublin · Down a set of dimly lit narrow stone steps, a vault underneath the church holds dozens of mummified remains, including an 800-year-old "Crusader" whose finger you are allowed to touch.

Wallabies of Lambay

Lambay Island · Despite being 10,000 miles (16,000 km) from their native Australia, a group of wallabies has made Lambay Island home for the last 25 years after being relocated by the Dublin Zoo.

NORTHERN IRELAND

THE VANISHING LAKE

BALLYCASTLE, ANTRIM

East of the seaside town of Ballycastle, on the side of the coastal road, is a lake—sometimes. When you get there, it may be gone. But it will come back.

Loughareema, also known as "the vanishing lake," sits on a bed of porous limestone with a "plug hole" that attracts peat. When enough peat accumulates in the hole, it prevents drainage, causing the water level to rise. When the peat dislodges, the lake empties—sometimes disappearing in a matter of hours.

Loughareema Road (by Ballypatrick Forest), Ballycastle. The lake is a 2-hour bus ride from Belfast. Be prepared for a dry chalk bed, an expansive lake, or anything in between. Ⓝ 55.157084 Ⓦ 6.108058

The vanishing lake of Ballycastle in its unvanished state.

THE GIANT'S CAUSEWAY

BUSHMILLS, ANTRIM

With its thousands of interlocking hexagonal columns that rise vertically like steps, the Giant's Causeway is a geological oddity that looks distinctly man-made.

Volcanic activity created the unusual formation near the start of the Paleogene period (23–65 million years ago) when molten basalt came into contact with chalk beds and formed a lava plateau. When the lava cooled quickly, the plateau contracted and cracked, forming 40,000 columns of varying heights that look like giant stepping stones. The largest stand almost 36 feet tall (11 m).

According to legend, an Irish giant by the name of Fionn mac Cumhaill constructed the causeway so he could skip over to Scotland to defeat his Scottish counterpart, Benandonner. While in transit to Scotland, Fionn fell asleep, and Benandonner decided to cross the causeway to look for his competitor. To protect her slumbering husband, Fionn's wife gathered him up and wrapped him up in cloth to disguise him as a baby. When Benandonner made it to Northern Ireland, he saw the large infant and could only imagine how big Fionn must be. Frightened, Benandonner fled back to Scotland—but the causeway remained.

44 Causeway Road, Bushmills. The causeway is an hour by car from Belfast or 3 hours by bus, which runs along a scenic route. Ⓝ 55.240807 Ⓦ 6.511555

ALSO IN NORTHERN IRELAND

Skate 56 at the Belfry

Newcastle · A church turned indoor skate park.

Peace Maze

Castlewellan · One of the world's largest hedge mazes celebrates peace in Northern Ireland.

The hexagonal basalt columns of the Giant's Causeway are the stuff of Celtic legend.

SCOTLAND
FINGAL'S CAVE

OBAN, ARGYLL AND BUTE

Like something out of an epic fantasy novel, Scotland's Fingal's Cave is a 270-foot-deep, 72-foot-tall (82 x 22 m) sea cave with walls of perfectly hexagonal columns. Celtic legends held that the cave was once part of a bridge across the sea, built by giants to fight one another. Science says it was formed by enormous masses of lava that cooled so slowly that they broke into long hexagonal pillars, like mud cracking under the hot sun.

When naturalist Sir Joseph Banks rediscovered the cave in 1772, it quickly captured people's imagination and inspired the work of artists, writers, and musicians. Composer Felix Mendelssohn wrote an overture about the cave in 1830, the same year painter J. M. W. Turner depicted it on canvas. Thus was born a Romantic-era tourist site that is just as entrancing today.

Get a train from Glasgow to Oban, where you can take a ferry to Craignure, located on the Isle of Mull. A bus will take you to Fionnphort for a boat tour to Staffa.
Ⓝ 56.433889 Ⓦ 6.336111

The cave's contours have provided artistic inspiration for everyone from Jules Verne to Pink Floyd.

ALSO IN SCOTLAND

The Ruins of St. Peter's

Cardross · This hulking skeleton of a modernist seminary was completed in 1966 and abandoned in the 1980s.

Yester Castle

Gifford, East Lothian · A castle that opens into a subterranean vaulted "goblin hall" from the 13th century.

The Dunmore Pineapple

Dunmore Park · This house with a top shaped like a giant pineapple, a symbol of hospitality and affluence, was built in the late 1700s and is now available as a vacation rental.

Dog Cemetery at Edinburgh Castle

Edinburgh · The final resting place of mascots and Scottish guards' loyal canine companions.

Greyfriars Cemetery Mortsafes

Edinburgh · Cages built over 19th-century burial sites to protect the dead from being disinterred by opportunistic body snatchers.

Holyrood Abbey Ruins

Edinburgh · A ruined 11th-century abbey built by King David I.

Britannia Panopticon Music Hall

Glasgow · The world's oldest surviving music hall.

Cultybraggan Camp

Perth · Built to hold the worst of the worst of Nazi war criminals.

Scotland's Secret Bunker

St. Andrews · A bunker that was built to shelter the politicians and "essential" people of Scotland in the event of nuclear attack.

Within these gardens are the keys to life and the universe.

GARDEN OF COSMIC SPECULATION

HOLYWOOD, DUMFRIES AND GALLOWAY

Among the daffodils and daisies of the Garden of Cosmic Speculation are black holes, Fibonacci sequences, fractals, and DNA double helixes.

Architectural theorist Charles Jencks and his late wife, Maggie Keswick, designed the 30-acre garden for their own property. Its aesthetic is guided by the fundamentals of modern physics, reflecting the shapes and patterns of the unfolding universe. Begun in 1988, the garden took almost 20 years to build, during which time Keswick succumbed to cancer. Jencks continued the project in her memory, occasionally altering designs in response to shifts and breakthroughs in scientific knowledge. (The Human Genome Project inspired the DNA Garden section, with its plant-threaded double helix.)

Holywood, 5 miles (8 km) north of Dumfries. The garden is open to the public one day a year, during the first week of May. Managed by the Scotland's Gardens Scheme, the yearly event helps raise money for Maggie's Centers, a cancer foundation named after Jencks's late wife. Ⓝ 55.129780 Ⓦ 3.665830

DNK

RUS

55°

0°   5°   10°   15°   20°

ENG

AMSTERDAM

Electric Ladyland Museum of Fluorescent Art

Micropia

POL

The Impaled Stork

Eisinga Planetarium

Giethoorn

Teufelsberg Spy Station

★ **BERLIN**

Thousand-Year Rose

Teylers Museum   **AMSTERDAM** ★

THE HAGUE ★ **NETHERLANDS**

The Externsteine

Castle of Wewelsburg

GERMANY

BRUSSELS ★

Tower of Eben-Ezer

Folx-les-Caves

The Mundaneum   **BELGIUM**

CZE

VIENNA

Kugelmugel

Esperanto Museum

50°

ENGLISH CHANNEL

LUX

SVK

Kaspar Hauser

PARIS ★

Hollow at La Meauffe

René de Chalon

Optical Telegraph

Space Travel Museum

★ **VIENNA**

Eisbachwelle

Saint Munditia

Gallery of Beauties

Hellbrunn Palace

World of Ice Giants

AUSTRIA

HUN

The Rat King

Bruno Weber Skulpturenpark

Astronomical Clock

FRANCE

Le Musée des Grenouilles

BERN ★
Child-Eater of Bern

SWITZERLAND

Abbey Library of Saint Gall

SVN

Passetto Di Borgo

45°

Oradour-sur-Glane

Villa de Vecchi

Povéglia

Temples of Damanhur

Wooden Books

ITALY

Vatican City

Pope Leo's Bathroom

Museum of Holy Souls

Le Palais Idéal

ROME

ADRIATIC SEA

Solar Furnace

AND

La Specola

Galileo's Middle Finger

Civita di Bagnoregio

Park of the Monsters

★ **ROME**

SPA

Corsica

Secret Cabinet of Erotica

MEDITERRANEAN SEA

Sardinia

TYRRHENIAN SEA

40°

10°

N

Miles
0   100   200

Kilometers
0   100   200

Capuchin Catacombs

Sicily

IONIAN SEA

5°

15°

WESTERN EUROPE

Greece and Cyprus, *see page 48*
Portugal and Spain, *see page 66*

TUN

DZA

MLT

The interior of this spherical state houses a museum on the history of the micronation.

AUSTRIA

REPUBLIC OF KUGELMUGEL

VIENNA

The Republic of Kugelmugel is one of many "micronations." These independent states, often established as art projects, social experiments, or simply for personal amusement, are not recognized by international governments.

When creative urges compelled artist Edwin Lipburger and his son to build a spherical studio on their farm in 1970, they didn't realize that local laws prohibited the building of spherical structures. In an attempt to protect their creation from demolition, the Lipburgers created their own Kugelmugel township, complete with self-made street signs around the building. Later, as the legal dispute escalated, Edwin Lipburger attempted to declare the structure its own federal state—the Republic of Kugelmugel—even going as far as to issue his own stamps and currency and refusing to pay taxes. The elder Lipburger was eventually sent to jail for 10 weeks in 1979 (he was later pardoned by the Austrian president).

In the early 1980s, Austria's culture minister suggested Kugelmugel be moved to the Vienna Prater amusement park. The Lipburgers went along with the idea, having been promised access to tap water, electricity, and sewers. But these facilities

The micronation of Kugelmugel was founded to protect an artist's space from demolition.

were never provided, which resulted in an ongoing dispute with the city of Vienna.

Edwin Lipburger died in 2015 and his son took over the presidency. The building now serves as an art gallery and is open for special exhibitions and performances. A proud "micronation," it counts about 600 people as citizens, though not all citizens can fit in the space at once.

2 Antifaschismusplatz, Vienna. Get a bus to Venediger Au. Kugelmugel is located in Vienna's famous Prater park, depicted in Carol Reed's classic film *The Third Man*, and is found at the western edge of the park in the shadow of a roller coaster. Ⓝ 48.216234 Ⓔ 16.396221 ➻

➼ Other Micronations

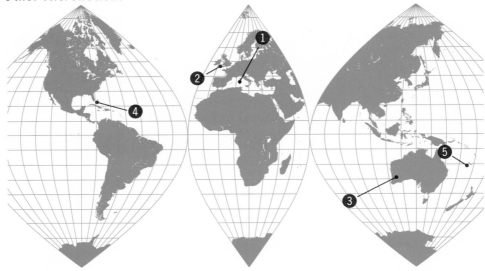

1 NIMIS, LADONIA

Nimis, a mountainous, multi-towered sculpture made of 70 tons of driftwood planks, is the main attraction of the micronation of Ladonia. It is also the only reason Ladonia exists at all. The ersatz country—two-thirds of a square mile (1.7 square km) in size, located on a Swedish peninsula protruding into the Kattegat—was established in 1996 following a long court battle between artist Lars Vilks and the Swedish government. Vilks began building Nimis in secret in 1980. Far from civilization and only fully visible from the water, it went unnoticed by Swedish authorities for two years until they discovered it and declared it would have to be destroyed. (The land is part of a nature reserve, where it is forbidden to build structures.) Goaded, Vilks ignored the announcement, sold Nimis to the artist Cristo, built a similarly sized stone-and-concrete sculpture called *Arx*, and decided to take control of the area and secede from Sweden. The nation of Ladonia was born.

Today, Ladonia claims over 15,000 citizens, all of whom reside outside its borders in accordance with the nation's nomadic lifestyle policy. Citizenship is free and requires an online application, but if you would like to become part of Ladonia's nobility, you will need to pay $12 and send an email informing the administrators of your preferred title.

The Ladonian flag is green, with a faint white outline of the Nordic cross—a design chosen because it is what the blue-and-yellow Swedish flag would look like if it was boiled. Taxes are payable, but money is not accepted. Instead, citizens must contribute some of their creativity. The citizenship application process caused confusion among 3,000 Pakistanis who applied for immigrant status with the intent to live in Ladonia, only to be told it was not possible to move there.

Though residing in Ladonia is prohibited, visiting is not.

The micronation of Ladonia is known for its artisanal wooden towers.

2 THE PRINCIPALITY OF SEALAND

Sealand came into being in 1967, when Roy Bates commandeered a former military sea fort off the east coast of England with plans to broadcast a pirate radio station. When members of the British Royal Marines approached the fort, Roy's 14-year-old son Michael fired warning shots at their boat. The Bates duo appeared in court on firearms charges, where the judge ruled the case invalid due to the army fort being outside British jurisdiction. The father and son returned to Sealand, where Roy dubbed himself prince of the fledgling micronation.

In 1978, Bates the elder and his wife were in England when Alexander Achenbach, self-described prime minister of Sealand, assembled a team to storm the fort by air and sea to attempt a coup. He captured the prince Michael Bates and held him hostage, only for Roy to return later—with armed assistants, in a helicopter piloted by a former James Bond–franchise stuntman—and capture Achenbach and his crew. Roy charged Achenbach with treason, keeping him captive for several weeks until Germany sent a diplomat to Sealand to negotiate Achenbach's release. Achenbach has since established a remote rebel government in exile: a micro-micronation.

3 PRINCIPALITY OF HUTT RIVER

When the Australian government imposed a restrictive wheat crop production quota in late 1969, wheat farmer Leonard Casley was infuriated. With fields full of wheat waiting to be harvested and nowhere to send the sizable yield, he fought to change the government's new agricultural policy. The government was unmoved. Casley then took the next logical step: secession.

The Principality of Hutt River was founded in April 1970, when Casley informed authorities that he and his family would no longer be under Australia's jurisdiction. His one-page letter name-checked the Magna Carta, English common law, the Atlantic Treaty, and the Charter of the United Nations.

During the 1970s, relations between Hutt River and Western Australia were strained. The national postal service refused to deliver to the area, and the Australian Tax Office repeatedly demanded the Casleys pay taxes, prompting the self-proclaimed Prince Leonard I to declare war on Australia in 1977. A few days later, the prince sent official notification that the war was over. There were no deaths, no injuries, and no acknowledgment from Australia that the conflict ever existed.

Conch Republic passports are not valid for international travel.

4 THE CONCH REPUBLIC

"We seceded where others failed" is the motto of this micronation in Key West, Florida, formed in 1982 in response to a new border patrol checkpoint that inconvenienced residents and slowed tourism. Soon after seceding, the Conch Republic engaged in a one-minute war with the United States, during which newly proclaimed Prime Minister Dennis Wardlow thwacked a man in a naval uniform on the head with a stale loaf of bread. The PM then surrendered and immediately sought a billion dollars in foreign aid from the US. He has yet to receive it.

Although Conch Republic passports are not valid for international travel, the micronation's website claims a Conch citizen traveling in Guatemala avoided certain death by showing his Conch passport to armed revolutionaries instead of his American one. Supposedly, the rebels dropped their weapons and treated the visitor to several rounds of tequila.

5 REPUBLIC OF MINERVA

Las Vegas real estate mogul Michael Oliver dreamed of an island utopia: a tax-free, welfare-free, government-intervention-free haven for 30,000 people, supported by fishing, tourism, and other "unspecified activities."

Oliver targeted two unclaimed submerged atolls south of Fiji and Tonga as the location for the new state. International law states that islands can only be claimed if they are a foot (.3 m) above the high tide point. The atolls became submerged at high tide, therefore no one owned them. In 1971, Oliver shipped in barges of sand, dumped them on the reefs, and staked his claim, naming it the Republic of Minerva.

This bold colonization soon caught the attention of Tongan king Taufa'ahau Tupou IV, who set sail for Minerva accompanied by nobles, cabinet ministers, soldiers, police, and a brass band. Upon arrival, the motley crew tore down the Minervan flag and claimed the land in the name of Tonga.

ESPERANTO MUSEUM

VIENNA

When Ludwig Lazarus Zamenhof invented the Esperanto language in the 1870s, his goal was to ease communication between people of different nationalities. Esperanto, a hybrid of Romance, Germanic, and Slavic languages, is documented and studied at this museum, along with around 500 other constructed languages, or "conlangs."

On display is an impressive array of Esperanto objects: sodas and toothpaste containers with Esperanto labels; novels written in Esperanto; language manuals; and 19th-century photographs of "planned-language" pioneers. At its peak, Esperanto had as many as two million speakers.

On the eve of WWI, Esperantists still dreamed of uniting the world behind a common language.

Esperanto ranks among the top 200 most-spoken languages out of over 6,000 left on the planet. There are estimated to be a thousand native Esperanto speakers, who learned the language as children; among them is the American financier George Soros.

If the museum piques your interest in learning Esperanto, take note that becoming conversant in the language entitles you to join the Esperantist passport service, a worldwide directory of people willing to host Esperanto speakers in their homes free of charge.

Herrengasse 9, Vienna. Take the U-Bahn to Herrengasse. ℕ 48.209474 🄴 16.365771 ➡

➡ How to Say "Crazy" in Toki Pona

More than 900 constructed languages have been invented since the 13th century. Some, like Esperanto and Volapük, were created with the ambitious goal of becoming a universal lingua franca. Others are meant to test the contentious Sapir-Whorf hypothesis, which holds that a person's worldview is shaped by the vocabulary and syntax available in his or her language.

Kala

Kasi

Toki Pona, created by Canadian linguist Sonja Elen Kisa in 2001, is a language for minimalists. Consisting of just 123 words, the language is meant to reflect a Zen outlook on life. Toki Pona combines simple words to create complex ones—such as joining the words for "crazy" and "water" to create the word for "alcohol." The language's two main features—a restricted vocabulary and the linking of root words—also occur in Newspeak, the fictional language used in George Orwell's dystopian novel *1984*.

Láadan is American science-fiction writer Suzette Haden Elgin's experimental answer to the feminist hypothesis that existing languages are inadequate for conveying the breadth of female experience. The language, created in 1982, includes words like *radiidin*, defined as "a time [that is] allegedly a holiday but actually so much a burden because of work and preparations that it is a dreaded occasion; especially when there are too many guests and none of them help."

French author and musician François Sudre began working on Solresol in the 1820s. The language is based on the seven syllables that correspond to sounds on a musical scale: *do*, *re*, *mi*, *fa*, *sol*, *la*, and *si*. Every word is composed of one or

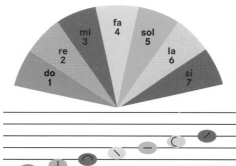

Sentences in Solresol can be performed on a violin, spoken as a series of numbers, or communicated using the colors of the rainbow.

more of those sounds—for example, *si* means "yes" while *dofalado* means "sincerity." There are 2,668 words in all.

Since the base sounds correspond to musical notes, Solresol messages can be communicated through musical instruments. Each of the seven base sounds also corresponds to a color of the rainbow. In his 1902 book on Solresol grammar, Boleslas Gajewski suggested messages be relayed "at night, by shooting rockets of each of the seven colors . . . always separating every syllable as needed, then pausing briefly between every word."

Listen for the archbishop ghost's cackling on the wind as water bursts from his old prank fountains.

TRICK FOUNTAINS OF HELLBRUNN PALACE

SALZBURG

The Prince-Archbishop Markus Sittikus von Hohenems was quite the prankster. When building Hellbrunn Palace, his summer home, the 17th-century Salzburg monarch booby-trapped its gardens with trick fountains that would spray unsuspecting guests as they toured the grounds. At each spot was a patch that remained dry—that is where the prince would stand while his visitors got soaked.

Nearly 400 years later, the palace remains almost completely unchanged, except for the 1750 addition of the mechanical theater, a water-powered diorama of a Baroque town with 200 miniature townspeople moving to organ music.

Fürstenweg 37, Salzburg. Hellbrunn is open from April to early November. Dress appropriately—you will get wet. Ⓝ 47.763132 Ⓔ 13.061121

ALSO IN AUSTRIA

Eggenburg Charnel

Eggenburg · Artfully arranged skeletons sit at the bottom of this 14th-century pit.

Fucking, Austria

Fucking · American soldiers discovered Europe's best one-liner at the end of World War II. Since then, the Fuckingers, as residents of the town are called, have had to deal with hordes of visitors posing in front of their street signs.

House of Artists

Gugging · In this home for the mentally ill, the walls are open canvases for patients to paint their frustrations, fears, and hopes.

Franz Gsellmann's Weltmaschine

Kam · This behemoth of whirring, spinning, turning parts includes toy gondolas, a xylophone, miniature windmills, a spaceship, and an oxygen tank—created by an untrained farmer.

Minimundus

Klagenfurt · A park filled with famous buildings from around the world built on a 1:25 scale.

Kremsmunster Observatory

Kremsmunster · Founded as a monastery in the 8th century and upgraded to a five-story "mathematical tower" in the 1750s, this weather observatory has been operational for over 250 years.

Dom Museum's Kunst und Wunderkammer

Salzburg · This lovingly re-created and restored collection of curiosities once belonged to the Archbishop Wolf Dietrich.

Starkenberger Beer Resort

Starkenberg · Lager lovers can literally immerse themselves in one of seven 13-foot (4 m) pools of warm beer, each containing some 42,000 pints. Cold beer is provided for drinking.

World of the Ice Giants

WERFEN, SALZBURG

At 26 miles (42 km) long, Eisriesenwelt, or "World of the Ice Giants," is the largest ice cave in the world. It is filled with naturally formed sheets of dripping, curving ice, the result of melting snow from the surface flowing into the cave through cracks and crevices and solidifying when it meets the freezing temperatures. Bright magnesium lights illuminate the ice formations, making the scene even more dramatic.

Getreidegasse 21, Werfen. Eisriesenwelt is 25 miles (40.3 km) south of Salzburg. The cave is open to visitors from May to October. Bring warm clothing; the caves are below freezing, even in summer. Ⓝ 47.50778 Ⓔ 13.189722

Handheld lamps illuminate the dripping formations in the world's largest ice cave.

BELGIUM

Tower of Eben-Ezer

BASSENGE, LIÈGE

This seven-story stone tower—topped with a bull, a lion, an eagle, and a sphinx—looks medieval, but harks from a more recent time. Beginning in 1951, Robert Garcet spent over a decade building the Tower of Eben-Ezer by hand, as a totem to peace and the pursuit of knowledge.

A devotee of the Bible, numerology, and ancient civilizations, Garcet used symbolic dimensions when designing the building. The tower is 33 meters tall (to represent Jesus's age when he died) and topped with a quartet of statues (matching the four horsemen of the apocalypse). Each floor measures 12 by 12 meters—12 being the number of Jesus's disciples.

The interior walls are filled with Garcet's artwork, which depicts scenes from the apocalypse, biblical quotes, and Cretaceous dinosaurs. Climb the spiral staircase to the rooftop and you'll emerge from under a winged lion to see panoramic views of the Belgian countryside.

4690 Eben-Emael, Bassenge. A train from Brussels takes about 2 hours. Ⓝ 50.793317 Ⓔ 5.665638

The hand-built turreted Eben-Ezer Tower was inspired by the Bible and ancient civilizations.

THE MUNDANEUM

MONS, HAINAUT

The Mundaneum was, to put it mildly, an ambitious undertaking. Belgian lawyer Paul Otlet and Nobel Peace Prize winner Henri LaFontaine established the project in 1910 with the aim of compiling the entirety of human knowledge on 3 × 5-inch index cards. The collection was to be the centerpiece of a "world city" designed by architect Le Corbusier, forming a nucleus of knowledge that would inspire the world with its libraries, museums, and universities.

To address the daunting task of arranging bits of paper into a coherent compendium of world history, Otlet developed a system called Universal Decimal Classification. Over the next few decades, a growing staff created and catalogued over 12 million cards summarizing the contents of books and periodicals. Having assembled this wealth of knowledge, Otlet began offering a fee-based research service. Queries came in by mail and telegraph from around the globe at the rate of 1,500 per year.

With the paper-based system becoming cumbersome by 1934, Otlet hoped to move on to another system: a mechanical data cache accessible via a global network of what he termed "electric telescopes." To his dismay, the Belgian government had little enthusiasm for the idea. With World War II looming and priorities elsewhere, the Mundaneum moved to a smaller site, eventually ceasing operations after years of financial instability. The final blow came during the Nazi invasion of Belgium, when soldiers destroyed thousands of boxes filled with index cards and hung Third Reich artwork on the walls.

Otlet died in 1944, his Mundaneum and world city mere memories. He is now regarded as one of the forefathers of information science—his vision of a globally searchable network of interlinked documents anticipated the World Wide Web.

The remains of the Mundaneum—books, posters, planning documents, and drawers with original index cards—are now on display at the Musée Mundaneum in Mons.

76, rue de Nimy, Mons. The Mundaneum is a 15-minute walk from the Mons train station.
Ⓝ 50.457674 Ⓔ 3.955428

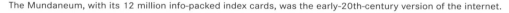

The Mundaneum, with its 12 million info-packed index cards, was the early-20th-century version of the internet.

THE GROTTOES OF FOLX-LES-CAVES

ORP-JAUCHE, WALLOON BRABANT

Down a narrow staircase, 50 feet below the small town of Folx-les-Caves, are nearly 15 acres of human-made caves. Carved from tuff, or compressed volcanic ash, they were probably dug during the Roman era or the Middle Ages. In 1886, farmers began growing mushrooms in the cold, dark grottoes—the fungi grow there to this day.

Prior to the mushroom era, the cave served as a refuge for bandits on the lam. Words and names chiseled into the stone—graffiti from a pre–spray can age—provide evidence of its past inhabitants. One of the most famous was Pierre Colon, an 18th-century thief who robbed passing merchants and shared his spoils with the poor. Colon was sent to jail, but escaped after his wife pulled the now-cliché move of sneaking him a metal file inside a cake. The legend of Colon is celebrated at Folx-les-Caves every year in early October with a festival of music, dance, and food.

35, rue Auguste Baccus, Orp-Jauche. Orp-Jauche is about an hour east of Brussels. Ⓝ 50.669216 Ⓔ 4.941959

ALSO IN BELGIUM

Caves of Remouchamps	Plantin-Moretus Museum of Printing	The Atomium	The Musical Instrument Museum	Dr. Guislain Museum
Aywaille · Take a 90-minute boat ride along the world's longest subterranean river. If you're lucky, you'll spot a translucent shrimp.	*Antwerp* · A 16th-century publishing house–cum-museum features two of the oldest printing presses, a Gutenberg Bible, and the world's only copy of the original Garamond punches and matrices.	*Brussels* · Enjoy the view from the top of an iron crystal structure magnified to 165 billion times its size, built for the 1958 Brussels World's Fair.	*Brussels* · A three-floor wonderland displays over 1,500 instruments of all shapes and sizes. Check out the componium, the first automatic instrument capable of improvisation.	*Ghent* · Housed within an operating mental institution, this museum mixes art and education in its mission to educate the public about psychiatric care.

FRANCE

HOLLOW AT LA MEAUFFE

LA MEAUFFE, NORMANDY

Holloways, which appear like deep trenches dragged into the earth, are centuries-old thoroughfares worn down by the traffic of time. In Europe, most of these sunken lanes go back to Roman times, or as early as the Iron Age.

These deep-recessed roads were naturally tunneled into the soft ground by years of footsteps, cart wheels, and animal hooves. Water flowing through the embankments like a gully further molded the paths into rounded ditches that have sunk as much as 20 feet lower than the land on either side. In some cases, trees rise up from the banks flanking the narrow path and reach toward each other to form a canopy over the road, making the holloway look like a tunnel running through the thick greenery.

Holloways are especially common in the *bocage*, or "hedgerow," landscape around Normandy, where the countryside is divided into small fields enclosed by sunken lanes and high hedges. Like many sunken roads, the trench-like holloway in La Meauffe was used as a shelter during times of war. During World War II, the La Meauffe hollow was a defensive strongpoint for the German army, providing perfect cover from the advancing American troops. The limited visibility of the terrain caused the Americans to suffer heavy losses during the attack, leading US soldiers to call the road in La Meauffe "Death Valley Road."

The holloway in La Meauffe is located off the main road on the southern edge of town. Ⓝ 49.174619 Ⓦ 1.112959

Many who walk through holloways don't realize they are retracing ancient steps.

Oradour-sur-Glane

Oradour-sur-Glane, Limousin

The village of Oradour-sur-Glane has stood in ruins since 1944. Among the scorched, crumbling buildings are the possessions of people who lived here over 70 years ago: burned shells of cars; sewing machines; bed frames; the skeleton of a stroller. All sit quietly, at the mercy of nothing but weather and time.

On June 10, 1944, in response to suspected Resistance activity in the region, the Nazi SS stormed Oradour-sur-Glane and ordered every resident to assemble in the village square. The unit then led the men to barns and sheds, where they had set up machine guns. They locked the women and children in the church, and set the

The burned-out remains of a village left untouched for 70 years.

building on fire. Anyone who tried to escape through a window was met with a hail of gunfire.

Within hours, the Waffen-SS had murdered 642 residents of Oradour-sur-Glane. Satisfied, they left, but not before torching every structure in the village.

When World War II ended, French president Charles de Gaulle declared that while a new Oradour-sur-Glane would be built next to the original, the old one must be kept in ruins as a reminder of the atrocities of war. Apart from the addition of signs, plaques, and a museum, the ghost village is untouched. A sign above the entrance reads *Souviens-Toi*: "Remember."

**Oradour-sur-Glane is a 30-minute drive west of Limoges. On June 10, wreaths are laid in the village's ruined church to mark the anniversary of the attack.
Ⓝ 45.931233 Ⓔ 1.035125**

The Chappe Optical Telegraph

Saverne, Bas-Rhin

Messages needed to travel swiftly across the country during the French Revolution, and Claude Chappe's optical telegraph was just the device for the job.

In 1791, Chappe debuted his chain of stone towers—topped with 10-foot (3 m) poles and 14-foot (4.3 m) pivoting crossbeams, and spaced as far apart as the eye could see—on the Champs-Élysées. He also created a language of 9,999 words, each represented by a different position of the swinging arms. When operated by well-trained optical telegraphers, the system allowed messages to be transmitted up to 150 miles (241.4 km) in two minutes.

The French military saw the value of Chappe's invention and built lines of his towers from Paris to Dunkirk and Strasbourg. Within a decade, a network of optical telegraph lines crisscrossed the

nation. When Napoleon seized power in 1799, he used the optical telegraph to dispatch the message "Paris is quiet and the good citizens are content."

Renovated in 1998, the optical telegraph next to the Rohan Castle in Saverne functioned as part of the Strasbourg line from 1798 until 1852. It is one of several remaining relay points in the system that can still be visited today.

Rohan Castle, place du Général de Gaulle, Saverne. The castle is a 5-minute walk from the Saverne train station. Ⓝ 48.742222 Ⓔ 7.363333

A key to the letter and number positions in the Chappe signaling system.

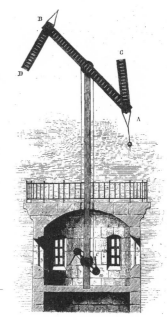

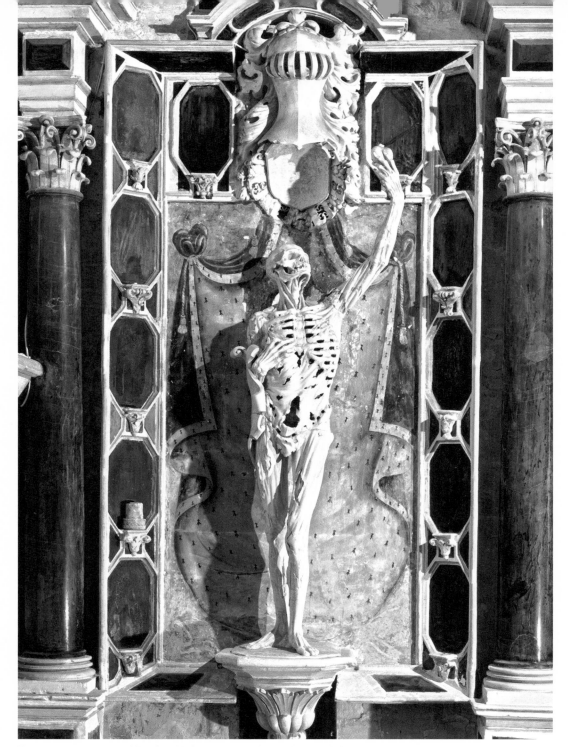

The postmortem statue of René de Chalon once held the man's own dried heart.

THE ROTTING BODY OF RENÉ DE CHALON

BAR-LE-DUC, LORRAINE

Saint-Étienne church, in the city of Bar-le-Duc, is home to a statue of a rotting corpse. Visible musculature and skin hang in flaps over the hollow carcass. The exposed skull looks toward a raised left hand, which once held the dried heart of René de Chalon, the 16th-century prince the statue depicts. (The heart is believed to have gone missing sometime around the French Revolution.)

The life-size sculpture by Ligier Richier is part of the "transi" Renaissance art form—stone sculptures of rotting bodies that served as a reminder of temporary flesh and eternal afterlife.

Saint-Étienne church, place Saint-Pierre, Bar-le-Duc. The train ride from Paris to Bar-le-Duc takes two and a half hours. Ⓝ 48.768206 Ⓔ 5.159390

The many dials of what may be the most complicated horological device ever constructed.

THE ASTRONOMICAL CLOCK OF BESANÇON CATHEDRAL

BESANÇON, FRANCHE-COMTÉ

Besançon Cathedral, located in the center of France's 19th-century clock-making capital, is home to a 19-foot-tall (5.8 m) clock with 30,000 pieces. It is one of the most complicated horological devices ever made. Installed in 1860, the clock shows the local time in 17 places around the world, as well as the time and height of the tides in eight French ports, a perpetual calendar with leap-year cycles, and the times of sunrise and sunset.

Besançon Cathedral, rue du Chapitre, Besançon. The cathedral is a leisurely stroll from the main train station.
Ⓝ 47.237829 Ⓔ 6.024054

ALSO IN FRANCE

The Chapel Oak

Allouville-Bellefosse · A tiny chapel built inside the trunk of the oldest known tree in France.

Mechanical Dragon Clock

Blois · A huge golden mechanical dragon emerges from a villa at regular intervals like a cuckoo clock—a homage to magician and automaton creator Jean Eugène Robert-Houdin.

La Maison Picassiette

Chartres · A home covered top to bottom in mosaics, made by a frustrated grave sweeper.

Douaumont Ossuary

Douaumont · The remains of nearly 130,000 soldiers fill this World War I memorial.

Secret Passages of Mont Sainte-Odile

Ottrott · An ancient fortified monastery contains hidden passages and has a history of mysterious book thefts.

Collégiale de Saint-Bonnet-le-Château

Saint-Bonnet-le-Château · Thirty mummified bodies, believed to be Catholic nobles killed by a Protestant leader in 1562, were rediscovered in 1837, when one of the vaults of this church was opened.

MUSÉE DE LA CHASSE ET DE LA NATURE

PARIS

Housed in the 17th-century Hôtel de Guénégaud, the Museum of Hunting and Nature displays stuffed animals, the decorated vintage weapons that killed them, and artwork depicting the chase.

One room features a large wooden cabinet dedicated to a variety of hunted animals. Concealed in the cabinet drawers are bronze casts of each animal's footprints and droppings, depictions of its natural habitat, and a poem dedicated to the creature. A room on the second floor has a ceiling entirely covered in the feathers and heads of five owls.

In the dramatic room packed with taxidermy specimens from around the world, look out for one item that's not like the others: a wall-mounted, animatronic moving albino boar head that speaks perfect French. **62, rue des Archives, Paris. Take the Métro to Rambuteau.** Ⓝ **48.857127** Ⓔ **2.354125**

Graffiti, moss, and abandoned train tracks form a loop around Paris's outer arrondissements.

THE PETITE CEINTURE

PARIS

The "little belt" railway ringing central Paris served urban travelers from 1862 to 1934, connecting five main stations owned by five different rail companies. The 20th-century expansion of Paris and the Métro eventually made the circular railway obsolete.

Stretches of the track now exhibit a quirky blend of idyllic nature and gritty urban life—plants and flowers grow between the rails against a backdrop of vivid graffiti and street art. Bridges, tunnels, and the original tracks remain mostly untouched, hidden just beyond the streets and neighborhoods of the outer arrondissements.

The Petite Ceinture is still owned by SNCF, France's national railway company, so a walk along the rails is considered trespassing. That said, an innocent bit of urban exploration is unlikely to land you in trouble. If you're strictly by the book, visit the section between the Porte d'Auteuil and the Gare de la Muette—in 2008, the area was classified as a nature trail open to pedestrians.

There are multiple points of entry to the Petite Ceinture. For the most scenic walks, access the track from the Balard, Porte de Vincennes, Porte Dorée, or Buttes Chaumont Métro stop. Ⓝ **48.821375** Ⓔ **2.342287**

RUINS OF LE JARDIN D'AGRONOMIE TROPICALE

PARIS

In 1907, as part of the city's Colonial Exposition, a set of pavilions and artificial villages went on display on the edge of the Bois de Vincennes. Each pavilion exhibited indigenous food, plants, products, and resources from colonies of the French empire. Each village mimicked its Asian or African original. Visitors didn't even need to imagine the actual people who lived in these far-off lands—as with the plants, Paris imported them and put them on display.

Throughout the summer, Indochinese, Malagasy, Congolese, Sudanese, and Tuareg people lived in re-created "typical" environments in the park. They wore costumes, sang, and danced for curious audiences. French visitors swarmed to the garden to observe the human zoo.

When cooler weather set in, the residents returned to their homes in the colonies, and the site began to decay. The land, owned by the French state, remained closed to the public until the city of Paris purchased it and reopened the site in 2007.

The five villages are still there, but the buildings—today fenced off from visitors—have almost completely collapsed. Most have trees and other vegetation growing from within. The plants inside the greenhouses have been left to rot. A few key pieces of architecture, such as a Chinese gate, are the only hints of the garden's former role as a tribute to the "glories" of colonial expansion.

45, avenue de la Belle-Gabrielle, Paris. The garden is a 10-minute walk from the Nogent-sur-Marne train station. ℕ 48.841007 Ⓔ 2.465697 ➤➤

➤➤ Humans on Display

Visitors to New York's Bronx Zoo in September 1906 were surprised to see a new mammal on display at the Monkey House. Imported from the Belgian Congo, he stood just under 5 feet (1.5 m), weighed 103 pounds (46.7 kg), and went by the name Ota Benga. Crowds jostled to watch him bare his sharp teeth and play with his cagemates: a parrot and an orangutan. The zoo had never displayed a member of his species before; Ota Benga was a human being.

The public exhibition of people was once a way for the world's colonizers to show off the "exotic" inhabitants of the countries they owned. Plucked from their homes like living souvenirs, indigenous people performed a counterfeit version of their daily lives at world's fairs, carnivals, and, most outlandishly, in zoos.

The producers of these shows showed little regard for the indigenous culture of the performers, preferring to create a version that would shock, amaze, and sell tickets. In 1882, Canadian theater agent Robert Cunningham visited Australia to recruit for a P. T. Barnum tour of "savage tribes." Cunningham selected nine Aboriginal men

and women from seven separate communities. Each community spoke a different traditional language, and only two of the nine spoke English.

Promotional posters described the performers as "tattooed cannibal black trackers and boomerang throwers." They danced, sang, and engaged in mock fights at exhibitions across America and Europe. Within two years, five of the nine were dead.

The case of Saartjie Baartman is a particularly unsettling example of 19th-century Europe's fascination with so-called primitives. In 1810, the young Khoisan woman was living in South Africa when William Dunlop, a visiting British doctor, persuaded Baartman to accompany

him to London. There, he exhibited her naked in a cage, demanding she walk, sit, or stand so that onlookers could get a better view of her large buttocks and genitalia. Anthropologists used her physical proportions as evidence that the white race ranked highest. Known as the Hottentot Venus, Baartman died at the age of 26.

As for Ota Benga, a flurry of complaints, particularly from African American ministers, brought about his release from the zoo. Freedom from the animal cage, however, did not mean happiness. After a stay at an orphanage, Benga had his teeth capped, went to school, and was working at a tobacco factory when World War I destroyed his

Ota Benga

dreams of returning to the Congo. Miserable at the thought of never going home, Benga chipped the caps off his teeth and fatally shot himself. He was 32.

DEYROLLE TAXIDERMY

PARIS

Exotic taxidermy, entomology, and natural history collections displayed in antique wooden cases and glass bell jars have made this store a destination for Parisians ever since it opened in 1881.

In 2007, many of the animals were reduced to blackened fragments after a fire tore through the store. With the help of artists and collectors worldwide, the shop is back in business, and today houses everything from stuffed house cats to polar bears among its 19th-century decor. Though some of the more exotic animals are not for sale, you can always borrow a lion for a party—almost everything in the store is available for rent.

46, rue du Bac, Paris. Take the Métro to rue du Bac.
Ⓝ 48.856444 Ⓔ 2.326564

A motionless menagerie with representatives from every corner of the animal kingdom.

Argonaute Submarine

19th Arr. • It was the dawn of the Space Age, but in 1958 the pride of the French military was one of the most modern submarines of any fleet, silently cruising the briny dark.

The "I Love You" Wall

Clignancourt • Covered in 612 cobalt-blue lava tiles and sprinkled with the red of a few broken hearts, this Montmartre wall expresses love via 250 languages.

Le Louxor Palais du Cinéma

10th Arr. • This stylish Egyptian Revival theater, inspired by Theda Bara's 1917 *Cleopatra*, is perhaps the oldest surviving movie palace in Paris, abandoned for decades and restored in 2013 to its former glory.

Parc Monceau

L'Europe • The ruins of the Duke of Chartres's 18th-century bucolic fantasy, where camels once roamed alongside faux Dutch windmills and Italianate vineyards, lend the landscape of this scruffy royal folly an airy charm.

UCJG and the World's Oldest Basketball Court

9th Arr. • Most weekend b-ballers who bang the backboards in the basement of this UCJG youth hostel (the French YMCA) have no idea that they are playing on the oldest basketball court in the world. Built in 1892, it is an almost exact copy of the Springfield, Massachusetts, court (long gone), where the game was invented.

Museum of Vampires and Legendary Creatures

Les Lilas • An eccentric scholar of the undead, and a rabid collector of their trappings, has assembled this macabre collection that includes Dracula toys, a vampire-killing kit, and a merry clutter of vampiric books, art, and literature.

A Parisian guide to the language of love.

Museum of Arts and Crafts

3rd Arr. • One of the world's great collections of mechanical instruments was founded by anti-cleric French revolutionaries, so it is no small irony that these engineering wonders and contraptions are now partly housed in the former abbey church of Saint Martin des Champs.

Medici Column

Les Halles • Standing extra tall yet supporting nothing, the nine-story Medici Column in front of the Paris Commodities Exchange hides a secret spiral staircase that runs to a viewing platform at the top, originally built so Catherine de Medici's personal astrologer could contemplate the stars.

The Duluc Detective Agency

Les Halles • Noirish green neon marks the location of one of France's oldest private detective agencies, a favorite spot for film directors and fans of gumshoe novels alike, with a Prussian-blue door and an engraved plain brass plaque that reads: DULUC. INVESTIGATIONS. 1ST FLOOR.

The Relic Crypt of St. Helena at Église Saint-Leu-Saint-Gilles

Les Halles • Forgotten by most Parisian Catholics but venerated by the Russian Orthodox community, this little-known reliquary holds stolen pieces of Constantine's mother, who helped spread Christianity throughout the Roman Empire in the 4th century, even as most of her body still rests in Rome.

House of Nicolas Flamel

Le Marais • This former home of the legendary 15th-century alchemist dates to 1407, making it the oldest stone house in Paris. Covered in strange and arcane symbols of magical transformation, it may hold the secret of turning tin into gold.

Paris Sewer Museum

Rive Gauche • What Victor Hugo described as "fetid, wild, fierce" has been a tourist destination since 1889, when the tangled labyrinth under the streets of Paris was first opened to the public to show off a marvel of French engineering.

Gustave Eiffel's Secret Apartment

Faubourg St-Germain • At the top of the city's iconic tower is a jewel box of an apartment built by Eiffel to entertain the elite science community of Paris—and to make everyone else jealous.

Bird Market

Île de la Cité • Every Sunday, when most of the flower vendors take their one day off, stacks of cages fill the market square on Île de la Cité near Notre-Dame, with pets and livestock of the chirping, squeaking, and flapping varieties.

Chapelle de Saint Vincent de Paul

7th Arr. • Behind the stark facade of a chapel near the Luxembourg Gardens at the top of a carved double staircase guarded by marble statuary, the robed bones and waxy remains of the 18th-century Saint Vincent de Paul rest in a glass-fronted solid silver reliquary, looking like he was just caught napping.

Arènes de Lutèce

St. Victor • Paris is the city of Notre-Dame, of Hemingway and Harry's Bar, of la Belle Époque and Marie Antoinette, but the Roman past of Paris is often overlooked. To wit, the quiet remnant of an ancient Lutetian amphitheater that has no guard, no entry fee, and few tourists.

Louis XIV's Globes

Quartier de la Gare • Created during the 17th-century reign of the Sun King, the golden age of French art, literature, and geographical exploration, two exquisite globes—the Earth and the cosmos—each 20 feet (6 m) in diameter, glow overhead at the National Library.

MUSÉE FRAGONARD

PARIS

At the Musée Fragonard, human fetuses dance a jig alongside a ten-legged sheep while a skinless horseman of the apocalypse looks on.

Founded in 1766 as a veterinary school with a private collection, the museum has rooms devoted to anatomy, physical abnormalities, articulated animal skeletons, and disease. However, by far the most striking room is the collection of *écorchés*, or "flayed figures," created by Honoré Fragonard.

Louis XV appointed Fragonard as a professor at the first veterinary school in Lyon, and it was there that he began skinning and preserving animal and, later, human corpses. Though he intended for his *écorchés* to be used as educational

A deceased rider flogging a dead horse.

tools, Fragonard arranged many of his figures into theatrical poses, creating eerie posthumous narratives. The horseman of the apocalypse, inspired by Albrecht Dürer's painting, is the most notable example. A skinless corpse with dried, varnished muscles and unnerving glass eyes sits astride a similarly preserved horse caught mid-gallop, the thick arteries of its neck filled with red wax. Reins loop from the animal's mouth and over the rigid sinew of the rider's hands.

Fragonard worked in Lyon for six years before his flayed figures began frightening the townspeople. He was dismissed from the institution amid accusations of insanity. The public would not see the flayed figures until the Fragonard museum opened in 1991.

7, avenue de Général de Gaulle, Maisons-Alfort. Located in a suburb of Paris, you can take the Métro to École Veterinaire de Maisons-Alfoet. Ⓝ 48.812714 Ⓔ 2.422311

ALSO IN AND NEAR PARIS

Dans le Noir?

Paris · Dine in complete darkness at this restaurant chain, staffed by vision-impaired waiters.

Eyewear Museum

Paris · A tiny museum filled with hundreds of famous spectacles.

The French Freemasonry Museum

Paris · Peek into the private world of a secret society.

Musée de la Contrefaçon

Paris · A museum dedicated to French counterfeits, from pens to pants.

Musée de la Magie

Paris · A museum documenting the history of magic in what was once the basement home of the Marquis de Sade.

Musée d'Anatomie Delmas-Orfila-Rouvière

Paris · France's largest collection of human anatomy specimens sits hidden from the public.

Musée des Arts et Métiers

Paris · The national museum of scientific and industrial instruments is home to Foucault's actual pendulum.

The Dog Cemetery

Asniéres-sur-Seine · Pay your respects at a pet graveyard dating back to the late 1800s.

Absinthe Museum

Auvers-sur-Oise · Learn about the history of the delicious beverage called the "green fairy"—purported to cause wild hallucinations and insanity, and banned in Europe and the US for nearly 100 years.

Musée Mondial de l'Aérostation

Balleroy · The top floor of this chateau is dedicated to the sport of ballooning.

Heart of the Dauphin

Saint-Denis · A crystal jar in the Basilica of Saint-Denis holds Louis XVI's small, withered heart.

THE RAT KING

NANTES, PAYS DE LA LOIRE

Tucked away in the Natural History Museum among the beautiful taxidermy birds, sparkling minerals, and mammal skeletons is a very rare and possibly forged specimen: a rat king. Folklore holds that rat kings are formed when a group of rats get their tails inextricably entangled. Thus trapped, the creatures spend the rest of their lives intertwined and unable to move, relying on other rats to bring them food.

The Nantes rat king specimen, found in 1986, consists of nine conjoined rodents preserved in alcohol. It is one of only a handful of cases ever found, though scientific consensus considers its natural occurrence unlikely.

12, rue Voltaire, Nantes. Take the tram to Médiathèque. Ⓝ 47.212446 Ⓦ 1.564685

LE PALAIS IDÉAL

HAUTERIVES, RHÔNE–ALPES

It all began with a postman and a pebble. Walking on his usual mail delivery route in 1879, Ferdinand Cheval tripped on a rock, took note of its peculiar shape, and had a sudden vision of a grand palace made of irregular stones. For the next 33 years, he worked diligently to transfer the palace from his imagination into reality, collecting rocks in a wheelbarrow while delivering letters. An oil lamp guided his hands as he worked on the palace alone at night. Not once did he ask for, or receive, help.

Cheval had little formal education and no architectural experience. His stone, cement, and wire creation was inspired by different styles and eras, incorporating Chinese, Algerian, and Northern European designs. The final castle (the Ideal Palace) is a magnificent tangle of grottoes, flying buttresses, and animal statues. A shrine holds the wooden wheelbarrow that carried the rocks.

When French authorities denied Cheval's wish to be buried within his palace, he was undeterred. Over the next eight years, the 80-year-old self-made architect built his own magnificent vault in the local cemetery, in a style closely resembling that of Le Palais Idéal. He died a year after its completion.

8, rue du Palais, Hauterives. The palace is 30 miles (48.3 km) south of Lyon, just off D538. The nearest train station is St. Vallier sur Rhône, a 45-minute ride from Lyon. Ⓝ 45.255889 Ⓔ 5.027794

"Everything you can see, passer-by, is the work of one peasant" is inscribed on the wall of this postman's castle.

The Odeillo solar furnace benefits from southern France's 2,400 hours of sunshine per year.

WORLD'S LARGEST SOLAR FURNACE

ODEILLO, LANGUEDOC-ROUSSILLON

A solar furnace uses a large concave surface of mirrors to reflect sunlight onto a focal point the size of a cooking pot. The temperature at this point may reach above 6,000°F (3,315°C), enough to generate electricity, melt metal, or produce hydrogen fuel.

The world's largest solar furnace is located in Font-Romeu-Odeillo-Via, a commune in the sunny Pyrenees mountains on the French-Spanish border. Operational since 1970, it uses a field of 10,000 ground-mounted mirrors to bounce the sun's rays onto a large concave mirror that shows a distorted, upside-down reflection of the countryside. Tours of the site include workshops and demonstrations on renewable energy and the solar system. **Grand Four Solaire d'Odeillo, 7, rue du Four Solaire, Font-Romeu-Odeillo-Via. The Odeillo station, a 15-minute walk from the furnace, is a stop on the scenic Little Yellow Train route. The train, which has two open-air carriages, offers splendid views of valleys, mountains, and Villefranche-de-Conflent, a fortified medieval town. Ⓝ 42.494916 Ⓔ 2.035357**

ALSO IN SOUTHERN FRANCE

Bugarach Mountain

Bugarach · New Agers believe this mountain spaceship is home to aliens who will rescue them in the apocalypse. It was the site of much hubbub around the Mayan Doomsday predictions of 2012.

Nude City

Cap d'Adge · Nudity is legal and common at this family-style resort. Roughly 40,000 daily visitors dine, shop, and stroll naked.

Carriolu Miniature Village

Carriolu · French cheese maker Jean-Claude Marchi has built a meticulously detailed mini village out of pebbles.

GERMANY

SPACE TRAVEL MUSEUM

FEUCHT, BAVARIA

Dedicated to space technology, the Hermann Oberth Space Travel Museum celebrates the zeal and inventiveness of its namesake—one of the forgotten founding fathers of rocketry and modern astronautics.

Born in 1894, Hermann Oberth was interested in astronomy from a young age. After reading Jules Verne's *From the Earth to the Moon* at 11, Oberth began sketching designs for rockets. By the time he was 14, Oberth designed plans for a recoil rocket that could propel itself through space by expelling exhaust gases from its base. (This was quite the achievement for an adolescent, considering rockets meant for manned spaceflight would not exist for another 50 years.)

After studying physics, aerodynamics, and medicine at universities in Munich, Heidelberg, and Göttingen, Oberth published a 429-page tome in 1929 titled *The Rocket into Interplanetary Space*. The book had a global impact. It would have been a banner year for the newly esteemed scientist if not for the fact that he lost vision in his left eye while building a flying rocket model for Fritz Lang's sci-fi film *Woman in the Moon*.

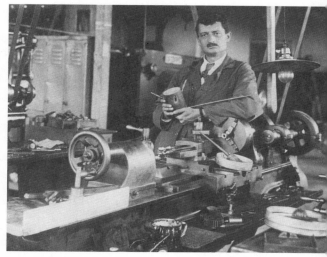

Hermann Oberth, the father of space travel, surrounded by his tools of the trade.

Exhibits inside the museum include a Cirrus rocket and a Kumulus rocket developed in the 1960s and launched outside of Cuxhaven, Germany. A Swiss Zenit sounding rocket is also on display. **Pfinzingstraße 12-14, Feucht. Feucht is a short S-Bahn ride from Nuremberg. Ⓝ 48.136607 Ⓔ 11.577085**

THE GALLERY OF BEAUTIES

MUNICH, BAVARIA

Though married at 24, 19th-century Bavarian King Ludwig I kept 36 beautiful young women in his palace. Whenever he became enamored with a female acquaintance's good looks, a portrait of her placid face would soon appear in the Schönheitgalerie, his Gallery of Beauties, a room in the south pavilion of Nymphenburg Palace—literally, Nymph Castle. The women are captured in their late teenage years or early twenties, milk-skinned and serene.

The standout work of the collection is Helene Sedlmayr, a doe-eyed, dark-haired shoemaker's daughter who gave toys to Ludwig's children. Her beauty was particularly appreciated by the king's valet, Hermes Miller, with whom she went on to bear 10 children.

Included in this collection are portraits of the adventurous aristocrat Jane Digby and Irish performer Lola Montez.

Schloss Nymphenburg, Eingang 19, Munich. The vast palace is 20 minutes via tram from central

Munich toward Amalienburgstraße. Ⓝ 48.136607 Ⓔ 11.577085

Saint Munditia's bejeweled right hand holds a glass container filled with dried blood.

JEWELED SKELETON OF SAINT MUNDITIA

MUNICH, BAVARIA

St. Peter's Church, or "Alter Peter," as locals call it, is Munich's oldest church, dating from before the city's founding in 1158. A quarter of the way down the aisle is a glass coffin bearing the skeleton of Saint Munditia, who met her end courtesy of a hatchet beheading in 310 CE. Once kept hidden in a wooden box, she was carefully adorned and put on display in 1883. Glass eyes stare out from the sockets of her sepia-stained skull. Jewels conceal her rotted teeth. And her skeleton, covered in gold and gemstones, has been reunited with her skull.

St. Peter's Church, Rindermarkt 1, Munich. St. Munditia's feast day, November 17, is celebrated with a candle procession at the church. Ⓝ 48.136497 Ⓔ 11.575672 ➻

➻ Jeweled Saints: Adorned and Adored

The practice of adorning the skeletons of saints with gold and jewels began after the rediscovery of the Roman catacombs in 1578. It was a tumultuous time for the Catholic Church—Protestant reformers had revolted against Catholic practices perceived as corrupt. In addition to criticizing papal authority, transubstantiation, purgatory, and confession, they also found the Church's excessive veneration of saints and their relics to be contrary to the teachings of scripture.

In response to the Protestant Reformation, a meeting of Catholic cardinals known as the Council of Trent set out to restore the Church's image. One way of doing that was to double down on the spiritual power of saints and their relics. The Church took the skeletons of supposed saints from the Roman catacombs and displayed them dressed in the elaborately embellished clothing of the time. This practice continued until the mid-1700s, and the jeweled skeletons found their final resting places in churches around Europe.

Waldsassen Basilica, a Bavarian church near the Czech border, has the largest collection of jeweled saints: There are 10 of them lining the aisles, each one in a glass case. Known as the Holy Bodies, the skeletons, many of whom have their teeth intact, seem to smile beatifically.

EISBACHWELLE

MUNICH, BAVARIA

If you're riding the Munich subway in the dead of winter and see someone carrying a surfboard, chances are they're headed for the Eisbach, the artificial stream that runs through Munich's biggest park, the English Gardens. A 1.5-foot (.5 m) stationary wave at a spot beneath a bridge was once an occasional sight, until local surfers funneled it into a more forceful, permanent swell by installing planks of wood on the sides of the river. Now the site attracts wave riders who brave the cold water—39°F in winter, 60°F in summer—to surf the stream in front of a crowd of onlookers.

So popular is the site that queues of surfers form on both sides of the narrow river. Those who pioneered the permanent wave have become disgruntled by the mass influx of newcomers, but their real ire is reserved for beginner surfers who lack the skills to conquer the wave.

Surf's up at the English Gardens, where there's an artificial wave beneath the bridge.

Englischer Garten, Munich. Get the U-Bahn to Lehel. The surf spot is at a river bridge in the English Gardens, just north of the Haus der Kunst art museum. There are three stationary waves in Munich, some better suited to beginners. Ⓝ 48.173644 Ⓔ 11.613079

THE EXTERNSTEINE

HORN-BAD MEINBERG, NORTH RHINE-WESTPHALIA

The Externsteine, translated variously as "stones of the ridge" or the "Star Stones," are a collection of limestone formations that jut up from the ground in a forest south of Detmold. There is scant evidence of the site's historical significance, but among the few known facts is that Christian monks carved stairs and reliefs into the stones during the late 8th century.

Between neo-pagans and neo-Nazis, the Star Stones have drawn quite the curious cult following.

The Externsteine were a blank slate for Heinrich Himmler and his Third Reich comrades. Himmler was the head of the Nazi's occult division, *Ahnenerbe*, a pseudo-scientific think tank devoted to finding or—in many cases—fabricating an honorable and far-reaching Germanic history. The society identified the Externsteine as an important location of ancient Teutonic activity.

Today, the site remains a pilgrimage spot for both neo-pagans and neo-Nazis. Visit on the summer solstice, or on the celebration of Walpurgisnacht, and you will encounter a motley crew of pagans, hippies, mystics, and skinheads—each group celebrating for different reasons.

Externsteine Strasse, Horn-Bad Meinberg. Walpurgisnacht is April 30. Ⓝ 51.867376 Ⓔ 8.918495

ALSO IN SOUTHERN GERMANY

German Butcher Museum

Böblingen · Learn about the development and history of animal butchery. A strong stomach is recommended.

Heidelberg Thingstätte

Heidelberg · A Nazi-built stone amphitheater sits on a hill littered with ancient burial grounds.

Nördlingen

Nördlingen · A town centered in a 15-million-year-old meteor crater in the Bavarian countryside.

European Asparagus Museum

Schrobenhausen · A museum dedicated to Germany's favorite "royal vegetable."

The Pig Museum

Stuttgart · An old slaughterhouse turned museum with 25 themed rooms dedicated to pig paraphernalia.

THOUSAND-YEAR ROSE

HILDESHEIM, LOWER SAXONY

The wild shrub that sprawls up a wall of Hildesheim Cathedral is believed to be the oldest living rose bush in the world.

Local tradition says that in 815, Louis the Pious, King of the Franks, was on a hunting trip in Hildesheim and stopped to observe Mass. After departing the area, he realized he had left behind a relic of the Blessed Virgin Mary. When he returned to retrieve it, he discovered the relic caught in a dog rose bush that refused to let it go. Taking this as a sign of the divine, Louis had a chapel built around the rose bush. In the 11th century, the chapel was expanded into what became Hildesheim Cathedral.

Even if the rose bush isn't as old as its legend, it has certainly proven its resilience. In March 1945, Allied bombs leveled the cathedral. Somehow, out of the burning rubble, the rose bush flourished once more. It now covers a courtyard-facing outer wall of the cathedral, which was rebuilt between 1950 and 1960.

Domhof 17, Hildesheim. To see the roses in bloom, visit at the end of May. Ⓝ 52.148889 Ⓔ 9.947222

The man at left represents an adult Kaspar, while the boy on the right is Kaspar as a "feral child."

KASPAR HAUSER MONUMENT

ANSBACH, BAVARIA

On a quiet street in the small city of Ansbach stand two statues depicting the same person, one as a boy and the other as a young man. They are both the mysterious Kaspar Hauser. The odd, unresolved story of Kaspar Hauser began on a May afternoon in 1828, when a stumbling, squinting teenage boy appeared on the streets of Nuremberg. The 4-foot-9-inch (1.4 m) Hauser had pale, soft skin, spoke few words, and seemed ill at ease among people. He carried two letters: The first, dated 1828, detailed the hardships that forced the author—apparently the boy's caretaker—to give him up. It ended with an ultimatum: Take care of Kaspar or, if you cannot, kill him. The second letter, dated 1812, was written by Kaspar's mother. She stated that she could not afford to raise him, and wished for him to join the Sixth Cavalry regiment at Nuremberg when he turned 17. A close examination of the letters revealed they were likely written by the same person in a single session.

The mystery deepened when Kaspar began talking. He claimed he spent his childhood locked in a small, dark cell, never seeing another person and waking up each day to a meal of bread and water. His only human contact, he said, was with a man who arrived shortly before his release and taught him to say the phrase, "I want to be a cavalryman like my father."

Frederich Daumer, a professor, took Kaspar into his home and taught him to speak German, ride horses, and draw. The boy began to adapt to his environment, but his highly attuned senses and inexplicable sensitivity to magnets and metal caused him anguish. One day in October 1829, Kaspar appeared with a bleeding wound on his forehead, claiming he had been attacked by a hooded man while using the bathroom. The man was never found, and the wound may have been self-inflicted. A few months later, a gunshot rang out from Kaspar's room. Again he was discovered with a wound to the head, this time declaring he had knocked a pistol to the ground, where it accidentally went off.

Kaspar's strange life ended in 1833, when he sustained a fatal stab to the chest at the palace gardens in Ansbach. He said a man had lured him there and lunged at him. He even provided a note from the assailant.

Platenstrasse, Ansbach. Kaspar Hauser is also remembered at Ansbach's Markgrafenmuseum, which exhibits the bloodstained clothes he was found in, the two letters, and some of his personal belongings. Ⓝ 49.302248 Ⓔ 10.570951

ALSO IN NORTHERN GERMANY

Museum of Letters, Characters, and Typefaces

Berlin · Peruse a collection of typographical objects, such as giant letterforms taken from signs. The museum recycles and archives letters from a wide range of languages and fonts.

Spreepark

Berlin · A now-abandoned amusement park that was closed when the owner was discovered smuggling drugs in the park's rides. These days it's popular with urban explorers.

Spicy's Spice Museum

Hamburg · A museum dedicated to spices from around the world, the only one of its kind.

Karl Junker House

Lemgo · Visit the single intricate masterpiece of a posthumously diagnosed schizophrenic architect.

CASTLE OF WEWELSBURG

WEWELSBURG, NORTH RHINE–WESTPHALIA

SS leader Heinrich Himmler had grand, horrifying plans for the triangular Renaissance castle in Wewelsburg. In 1934, undeterred by its dilapidated state, he signed a 100-year lease on the property. His mission was to turn the castle into an SS training center, where young Aryan minds could study Nazi-skewed versions of history, archaeology, astronomy, and art.

The SS redesigned the castle, incorporating swastikas and occult symbols, and used slave laborers from the nearby Niederhagen and Sachsenhausen concentration camps to bring the plans to life. Nazi-approved artwork and historical objects decorated the halls. The focal point of the redesign was a circular chamber known as "the crypt," which featured an eternal flame at the center of the room surrounded by 12 seats—an allusion to the Knights of the Round Table. On the ceiling was a large swastika.

As construction continued, Himmler expanded his vision for the town of Wewelsburg. From 1941, he began to view it as the future center of the new world order, with the castle anchoring a village populated exclusively by SS leaders. But in spite of all the plans, SS training never took place at Wewelsburg. The SS did conduct meetings there, and it functioned as a venue for SS officer marriage consecrations—although prospective spouses had to provide genealogical documentation proving their Aryan heritage before the ceremony could take place.

After Germany's 1943 defeat at Stalingrad—regarded as the beginning of the end for the Nazis—construction at Wewelsburg halted. On March 30, 1945, a month before Hitler's suicide, Himmler ordered SS major Heinz Macher to destroy the castle. The US Third Infantry Division arrived the next day to discover the castle interior in ashes. Only the outside walls remained.

Today, Wewelsburg is one of Germany's largest youth hostels, offering 204-bed accommodations and team-building programs for schoolchildren. A museum at the entrance presents the history of the SS and pays tribute to its victims.

From Nazi indoctrination center to youth hostel.

Burgwall 19, Büren. From the Paderborn train station, take a half-hour bus ride to Büren-Wewelsburg. The castle is a three-minute walk from the Schule/Kreismuseum stop. Ⓝ 51.606991 Ⓔ 8.651241

THE IMPALED STORK

ROSTOCK, MECKLENBURG–VORPOMMERN

Until the 19th century, the annual disappearance of white storks each fall puzzled European bird-watchers. Aristotle thought the storks went into hibernation with the other disappearing avian species, perhaps at the bottom of the sea. Other hypotheses argued that the disappearing birds flew to the moon to escape the cold weather.

In 1822, a stunning piece of evidence proved key to solving the mystery of the disappearing birds. A white stork, shot on the Bothmer Estate near Mecklenburg, was discovered with a 2.5-foot-long (80 cm) Central African spear embedded in its neck. Remarkably, the stork had flown all the way from its equatorial wintering grounds in this impaled state.

The stork now sits alongside the 60,000 other specimens of water animals, mollusks, birds, and insects of the University of Rostock's Zoological Collection. **Universitätsplatz 2, Rostock. Get a bus or tram to Lange Straße and walk two blocks south.** Ⓝ **54.087436** Ⓔ **12.134371**

ALSO IN NORTHWESTERN GERMANY

Wunderland Kalkar

North Rhine-Westphalia · An amusement park built on the grounds of an unused nuclear reactor.

The Wuppertal Suspension Railway

Wuppertal · The name of the world's oldest monorail system translates as "floating railway."

TEUFELSBERG SPY STATION

BERLIN

Atop a mountain in Grunewald forest, about six miles (9.7 km) west of central Berlin, are two sphere-topped cylindrical towers wrapped in shredded white canvas. During the Cold War, these buildings were part of Field Station Berlin, an NSA listening post where spies could tune in to Soviet radio frequencies.

The spy station was constructed on top of a Nazi military college complex built in 1937. Intended to be part of Hitler's World Capital Germania—a revitalized National Socialist version of Berlin—the college was abandoned at the start of World War II.

Postwar, the rubble of bombed-out Berlin was trucked to Grunewald and dumped on top of the college in a pile that topped 37 stories and became known as Teufelsberg: German for "Devil's Mountain." In 1963, the NSA station began operating in newly constructed buildings at the top of Teufelsberg, its satellite antennae positioned in prime hilltop spots concealed by the canvas-covered spheres.

The station became a key surveillance post for American and British intelligence officers studying the goings-on in East Germany. After the fall of the Berlin Wall, the station was abandoned. In 1996 the site was sold to property developers Hartmut Gruhl and Hanfried Schütte, who envisioned a bold transformation involving luxury apartments, a hotel, and a restaurant. Those plans, however, have not materialized. The former NSA station is still at Teufelsberg, and has attracted artists looking to make their mark on its abandoned walls and, eventually, establish an official residency. **From Berlin, get a train to the Heerstrasse S-bahn stop and walk to Teufelsberg via the Teufelsseestrasse. You may encounter tour guides at the gate asking you for a fee to visit the site.** Ⓝ **52.497992** Ⓔ **13.241283**

The tattered mountaintop remains of an NSA listening post, built atop a Nazi military college.

Subterranean Berlin

Brunnenstraße • A tour of the subterranean city meanders through abandoned tunnels, old hospitals, secret bunkers, World War II munitions storage, and Cold War–era passages used by dissenters to escape from East Germany.

Spandau Citadel

Spandau • Since the 12th century, a strategic island at the convergence of the Havel and Spree Rivers has been the site of a protective citadel, including the 13th-century Julius Tower (Berlin's oldest building) and 16th-century fortifications. Today it is the site of tours, concerts, and festivals.

Gardens of the World

Marzahn-Hellersdorf • Once an austere East Berlin park, it was transformed in 1987 into Gardens of the World, 50 acres of lush gardens: Chinese, Japanese, Italian, some even biblical and fairy-tale themed.

Tieranatomisches Theater

Mitte • Within this elegant anatomical theater built in 1789, with its airy dome and pastoral murals, live the ghosts of thousands of animals dissected under the orders of King Frederick William II, who thought the four-legged citizens of Berlin needed a dedicated veterinary school.

Designpanoptikum

Mitte • A shrine to the art of repurposing, recycling, and reengineering, the Surrealist Museum of Industrial Objects is a disorienting trip through an inspired world of retro machinery, strange medical devices, and a few interesting things stolen from the salvage yard.

Tajikistan Tearoom

Mitte • A little taste of Tajikistan dropped into a central Berlin courtyard, this classic Central Asian–style tearoom is plush with

These murals embody Berlin's hopes for a brighter future.

Persian carpets, pillows, and tapestries, and beckons you to lounge on the floor to enjoy a traditional tea service.

Hall of Mirrors, Clarchens Ballhaus

Mitte • Off to the side of Clarchens Ballhaus, an ornate dance hall that predates World War I, there is a mirrored ballroom that still carries the spirit of Wilhelmine Germany, moody and evocative, and mostly undiscovered, even by locals.

Monsterkabinett

Mitte • A design and robotics studio is the creation of the curiously named Dead Pigeon Collective, where unnerving human-androids turn dreams into nightmares and nightmares into eerie performance art.

Hohenzollern Crypt

Mitte • Beneath the Baroque cathedral in the center of the city is the world's largest collection of dead Prussian royals, sent into the afterlife in gorgeous gilded style.

DDR Museum

Mitte • An interactive museum of life in Deutsche Demokratische Republik (East Germany), including a cafeteria that serves up Soviet bloc cuisine, televisions that play crusty old state-run programming, and telephones that are very likely recording your conversations.

Hatch Sticker Museum

Friedrichshain • Opened in 2008 as a side venture for Oliver Baudach and his skate-cult sticker company, this small museum pulls mostly from Baudach's own bottomless pile of vintage street slicks.

Hansa Studios

Mitte • A magnificent recording studio, with glowing chandeliers, gleaming parquet floors, and original 1913 architectural detail, where some of the all-time greats have laid down tracks—from Iggy Pop to David Bowie to R.E.M.

Museum of Things

Friedrichshain • Brimming with more than 35,000 vintage and unique pieces, this collection, including kitchenware, advertisements, packaging, and children's toys, honors the beauty and simplicity found in 20th-century design.

East Side Gallery

Friedrichshain • An open-air art gallery of more than a hundred murals, these reclaimed vestiges of the Berlin Wall have been transformed by a cadre of international artists from the once-impenetrable symbol of suppression into a stark reminder of the fragility of freedom.

Ramones Museum

Friedrichshain • Blitzkrieg bop your way through a collection of more than 500 pieces of memorabilia, including posters, signed photos, ticket stubs, and pants worn by actual Ramones.

Schwerbelastungskörper

Tempelhof • Originally built in the early 1940s to test some Nazi theories of construction, this enormous concrete cylinder—now sunk into Berlin's cushy soil—is too massive to demolish. Visitors can tour its interior rooms.

Tempelhof Field

Tempelhof • The city's oldest airfield hasn't seen a plane come or go for nearly a decade, but its new life as a public park raises hopes of preserving the airport's old Art Deco–style buildings, a throwback to a time when air travel was still in its glamorous youth.

Peacock Island

Berlin-Wannsee • Sitting within the rush of the River Havel, this quiet oasis is scattered with abandoned buildings, flocks of wild peacocks, and the facade of a fairy-tale castle built by a Prussian king.

GREECE

MARKOPOULO SNAKE FESTIVAL

CEPHALONIA, MARKOPOULO

For Orthodox Christians, August 15 commemorates the Assumption of the Virgin Mary into Heaven. At the small hamlet of Markopoulo, on the island of Cephalonia, villagers celebrate by collecting snakes in bags and bringing them to the church, where they slither over feet, flick their tongues at a portrait of Mary and Jesus, and are placed on children's heads for good luck.

The combination of Jesus's mother and snakes—a Biblical symbol for evil and corruption—may seem incongruous, but the basis for the tradition is a miracle said to have taken place on the island in 1705. That year, the story goes, nuns in the village convent escaped a gang of attacking pirates when the Virgin answered their prayers to be transformed into snakes.

The serpents—a small, non-threatening species known as the European Cat Snake—now make an annual appearance at the Church of Virgin Mary, where the faithful and the curious gather to handle them. According to locals, the snakes only emerge in the days leading up to the festival, and cannot be found on the island during the rest of the year.

Markopoulo is a one-hour flight west of Athens. Ⓝ **38.080451** Ⓔ **20.732007**

Festival snakes are non-poisonous and defanged prior to church service.

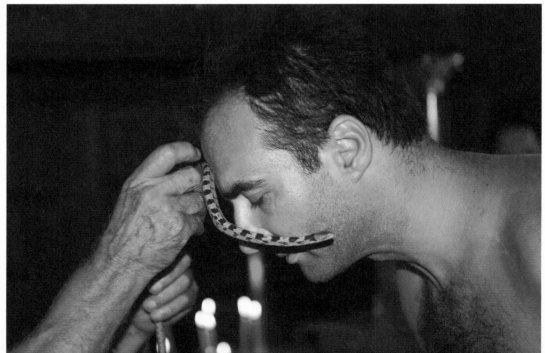

ANTIKYTHERA MECHANISM

ATHENS

An ancient geared computer that uses technology lost in antiquity.

This corroded lump of bronze with a dial on top doesn't look like much, but its discovery forced a complete reevaluation of ancient Greek engineering.

The Antikythera mechanism was part of a shipwreck that lay in the Mediterranean for over 2,000 years. In 1900, sponge divers happened upon the fragmented vessel and hauled the mechanism from the depths. It sat in a museum for 50 years before historians took a serious look at it. They discovered a device of unprecedented complexity.

The machine was built between 150 and 100 BCE, and with over 30 gears hidden behind its dials, it is easily the most advanced technological artifact of the pre–Common Era. Regarded as the first known analog computer, the mechanism can make precise calculations based on astronomical and mathematical principles developed by the ancient Greeks.

Although its builder's identity and its purpose on the ship remain unknown, after a century, scientists are beginning to understand how the device worked. To use the instrument, you would simply enter a date using a crank, and, when the gears stopped spinning, a wealth of information would appear at your fingertips: the positions of the sun, moon, planets, and stars, the lunar phase, the dates of upcoming solar eclipses, the speed of the moon through the sky, and even the dates of the ancient Olympic games. The mechanism's calendar dial could compensate for the extra quarter day in the astronomical year by turning the scale back one day every four years— impressive, given that the Julian calendar, which was the first in the region to include leap years, was not introduced until decades after the instrument was built.

Today, the Antikythera mechanism is housed in the Bronze Collection of the National Archaeological Museum of Athens. The device continues to reveal its secrets to the researchers of the Antikythera Mechanism Research Project, an international effort supported by various universities and technology companies.

National Archaeological Museum, 44 Patission Street, Athens.
Ⓝ 37.989906 Ⓔ 23.731005

THE EASTER ROCKET WAR

VRONTADOS, CHIOS

Since the 19th century, a pair of rival parishes on the Greek island of Chios have celebrated Easter not with bunnies or eggs, but by launching thousands of flaming projectiles at each other's churches while congregants observe Mass inside. When the sun sets on the eve of Orthodox Easter Sunday, members of the Angios Marcos and Panaghia Ereithiani churches— wearing bandanas over their faces to filter out the sulfur-scented smoke—launch the cigar-size rockets from ramps. They blaze across the sky in quick succession, leaving trails of vapor and providing a cacophonous accompaniment to the solemn Easter services taking place within the churches. (Wire mesh protects the windows from damage.)

The origin of this Orthodox yet unorthodox tradition is uncertain, but likely began as an act of defiance against Ottoman occupiers during the 19th century. (There are two versions of the story: In the first, the islanders launched rockets after the Turks confiscated their cannons. In the second, locals fired rockets at each other's churches to keep the Turks away during Easter services.)

On the island of Chios, Easter is a time of rebirth, renewal, and shooting rockets at churches.

The aim of the proceedings is to hit the opposing church's bell tower, but this goal is largely irrelevant given the general chaos of the evening—neither side is declared the victor of the "war," and an annual rematch is assured.

Orthodox Easter is in April or early May. From Athens, the island of Chios is a 45-minute flight or 7-hour ferry ride.
Ⓝ 38.370981 Ⓔ 26.136346

MOUNT ATHOS MONASTERY

MOUNT ATHOS

On a mist-shrouded peninsula east of Thessaloniki is a place where time has stalled and the rules of the modern world do not apply. Mount Athos, known to Greeks as the "holy mountain," is the home of Eastern Orthodox monasticism. Self-governed, and running on Byzantine time—in which the day begins at sunset—Mount Athos accommodates 1,500 monks within its monasteries, most of which were built during the 10th century. The monks' sole purpose in life is to become closer to God.

The monks believe that achieving complete oneness with Jesus Christ is only possible after death, but preparations are made throughout life. Every waking hour is spent praying or reflecting in silence. Monks, who wear long black robes to signify their death from the surrounding world, live in one of 20 communes, or, for those who prefer greater solitude, in cloisters or cells. There are eight hours of church services every day, beginning at 3 a.m. When not at church, monks pray individually, their lips moving silently under their long beards.

Women are forbidden from visiting or living on Mount Athos in accordance with the belief that a female presence would alter the social dynamics and impede the monks' journey toward spiritual enlightenment. The absence of women, according to the monks, also makes it easier to live a life of celibacy. Finally, there is the consideration for the Virgin Mary. According to Athonite tradition, Mary was blown off course during her journey to Cyprus and landed on Mount Athos, where she converted its pagan tribes to Christianity. Banning women from the peninsula allows the Virgin to be revered as the only female influence.

Male visitors to Mount Athos are permitted to attend church services, dine with the monks, and stay overnight in one of the monasteries. A calm, pious demeanor is expected, as most travelers are Orthodox pilgrims seeking spiritual refuge and community. **To visit Mount Athos, you'll need to obtain written permission from the Holy Executive Bureau in Thessaloniki. The office issues about 10 permits per day to non-Orthodox travelers, and 100 to Greek Orthodox travelers—contact them six months in advance to avoid disappointment. From Thessaloniki, get a bus to the small village of Ouranoupolis, where a ferry will transport you to Mount Athos. Female travelers may view the verdant hills and ancient monasteries from a distance on a boat tour. ℕ 40.157222 Ⓔ 24.326389**

Built on the side of a cliff, this isolated sanctuary is home to 1,500 monks.

CYPRUS

VAROSHA BEACH RESORT

FAMAGUSTA

Turquoise water, golden beaches, and signs illustrated with a gun-toting soldier that read "Forbidden Zone"—this is the resort town of Varosha.

Since 1974, the northern and southern parts of Cyprus have been divided by the "Green Line," a UN buffer zone that splits the country into the Greek-controlled south and the Turkish-controlled north. The division happened amid much violence: After the Greek military junta backed a coup against the Cypriot government, Turkey invaded Cyprus from the north, forcibly expelling hundreds of thousands of Greek Cypriots and driving them south. Turkish Cypriots in the south abandoned their homes and headed north.

In the early '70s, Famagusta, a town 2 miles north of the Green Line, was the top tourist destination in Cyprus. Its beachside

**YASAK BÖLGE
GİRİLMEZ**

FORBIDDEN ZONE

ZONE INTERDITE

VERBOTENE ZONE

ΑΠΑΓΟΡΕΥΜΕΝΗ ΖΩΝΗ

Varosha quarter, dotted with high-rise hotels, played host to moneyed movie stars like Elizabeth Taylor and Brigitte Bardot. In the wake of the Turkish invasion, its 39,000 residents fled, and Varosha became a ghost town. It has remained enclosed in barbed wire, uninhabited, and under the control of the Turkish military ever since. Left unmaintained for decades, buildings are slowly collapsing.

Just a few feet north of the fenced-off zone is the Arkin Palm Beach Hotel, a newly renovated resort where visitors can sip Caribbean-inspired cocktails beside the lagoon-shaped pool while gazing at the crumbling balconies of the decayed resort next door.

Varosha is closed to the public but visible through barbed-wire fences from the Arkin Palm Beach Hotel area. Photographing the town is forbidden, and soldiers will stop you if they suspect you are carrying a camera. ⓝ 35.116534 ⓔ 33.958992

ITALY

VILLA DE VECCHI

CORTENOVA

Just east of Lake Como, at the foot of the forested mountains of Cortenova, sits a house that's said to be haunted. Villa de Vecchi, also called the Red House, Ghost Mansion, and Casa delle Streghe (House of the Witches), was built between 1854 and 1857 as a summer residence for Count Felix de Vecchi. Within a few short years of its completion, the house witnessed an inexplicable string of tragedies that would forever cement its gothic legacy.

De Vecchi, head of the Italian National Guard and a decorated hero, set out to build a dream retreat for his family with the help of architect Alessandro Sidoli. A year before the villa was complete, Sidoli died. Many would later view this as the first ill omen. Nevertheless, the count and his family made Villa de Vecchi their home during the spring and summer months, and by most accounts lead an idyllic—if brief—existence.

The great mansion boasted a blend of Baroque and classical Eastern styles and was outfitted with all the modern conveniences of the time, including indoor heating pipes and painstakingly detailed frescoes and friezes. A larger-than-life fireplace presided over the main parlor, where a grand piano stood at the ready. Extensive gardens, promenades, and an equally impressive staff house rounded out the already picturesque surroundings.

Legend has it that in 1862 the count returned home to find his wife murdered and daughter missing. The count commandeered a lengthy, unsuccessful search for his daughter before dying by suicide that same year. The house made the rounds of owners and prospective buyers, but by the 1960s was left permanently uninhabited.

While the natural elements began their assault early on, the majority of the abandoned house's irreversible damage has been done by humans. Graffiti covers the walls, and anything capable of

This decaying mansion may be Italy's most haunted house.

being vandalized has been given its due makeover. The grand piano, once said to be played at night by a ghostly entity, has since been smashed to pieces, though some locals claim that music can still be heard coming from the house. **Note that if you enter the house, you are trespassing. An upper floor has collapsed, the stairs are rapidly deteriorating, and bits of ceiling regularly rain down. For safety's sake—not to mention law-abidance—it's best to admire from afar. ⓝ 46.003464 ⓔ 9.387661**

Underground Temples of Damanhur

BALDISSERO CANAVESE, PIEDMONT

From 1978 to 1992, members of the Damanhur commune dug into the mountain where they lived, tapping into what they believed were energy lines connecting the Earth to the cosmos. Their excavations were done in secret: Having neglected to secure planning approval, they had to conceal their work from authorities. But the world would eventually find out.

Led by philosopher, writer, and painter Oberto Airaudi, the "eco-society" of Damanhur began in 1975 with about 24 members. Billing itself as a "laboratory for the future of humanity," Damanhur is based on neopagan and New Age beliefs with emphasis on creative expression, meditation, and spiritual healing. Residents adopt animal and plant names (such as "Sparrow Pinecone") and live in "nucleo-communities" of 20 people in the foothills of the Alps, 30 miles (50 km) north of Turin.

Some former members have railed against Damanhur's sunny-spirited collective, describing it as a cult. It was an ex-Damanhurian who tipped off police about the unauthorized underground construction. When three officers and a public prosecutor arrived to conduct an early-morning raid, they were astonished. Beneath a humble farmhouse, behind a secret door, was a collection of temples spanning five levels.

Damanhur citizens had spent a decade and a half working around the clock in shifts to excavate 8,500 cubic meters of earth and rock. They decorated each hall and hallway in a different theme, with murals, stained-glass windows, mirrors, and mosaics. The New Age, 1970s-style artwork depicts everything from the history of the universe to a forest of endangered animals to the International Space

Minimalism is frowned upon among the interior decorators of Damanhur. Above, the Hall of Earth; below, a section map of the underground temple.

Station. The perimeter of one of the circular rooms is cluttered with sculptures, due to the directive that each member of the community must carve a statue in their own likeness.

Boggled by the unexpected beauty of the subterranean halls, Italian police granted a retroactive construction permit. The eco-society, which currently numbers around 1,000 members, now welcomes visitors to its temples. **Via Pramarzo, 3 Baldissero Canavese.** Ⓝ **45.417763** Ⓔ **7.748451**

1. Hall of Mirrors
2. Hall of Spheres
3. Hall of Metals
4. Earth Hall
5. Water Hall
6. Blue Temple
7. Labyrinth

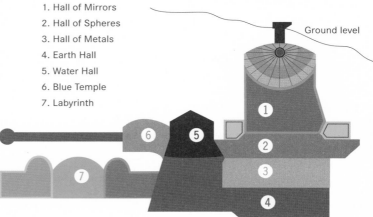

Poveglia Island

VENICE, VENETO

The first challenge to visiting Poveglia is finding someone to take you there. It won't be easy. Poveglia, an island just south of Venice, is strictly off-limits for both locals and tourists. Given the island's history, you'll understand why.

For much of its history, Poveglia was a dumping ground for the diseased, the dying, and the mentally ill. In the early 15th century, it functioned as a quarantine island. Those afflicted by one of the many Black Death plagues over the years were taken to the island, along with disease-ridden corpses. The living shivered and vomited blood while the hundreds of dead were thrown into plague pits and burned. The remains of an estimated 160,000 people are mixed into the island's soil.

In 1922, a psychiatric hospital opened on the island. Legend holds that one of the doctors, a sadistic man fond of experimenting on his patients, threw himself off the bell tower after being haunted by the spirits of those who had died on the island. (In some versions of the story, he is thrown off the tower, presumably by a vengeful patient.)

The hospital building, which closed in 1968, is still on Poveglia, covered in scaffolding and colonized by trees. Rusting bed frames, rotting wood beams, and pieces of the ceiling litter the floors. In the bushes surrounding the hospital are the rectangular metal grates that were once fastened to every window to keep patients trapped in their rooms.

Officially, Poveglia is off-limits to visitors. Unofficially, you might find a Venetian boat operator to take you—especially if you're willing to offer a heap of euros.
Ⓝ 45.381879 Ⓔ 12.331196

Poveglia, also known as Plague Island, is guarded by an octagonal fort.

Wooden Books of Padua University

SAN VITO DI CADORE, VENETO

 Each book in this 56-volume collection tells the story of a particular tree—not in words or illustrations, but with parts of the tree itself.

The wooden books of Padua University date from the late 1700s and early 1800s. While most books are made from wood—pulped into paper—these are different: The back and front covers are made from the tree's wood, its spine made of the tree's bark. Inside are samples of the tree's leaves, twigs, flowers, seeds, and roots. Each book is accompanied by a handwritten piece of parchment with a key explaining the contents.

41 Via Ferdinando Ossi, San Vito Di Cadore. Ⓝ 46.453240 Ⓔ 12.213190

La Specola

FLORENCE, TUSCANY

In 18th-century Florence, artists created wax anatomical models to show medical students what lay beneath the skin of the human body. The model-making process was labor-intensive and began with an artist pressing plaster against the individual organs of a recently dissected cadaver to create a cast. Wax was poured into the molds and each organ painted and varnished. All the body parts were then assembled into a wax torso and overlaid with muscles and membranes, which were either simulated with thread or painted on.

The works were so uncannily realistic—red, glistening muscles lying taut against knobby bones, encased in an intricate web of veins—that it was felt the artistry deserved a wider audience. In 1775, the figures were put on display at La Specola, a natural history and zoology museum established by the Medici family. Over several

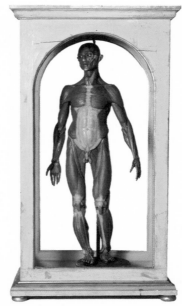

The exhibits at La Specola show what's under the skin.

generations, the Medicis amassed a vast quantity of fossils, minerals, animals, and plants. When La Specola opened, it became the first publicly accessible science museum in Europe. It featured consistent hours of operation, tour guides, and guards.

Today, La Specola contains 34 rooms filled with human and animal wax models, zoological specimen taxidermy, and medical instruments. Particularly compelling are the wax anatomical Venuses—naked women posed in demure but erotic poses, with their abdominal skin removed and rib cages pulled open to expose the organs underneath. The Marquis de Sade, who harbored a taste for sexual violence, had a particular fondness for these sensuous, gutted female forms.

Via Romana, 17, Florence.
Get a bus to San Felice. For a picturesque chaser, visit the Pitti Palace and the Boboli Gardens afterward—they're right next door.
Ⓝ 43.764487 Ⓔ 11.246972 ➥

Galileo's Middle Finger

FLORENCE, TUSCANY

It's hard to think of a more fitting tribute to the ever-defiant Galileo than displaying his middle finger in a goblet accented with gold.

Ninety-five years after Galileo's death in 1642, Anton Francesco Gori, a Florentine priest and scholar, pocketed the astronomer's finger while transporting his remains from their humble original grave to a monumental tomb. The relic was exhibited at Florence's Laurentian Library until 1841, when the town's Natural History Museum—also home to the anatomical and zoological museum La Specola—snatched it up for display.

In 1927, the finger landed at its current resting place—a museum devoted to scientific instruments that was renamed the Galileo Museum in 2010. The middle digit is the only human fragment among the institution's telescopes, meteorological instruments, and mathematical models. It is mounted vertically in a goblet on a column with a commemorative inscription:

This is the finger, belonging to the illustrious hand that ran through the skies, pointing at the immense spaces, and singling out new stars.

Whether the middle finger points upward to the sky, where Galileo glimpsed the glory of the universe and saw God in mathematics, or if it sits eternally defiant to the church that condemned him is for the viewer to decide.

Galileo Museum, Piazza dei Giudici 1, Florence. Ⓝ 43.767734 Ⓔ 11.255903

⇒ Other Medical Museums of Europe

JOSEPHINUM
VIENNA
Established in 1785, the Josephinum contains over 1,000 wax models, including anatomical Venuses and a heart seemingly floating under a glass dome.

THE NARRENTURM
VIENNA
The Narrenturm, or Fool's Tower, was built in 1784 to house psychiatric patients suffering from such maladies as ecstasy, melancholy, and delirium tremens. Now the circular building, nicknamed the "pound cake" by locals, is an anatomy and pathology museum.

MUSEUM BOERHAAVE
LEIDEN, NETHERLANDS
In one jar, a child's arm protrudes from a lacy sleeve, its fingers suspending the vascular tissue of an eye as though it were a yo-yo. In another jar nearby, a lily-white pig with a deformed head floats. Walking through this museum's exhibit halls filled with artifacts from the anatomical, medical, and scientific history of the Netherlands, visitors are pointedly reminded of their mortality.

To hammer the point home, articulated skeletons even carry flags inscribed with Latin phrases: *pulvis et umbra sumus* ("we are but dust and shadow"), *vita brevis* ("life is short"), and *homo bulla* ("man is a soap bubble"). An old operating theater, antique scientific instruments, and anatomical models of animals complete the experience.

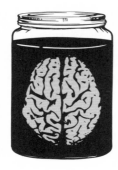

There are plenty of jarred human specimens at Museum Vrolik.

MUSEUM VROLIK
AMSTERDAM
During their many years as anatomy professors in the 1800s, father-and-son team Gerardus and Willem Vrolik amassed a formidable assortment of human abnormalities. Their specimens form the core collection of this museum. Willem focused largely on congenital anomalies such as conjoined twins and cyclopic fetuses. The infant specimens float, gently and ghostlike, in glass jars.

MUSEUM OF MEDICINE
BRUSSELS
Located beside the Museum of Cocoa and Chocolate, the Museum of Medicine features an entire wall of wax reproductive organs ravaged by sexually transmitted diseases.

MUSÉE DES MOULAGES
PARIS
Dedicated to "moulages"— wax models of body parts afflicted with disease—this warehouse of a museum has two spectacular floors of pus-filled, boil-covered, rash-afflicted skin.

MUSÉE DUPUYTREN
PARIS
After running out of money and being forced to close, this collection of thousands of anatomical wax models, pathological models, and abnormal physiological specimens sat neglected for 30 years before reopening in 1967.

HUNTERIAN MUSEUM
LONDON
The Hunterian Museum's collection includes half the brain of mathematician Charles Babbage, Winston Churchill's dentures, and the skeleton of a famed Irish giant.

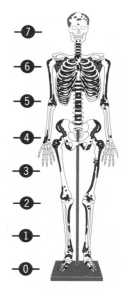

The 7'7" skeleton of Charles Byrne is on view at the Hunterian Museum.

ALSO IN TUSCANY

The Relic of St. Antoninus
Florence · View the mummified body of St. Antoninus, who died in 1459.

Hall of Maps
Florence · Fifty-four exquisite hand-painted Renaissance maps decorate these Medici rooms.

Santa Maria Novella Pharmacy
Florence · The oldest operating pharmacy in the world still sells balms, salves, and medicines made from 800-year-old recipes.

The Garden of Tarot
Grosseto · A garden filled with huge sculptures representing tarot cards.

Incorruptible St. Zita
Lucca · Here lies the 700-year-old naturally mummified body of a peasant girl–made-saint.

The Museum of Medieval Torture
San Gimignano · Housing an impressive array of medieval torture devices, such as spiked inquisitorial chairs, the museum seeks to combat torture around the world through education.

St. Catherine of Siena's Severed Head
Siena · A dismembered holy head stares out from her beautiful reliquary in St. Dominic Basilica.

POPE LEO'S BATHROOM

HOLY SEE, ROME

In 1516, Cardinal Bibbiena, a longtime friend and confidant to Pope Leo X, decided to redecorate the bathroom within the Vatican's Papal Apartments. Bibbiena's predilection for the ribald and the risqué inspired him to commission a series of erotic frescoes by another close friend, the artist Raphael. The panels depict Venus, Cupid, nymphs, and satyrs, all cavorting naked in the wild.

The ensuing centuries have brought censorship to the Stufetta del Cardinal Bibbiena (translated as "small heated room of Cardinal Bibbiena"). Scandalized residents of the Papal Apartments painted over sections of the artwork. A few of the panels remaining depict the naked Venus swimming, looking at herself in a mirror, and reclining between the legs of Adonis.

Though it wouldn't be sanctioned by today's Papal Apartment decorators, Raphael's erotic artwork was tame compared to the famed licentiousness of the Renaissance popes. Pope Alexander VI's Banquet of Chestnuts, which took place in the papal palace in 1501, is an example of the behavior that occurred within the Vatican walls. During the event, the pope brought in 50 women, auctioned off their clothing, and made them crawl naked on the floor to pick up chestnuts he and his guests had thrown. Members of the clergy and party guests were then encouraged to have sex with the women. The man who carried out the most "conquests" received a prize of clothing and jewels.

A bathroom in the Vatican's Papal Apartments was customized for 16th-century Cardinal Bibbiena, who had a penchant for the erotic arts.

Viale Vaticano, Rome. Get the Rome metro to Ottaviano–San Pietro–Musei Vaticani. Tours of the Stufetta del Bibbiena are rarely granted—now is the time to exploit any Vatican connections. As an alternative, visit the Borgia Apartments, the scene of Alexander VI's aforementioned banquet. Ⓝ 41.903531 Ⓔ 12.456170

The pope's secret escape route runs along a stone wall in Rome.

PARK OF THE MONSTERS

BOMARZO, LAZIO

The stone sculptures in the Parco dei Mostri emerged from the tormented mind of 16th-century Italian prince Pier Francesco Orsini. Pier endured a brutal war, saw his friend killed, was held for ransom for years, and returned home only to have his beloved wife die. Seeking a way to express his grief, Orsini hired architect Pirro Ligorio to create a park that would shock and frighten its visitors.

The park exhibits the 16th-century Mannerist style—an artistic approach that rejected the Renaissance's elegance and harmony in favor of exaggerated, often tortured expressions and a mishmash of mythological, classical, and religious influences. Its wretched sculptures—including a war elephant attacking a Roman soldier, a monstrous fish head, a giant tearing another giant in half, and a house built on a tilt to disorient the viewer—caught the attention of Salvador Dalí, who visited in 1948 and found much to inspire his surrealist artwork.

The Parco dei Mostri is the monster-filled garden of a grief-stricken prince.

A trip to the park is not complete without a walk up the stone stairs leading into the "Mouth of Hell"—the face of an ogre captured midscream. Walk into its gaping maw, inscribed with "all reason departs," and you'll find a picnic table with benches. **Localita Giardino, Bomarzo. From Rome take a train to Orte Scalo, where you can switch to a bus to the gardens.** Ⓝ **42.491633** Ⓔ **12.247575**

PASSETTO DI BORGO

ROME

To the casual eye, the Passetto di Borgo looks like a plain old fortification, but its stone walls hide a passageway that several popes have used as an emergency escape route. Construction of the wall dates back to 850, with Pope Nicholas III overseeing the creation of its current form in 1277. Pope Alexander VI finished the wall in 1492—and just in the nick of time. He used it to flee the invading French only two years later.

The most recent papal escape was in 1527, when Clement VII evaded the 20,000 mutinous troops of Charles V, who murdered most of the Swiss Guard on the steps of St. Peter's Basilica. Since then, the Passetto had languished in declining condition, closed to visitors but available to the reigning pope in case of a crisis. But in 2000, in honor of the pope's Jubilee year, the Passetto was renovated. It now opens to visitors for a limited time each summer.

Borgo Pio, 62, Rome. Take a tram to Risorgimento/ San Pietro. Ⓝ **41.903817** Ⓔ **12.460230**

ALSO IN ROME

The Criminology Museum

First assembled in 1837, this collection of prison paraphernalia, torture devices, and items showing the history of criminal anthropology only opened to the public in 1994.

Torre Argentina Cat Sanctuary

Hundreds of stray cats haunt the ruins of the famous Theater of Pompey, where the Roman emperor Julius Caesar was assassinated.

Vigna Randanini

This Jewish catacomb, used in the 3rd and 4th century CE, is one of only two open to the public.

THE MUSEUM OF HOLY SOULS IN PURGATORY

ROME

This small collection of hand-shaped burns imprinted onto prayer books, bedsheets, and clothing offers purported proof of communication between the dead and the living.

According to Catholic doctrine, the souls of the dead are stranded in purgatory until they atone for their sins. Their ascent to heaven can be hastened, however, through the prayers of the friends and family they've left behind.

Victor Jouet, founder of Rome's Church of the Sacred Heart, was supposedly inspired to build this purgatorial museum after a fire destroyed a portion of the church in 1898, leaving behind the scorched image of a face that he believed to be a trapped soul.

The collection of hand marks, which mostly date from the 18th and 19th centuries, are presented as the earthly manifestations of trapped souls reaching out from purgatory, pleading that their loved ones pray harder.

Lungotevere Prati 12, Rome. The museum is a 15-minute walk from the Lepanto metro stop. Ⓝ 41.903663 Ⓔ 12.472009

ALSO IN SOUTHERN ITALY

Manna of St. Nicholas	Blue Grotto	The Anatomical Machines of Cappella Sansevero	Catacombs of San Gennaro	Il Castello Incantato
Bari · The remains of St. Nicholas are said to excrete a sweet-smelling liquid known as manna. Every year on May 9, manna is collected in vials and sold to the public.	*Capri* · Once the personal swimming hole of Roman emperor Tiberius, this sea cave has an unearthly blue glow.	*Naples* · Anatomical models built on top of real human skeletons are just a part of the strange collection of mysterious 18th-century prince Raimondo di Sangro.	*Naples* · An underground proto-Christian burial site comprised of three adjacent cemeteries dating back to the third century CE.	*Sciacca* · Containing over 1,000 carved heads, the life's work of "village madman" Filippo Bentivegna can be found in this small garden.

THE SECRET CABINET OF EROTICA

NAPLES, CAMPANIA

Citizens of Pompeii and Herculaneum believed phalluses provided protection, prosperity, and good luck, and incorporated them into everything from furniture to oil lamps. Frescoes on the walls of homes depicted erotic encounters between wood nymphs and satyrs. Erotica was everywhere.

After the excavation of Pompeii and Herculaneum in the 19th century, the sexy *objets* were put on display in the National Archaeological Museum of Naples. But when the future king of the two Sicilies, Francis I, visited with his wife and young daughter in 1819, he was shocked by all the erotica. He ordered all the explicit items removed from view and locked in a secret cabinet, where access could be restricted to mature gentlemen of high moral standing.

All of this fuss only served to make the collection more famous. The whispered-about secret collection became a must-see for gentlemen during their grand tours of Europe. It remained off-limits to women, children, and the general public.

For a century and a half, the Gabinetto Segreto stayed out of sight, on view only during brief liberal periods under radical 19th-century general Giuseppe Garibaldi and again in the 1960s. The Gabinetto Segreto was finally opened to the public in 2000 and moved into a separate gallery in 2005.

Among dozens of stone penises, phallic wind chimes, and naughty mosaics, one item became the most notorious: *The Satyr and the Goat*. It is a detailed carving of a satyr in flagrante delicto with a female goat, her cloven feet pressed up against his chest as she gazes at him demurely.

19 Piazza Museo, Naples. Take the train to Museo. Ⓝ 40.852828 Ⓔ 14.249750

Sculptures of a more carnal nature are kept in a secret room at the National Archaeological Museum of Naples.

CIVITA DI BAGNOREGIO

CIVITA

The small town of Civita di Bagnoregio has survived for centuries clinging to its high perch, but as the soft clay on the edges of town erodes, its architecture is being lost.

Etruscans first built the tiny town on a tall column of volcanic rock over 2,500 years ago. Civita was accessible via a slim donkey path that led to the arch marking its only entrance. Over the centuries, the borders of the circular town became flush with the edges of the plateau as the tightly packed architecture filled the limited space atop the column. All the while, natural disasters weakened the clay foundations, which were already prone to erosion.

In the 20th century, many of Civita's thousands of residents began to move to more stable areas. Today, the donkey path has finally eroded away, and the only way to get to the village is via a steep footbridge. Supplies only make it up there via moped or three-wheeled *apès*, or "bees," tiny motorized pickup trucks. The population reaches about 100 during the summer months and drops into the single digits during winter. In 2013, the town began charging visitors an entry fee. Tourism has been profitable for the townspeople, but regular landslides continue to reduce the town's borders.

Buses runs from Orvieto, taking about an hour to reach Civita. Come in June or September to catch the wild donkey races. Ⓝ 42.627815 Ⓔ 12.114002

There is only one path into this ancient village built on volcanic tuff.

CAPUCHIN CATACOMBS

PALERMO, SICILY

Eight thousand corpses in various states of decay inhabit the musty, ill-lit Capuchin catacombs of Palermo. Originally intended exclusively for monks, the passageways were expanded over time to make room for prominent locals who paid to be buried in the holy vaults. Separated according to age, sex, occupation, and social status, the mummified bodies are arranged in open coffins, hung from the walls in narrow corridors, and piled on shelves.

In the "Chapel of the Virgins," girls whose families declared them virgins are displayed in faded and tattered white dresses under the inscription: "We follow the lamb wherever he goes; we are virgins."

They wear their best clothes, but their appearance is marred by caved-in noses, empty eye sockets, and sunken cheeks. Many have wide-open mouths—due to a combination of decomposing facial ligaments and gravity—making them look as though they are silently screaming.

The particularly dry atmosphere of the catacombs facilitated the natural mummification of the bodies. Priests would lay the dead on shelves and allow them to drip until they were completely depleted of bodily fluids. When a year had passed, they would rinse the dried-out corpses with vinegar before re-dressing them in their best attire and sending them to their designated eternal room.

The oldest corpse in the collection is that of Silvestro da Gubbio, a friar who passed in 1599. The most recent, Rosalia Lombardo, was only two years old when she was embalmed in 1920 after dying of pneumonia. The embalming procedure has kept Rosalia looking so well preserved that she has been dubbed "Sleeping Beauty."

The catacombs also serve as a macabre history-of-fashion museum, charting the rise and fall of Palermo high style from the 1600s to the 1920s. Though the mummies once sported glass eyes, they disappeared during World War II when American GIs came through town and plucked them out as souvenirs.

1 Piazza Cappuccini, Palermo. The monastery is a 25-minute walk from Palermo's central railway station.
Ⓝ **38.116191** Ⓔ **13.362122** ➤

You're always in the company of monks in the halls of the Capuchin catacombs.

➤ Other European Ossuaries

CHURCH OF ST. URSULA COLOGNE
The Golden Chamber of this church is filled with the bones of hundreds of virgin martyrs—or so the story goes.

SANTA MARIA DELLA CONCEZIONE DEI CAPPUCCINI ROME
A stunning memento mori, this 6-room crypt is full of intricate wall frescoes, arches, and ceiling decorations made entirely from human bones—the remains of some 4,000 Capuchin friars who died between 1528 and 1870.

CHURCH OF ST. PETER AND PAUL MELNÍK, CZECH REPUBLIC
When a plague epidemic swept through Melník in the 1520s, creating a huge demand for additional burial ground, the remains of 15,000 people laid to rest in the surrounding cemeteries were disinterred, cleaned, and dumped into this church vault.

In the 1780s, when ossuaries were declared a health risk, the vault was bricked up and forgotten about until 1913, when Czech anthropologist Jindřich Matiegka reopened the entrance to the crypt and began arranging the bones into orderly piles and decorative patterns. Thousands of bones form a large cross adorned with a palm frond, while skulls are arranged into heart shapes. A long, deep tunnel made mostly of leg bones represents Christ's resurrection.

SEDLEC OSSUARY KUTNÀ HORA, CZECH REPUBLIC
Sedlec Ossuary, also known as the Bone Church, has a stunning centerpiece: a chandelier made from every bone in the body.

SAN BERNARDINO ALLE OSSA MILAN
This ossuary was built in 1210 after the local graveyard became overfilled with deceased patients from the nearest hospital.

FONTANELLE CEMETERY CAVES NAPLES
In the early 20th century, a cult emerged around the skulls at this crypt. Visitors brought them flowers and offerings, made wishes, and asked for luck—the most sought-after skulls were those with a gift for predicting the winning lottery numbers.

Artwork, artifacts, and fossils are displayed in beautiful old glass cases and bell jars.

NETHERLANDS

TEYLERS MUSEUM

HAARLEM, NORTH HOLLAND

Lit only by sunlight since it opened in 1784, Teylers Museum is the first and oldest museum in the Netherlands. Its core collection of fossils, paintings, and scientific instruments was originally a "Cabinet of Curiosities" belonging to Pieter Teyler van der Hulst, a Haarlem banker and silk manufacturer. In his will, the Enlightenment-era thinker decreed that his money be used to establish a foundation for science and the arts. This museum is the glorious result.

The current collection includes 25 Michelangelo artworks, drawings by Rembrandt, fossils, beautifully illustrated 18th-century natural history books, coins from the 1600s, and the world's largest electrostatic generator. Its main building—an airy, oval-shaped room with lofty glass-vaulted ceilings—is as eye-catching as the curiosities within.

Spaarne 16, Haarlem, Noord Holland. Buses stop in front of the museum. Ⓝ 52.380256 Ⓔ 4.640391

EISINGA PLANETARIUM

FRANEKER, FRIESLAND

Dutch priest Eelco Alta induced mass panic in 1774 when he predicted that an upcoming alignment of the moon, Mercury, Venus, Mars, and Jupiter would send Earth hurtling into the sun. To calm the widespread fear and demonstrate the varying speeds of the solar system's orbits, amateur astronomer Eise Eisinga installed a planetarium in his living room ceiling. It has been operating ever since, making it the oldest functional planetarium in the world.

Construction took seven years. Eisinga carved the era's six known planets—Mercury, Venus, Earth, Mars, Jupiter, and Saturn—out of wood, painted them gold, and suspended them from the ceiling. A pendulum sends the planets traveling, in real time, in concentric orbits around a painted sun. The ceiling is painted cerulean blue to represent the sky, and shows zodiac constellations and the points at which each planet is farthest from the sun. Hidden in the roof space above the ceiling are 60 wheels and gears that keep the whole thing moving.

Just as Eisinga was putting the finishing touches on his planetarium in 1781, scientists made a discovery: Uranus. But the astronomer did not alter his creation, as adding the planet would have destroyed the planetarium's 1:1012 scale.

Eise Eisingastraat 3, 8801 KE Franeker. Get a bus to Theresia. Ⓝ 53.187335 Ⓔ 5.543735

The world's oldest functional planetarium packs our whole solar system into a cozy-looking Dutch house.

GIETHOORN

GIETHOORN, OVERIJSSEL

With its thatched cottages, narrow canals, and 180 wooden bridges, the village of Giethoorn (population 2,600) is a tiny, bucolic version of Venice.

The village got its canals during the 16th century, when the inhabitants began digging troughs to extract peat for fuel. The resulting four miles of waterways, along with the adjacent footpaths, constitute Giethoorn's transport network—the village has no roads, so all travel is via boat or bike or on foot. The preferred way to ride along the canals is by "whisper boats," whose motors are quiet enough to maintain the tranquil atmosphere.

Giethoorn is a 90-minute drive from Amsterdam. To sightsee around the village, join a boat tour or rent your own flat-bottomed vessel. Ⓝ 52.740178 Ⓔ 6.077331

Bask in the glow of the Summer of Love between Electric Ladyland's fluorescent walls.

ELECTRIC LADYLAND MUSEUM OF FLUORESCENT ART

AMSTERDAM

In a small, dark basement just a five-minute walk from the Anne Frank house is a museum as psychedelic as the Jimi Hendrix album of the same name. The Electric Ladyland Museum of Fluorescent Art begins with a walk through a neon-accented "participatory environment," where the walls glow when the lights go off. Then comes the cabinet of fluorescent minerals. These gray lumps of rock turn brilliant colors when viewed under an ultraviolet bulb.

Museum founder Nick Padalino will gladly talk you through the many kinds of fluorescence being demonstrated under the varying wavelengths of light. Most surprising is the display of common items, such as coconuts, seashells, Depression-era glass, and lentils, rendered in glowing fluorescent hues.

Tweede Leliedwarsstraat 5, Amsterdam. For maximum psychedelic effect, listen to Hendrix's "All Along the Watchtower" while exploring the curves and contours of the immersive environment. Ⓝ 52.375602 Ⓔ 4.882301

Micropia museum makes the invisible observable.

MICROPIA

AMSTERDAM

Micropia is a zoo with a small difference: The creatures it keeps are invisible to the naked eye. Established in 2014, Micropia sheds light on the molds, yeasts, bacteria, viruses, and other microorganisms that are ever-present but go largely unnoticed.

The "Meet Your Microbes" exhibit highlights a few of the 100 trillion microorganisms living in and around your body, while the kissing exhibit ratchets up the romance by illustrating the massive exchange of bacteria that occurs when two people make out. **Artisplein, Plantage Kerklaan 36–38, Amsterdam. Eat something before visiting the museum—the animal feces exhibit has a tendency to quell the appetite. Ⓝ 52.367668 Ⓔ 4.912447**

ALSO IN THE NETHERLANDS

Bijbels Museum

Amsterdam · A collection of biblical memorabilia that includes an exact copy of a Dead Sea scroll. If the Bibles bore you, you can also check out the best-preserved 17th-century kitchen in the Netherlands.

The Hash, Marihuana, Hemp Museum

Amsterdam · Learn about old tools and implements for turning hemp into rope, paper, and clothing, as well as a variety of ancient and not-so-ancient smoking devices.

The Torture Museum

Amsterdam · Inspect torture methods and devices from the past, including the rack, the skull cracker, and the heretic fork.

The Three-Country Labyrinth

Vaals · Seventeen thousand hornbeam shrubs constitute Europe's largest outdoor shrub maze, located where the Netherlands, Belgium, and Germany meet.

Cigar Band House

Volendam · Mosaics created from 11 million cigar bands cover the walls of this house.

ATLANTIC OCEAN

PORTUGAL

SPAIN

● Baby Jumping Festival

★ Don Justo's
Self-Built Cathedral
MADRID

● Boulders of Monsanto

● Mafra Palace Library
Quinta da Regaleira
★ **LISBON**
Preserved
Head of
Diogo Alves

Balearic
Islands

Miles
0   50   100

0   50   100
Kilometers

N

● El Caminito del Rey

Ronda ●

MEDITERRANEAN
SEA

ATLANTIC
OCEAN

CANARY ISLANDS

The Whistling Island

Miles

Kilometers

PORTUGAL AND SPAIN

PORTUGAL
THE BOULDERS OF MONSANTO

MONSANTO, IDANHA-A-NOVA

The hillside medieval village of Monsanto was built around a pile of massive granite boulders. The placement of the rocks determined the shapes of the winding streets and the architecture of the stone houses—instead of trying to move the boulders, the villagers used them as walls, floors, and even roofs. From a distance, some of the houses look like they've been crushed by giant falling rocks.

Other than the odd plastic chair and air-conditioning unit, Monsanto has retained its medieval appearance. It is now home to about 800 people. There are no cars—the preferred mode of transport through the narrow cobbled streets is donkey. **Trains run from Lisbon to Castelo Branco. From there, it's a short bus ride northeast to Monsanto.**
Ⓝ 40.031970 Ⓦ 7.0713570

Residents of Monsanto literally live under a rock.

Masonic initiations are rumored to have taken place in this 88-foot-deep well.

Quinta da Regaleira

Sintra, Grande Lisboa

Eccentric millionaire António Augusto Carvalho Monteiro designed this palace in 1904 as a monument to his diverse interests and secret affiliations. The five-story hilltop mansion mixes Roman, Gothic, Renaissance, and Manueline styles. Its surrounding gardens are a fantasy land of grottoes, fountains, statues, ponds, underground tunnels, and a deep, moss-covered "Initiation Well," believed to be the former site of Masonic rituals. The architecture hides shapes and symbols relating to alchemy, Masonry, the Knights Templar, and the Rosicrucians. A Roman Catholic chapel in front of the palace depicts Catholic saints, but also pentagrams, which are often used in occult religions.
Avenida Barbosa du Bocage 5, Sintra. Make sure to bring a flashlight. Ⓝ 38.812878 Ⓦ 9.369541

◀ .... ▶

Also in Portugal

Carmo Convent Ruins	**Bone Chapel**	**Drowned Village**	**Bussaco Palace Hotel**
Lisbon · The legacy of a massive earthquake that nearly wiped Lisbon off the face of the earth in 1755.	*Faro* · A small chapel built out of human bones and decorated with a golden skeleton.	*Vilarinho da Furna* · A submerged village appears when the water levels drop at a nearby dam.	*Luso* · A majestic resting place for dreamers nestled in a fairy-tale forest.

Mafra Palace Library

MAFRA

The magnificent library within the 18th-century Palace of Mafra is a national architectural treasure, and it ranks among the finest libraries in Europe. It is also home to a colony of bats that patrol the stacks each night in search of book-eating pests.

The wooden shelves of the rococo-style library are lined with thousands of valuable old volumes. These books are fragile, however, and bookworms, moths, and other insects can wreak havoc on their delicate pages. Most libraries control such pests with more traditional techniques like fumigation or irradiation, but the Mafra Palace Library deploys its own very special force of winged protectors.

During the day, the bats sleep behind the bookcases or out in the palace garden. Then at night, after the library has closed, these tiny flying mammals swoop between the stacks, hunting down book-eating bugs.

This nocturnal feasting has been going on for centuries, perhaps as far back as the creation of the library itself. The bats, however, do come with one disadvantage: the copious covering of droppings they expend upon the floors, shelves, and furniture each and every night. To combat this, library workers cover the furniture before they leave and spend their mornings carefully cleaning the marble floors to erase all evidence of the previous night's excreta.

The Mafra Palace Library is open to researchers and scholars. An appointment is recommended. Because the bats only hunt at night, you're unlikely to see them during visiting hours. To catch a glimpse, stand just outside the library at night and wait for them to make their exit. It's also sometimes possible to hear the bats making sounds from their roosts inside the library on rainy, wet days.
Ⓝ 38.936976 Ⓦ 9.325933

A colony of tiny bats protects the 36,000 volumes of this library.

PRESERVED HEAD OF DIOGO ALVES

LISBON

The head of a 19th-century Portuguese serial killer is alarmingly well preserved in a jar at the University of Lisbon. Upon entering the anatomical theater at the Faculty of Medicine, the first thing you'll notice is the lone pickled head, up on a shelf next to a diaphonized hand. It's yellow, peaceful-looking, and somewhat akin to a potato. It's the head of Diogo Alves, whose claims to fame include being both Portugal's first serial killer and the last man to be hanged.

At least one half of each claim is true. Alves was a serial killer, indeed, but not the first. And he was not the last man to be executed—at least six more followed him to the gallows before Portugal ruled out capital punishment. So why, then, is Alves's head in a jar?

Timing, most likely. Alves was executed in 1841, as phrenology was just beginning to rear its ugly head in Portugal. Today, we recognize phrenology as a pseudoscience but at the time, people believed that personality traits—criminal propensity included—could be felt and measured right on the individual's skull. It's no surprise, then, that a notoriously wicked corpse would draw the attention of Portugal's budding band of phrenologists, who requested Alves's head be severed and preserved for posterity, so the source of his criminal urges could be studied in depth.

This part of the university is only open to students and is not typically accessible by the public.
Ⓝ 38.746963 Ⓦ 9.160439

SPAIN

DON JUSTO'S SELF-BUILT CATHEDRAL

MEJORADA DEL CAMPO, MADRID

Don Justo Gallego Martinez, a former monk with no experience in architecture, construction, or engineering, has been building this cathedral out of recycled and donated materials since 1961. There has never been a formal plan for the building—the design is influenced by St. Peter's Basilica but has changed over the years according to Don Justo's shifting inspirations.

The project began when Don Justo contracted tuberculosis and had to leave the monastic order. With his health in a perilous state, he prayed to the Virgin Mary and vowed to create a shrine in her honor should he survive. Though he never received an official construction permit, the recovered Don Justo devoted himself to building a church that now stands 13 stories tall. Oil drums, paint buckets, scrap metal, and bricks salvaged from a nearby brick factory are all pasted together with thick layers of concrete to form the walls and spires.

Don Justo receives occasional assistance from his nephews and volunteers, but the bulk of the work is done with his own hands. The cathedral is about 10 to 15 years away from completion—a problem, considering its chief builder is in his 90s. The fate of the building is up in the air. As an unapproved construction it could well be razed, eliminating the life's work of a most determined man.

Don Justo, a monk with no formal architectural training, inside his massive cathedral.

Calle del Arquitecto Gaudí, 1, Mejorada del Campo, Madrid. Take a bus from Conde de Casal to Calle de Arquitecto Antoni Gaudí. Ⓝ 40.394561 Ⓦ 3.488481

CLIFF-TOP CITY OF RONDA

RONDA

The city of Ronda is perched high atop the two cliffs of El Tajo Canyon as though a fissure opened and swallowed its center.

Romans established a settlement at Ronda nearly two centuries before the reign of Julius Caesar, and it has since survived through several invading forces. The walls of the canyon are sheer drops to the Guadalevín River over 330 feet (100.6 m) below—that's more than the height of the Statue of Liberty and her pedestal. Ronda's white stone buildings teeter on the very edge of the chasm.

Connecting the two parts of the city are three bridges: the Roman Bridge, the Arab Bridge, and the New Bridge. The first two are so called to recognize the regimes that built them. The (not so) New Bridge was completed in 1793 by the town's Spanish inhabitants. The bridges are impressive feats of stonework with ornate roofs above and massive columns that reach down into the canyon.

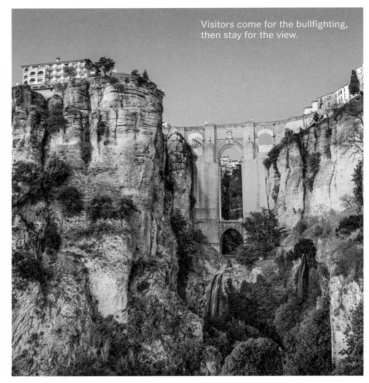

Visitors come for the bullfighting, then stay for the view.

Trains from Málaga (2 hours), Madrid (4 hours), or Granada (3 hours) make for a peaceful and scenic path to Ronda. If you're not into bullfighting, skip the popular 18th-century bullring and visit the Arab Baths instead. Ⓝ 36.740529 Ⓦ 5.164396

The devil men of the Baby Jumping Festival can leap four infants in a single bound.

BABY JUMPING FESTIVAL

CASTRILLO DE MURCIA, CASTILE AND LEÓN

Every year, 60 days after Easter, the village of Castrillo de Murcia invites men dressed as devils to leap over some babies.

The annual Baby Jumping Festival takes place in the streets, where infants are laid out on mattresses in rows of two or three. As the babies fidget and squirm in front of an audience gathered on the sidewalks, men dressed in bright yellow suits and devilish masks come charging down the street. They threaten bystanders with whips and act in a menacing manner before jumping over the rows of infants as though they were hurdles on a running track.

Dating back to the 17th century, the festival is intended to absolve infants of evil. Any child born within the last year is eligible to serve as a human hurdle for the devil men. Castrillo de Murcia. A taxi from Burgos is the best way to reach the tiny town. Ⓝ 42.358769 Ⓦ 4.060704

With its newly upgraded railings and walkway, the path is safer but no less scary.

El Caminito del Rey

EL CHORRO, MÁLAGA

Until its renovation in 2015, El Caminito del Rey was the most dangerous stroll in Spain. A 3-foot-wide (1 m) concrete path along the steep walls of the El Chorro gorge, 350 feet (107 m) above ground and over a century old, was riddled with large, ragged holes. In some sections the concrete had completely fallen away, leaving only the 3-inch-wide steel supporting beams. Mountain climbers and hikers would come here for the ultimate thrill, inching along the beams in windy conditions and trying not to look down.

The pathway was built in 1901 to provide a faster commute for workers at the nearby hydroelectric plant. Twenty years after the pathway was completed,

King Alfonso XIII crossed it for the inauguration of a new dam. This led to the pathway's modern name, "The King's Little Pathway," or "El Caminito del Rey." Soon after the king's crossing, however, the structure fell into disrepair.

Having recognized the perilous nature of the path—three men fell to their deaths in 2000—the local government began restoring it in 2014, laying down a new trail and adding handrails. Though the pathway no longer presents a potentially lethal challenge to hikers, it is just as thrilling to stand on—especially when the wind picks up.

Trains run infrequently from Alora to El Chorro—a taxi is your best bet for the 8-mile (13 km) journey north. Bring a sturdy harness, and clip onto the wire.
Ⓝ 36.729388 Ⓦ 4.442312

ALSO IN SPAIN

Josep Pujiula Labyrinth

Argelaguer · One man's self-made wonderland, this labyrinth was created from the natural landscape.

Chocolate Museum

Barcelona · Don't miss the chocolate model of the Sagrada Família.

The Hearse Museum

Barcelona · Displaying the finest in cadaver transportation.

Setenil de las Bodegas

Cádiz · This town built into the cliffs, possibly in the first century CE, still has a few thousand inhabitants.

Potty Museum

Ciudad Rodrigo · Thirteen hundred chamber pots are displayed by a single eccentric collector.

World's Biggest Chair

Cordoba · What better place to take a seat?

Castellfollit de la Roca

Girona · A small village situated on a narrow volcanic escarpment.

Los Jameos del Agua

Las Palmas · This partially collapsed lava tube and cave system is complete with concert hall, underground pond, and unique albino crabs.

Cave of the Moon

Madrid · Mysterious Spanish catacombs of unknown origin.

The Gala Dalí Castle

Púbol-la Pera · After buying this medieval castle, Salvador Dalí filled it with sculptures of spindly-legged elephants, busts of Richard Wagner, and a throne for his wife.

San Romà de Sau

Sau · This Romanesque tower appears when water levels drop.

Museo de las Brujas

Zugarramurdi · A museum dedicated to the Spanish occult.

THE WHISTLING ISLAND

LA GOMERA, CANARY ISLANDS

The whistles that echo across the valleys of La Gomera are not mere noise, but conversation. The tiny island's inhabitants speak to one another in Silbo, a wordless language that relies on pitch variation to communicate meaning.

Silbo originates with the Guanches, the first inhabitants of La Gomera, who spoke a tonal language with a simple structure. When Spanish settlers arrived in the 16th century, the Guanches adapted their simple Silbo to the Spanish dialect, creating the more complex version that is used today. To non–Silbo speakers, the sound is like birdsong. More distinctive is the method of speaking: Gomerans insert a finger or a knuckle of one hand into their mouths to make the sounds, while the other hand cups the side of the face to focus sound in the direction of the listener.

Fearing the extinction of the language, Gomerans made Silbo a mandatory part of the island's elementary school curriculum in 1999.

Get a ferry from Los Cristianos in Tenerife to San Sebastian de la Gomera. Ⓝ 28.103304 Ⓦ 17.219358

SWITZERLAND

BRUNO WEBER SKULPTURENPARK

DIETIKON, AARGAU

This park of monsters and mythical creatures is a glimpse into the marvelous mind of Bruno Weber. The Swiss sculptor began building his collection of oversize, exotic animals in 1962 to celebrate the power of imagination in his increasingly modernized hometown of Dietikon. Serpents, winged dogs, caterpillars, and mythological creatures surround a gothic, fairy-tale castle with an 82-foot tower—home to Weber and his wife, Mariann Godon, for decades.

When he was 75, Weber spoke of his plans to build a water garden in the park—a playful place with mosaic-coated sculptures, fountains, and ponds that would provide the finishing touch. Unfortunately, he died in 2011 at the age of 80, unable to complete his final flourish. Picking up his designs, Godon stepped in and finished the garden. It opened to the public six months after Weber's death.

Zur Weinrebe, Dietikon. The park is open on weekends from April to October. Get the train from Zurich to Dietikon, then hop on the bus to Gjuchstrasse. From there, it's a 15-minute walk to the sculpture park. Ⓝ 47.405469 Ⓔ 8.381182

A row of openmouthed, sharp-toothed creatures lines the roof at Bruno Weber Skulpturenpark.

THE CHILD-EATER OF BERN

BERN

Atop a blue and gold column in the middle of Bern sits an ogre, his jaw gaping and teeth bared as he happily eats a baby. He is Kindlifresser—"the Child-Eater"—and hoisting his sack of ready-to-eat-babies, he forms the centerpiece of one of the oldest fountains in the city.

Kindlifresser's origins are contentious. He may represent Kronos, the Titan king who, according to Greek mythology, ate five of his children. Another unfortunate possibility is that the Child-Eater reflects the 16th-century belief that Jews murdered children to use their blood for religious rituals. Kindlifresser's yellow pointed hat is strikingly similar to the headwear that Jews wore at the time.

Whatever the Child-Eater of Bern was originally intended to represent, his wide eyes, grotesque face, and sack of screaming infants makes for a unique city fountain.

Kornhausplatz, Bern. Take a bus to the medieval Zytglogge tower. You'll find the Kindlifresser in the middle of the plaza.
Ⓝ 46.948652 Ⓔ 7.447435

ALSO IN SWITZERLAND

H. R. Giger Museum

Gruyères · The bizarre visions of surrealist artist H. R. Giger are tucked away in a medieval Swiss city.

Collection de l'Art Brut

Lausanne · See artwork by loners, prisoners, and the criminally insane.

St. Maurice's Abbey

Saint Maurice, Valais · An abbey built on the ruins of Roman catacombs.

The Henkermuseum

Sissach · An extensive private collection of authentic medieval devices used for human torture and execution.

Maison d'Ailleurs

Yverdon-les-Bains · A museum of science fiction, utopias, and extraordinary journeys.

Evolver

Zermatt · See the Matterhorn Mountain from another angle on this spiral alpine-view construction.

Medizinhistorisches Museum

Zürich · A museum of medical history featuring Zurich's only authentic 14th-century plague doctor's uniform.

Bern's baby-eater with his sack full of appetizing infants.

LE MUSÉE DES GRENOUILLES

ESTAVAYER-LE-LAC, FRIBOURG

This museum in the medieval town of Estavayer-le-Lac caters to two interests: frogs and guns. During the mid-1800s, eccentric artist François Perrier stuffed over 100 frogs and arranged them into tableaux replicating everyday scenes, including a barber shop, a billiards game, a feast, and a rousing round of dominoes. The taxidermy scenes are in a room attached to an armory that contains firearms and military equipment dating from the Middle Ages through the early 20th century. The relationship between the two rooms, if any, is not clear.
Rue du Musée 13, Estavayer-le-Lac. The museum is near the south shore of Lake Neuchâtel. The train from Fribourg takes around 40 minutes. Ⓝ 47.405469 Ⓔ 8.381182

ABBEY LIBRARY OF SAINT GALL

ST. GALLEN

There are beautiful old libraries, and then there is the Abbey Library of Saint Gall.

Architectural plans for the library attached to the main church date back to the 9th century. As the abbey grew over the years, so did its library, and soon the site became known for its collection of illuminated manuscripts and writings, as well as becoming a leading center for science and Western culture between the 6th and 9th centuries.

In the mid-18th century, the world-renowned collection was moved to a new library space lavishly decorated in rococo style. Elaborate artworks were installed in the ceiling, framed by flowing, curved moldings. The wooden balconies are adorned with flowering shapes and designs.

Today, the library at St. Gall is still considered nearly unrivaled in its beauty. It also holds one of the more important collections in the world, covering 12 centuries. The collection even includes the first example of an architectural plan on parchment. **Klosterhof 6B, St. Gallen. The abbey is a short walk from St. Gallen Spisertor train station. Before touring the library, you'll be given slippers to slide over your shoes to protect the floors. Ⓝ 47.423348 Ⓔ 9.376754**

One of Europe's oldest surviving libraries is also one of its most beautiful.

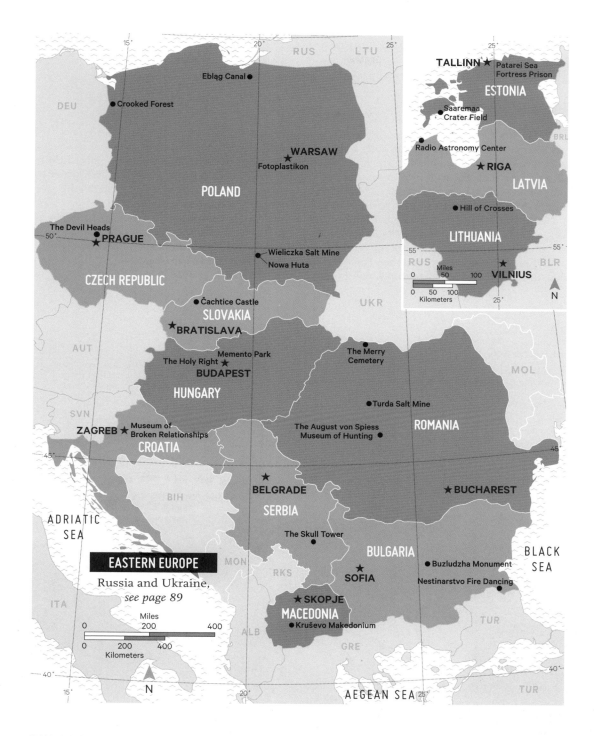

EASTERN EUROPE

Russia and Ukraine,
see page 89

BULGARIA

NESTINARSTVO FIRE DANCING

BALGARI, BURGAS

A tiny village tucked into the southeast corner of Bulgaria, Balgari is the only place in the country where Nestinarstvo, an annual fire-dancing ritual, is performed in its authentic form. The tradition is one in which dancers known as *nestinari* dance barefoot on smoldering embers to encourage fertility and good health.

Nestinarstvo is an amalgam of pagan and Christian practices. Dancers, many of whom enter a trancelike state prior to the ritual, carry icons of the saints as they step onto the circle of embers. Surrounded by an audience of villagers, they walk back and forth to the beat of a drum, shouting prophesies over the sound of bagpipes.

The dance is performed each year on June 3, the feast day of saints Constantine and Helena.

Ⓝ 42.087878 Ⓔ 24.729355

BUZLUDZHA MONUMENT

KZANLAK, STARA ZAGORA

On a remote mountain in Bulgaria sits a smooth, gray, disc-shaped monument that wouldn't look out of place in a schlocky sci-fi film. A red star on the monolith beside it, however, reveals its true origins: In the 1970s, 6,000 workers spent seven years constructing the building as a tribute to Communism. (A compulsory donation from every Bulgarian citizen provided funding for the project.)

When the Bulgarian Communist Party surrendered its political monopoly in 1989, and Bulgaria began the transition toward democracy, the Buzludzha site quickly lost its relevance. Vandals soon attacked the abandoned monument, destroying its interior artwork. The concrete structure remains, but a visit is more likely to inspire anti-communist sentiment than celebrate the wonder of socialism. A message painted in big red letters over the doorway reads: FORGET YOUR PAST.

Approximately 7 miles down a side road from the Shipka Pass in the Balkan Mountains.
Ⓝ 42.735819 Ⓔ 25.393819 ➤➤

ALSO IN BULGARIA

Belogradchik Rocks

Belogradchik · Bizarrely shaped rock formations are named after equally bizarre legends—the Dervish, the Rebel Velko, the Schoolgirl.

Kaliakra Transmitter

Bulgarevo · A massive, unfinished broadcasting station stands as a monument to the fall of Communism.

Though it looks like a sci-fi movie set, Buzludzha is a homage to the Bulgarian Communist movement.

➟ Other Brutalist Monuments of the Former Yugoslavia

During the 1960s and 1970s, Yugoslavian president Josip Broz "Tito" ordered the construction of these monuments to honor the Communist Party and commemorate the battle sites of World War II. All are made from concrete and designed in the Brutalist style—an imposing architectural movement popular with Socialist countries for its raw, imposing aesthetic.

MRAKOVICA MEMORIAL: MONUMENT TO THE REVOLUTION
Kozara, Bosnia and Herzegovina

KOSOVSKA MITROVICA MONUMENT
Kosovska Mitrovica, Kosovo

KOLAŠIN MONUMENT
Kolašin, Montenegro

MONUMENT TO THE REVOLUTION OF THE PEOPLE OF MOSLAVINA
Podgarić, Croatia

VALLEY OF THE HEROES MONUMENT
Tjentište, Bosnia and Herzegovina

MONUMENT TO THE UPRISING OF THE PEOPLE OF KORDUN AND BANIJA
Petrova Gora, Croatia

JASENOVAC MONUMENT
Jasenovac, Croatia

THE THREE FISTS AT BUBANJ MEMORIAL PARK
Niš, Serbia

The Tjentište War Memorial in Bosnia commemorates World War II's Operation Fall Schwarz.

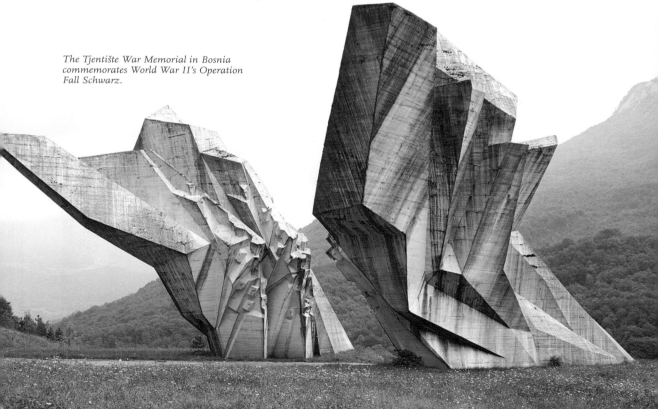

CROATIA
MUSEUM OF BROKEN RELATIONSHIPS

ZAGREB

When Croatian artists Olinka Vištica and Dražen Grubišić's four-year romance came to an end in 2003, the former couple joked that they would have to set up a museum to display all the objects they had shared. Three years later, they opened the Museum of Broken Relationships.

The institution contains a fascinating gathering of former tokens of affection. Besides the standard teddy bears and letters, the collection also includes a tiny bottle filled with tears, an ax, airsickness bags, and a prosthetic leg. While some of the items are tragic—a woman used the ax to smash her ex-girlfriend's furniture—some are sweet. The airsickness bags are from flights during a long-distance relationship, and the prosthetic leg came from a man who fell in love with his physical therapist.

Ćirilometodska 2, Zagreb. Get the funicular to avoid a steep hill-climb. If you are in the wake of a recently ended relationship, you are welcome to donate an object for exhibition. Ⓝ 45.815019 Ⓔ 15.973434

ALSO IN CROATIA

Goli Otok Prison

Goli Otok · Island gulag shut down in 1988.

Birthplace of Tesla Museum and Memorial Center

Smiljan · Learn about both Nikola Tesla's country upbringing and the scientific discoveries of his adult life.

Sea Organ

Zadar · Random harmonic sounds emerge when the wind and waves hit the tubes of this architectural object.

CZECH REPUBLIC
THE DEVIL HEADS

ŽELÍZY

A disturbing sight awaits hikers exploring the forest above the village of Želízy in Czechia. Looking out over the Kokořínsko nature reserve, two enormous demonic faces carved from the native stone stare back with empty eyes.

Created by the renowned Czech sculptor Václav Levý in the mid-19th century, the nearly 30-foot-tall sandstone heads are known as *Certovy Hlavy*, or "the Devil Heads," and they have been a local attraction for generations. Now suffering slightly from the ravages of time and weather, the monstrous faces have grown less distinct over time—but no less creepy. It's possible to see the Devil Heads from the street or by hiking about 0.3 miles up the relatively steep hill to reach the sculpture. There are several other stone works created by Václav Levý in the area, including artificial caves carved into nearby rock faces. Ⓝ 50.420551 Ⓔ 14.464792

The heads are best explored during the daytime—with a friend.

ALSO IN THE CZECH REPUBLIC

Capuchin Crypt

Brno · The final resting place for 24 perfectly preserved Capuchin monks, arranged in neat rows in front of a large wooden cross.

Křtiny Ossuary

Křtiny · This small-town ossuary features a dozen skulls painted with black laurel leaves.

Alchemy Museum

Kutná Hora · A museum and underground laboratory of modern-day alchemist Michal Pober. Filled with cauldrons, vials, potions, poster board explanations, and life-size dioramas.

Communist Clock

Olomouc · Originally built in the early 15th century, this clock was reconstructed in the Social-Realism style, featuring figures of Communist workers.

Church of St. John of Nepomuk

Žd'ár nad Sázavou · A Gothic Baroque pilgrimage site containing the incorruptible 14th-century remains of the Czech Republic's national saint.

ESTONIA
SAAREMAA CRATER FIELD

KAALI, SAAREMAA

Opinions vary on when it happened, but at some point between 5600 BCE and 600 BCE, a large meteor entered the atmosphere, broke into pieces, and slammed into the forest floor of the island of Saaremaa. The heat of the impact instantly incinerated trees within a 3-mile radius (5 km).

A mythology developed around the nine craters clustered in Saaremaa. Water gathered in the largest cavity—a 360-foot-wide, 72-foot-deep basin (110 x 22 m)—now regarded as a sacred lake. Iron Age inhabitants built a stone wall around it, and the discovery of silver and animal bones during archaeological excavations in

A 360-foot-wide basin created by a blazing meteorite.

the 1970s suggests the lake was a site for animal sacrifice and pagan worship. Some of the site's animal remains were dated to the 1600s, long after the church forbade such rituals.

The crater field now features a meteor museum, souvenir shop, and hotel offering a buffet breakfast and sauna. **Kaali Küla, Pihtla vald, Saaremaa. Ⓝ 58.303309 Ⓔ 22.70604**

PATAREI SEA FORTRESS PRISON

TALLINN, HARJU

Patarei, a sea fortress built in 1840, housed inmates in its cold, dark confines from 1919 until 2002. The prison has been left virtually untouched since it closed, right down to used cotton swabs in the operating room and pictures of women torn from magazines plastered on the cell walls. Rusting wheelchairs, flaking paint,

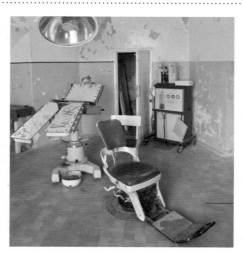

and dust-covered, neatly made beds provide a creepy atmosphere exceeded only by the musty, dimly lit Hanging Room.

Patarei is available to rent for weddings and parties. **Kalaranna 2, Tallinn. The fortress is open from May to October. Take a bus to Kalamaja. Ⓝ 59.445744 Ⓔ 24.747194**

The walls of the prison's dentistry room are painted a soothing blue.

HUNGARY
MEMENTO PARK

BUDAPEST

When Communism collapsed in Hungary in 1989, the city of Budapest was left with dozens of public monuments that celebrated the fallen regime. Rather than destroy these socialist relics, the city decided to banish them to the suburbs.

Twenty minutes outside the Budapest city center, Memento Park is the final resting place of more than 40 Communist-era statues and plaques. The open-air park displays the outcast monuments in a neutral setting, neither making a mockery of them nor honoring them.

Perhaps the most curious item in the park is a full-scale replica of Stalin's giant boots. A huge 26-foot statue of the Soviet leader once stood at Felvonulási Square in central Budapest, serving as a rallying point and parade route for the Communist regime. On October 23, 1956, Hungarians revolted against the regime and pulled down the huge statue, leaving only Stalin's massive boots behind. Though the revolution was brutally crushed, the replica of the footwear now serves as a memorial to those lost in the uprising—and a reminder of life behind the Iron Curtain.

The park can be reached from Budapest via public transport; the trip is approximately 25 minutes by metro or bus. Ⓝ 47.426346 Ⓝ 18.998732

ALSO IN HUNGARY

Electronic Museum

Budapest · This electric curiosity museum features Tesla coils and a Van de Graaff generator, housed in an old transformer station.

Golden Eagle Pharmacy Museum

Budapest · An alchemy museum that began its life in 1896 as a private collection of pharmaceutical oddities.

Taródi Vár Castle

Sopron · One dedicated family hand-built this 20th-century castle based on a medieval design.

The park encourages visitors to remember the past by confronting its relics.

CITY GUIDE: More to Explore in Budapest

Columbo Statue

Szent István krt • Installed in 2014, TV's iconic rumpled detective, along with his droopy basset hound, is immortalized as a permanent (and puzzling) bronze statue in a glitzy shopping district a couple of blocks from the Danube.

FlipperMuzeum

Józsefvóros • A museum devoted to classic pinball machines in the basement of an unremarkable apartment building, it's packed with rows of rare and vintage machines, including well-worn favorites and an example of the only known Hungarian-made machine.

The Timewheel

Zugló • This art installation in a public park near Vajdahunyad Castle is possibly the world's largest hourglass. It takes an hour to turn it and a complex system of steel cables—a big job for the team of four men who reset it every New Year's Eve.

Vajdahunyad Castle

Zugló • Built in 1896 as part of Hungary's millennial celebration, a temporary exhibition of a Transylvanian-style castle—flimsily glued and nailed together from cardboard and wood—was such a crowd favorite that in 1904 it got an upgrade to real stone.

The Hall of Hunting

Zugló • Inside the mock-Transylvanian Vajdahunyad Castle, to the Agricultural Museum exhibits of forestry, fishing, and taxidermied birds and stags, and up a stately stone staircase is the Hall of Hunting, a cathedral of antlers, horns, and hooves.

Gyermekvasút

Széchenyi-hegy • To say that Gyermekvasút is the world's largest children's railway is true, but it doesn't tell the whole story: Other than the engineer, this 7-mile-long (11.2 km) narrow-gauge rail line is run entirely by children aged 11 to 14.

Shoes on the Danube Promenade

Inner City • A moving place for reflection, this trail of iron footwear is a haunting memorial to the thousands of Jews who were told to remove their shoes before being executed along the river during World War II.

György Ràth Museum

Terézváros • A wealthy Hungarian optometrist's impressive and charming collection of more than 4,500 Asian curiosities, all collected during the late 19th and early 20th centuries, is a branch of the Museum of Applied Arts, and home to their conservation workshop.

The Hungarian Geological Institute Building

Istvánmezó • Home to the Hungarian Geological Institute (the country's oldest scientific research body) and the Geological Museum of Budapest, the building is a glorious example of the homegrown Art Nouveau architectural movement known as *Szecesszió* (Secessionist).

Hospital in the Rock Museum

Castle Hill • Under Buda Castle Hill, squirreled away in the network of tunnels and caverns, there is a museum dedicated to preserving the history (and occasional grisly tableaus) of a secret military hospital, now staffed mostly by stiff-backed uniformed mannequins.

The Castle Hill Funicular

Castle Hill • Built in 1870 at the entrance to the Széchenyi Chain Bridge, the span that originally connected Buda and Pest, this spectacular incline railway was almost lost to relentless shelling during World War II.

Semmelweis Medical History Museum

Castle Hill • With rare wax medical models by Clemente Susini (including a supine Venus), early X-ray machinery, and an array of anatomical curios, this museum (housed in the former childhood home of Ignác Semmelweis, a medical pioneer) spans the history of medicine, surgery, and pharmaceuticals.

Red Ruin

Inner City • The decor of the city's only Communism-themed "ruin bar" takes pride in presenting its own subversive poke in the eye to Cold War propaganda, with plenty of good beer, a whole lot of comic book design, and a tall glass of post-Communism political punditry.

For Sale Pub

Inner City • Covered floor to ceiling in stalactites and stalagmites of cardboard, napkins, and business cards, every possible surface of this cozy bar is available for patrons to post their drawings, notes, photos, and drunken come-ons.

The Garden of Philosophy

Tabán • Up high on Gellért Hill, in a park designed to quiet the mind, somber black statues of the world's most influential religious figures (Abraham, Jesus, Buddha, Lao-tzu, and Akhenaton) convene around a small metal ball under the watchful eyes of Gandhi, Bodhidharma, and Saint Francis of Assisi—representing the confluence and progression of human culture.

The Citadella

Gellérthegy • Occupied by the Austrian army, the Nazis, and then the Soviets, the imposing remains of this fortress have a story as tangled and complex as the city itself. Once a symbol of occupation and oppression, the citadel now represents the city's ability to transform itself.

Zwack Unicum Museum

Ferencváros • Often called the Hungarian national spirit, the Zwack family's eccentric liqueur, Unicum, dates to the late 18th century and gets the full museum treatment next door to their distillery. Exhibits include the 200-year history of the unique herbal concoction, plus Central Europe's largest collection of miniature alcohol bottles.

This installation commemorates Hungary's 2004 inclusion into the European Union.

THE HOLY RIGHT

BUDAPEST

Stephen I, Hungary's first king, died in the 11th century, but a piece of him lives on in Budapest: his right hand.

Talk of healing miracles occurring at Stephen's tomb led to his canonization in 1083. The exhumation of the former king's body revealed an incorruptible right arm—"incorruptible" being the Catholic belief that divine intervention can prevent the posthumous decay of saintly bodies.

Over the ensuing centuries, the king's detached arm passed through multiple countries and owners. During the 13th-century Mongol invasion of Hungary, it was sent to Dubrovnik, Croatia, for safekeeping by the Dominican monks. It was probably at this time that the right hand, or "Holy Right," was severed from the rest of the arm. (Dividing saintly body parts was a common practice at the time. Portions of a relic were often sent to churches in neighboring countries in order to prevent squabbling and political unrest.)

Today, the Holy Right, known to sacrilegious young Hungarians as "the monkey paw," resides in an ornate golden reliquary in Budapest's Basilica of St. Stephen. Drawn into a tight fist and clutching precious jewels, the hand—now shrunken and yellowed—still manages to look strong and defiant. **Szent István Bazilika, Szent István tér 1, Pest.**

Stephen I's desiccated, saintly hand sits in a gilded case at the Budapest basilica named for him.

St. Stephen's Basilica is a block west of the Bajcsy-Zsilinszky stop on the metro. To see the hand, go to the back left of the basilica and put a 100 forint coin in the slot. A light will illuminate the Holy Right for about 30 seconds. Ⓝ 47.500833 Ⓔ 9.053889

LATVIA

RADIO ASTRONOMY CENTER

IRBENE, VENTSPILS

Until 1993, the 105-foot (32 m) radio antenna in the remote forests of Irbene was a top-secret piece of espionage equipment. Members of the Soviet military, who lived in a purpose-built housing complex nearby, used the dish to monitor communications between NATO countries during the Cold War.

Following the restoration of Latvia's independence in 1991, Soviet troops gradually withdrew from the country. Before departing Irbene, soldiers took the time to attack the radio equipment, pouring acid into motors, cutting cables, and hurling pieces of metal into the antenna's mechanisms.

A radio tower once used by spies, now operated by astrophysicists.

Though severely damaged, the big dish—the eighth-largest in the world—survived. In July 1994, the Latvian Academy of Sciences took over the site, spending three years conducting repairs and reconfiguring the antenna to operate as a radio telescope for astronomical studies. The academy's Ventspils International Radio Astronomy Center division now uses the telescope to observe cosmic radiation and debris.

Irbene's Soviet past is evident in its military ghost town, a collection of crumbling concrete blocks filled with the abandoned possessions of its former inhabitants. **Ances Irbene. LV-3612, Ventspils rajons. Irbene is a three-hour bus ride west of Riga, the capital. From there, it's 20 miles (32 km) north to Irbene. The astronomy center offers tours, which include a climb of the telescope. Ⓝ 57.558056 Ⓔ 21.857778**

LITHUANIA
HILL OF CROSSES
MEŠKUIČIAI, ŠIAULIAI

Crosses have been accumulating on this small hill since the 14th century, when Teutonic Knights of the Holy Roman Empire occupied the nearby city of Šiauliai. New crosses tend to appear during periods of occupation or unrest as symbols of Lithuanian independence. This was particularly evident during a peasant uprising against Russian control in 1831, when people began placing crucifixes in remembrance of missing and dead rebels. By 1895, there were 150 large crosses on the site. In 1940, the number had grown to 400.

During the Soviet occupation, which lasted from 1944 to 1991, the Hill of Crosses was bulldozed three times. Each time, locals and pilgrims returned to put up more crosses. The site achieved worldwide fame when Pope John Paul II visited in 1993 to thank Lithuanians for their enduring symbol of faith.

There are now approximately 100,000 crosses on the hill. The faithful are welcome to add their contribution, in whatever form they wish—a crucifix made of Legos recently joined the collection.

The hill is 7 miles (11.3 km) north of Šiauliai, which is reachable by bus or by train from Vilnius. From Šiauliai, catch a bus bound for Joniškis and get off at the Domantai stop. From there, it's a brief walk to the crosses. Ⓝ 56.015278 Ⓔ 23.416668

ALSO IN LITHUANIA

Witches Hill

Curonian Spit · This outdoor sculpture trail features carvings of 80 Lithuanian folk heroes located at the site of annual midsummer celebrations.

Grūtas Park

Druskininkai · A sort of Soviet theme park and open-air museum, it features recreations of gulag prison camps complete with barbed wire and guard towers.

Devils' Museum

Kaunas · A collection of 3,000 artworks depicting the devil shows myriad interpretations of the dark lord.

100,000 crosses of all sizes are crammed together on a hill.

MACEDONIA
THE KRUŠEVO MAKEDONIUM

KRUŠEVO, KRUŠEVO

The space-age spherical building on a hill overlooking the medieval town of Kruševo resembles something between a *Star Wars* set piece and a giant virus. Neither of these remotely relates to the monument's solemn purpose: to commemorate the 1903 Ilinden uprising, when a group of Macedonians revolted against the Ottoman Empire in an attempt to establish an autonomous state. Eight hundred rebels took control of Kruševo on the night of August 2, renaming it the "Kruševo Republic."

The Kruševo Republic lasted 10 days before the Ottomans struck back. An 18,000-strong army stormed the town and quickly recaptured it, burning and plundering as they went. Despite the short life of the Kruševo Republic, Macedonians revere the leaders of the Ilinden Uprising, and August 2 is a national holiday. The Makedonium monument, built in 1973, is held in similar esteem, and it also appears on national currency. It contains stained glass skylights, a centerpiece that resembles an oversize gas burner, and the tomb of the uprisings' leader Nikola Karev.

The Makedonium is a space-age–style tribute to an early-20th-century uprising.

Kruševo is a two-hour drive south of Skopje. The monument is less than a mile from the center of town. Ⓝ 41.377404 Ⓔ 21.248334

POLAND
ELBLĄG CANAL

JELONKI

Due to drastic changes in elevation, the Elbląg Canal is broken up into short strips of water separated by stretches of land. In order to navigate this tricky waterway, an ingenious system of inclined planes was created to transform boats into railroad cars for the troublesome portions of the journey.

Stretching from Lake Drużno to Jeziorak Lake, the narrow course is the longest navigable canal in Poland. Yet it was nearly unusable until the mid-1800s, when the King of Prussia ordered a novel solution. As the canal is too long and steep to use traditional water locks, pairs of rail tracks are laid across the dry stretches. Giant water-powered cradles then lift the boats up out of the water,

A creative engineering feat that combines two modes of transportation.

place them on the tracks, and carry them across the ground to the next bit of sailing territory. The unique amphibious canal has been hailed as one of Europe's most impressive engineering marvels.

Today, the canal is still in use, though mostly as a tourist attraction. Boat tours run the length of the canal, a roughly 11-hour journey. If the entire trip is too lengthy, you can hop off about halfway through, but the full impact of this clever invention might be somewhat derailed. Ⓝ 54.028372 Ⓝ 19.594404

The salt mine's Chapel of St. Kinga features chandeliers made of salt.

WIELICZKA SALT MINE

WIELICZKA, LESSER POLAND

Miners at Wieliczka carved its rock salt deposits without interruption from the 13th century until the 1990s. Over the centuries, workers slowly turned the seven-level subterranean mine into a majestic salt city replete with life-size rock salt sculptures of saints, biblical wall reliefs, and tableaus depicting their daily lives.

In the early 1900s, the workers undertook their most ambitious project: an underground church named after Kinga, the patron saint of salt miners. The 331-foot-deep (101 m) St. Kinga's Chapel features a sculpture of Christ on the cross, depictions of scenes from the New Testament, a wall relief of Da Vinci's *The Last Supper*, and two altars. All are carved from salt. Hanging from the ceiling are five chandeliers that miners crafted by dissolving salt, removing its impurities, and reconstituting it into crystals as clear as glass.

Another memorable sight on the tour is the placid subterranean lake in the Józef Piłsudski Chamber, softly lit and overseen by a statue of Saint John Nepomucene—the patron saint of drowning. Take a moment of reflection before you bundle into a small, dark miners' cage with five other people for the long ascent back to the surface.

Take Danilowicza 10, Wieliczka. Located in a suburb of Kraków, the nearest railway station is Dworzec PKP Wieliczka-Rynek. Ⓝ 48.983039 Ⓔ 20.055731

THE CROOKED FOREST

NOWE CZARNOWO, WEST POMERANIA

At first, Gryfino Forest looks to be a run-of-the-mill field of trees. And then you see it: a group of 400 pines, each with a mysterious, dramatic bend close to the ground.

The trees' unusual but uniform "J" shape is likely the result of human intervention—probably farmers who manipulated the trees with the intention of turning them into curved furniture. The pines, planted in 1930, had around ten years of normal growth before being distorted. An alternative theory holds that regular flooding caused the unusual shapes.

Gryfino is a 30-minute train ride from the city of Szczecin. Ⓝ 53.214827 Ⓔ 14.474695

By hook or by crook, the Gryfino trees flourish.

NOWA HUTA

KRAKÓW

As Soviet occupying forces rolled into Poland toward the end of World War II, they found a country devastated by the ferocious fighting on the Eastern Front. Rebuilding was in order, and Moscow saw the opportunity not only to remake Poland's cities, but also to dramatically reshape Polish society while they were at it.

To achieve this, the Soviets set to work planning and building Nowa Huta, which was to be an ideal city representing a vision of a glorious Communist future. The project was approved in 1947, and construction of the urban experiment in social engineering began in 1949.

One of only two fully planned socialist realist cities ever built (the other being Magnitogorsk, Russia),

A Soviet model for proletarian paradise.

Nowa Huta was created as a bustling working-class enclave. Built on the outskirts of Kraków, for a population of 100,000, Nowa Huta was laid out in a sunburst pattern radiating from a monumental central square (Plac Centralny) and built in a stirring architectural style that combined Renaissance elegance with the grand, overwhelming scale typical of Soviet projects. Wide avenues were designed to halt the spread of fires. Trees lining those avenues were planted to absorb the impact of a nuclear blast.

Most important was Nowa Huta's intended status as a proletarian paradise. To that end, the city was built with a massive steel mill at one end (*Nowa Huta* means "New Steel Mill"). The Lenin Steelworks contained the largest blast furnace in Europe and employed 40,000 people at its height, with the capacity to produce 7 million tons of steel annually. It was an odd location for such a facility, given weak local demand for steel, but in this case—as in many others during the Cold War—the symbolism was more important than the logic.

Ironically, Nowa Huta later turned into an anti-Communist hub and was key in the Solidarity movement of the 1980s. Nevertheless, the city remains to this day one of the best examples of socialist realist architecture and urban planning.

Run to Nowa Huta from Kraków's city center. While you're there, swing by Lord's Ark church. Locals fought for 28 years for permission to build it, after which it became a powerful symbol of Polish resistance to Communist rule. Ⓝ 50.071703 Ⓔ 20.037883

ALSO IN POLAND

Live-in Salt Mine	UFO Memorial	Konstantynów Radio Tower	Skarpa Ski Jump
Bochnia · Poland's oldest salt mine has nearly everything you need to live underground forever, including a gymnasium and a spa.	*Emilcin* · Site of the most famous alleged UFO abduction case in Polish history.	*Płocki* · The shattered remains of what was once, at 2,120 feet (646 m), the tallest structure in the world.	*Warsaw* · Ruins of a late 1950s ski jump ramp can be found in the middle of the city.

FOTOPLASTIKON

WARSAW, MASOVIAN

Before movie theaters and motion pictures, the European public entertained itself with a visit to the "fotoplastikon." Invented in Germany in the late 19th century, the fotoplastikon, or Kaiserpanorama, is a cylindrical wooden structure with multiple viewfinders through which people can view illuminated stereoscopic photographs.

In the first half of the 20th century, there were around 250 fotoplastikon devices across Europe. Visitors sat at one of the pairs of goggles and watched, spellbound, as seemingly three-dimensional scenes from around the globe paraded past. Images of African deserts, American cities, and Arctic expeditions—all the stuff of fantasy in the pre-cinema,

pre-air-travel era—provided an escapist thrill and broadened people's perceptions of the world.

This Warsaw model, built in 1905, is one of only a few left that are still in working condition. It is equipped with 18 viewing stations and sits in the middle of a parlor plastered with old travel posters. **Aleje Jerozolimskie 51, Warsaw. The fotoplastikon parlor is 2 blocks from the Centrum metro station. Ⓝ 52.231374 Ⓔ 21.008064**

ROMANIA

THE MERRY CEMETERY

SĂPÂNȚA, MARAMUREȘ

At the Cimitirul Vesel, or "Merry Cemetery," over 600 colorful wooden crosses bear the life stories, dirty details, and final moments of the bodies that lie below. Displayed in bright, cheery pictures and annotated with limericks are the stories of almost everyone who has died in the town of Săpânța. Illustrated crosses depict soldiers being beheaded and a townsperson being hit by a truck. The epigraphs are surprisingly frank and often funny: "Underneath this heavy cross lies my mother-in-law . . . Try not to wake her up. For if she comes back home, she'll bite my head off."

The cemetery's unique style was created by a local named Stan Ioan Pătraș, who at the age of 14 had already begun carving crosses for the graveyard. By 1935, Pătraș was carving clever and ironic poems—done in a rough local dialect—about the deceased, as well as painting their portraits on the crosses, often depicting the way in which they died.

Pătraș died in 1977, having carved his own cross and leaving his house and business to his most talented apprentice, Dumitru Pop. Pop has spent the last three decades continuing the carving work, and has also turned the house into the Merry Cemetery's workshop-museum. Despite the occasionally darkly comic—or merely dark—tones of the crosses, Pop says no one has ever complained about the work:

It's the real life of a person. If he likes to drink, you say that; if he likes to work, you say that . . . There's no hiding in a small town . . . The families actually want the true life of the person to be represented on the cross.

Church of the Assumption, Săpânța. The cemetery is near the Romanian–Ukrainian border, at the intersection of routes 19 and 183. Ⓝ 47.97131 Ⓔ 23.694948

ALSO IN ROMANIA

Zoological Museum

Cluj-Napoca · The natural history museum housed within the Babes-Bolyai University appears as if it has been untouched for half a century. Scruffy, ratty taxidermy fills glass cases, hangs from the ceiling, and peers down from an off-limits second floor.

Decebal's Head

Orșova · While the 13-story bearded stone face overlooking the Danube River looks like something straight from Middle Earth, it is the recent creation of a Romanian businessman. When the ten-year sculpting process was completed in 2004, the head became the tallest rock sculpture in Europe.

Life and death are celebrated in equal measure at the Merry Cemetery.

THE AUGUST VON SPIESS MUSEUM OF HUNTING

SIBIU, SIBIU

Glass-eyed animal heads cover the dimly lit walls of this museum, a stark reminder of Romania's appeal to hunters. The country's large bear population is due to the policies of the country's last Communist leader, Nicolae Ceaușescu, who, after depleting the bear population in his own private reserve, made bear hunting illegal for everyone but himself and a few handpicked Communist Party members.

The measure protected many bears from slaughter, but Ceaușescu killed more than his fair share. Driven by a desire to hunt the biggest animals, he had bear cubs captured, fed a hearty diet, and then released back into the wild when they had fattened up. But the animals had grown so used to being fed by humans that they died hungry in the wild. Undeterred, Ceaușescu switched methods, ordering that the bears be fed raw meat and beaten with sticks to discourage attachment. The resulting aggressive bears were known to attack hikers and cars.

One of Ceaușescu's largest trophies—the skin and stuffed paws of a huge brown bear—is on display at this museum. The bulk of the trophies, however, hail from the 1,000-strong personal collection of Colonel August von Spiess, a fellow chaser of Carpathian bears and Romania's royal hunt master during the 1920s and '30s.

The dog was killed by the bear; the bear by the dog's owner. Both were mounted: one as trophy, the other as homage.

Strada Şcoala de Înot, Nr. 4, Sibiu. In the heart of Transylvania, Sibiu is a 5-hour train ride northwest from Bucharest. Ⓝ 45.786634 Ⓔ 24.146900

TURDA SALT MINE

TURDA, CLUJ

This former salt mine, excavated by hand and machine over hundreds of years, is now a subterranean fairground-cum-spa. Operational from the times of the Roman Empire until 1932, the mine closed for 60 years, reopening to the public in 1992. The microclimate—a steady 53°F (11.6°C) year-round, with high humidity and no allergens—is ideal for halotherapy, an alternative health treatment in which people with respiratory problems spend time breathing in the salt-infused air.

The current attractions in the 260 × 130-foot (80 × 40 m) space make it easy to pass the hours. They include a Ferris wheel, mini-golf course, bowling alley, and underground lake with paddle boats. To offset the darkness, bright lights hang vertically on strings from the 16-story ceiling, illuminating dripping stalactites with a blue-tinged glow.
Aleea Durgăului 7, Turda. Ⓝ 46.566280 Ⓔ 23.790640

Yellow rowboats bob on an underground lake, illuminated by hanging lights.

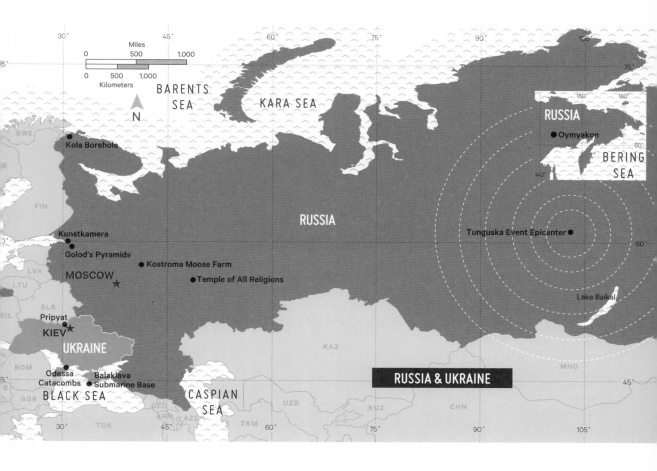

RUSSIA

ALEXANDER GOLOD'S PYRAMIDS

OSTASHKOV, TVER OBLAST

Aggression; osteoporosis; black-heads; dizziness; heartburn; depression; sterility; learning disabilities; arachnophobia: All these ailments and many more can be swiftly cured, according to Russian scientist and defense engineer Alexander Golod. The remedy is simple: pyramids.

Seizing upon a New Age belief that pyramids exude healing energy, Golod has built fiber-glass pyramids all over Russia. The largest of the structures is 15 stories tall, and located one hour

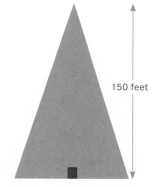

150 feet

Step inside to experience a number of health benefits. (The number likely being zero.)

outside of Moscow. People feeling unwell, run-down, or burdened by life's responsibilities are invited to step inside the pyramid and

experience the musty-smelling tranquility. They are then gently guided to the gift shop, where they may purchase pebbles, mini pyramids—which Golod claims emit a calming, energy-rebalancing force field—and bottled water that has been stored inside the pyramid, where it is said to acquire healing properties.

Despite Golod's claims of miraculous growth and recovery within the pyramids, no scientific body has ever confirmed that the building or its trinkets confer any tangible curative effect.

Ostashkov, Tver. The pyramid is by Lake Seliger, northwest of Moscow.
Ⓝ 57.140268 Ⓔ 33.128516

KUNSTKAMERA

ST. PETERSBURG

Peter the Great, who ruled Russia from 1682 to 1725, was interested in all things modern, scientific, and rational. During his reign—when not busy ordering the interrogation, torture, and death of his own son—Peter collected artwork, scientific books and instruments, fish, reptiles, insects, and human specimens. In 1714, he ordered that his collection form the foundation of a new museum in St. Petersburg. The institution, called the Kunstkamera, was the country's first museum, and aimed to show the world that Russia was a modern, scientific, secular country.

Peter the Great's 300-year-old collection of human body parts

Completed in 1654, the Gottorb Globe was the world's first planetarium.

is displayed on level two of the Kunstkamera. The focus of the collection, established in 1727, is on infant anatomy and disease—malformed fetuses, tumor-ridden stomachs, and jarred children's heads, preserved with care by 17th-century Dutch anatomist Frederik Ruysch.

Also on display is the skeleton of Nikolai Bourgeois, a 7-foot-2-inch (2.2 m) man who was Peter's assistant and a living exhibit at the museum, as well as a stuffed two-headed calf and the preserved fetuses of conjoined twins. The 32 human teeth neatly arranged into a grid were all extracted by Peter the Great, who found dentistry to be a rewarding hobby.

Universitetskaya Embankment, 3, St. Petersburg. Get a bus or trolleybus to Universitetskaya Naberezhnaya. ℕ 59.941568 Ⓔ 30.304588

ALSO IN RUSSIA

Kadykchan

Kadykchan · A Siberian ghost town built by gulag prisoners, many of whom later worked in the nearby coal mines.

Manpupuner Rock Formations

Komi · Enormous natural rock formations rise abruptly from the flat landscape surrounding them and tower over the Russian plateau.

Lena's Stone Pillars

Lena River · Evidence of ancient human life and fossils from mammoths, bison, and fleecy rhinos have been found at this remote forest of stone pillars.

Museum of Soviet Arcade Machines

Moscow · Founded by Russian university students in the basement of a technical school, the Museum of Soviet Arcade Machines features over 40 machines from the era, including video games, pinball machines, and collaborative hockey foosball.

House of Evgeny Smolik

Selo Irbeyskoe · Inspired by fairy tales, Evgeny Smolik turned his village home into a surreal palace with intricate wood carvings and fantasy-themed furniture.

Church on Blood in Honor of All Saints Resplendent in the Russian Land

Yekaterinburg · This Russian Orthodox church was built on the site where the last tsar and his family were shot by the Bolsheviks during the Russian Civil War.

TUNGUSKA EVENT EPICENTER

VANAVARA, KRASNOYARSK KRAI

On June 30, 1908, at 7:14 a.m., a powerful explosion shattered windows, knocked people off their feet, and leveled 80 million trees over 830 square miles (2,150 sq. km) of forest around Siberia's Podkamennaya Tunguska River basin. Initial speculation was that a meteorite had hit Earth, but subsequent investigations found no crater in the area.

Naturally, the mysterious nature of the Tunguska Event has given rise to a wealth of conspiracy theories. Among the more far-fetched culprits: a tiny black hole passing through the Earth; a UFO crash; and the testing of Nikola Tesla's secret "death ray." Today, the favored scientific explanation involves the midair explosion of a large meteoroid or comet. Indeed, it is the largest impact event in recent history.

Split, mangled, and felled trees are all still visible around the Tunguska site.

The closest village to the Tunguska Event epicenter is the town of Vanavara, located about 40 miles (65 km) southeast of the epicenter. ℕ 60.902539 Ⓔ 101.904508

KOLA SUPERDEEP BOREHOLE

MURMANSK, MURMANSK OBLAST

Until 1970, geologists could only theorize about the composition of the Earth's crust. That was the year Soviet scientists began drilling what would become the deepest hole in the world.

Engaged in a subterranean version of the Space Race, the USSR went all out to beat the US in a journey to the center of the Earth. While American researchers faltered with Project Mohole, a dig off the coast of Mexico that ran out of funding in 1966, their Russian counterparts took a more determined approach. From 1970 to 1994, their drill on the Kola Peninsula burrowed through

Kola hole is the deepest pit ever drilled in the name of science.

layers of rock, reaching an ultimate depth of 7.5 miles (12 km).

The most intriguing discovery made by the Kola borehole researchers was the detection of microscopic life-forms 4 miles (6.7 km) beneath the surface of the Earth. Usually fossils can be found in limestone and silica deposits, but these "microfossils" were encased in organic compounds that remained surprisingly intact despite the extreme pressures and temperatures of the surrounding rock. Drilling at Kola stopped in the early 1990s, but data from the dig is still being analyzed.

The hole is northwest of Murmansk on the Kola Peninsula, a few miles from the Norwegian border.
Ⓝ 69.396219 Ⓔ 30.608667 ➤➤

➤➤ World's Deepest Places

DEEPEST CANYON:
11,596 FT/3,534.5 M
Cotahuasi Canyon,
Peru

DEEPEST CAVE:
7,208 FT/2,197 M
Krubera Cave,
Abkhazia

DEEPEST HOLE:
39,600 FT/12,070 M
Kola Borehole,
Russia

DEEPEST MINE:
12,795 FT/3,900 M
TauTona Gold Mine,
South Africa

DEEPEST OPEN PIT MINE:
3,937 FT/1,200 M
Bingham Canyon
Mine, United States

Bingham Canyon Mine, United States

DEEPEST RAIL TUNNEL:
790 FT/240.8 M
Seikan Tunnel,
Japan

DEEPEST POINT UNDER WATER:
35,838 FT/
10,923.4 M
Mariana Trench,
Pacific Ocean

DEEPEST LAKE:
5,314 FT/1,619.7 M
Lake Baikal, Russia

DEEPEST FOUNDATION:
394 FT/120 M
Petronas Towers,
Malaysia

DEEPEST MAN-MADE POINT:
40,502 FT/12,345 M
Sakhalin-I Oil Well,
Russia

Exhibition of Achievements of National Economy

Ostankinsky • Bigger than the principality of Monaco and inspired by the great 19th-century expositions of London, Paris, and Chicago, this open-air market/ museum/amusement park works overtime to glorify Russian agriculture, industry, and technology, all on a grand scale.

Monument to the Conquerors of Space

Ostankinsky • Swooping up over the Memorial Museum of Cosmonautics is a 350-foot (107 m) Jetsonian curve of titanium, topped by a rocket blasting into the sky. At the base of the monument are stone bas-relief sculptures of scientists and cosmonauts.

Laika Monument

Airport • It took more than half a century, but in 2008 a monument to a famous space dog was finally unveiled at a space-training facility, where a plaque tells of her bravery and tragic contribution, and her fans can leave flowers and trinkets to say, "Good dog, Laika."

Laika was the first animal to orbit the Earth.

Aquarelle Train on the Moscow Metro

Sokolniki • With stations that look like something out of *Dr. Zhivago*, the Moscow Metro turns public transportation into an art form. Especially nice is the Aquarelle car on the Red Line, wrapped in a floral watercolor skin and with an interior like a museum gallery.

Padlock Tree Park

Yakimanka • In what started as a practical solution to the structural problem often caused by the weight of "love locks" on bridges, the city has installed rows of metal "trees" over Luzhkov Bridge and along the Moscow River, where newlyweds can lock, kiss, and toss their keys into the water to their bursting hearts' content.

Lubyanka

Meshchansky • This Neo-Baroque block of a building was the All-Russia Insurance Company before the KGB filled its long halls. Make an appointment to visit its propaganda-filled museum.

Lenin's Mausoleum

Tverskoy • The world's most famous "modern mummy," the embalmed remains of the leader of the Bolshevik Revolution, can be viewed by small groups inside an oddly stunted pyramid of a tomb.

The Old English Court Museum

Tverskoy • The 16th-century headquarters of the Mystery and Company of Merchant Adventurers, also known as the Muscovy Company, was the conduit for all official trade between England and Russia from 1551 to 1917.

Romanov Palace Chambers in Zaryadye

Tverskoy • Long before they became tsars of the Russian Empire, the Romanovs were just another bunch of aristocrats wrangling for power, and their centuries-old lineage can be traced to this fairly unpretentious family home in the historic Zaryadye district.

Tsar Bell

Tverskoy • It's never actually been rung, but it's still the largest bell in the world. Cast in 1735, the bell rests on some block stone behind the Ivan the Great Bell Tower at Red Square, and is so big it once served as a chapel where parishioners could enter through a "door," which was really a hole caused by the bronze cracking under its own weight.

Miniature Moscow—Capital of the USSR

Dorogomilovo • It's the tallest hotel in Europe, and all 34 stories sparkle and gleam like Stalin ordered, but on the first floor of the historic Hotel Ukraina is a vastly smaller achievement: a 3,200-square-foot (300 m) miniature Moscow, a diorama of the entire city, just as it looked in 1977.

A. N. Scriabin House

Arbat • The innovative and controversial composer Alexander Scriabin's home is reverently preserved and features a working version of his theosophical color keyboard.

Children Are the Victims of Adult Vices

Yakimanka • Along the Moscow River, there is a surreal 13-piece allegorical art installation by Mihail Chemiakin that depicts greed, poverty, and indifference as figurative corrupting influences over two angelic children.

Moscow Cats Theatre

Meshchansky • Be forewarned: This circus of cats and clowns is not without controversy and its wild popularity has brought some well-deserved scrutiny to the felines' welfare, but for anyone curious to see if cats can, in fact, be herded, this may be the place to go.

The Battle of Borodino

Meshchansky • The brutal clash between French and Russian forces at the 1812 Battle of Borodino is masterfully re-created on a 360-degree, 375-foot (115 m) panoramic canvas by artist Franz Roubaud, depicting the more than 250,000 troops and 70,000 casualties.

Fallen Monument Park

Yakimanka • This odd sculpture garden, also called Muzeon Park of Arts, dispassionately displays tossed-aside Soviet sculptural relics like busted-up statues of Stalins, Lenins, and the founder of the KGB, and a Soviet emblem that looks like an old James Bond prop.

State Darwin Museum

Academic • The world's first museum of evolution, its collection dates back to 1907, with dinosaur models (enhanced with a catchy dinosaur soundtrack), dioramas of wildlife from the North Pole to the South, and only slightly moth-eaten taxidermy by master stuffer Filipp Fedulov.

Lenin Hills Museum–Reserve at Gorki Leninskiye

Gorki Leninskiye • Outside of Moscow's city limits is the palatial, very unproletariat final home of Vladimir Lenin, preserved as a museum to his memory, with his Rolls-Royce Silver Ghost (outfitted with tank treads), a reproduction of his Kremlin office, his death mask, and a plaster cast of his stroke-plagued hands.

OYMYAKON

OYMYAKON, SAKHA REPUBLIC

Located just a few hundred miles from the Arctic Circle, the Siberian village of Oymyakon is the coldest permanently inhabited place on Earth.

Every January, the fur-swaddled citizens of Oymyakon endure average daily highs of −47°F (−43.9°C), with nighttime temperatures plummeting to around −60°F (−51.1° C). The lowest temperature ever recorded was −90°F (−67.8°C) in 1933.

Oymyakon's 500 residents live on a diet of mostly reindeer and horse meat because the frozen

Welcome to the coldest town on Earth.

ground makes it difficult to grow crops. Cars are hard to start because the axle grease and fuel tanks freeze, and batteries lose life at an alarming speed.

Summer, however, brings relief. Temperatures can even reach the 70s (20s C) during July.

Trips to Oymyakon start with a flight to Yakutsk, capital of the Sakha Republic and the coldest major city in the world. From there it's about 20 hours of driving to get to Oymyakon. It's best to travel with a local who has a car well-equipped to handle the chill.
Ⓝ 63.464263 Ⓔ 142.773770

TEMPLE OF ALL RELIGIONS

KAZAN

The colorful Temple of All Religions, or Universal Temple, is a mishmash of architectural flourishes culled from most of the major world religions.

Established by philanthropist Ildar Khanov in 1992, the site is not a chapel in the traditional sense, but a center meant to stand as a symbol of religious unity. Khanov, an advocate for rehabilitation services for substance abusers, built the center with the help of patients he met through his work.

The exterior of the temple looks almost like something out of Disneyland's It's a Small World ride, with a Greek Orthodox dome here and a Russian minaret there. There are design influences from Jewish synagogues and Islamic mosques, along with a number of spires and bells. All in all, the temple incorporates architectural influences from 12 religions in a bright cacophony of devotion.
Arakchinskoye Shosse, 4, Kazan. Khanov died in 2013, but his associates continue to work on the temple. You can get there by bus and train from Kazan's railway station. Ⓝ 55.800620 Ⓔ 48.974999

A colorful symbol of religious unity and rehabilitation.

KOSTROMA MOOSE FARM

SUMAROKOVO

In the early 1930s, the USSR set its sights on spreading Communism throughout the world—on mooseback. Large, strong, and agile even through deep snow, the moose seemed to be the perfect animal for the Soviet cavalry. And so began the quest to domesticate the northern moose.

In secret moose husbandries, the wild animals were trained to carry armed riders and not be gunshy. There were attempts to mount pistols and shields to the antlers of the bull moose. In the end, military moose never took off, but these taming efforts did lead to the rise of modern experimental moose farms, where semi-domesticated moose are still raised today.

The early farms were tasked with raising moose for milk, transportation, and to nourish the hungry populace. Despite initial apprehension, it turned out these gentle giants were just as easily milked as cows. However, raising them for meat proved prohibitively expensive, and moose, being clever creatures, wouldn't be led easily to slaughter. Despite years of brutal and bloody experimentation, it was discovered that it's difficult to make a moose do what a moose does not wish to do, and efforts to fully control them were abandoned.

The Kostroma Moose Farm opened in 1963 with a new approach, known as free-range moose ranching. The tamed moose roam the forest but return to the farm by choice, recognizing it as a reliable food source and safe place to give birth. The Kostroma Farm started with just two calves and has since been home to over 800 moose. It's also functioned as a scientific research facility, but today the farm's primary functions are producing moose milk for medical treatments, harvesting antler velvet for pharmaceutical purposes, and providing a place for tourists to visit these fascinating creatures.

The Kostroma Moose Farm is located near the village of Sumarokovo, about 12 miles (20 km) east of the city of Kostroma. The facility is open daily and guided tours can be purchased. The best time to visit is in early spring, when the first calves are born. Ⓝ 57.664955 Ⓝ 41.114243

SERBIA

THE SKULL TOWER

NIŠ, NIŠAVA

The skull tower is the grim product of the 1809 Battle of Čegar, a turning point during the First Serbian Uprising against the Ottoman Empire. Desperate in the face of certain defeat, rebel commander Stevan Sinđelić fired into a gunpowder keg, annihilating his entire army as well as the enemy soldiers who had flooded the trenches.

Angered by Sinđelić's actions, Turkish commander Hurshid Pasha ordered the mutilation of the dead rebels' bodies. Their skins were peeled off their decapitated heads, stuffed with straw, and sent to the imperial court in Istanbul as proof of Turkish victory.

The 952 skulls left behind were used as building blocks for a 15-foot-tall (4.6 m) tower constructed at the entrance of the city. Sinđelić's skull sat at the top. The gruesome construction left a deep scar in the national psyche but did not deter its citizens from fighting for freedom from the Ottoman Empire. The Serbs rebelled again in 1815, this time successfully, driving off the Turks and winning independence in 1830.

In the years immediately following the construction of the tower, the families of deceased rebels chiseled away some of the skulls in order to give them proper funerals. Today, only 58 skulls remain. They are surrounded by hundreds of cavities, each one representing a person who died in battle. A chapel was

A grisly reminder of the sacrifices made for Serbian independence.

built around the tower in 1892 to shield it from the elements.

Dušana Popovića, Niš. Niš is a 3-hour bus ride from Belgrade. Ⓝ 43.311667 Ⓔ 21.923889

ALSO IN SERBIA

Devil's Town Rock Formation

Kuršumlija · The 202 bizarre rock spires of Đavolja Varoš are the result of natural soil erosion.

Bridge of Love

Vrnjačka Banja · Young couples have sealed their love by attaching a padlock to this bridge since World War I.

Red Cross Concentration Camp

Niš · A Red Cross memorial museum sits at the site of the Niš concentration camp that operated for four years during World War II. The museum celebrates a daring, 105-person prison escape in February 1942.

SLOVAKIA

ČACHTICE CASTLE RUINS

ČACHTICE

Four hundred years ago, Hungarian countess Elizabeth Báthory died in a closed room inside this castle. Known as the Blood Countess, she was imprisoned for unimaginable acts, which included torturing and killing hundreds of girls, and, if the legends are to be believed, bathing in their blood.

In 1610, after multiple tip-offs from locals that terrible things were happening inside Čachtice Castle, King Matthias II ordered the collection of testimonies and evidence, but Báthory was never convicted. In return for avoiding a trial, the family waived the king's debts. There is no way of knowing how many girls Báthory killed. Estimates of the number of victims ranged from 50 to over 600. Little is left of the building today, but there is enough to evoke the blood-soaked walls and agonized screams of Báthory's torture chambers.

Home of the world's most prolific female serial killer.

The castle is on a hill overlooking Čachtice.
Ⓝ 48.725075 Ⓔ 17.760988

UKRAINE

ODESSA CATACOMBS

ODESSA, ODESSA OBLAST

Rusted mining equipment, World War II grenades, 19th-century wine barrels, and human remains are some of the things you might stumble upon during a journey into the labyrinthine Odessa catacombs.

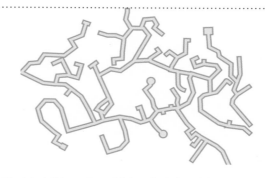

The labyrinthine nature of Odessa's dark and dusty catacomb tunnels has occasionally resulted in tragedy.

The estimated 1,500 miles (2,800 km) of passage that weave beneath the streets of the city were mostly dug by limestone miners in the early 1800s. When the mines were abandoned, they quickly became the preferred hideout of rebels, criminals, and eccentrics.

After the Soviets were forced out of Odessa during WWII, dozens of Ukrainian rebel groups stayed, hidden in the tunnels. While waiting for opportunities to strike they attempted to lead normal lives, playing chess and checkers, cooking, and listening to Soviet radio. Meanwhile, the Nazis tossed poison gas canisters into the catacombs and sealed random exits, hoping to trap or smoke out the rebels.

Today, only a small portion of the catacombs is open to the public as part of the Museum of Partisan Glory in Nerubayskoye, north of Odessa. The rest of the tunnel system is structurally unstable, partially flooded, and irresistible to urban spelunkers. Groups of explorers spend days underground, bringing headlamps, waders, and backpacks full of food and wine.

Occasionally, the subterranean parties turn deadly. In 2005, a group of Odessa teens spent New Year's Eve partying in the catacombs. In the drunken revelry, one of the girls became separated from the group and got lost in the catacombs. She spent three days wandering in the freezing cold and pitch-black darkness before she died of dehydration. Two years later, police were finally able to locate her body and retrieve it from the depths.

Most catacomb exploration begins in Nerubayskoye, a small town just northwest of downtown Odessa. Exploring the tunnels is not illegal, but it is not encouraged. It would be folly to venture into the catacombs without a guide. Choose carefully—your life is in their hands.
Ⓝ 46.546667 Ⓔ 30.630556 ➡

➺ Other Hidden Tunnels

METRO 2

The existence of Metro 2, the informal name of Moscow's alleged secret underground metro system, has never been proven, but KGB defectors, US intelligence, and former Russian ministers all say it's there. The system—said to be larger than Moscow's public metro network—was apparently constructed during Stalin's reign to evacuate leaders during periods of unrest. It continues to operate under the control of the Ministry of Defense, according to everyone except the KGB. In the mid-1990s, urban explorers claimed to have discovered an entrance to the system.

LONDON SEWER SYSTEM

Conditions were pretty grim in 1850s London. Its citizens used the River Thames as an open sewer, resulting in an ever-present stench and waves of cholera epidemics. It was time for the government to build a modern sewer system.

Construction of the 550-mile (885 km) network took place between 1859 and 1865. It incorporated the River Fleet, a major Thames tributary during Roman London that had been

An upside-down medusa head is part of the architecture at the Basilica Cisterns in Istanbul.

gradually forced underground by industrial development.

BASILICA CISTERNS OF ISTANBUL

Beneath Istanbul are hundreds of Byzantine cisterns—underground reservoirs built during the 5th and 6th centuries to store rainwater. The cathedral-like structures have an elegance that belies their utilitarian nature, and contain decorated arches, marble columns, and carvings of Medusa's head.

ANTWERP RUIEN

Antwerp's many natural ditches served as fortifications, trade routes, and open sewers from the 11th to 16th centuries. When the smell became too much to bear, the city asked each citizen to take responsibility for covering the ditches, or *ruien,* on their land. It took citizens 300 years to cover the ditches in a diverse range of materials that reflected their wealth, taste, and competence as builders. The underground ruien functioned as the city's sewers until the 1990s, when they were emptied in favor of a new network of pipes.

UNDERGROUND CITIES OF CAPPADOCIA

Cappadocia, a historic region in Turkey, contains a network of underground, multilevel cities sealed off from the world with large stone doors. Carved from volcanic rock around the 7th or 8th century BCE and connected by tunnels, the cities contained kitchens, wine cellars, wells, staircases, stables, and chapels. Early Christians used the cities as a hiding place to escape Roman persecution.

BALAKLAVA SUBMARINE BASE

BALAKLAVA, CRIMEA

Balaklava was a quiet fishing village until 1957, when the Soviet government suddenly wiped it from official maps in order to establish a secret submarine base. Working under Stalin's orders, military engineers created "Object 825 GTS," a seaside underground complex dedicated to housing and repairing naval submarines, storing weapons and fuel, and acting as a safe bunker in case of nuclear attack.

Moscow subway workers spent long hours gouging out granite to build the rockbound complex. When the four-year construction process finished in 1961, Object 825 GTS boasted a 2,000-foot-long (607 m) canal capable of housing six submarines, a hospital, communication centers, food storehouses, and an ample arsenal of torpedoes, nuclear warheads, and rockets.

The construction of Object 825 GTS turned Balaklava into a military town with closed borders. Residents—almost all of whom worked at the base—were not even permitted to receive visits from family members.

The submarine base remained secret and operational until 1993, when post-Soviet conditions rendered it unnecessary. In 2004, the base opened to the public as a naval museum. The submarines are gone, but the long stone corridors, dark canals, and a few leftover missiles provide plenty of Cold War atmosphere.

Mramornaya Street, Balaklava. The submarine base turned museum is in Balaklava Bay—get a bus from Sevastopol. Ⓝ 44.515236 Ⓔ 33.560650

Pripyat

PRIPYAT, KIEV OBLAST

Pripyat's clocks all read 11:55. That's the moment when, on April 26, 1986, the electricity was cut following a meltdown at the Chernobyl nuclear reactor. A day later, Pripyat residents received the following evacuation announcement:

For the attention of the residents of Pripyat! The city council informs you that due to the accident at Chernobyl Power Station in the city of Pripyat, the radioactive conditions in the vicinity are deteriorating . . . Comrades, leaving your residences temporarily, please make sure you have turned the lights, electrical equipment, and water off, and shut the windows. Please keep calm and orderly in the process of this short-term evacuation.

Today Pripyat is a city of abandoned buildings with paint peeling away from the walls, falling in flakes onto dusty shoes, toys, and Communist propaganda posters. Outside the crumbling City Center Gymnasium, a rusting Ferris wheel sits beside a jumble of bumper cars. They are the lone remains of a carnival that was due to open on May 1, 1986.

This somber, silent city seems an unlikely vacation spot, but it is possible to tour the Chernobyl area. A government-issued day pass is obtainable in Kiev. It is deemed safe to walk around Pripyat for only a few hours at a time, and several precautions must be followed to avoid contamination. Visitors must be accompanied by a tour group and are forbidden from touching structures or placing anything on the ground within the exclusion zone. Arms, legs, and feet must be covered, and the trip ends with everyone being screened for radiation using a Geiger counter.

Visitors are free to take photographs, view the reactor from a distance of 100 meters, and even talk to the few remaining residents of Pripyat who disobeyed orders after the blast and returned to their radiation-contaminated homes.

In the 30 years since Pripyat was abandoned, plants and animals have begun to thrive despite the high levels of radioactivity. Tree roots burst through concrete floors, forests encroach on the roads, and animals, such as beavers, boars, wolves, and bears, long vanished from the area, have returned. Free of human influence, the area has a much greater biodiversity than it did before the disaster.

Guided tours are available and depart by bus from Kiev.
Ⓝ 51.405556 Ⓔ 30.056944

ALSO IN UKRAINE

Underwater Museum

Crimea · Over 50 busts of Communist and Socialist figures from the USSR are lined up on stone shelves underwater.

The Swallow's Nest

Gaspra · A castle-like home perched on the very edge of a cliff overlooks the Crimean Sea.

Monastery of the Caves

Kiev · This 1,000-year-old relic-filled cave also hosts an amazing museum of minuscule portraits, documents, and sculptures that must be viewed through a microscope.

Salo Museum

Lviv · A museum dedicated to one of Eastern Europe's essential ingredients: pure pig fat.

Eternity Restaurant

Truskavets · An eatery, run by the local funeral parlor, features the largest coffin in the world.

Pripyat's creaky Ferris wheel has been still since the Chernobyl meltdown caused the town to be abandoned in 1986.

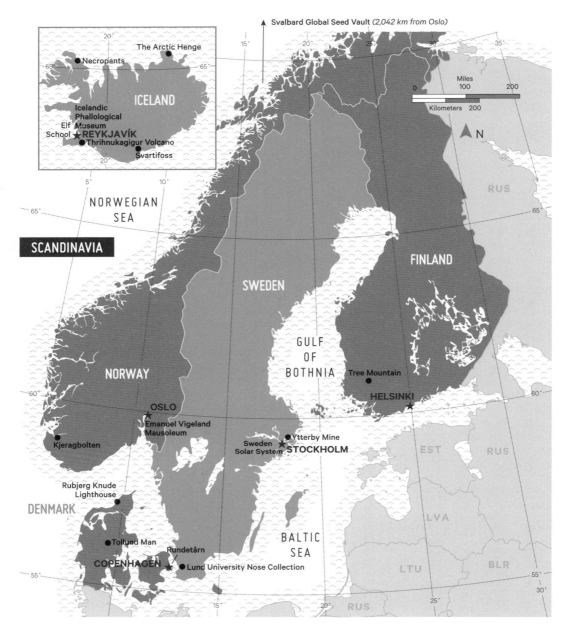

Svalbard Global Seed Vault *(2,042 km from Oslo)*

ICELAND

- Necropants
- The Arctic Henge
- Icelandic Phallological Museum
- Elf School
- ★ REYKJAVÍK
- Thrihnukagigur Volcano
- Svartifoss

NORWEGIAN SEA

SCANDINAVIA

RUS

SWEDEN

FINLAND

GULF OF BOTHNIA

NORWAY

- Kjeragbolten
- ★ OSLO
- Emanuel Vigeland Mausoleum

- Sweden Solar System
- ★ STOCKHOLM
- Ytterby Mine

- Tree Mountain
- ★ HELSINKI

EST

RUS

LVA

DENMARK

- Rubjerg Knude Lighthouse
- Tollund Man
- Rundetårn
- ★ COPENHAGEN
- Lund University Nose Collection

BALTIC SEA

LTU

BLR

RUS

Miles 0 100 200
Kilometers 200

N

DENMARK

RUBJERG KNUDE LIGHTHOUSE

LØKKEN, HJØRRING

The Rubjerg Knude lighthouse is slowly being swallowed by its surroundings. Built just off the North Sea in 1900, the 75-foot (23 m) tower is now half-buried in sand, the result of coastal erosion, wind, and shifting dunes.

For a few decades, the lighthouse keepers fought against the encroachment. They planted a perimeter of trees and shoveled sand from the courtyard. But it was a losing battle. More sand blew in, hampering views of the sea and forcing the lighthouse to cease

This lighthouse is losing its battle against shifting sands.

operations in 1968. The tower and its surrounding buildings stayed open as a museum and coffee shop until 2002, when the growing dunes threatened to overwhelm the entire operation.

All five of the surrounding buildings are now smothered in sand. The lighthouse, too, will soon be blotted out by the forces of nature. The site is now open for just a few weeks in the summer.

Fyrvejen, Løkken. Ⓝ 57.448989 Ⓔ 9.777089

RUNDETÅRN

COPENHAGEN

Danish for "round tower," Rundetårn is a cylindrical building, topped with a dome that contains Europe's oldest functioning observatory. Built in 1642—the year of Galileo's death—under the orders of King Christian IV, the tower originally contained a planetarium showing two versions of the solar system: the Galileo-approved, heliocentric model and Danish astronomer Tycho Brahe's geocentric interpretation.

Rundetårn is notable for its internal architecture—it contains no stairs, just a spiral brick path that winds around a central column seven and a half times. The unusual design had a practical purpose: Large, heavy scientific instruments needed to be transported to the top of the tower, and wagons made the job a lot easier.

The Rundetårn observatory operated in conjunction with the University of Copenhagen until 1861, when it was replaced by the new Østervold Observatory, built on the outskirts of town to avoid light pollution.

Rundetårn is now open to the public for stargazing and sightseeing. The tower is also the site of an annual spring unicycle race in which riders pedal up and down the spiral. The current record, set in 1988, is 1:48.7. **Købmagergade 52A, Copenhagen. Take the metro to Nørreport. Ⓝ 55.681964 Ⓔ 12.575691**

Come spring, the interior path of Europe's oldest observatory will be filled with unicyclists.

ALSO IN DENMARK

Skeletons of Æbelholt Abbey

Æbelholt · Explore the ruins of a 12th-century abbey and hear the legend of the miracle-working monk who once lived there.

Little Mermaid Statue

Copenhagen · Decapitated, mutilated, and then blown up, this beloved statue and national icon has seen her share of abuse.

Tobacco Museum

Copenhagen · Learn about the long history of tobacco smoking and its assorted paraphernalia and accoutrements.

Hair Jewelry

Frederikshavn · A local-history museum displays examples of the 19th-century fad for jewelry made from human hair.

World Map

Klejtrup Lake · Play mini-golf while traversing a perfect miniature version of the Earth.

Killed in the 4th century BCE and pulled from a peat bog in 1950, the Tollund Man still looks pretty good.

TOLLUND MAN

SILKEBORG, CENTRAL DENMARK

Walking around the Bjældskovdal bog in 1950, brothers Emil and Viggo Højgaard (along with Grethe, Viggo's wife) stumbled upon a body. Believing the man to be the victim of a recent killing, they called the police. Further investigations revealed that he had indeed been murdered—some 2,300 years earlier.

The Tollund Man was found curled in the fetal position with his eyes closed and a serene expression frozen on his face. The cold, acidic, oxygen-starved conditions of the peat bog had kept him remarkably well preserved. His hair, beard stubble, eyelashes, and toenails were all intact, and he was nude, but for a sheepskin cap and wide belt around his waist. A rope was wound tightly around his neck. The Iron Age man had been hanged, likely during a ritual sacrifice.

In 1950, it was not yet known how best to preserve discoveries like Tollund Man. Accordingly, only the head of the original specimen was kept intact. The rest of the body was subjected to various tests to determine his probable age (probably around 40, due to the presence of wisdom teeth and wrinkles) and the conditions surrounding his life and death. Among the details found: Tollund Man was 5 feet 3 inches (1.6 m) tall, his final meal was a gruel made from barley and flaxseed, and his "sacrificers" (read: killers) took the time to close his mouth and eyes after death.

Thousands of "bog bodies" have been discovered in sphagnum swamps across Northern Europe, but the Tollund Man remains the best preserved. His original head and reconstructed body now reside at the Silkeborg Museum. The rope used to end his life is still wrapped around his throat.

Silkeborg Museum, Hovedgårdsvej 7, Silkeborg. A slew of buses stops right in front of the museum.
ⓝ 56.164444 ⓔ 9.392778

FINLAND

TREE MOUNTAIN

YLÖJÄRVI, PIRKANMAA

This conical tree-covered hill was not created by nature: It's a planned work of art that was 14 years in the making. Artist Agnes Denes first proposed *Tree Mountain*, a human-made forest on a human-made hill, in 1982. Ten years later, the Finnish government announced that it had approved the project. From 1992 to 1996, 11,000 people each planted a tree on a specially sculpted mound of dirt. Together, the trees form an intricate pattern derived from a combination of the golden ratio and Denes's own pineapple-inspired design.

Each tree belongs to the person who planted it, and their descendants—the site is legally protected for 400 years.

Pinsiönkankaantie 10, Pinsiö. The hill is about a 3-hour drive from Helsinki. ⓝ 61.571030 ⓔ 23.477081

ALSO IN FINLAND

Helsinki University Museum

Helsinki · Browse the labyrinthine halls to discover a mid-19th-century Finnish pharmacy, a collection of brass cartography tools, and wax models of infant diseases.

Rock Church

Helsinki · Hidden inside a rocky outcrop is a late-1960s church with rock walls and a wraparound skylight that bathes the space in light. The church has excellent acoustics and hosts concerts.

Veijo Rönkkönen Sculpture Park

Parikkala · See eerie sculptures sporting real human teeth in one of the most important collections of contemporary folk art in Finland.

International Coffee Cup Museum

Posio · Explore worldwide coffee culture through a collection of almost 2,000 cups from over 80 countries.

The Lenin Museum

Tampere · One of the only permanent Lenin museums in the world is located at the Tampere Workers Hall, where Lenin and Stalin met for the first time in 1905.

ICELAND

ICELANDIC PHALLOLOGICAL MUSEUM

REYKJAVÍK

Sigurður Hjartarson began collecting phallic specimens in the 1970s, beginning with a "pizzle"—a whip made from a bull penis. Since then, his phallic collection has grown to enormous proportions.

The museum aims not merely to titillate, but to advance the "ancient science" of phallology, which examines how male genitalia have influenced history, art, psychology, and literature. Devoted to the study and appreciation of mammalian penises, its 280 specimens are drawn from a wide range of animals, including polar bears, badgers, cats, goats, seals, and even a blue whale, whose daunting 5-foot-7-inch (1.7 m) member is the largest in the collection.

The museum also oversees a small collection of *Homo sapiens* specimens, courtesy of men who bequeathed their genitals to the museum. One of the donors, an American, made a cast of his penis—which he dubbed "Elmo"—to be kept in the museum until the real item could be donated. Another, a 95-year-old man from Iceland, decided to contribute his penis so that it might be preserved as an eternal totem of his many youthful indiscretions.

In addition to biological specimens, the museum also features phallic artwork and objects. Following their silver medal win at the 2008 Summer Olympics, 15 members of Iceland's handball team provided casts of their penises. They are painted silver and displayed in a row behind glass—the phallic equivalent of being featured on a cereal box.

Laugavegur 116, 105 Reykjavík. The museum is near the Hlemmur bus station. Ⓝ 64.143033 Ⓦ 21.915643

THE ARCTIC HENGE

RAUFARHÖFN

Located in one of Iceland's most remote northern villages, the Arctic Henge is a colossal piece of stone construction that, when finished, will make Stonehenge look like amateur hour.

Started in 1996, the Arctic Henge project is a monument not only to the country's nordic roots, but also to some of the neo-pagan beliefs that have arisen in certain areas. The piece was inspired by the Eddic poem *Völuspá* (*Prophecy of the Seeress*), taking from it the concept of 72 dwarves who represent the seasons in the world of the poem.

In the Arctic Henge, 72 small blocks, each inscribed with a specific dwarven name, will eventually circle four larger stone monuments, which in turn will surround a central balanced column of massive basalt blocks. Each aspect of the deliberate layout corresponds to some aspect of ancient Norse belief, and when each piece of the monument is installed, visitors will be able to "capture the midnight sun" by viewing it through the various formations at different vantage points depending on the season.

So far, only the imposing central tri-column and one of the four larger gates have been constructed, along with a smattering of the smaller stones. **The Henge is a 90-minute drive from Húsavík, which is a prime spot for whale-watching. Ⓝ 66.462132 Ⓦ 15.962863**

The towering Artic Henge is still a work in progress.

NECROPANTS

HÓLMAVÍK

Some might say it's unseemly to exhume the corpse of a departed friend, flay his skin in one piece from the waist down, and wear that flesh as a pair of leggings. These people do not know about the rich tradition of necropants, a wealth-attracting good-luck garment.

According to the Museum of Icelandic Sorcery and Witchcraft, necropants were a real thing in 17th-century Iceland. The rules were complex. First, you had to get permission from a living man to use his skin after he died. When he kicked the bucket, you would wait around for the burial formalities to conclude, then approach the grave and start digging. Corpse exhumed, you cut around the waist and peeled the skin from the bottom half of the body, making sure to keep it all in one piece.

The next step was to steal a coin from a poor widow. This coin was placed into the scrotum of the necropants, where it would magically attract more money, leaving the wearer with a groin full of coins at all times. Once you'd had enough of the great wealth, or the necropants began to chafe, you would have to find another wearer to step into the magical leggings. In this way, the prosperity was passed down for generations.

A pair of necropants is on view at the museum, in a softly lit alcove, standing on a bed of coins.

Museum of Icelandic Sorcery and Witchcraft, Höfðagata 8-10, Hólmavík. The village of Hólmavík is a 4-hour bus ride from Reykjavík. Ⓝ 65.706546 Ⓦ 21.665667

Svartifoss: It's fun to say, and even more enchanting to visit.

SVARTIFOSS

KIRKJUBÆJARKLAUSTUR, SKAFTÁRHREPPUR

Svartifoss, meaning "black fall," is a modest waterfall in terms of height, width, and force, but its backdrop of black hexagonal columns makes for a rare and splendid sight. The columns are basalt crystals, formed from lava flows that cooled over centuries—the same process that created the textured walls of Fingal's Cave in Scotland. Parts of the crystals often break off and plunge into the river, so mind the sharp rocks at the base of the falls.

785 Fagurholsmyri, Skaftafell National Park, Kirkjubæjarklaustur. Bus services to Skaftafell run daily from Reykjavík during the summer months. Svartifoss is an hour-long hike from the Skaftafell National Park visitors' center. Ⓝ 64.020978 Ⓦ 16.981623

ALSO IN ICELAND

Santa's Workshop

Akureyri · The year-round home of Santa Claus—or a very convincing lookalike—is decked out like a gingerbread house.

Bjarnarhöfn Shark Museum

Bjarnarhöfn · Get a taste of Iceland's fishing industry and sample a famous Icelandic delicacy: fermented shark.

Blue Lagoon

Grindavík · A medicinal spa built around the discharge of a geothermal energy plant.

Jökulsárlón

Höfn · Iceland's largest lagoon is home to stunning multicolored icebergs.

THRIHNUKAGIGUR VOLCANO

BLUE MOUNTAINS COUNTRY PARK, BLÁFJÖLL

Volcanoes are usually best admired from a safe distance, but Iceland's Thrihnukagigur is so geologically unique, it is possible to go right inside the heart of the volcano. The three calderas of Thrihnukagigur have lain dormant for so long—the last eruption was over 4,000 years ago—brave visitors can actually descend into the volcano's colorful magma chamber.

An open elevator takes you over 600 feet down into the depths of the enormous crater, which is so large it could fit the Statue of Liberty in its entirety. Inside the cavern, you're met with a surreal sight. Rather than the jet-black obsidian you might expect, the craggy walls are covered in a gleaming, pearlescent rainbow of color that almost makes the cave look like it's composed purely of gems.

This is the only place on Earth where you can take a cable lift into the heart of a volcano, thanks to a strange natural phenomenon. Usually, after an eruption, the roiling magma cools and solidifies in place, effectively plugging the opening. But somehow the fathoms of magma that once boiled inside one of Thrihnukagigur's peaks sank back down into the earth, leaving behind a massive open cavern. **It is a moderate 2-mile hike (about 45 minutes long) through the Icelandic highlands to the Thrihnukagigur summit. Each excursion into the volcano's depths lasts around an hour or two. Ⓝ 63.998920 Ⓦ 21.697522**

A cable lift brings visitors to the glittering heart of a volcano.

ICELANDIC ELF SCHOOL

REYKJAVÍK

When Icelandic member of parliament Árni Johnsen escaped unharmed from a car crash in 2010, he knew whom to credit for his survival: elves. After rolling five times, the politician's SUV came to rest beside a 30-ton boulder. Johnsen, believing that multiple generations of elves called that boulder home, concluded that they used their magic to save him. When roadwork later required the removal of the boulder, he claimed it for himself, transporting it to his home to ensure the elves would continue to watch over him.

Johnsen's beliefs are not unusual. According to Icelandic folklore, thousands of elves, fairies, dwarves, and gnomes—collectively known as "hidden people"—live in rocks and trees throughout the country. It is no wonder, then, that the world's only elf school is located in Reykjavík.

Historian Magnús Skarphéðinsson, who has spent decades documenting people's encounters with elves, established the school in 1991. Classes focus on the distinguishing characteristics of Iceland's 13 varieties of hidden people. The school also offers five-hour classes for travelers, which include a tour of Reykjavík's elf habitats. Students receive a diploma in "hidden people research."

Skarphéðinsson has never seen an elf. His knowledge of their appearance and behavior comes from the hundreds of testimonies he has collected from people who claim to have made contact with hidden people.

Though Skarphéðinsson has devoted 30 years to the subject and considers himself the foremost authority on elves, he maintains a sense of humor about it all. At the end of class, he serves homemade coffee and pancakes and tells stories about the people who come up to him to say, "I swear I'm not on drugs, but I saw the strangest thing . . ."

108 Síðumúli, Reykjavík. Buses run on Suðurlandsbraut or Háaleitisbraut. Ⓝ 64.133062 Ⓦ 21.876143

NORWAY

SVALBARD GLOBAL SEED VAULT

LONGYEARBYEN, SVALBARD

A winter night in Longyearbyen lasts four months. In the ice-covered mountains, the darkness is broken only by a slim concrete building that emits a pale blue glow as it overlooks the 1,000-resident town. The simple structure offers no hint as to what's protected inside: a collection of seeds that could save humanity.

Humanity's genetic safe-deposit box

Due to the loss of genetic diversity among commercially cultivated crops, which tend to be grown from clonal monocultures, many worldwide food crops are at risk of disease. Mutated strains of fungus, or a new bacterium, could potentially wipe out an entire world crop in a matter of months, causing massive food shortages. The Svalbard Global Seed Vault was established by the Norwegian government in 2008 to function as a sort of genetic safe-deposit box.

The facility has the capacity to conserve 4.5 million seed samples. Under the current temperature conditions in the vault, which are similar to those in a kitchen freezer, the seed samples can remain viable to begin new crops for anywhere from 2,000 to 20,000 years.

Svalbard was chosen as the location because it is tectonically stable and its permafrost provides natural refrigeration in case of a power failure. There is no permanent staff at the seed bank, but it is monitored constantly using electronic surveillance. Access to the vaults, open only to employees, requires passing four locked doors protected by coded access keys.

To get to Longyearbyen, you'll need to fly from Oslo with a stopover in Tromsø. While the seed vaults are off limits, the building itself is a dramatic sight surrounded by banks of snow. Other winter activities in the tiny town include dog sledding and reindeer spotting. Ⓝ 78.238166 Ⓔ 15.447236

Is the Kjeragbolten stone stable enough to stand on? There's one way to find out.

Kjeragbolten

FORSAND, ROGALAND

Kjeragbolten is a boulder wedged in a mountain crevasse, 3,228 feet (984 m) above the ground. It is a favored spot for BASE jumpers, who hurl themselves from the cliff toward the spectacular fjord below. Visitors without vertigo are welcome to step onto the boulder for a unique photo opportunity—there are no fences restricting access.

Øygardstøl, Forsand. In the summer months, a bus runs from Stavanger to Øygardstøl, where the hike to Kjeragbolten begins. Wear appropriate shoes and be prepared for a steep climb. The journey takes about 3 hours each way. Don't step onto the rock if it's raining or damp. Ⓝ 59.022535 Ⓔ 6.581841

Also in Norway

Saltstraumen Maelstrom

Bodø · Behold the world's strongest whirlpool, created by a powerful tidal current.

Hessdalen AMS

Hessdalen · A remote research station devoted to unlocking the mystery behind floating lights seen in a Norwegian valley.

Mølen

Larvik · Found on Norway's largest stone beach, these 230 large man-made rock piles, or *mols*, are in fact an ancient cemetery dating back to 250 BCE.

Steinsdalsfossen Waterfall

Norheimsund · Walk the path behind this 164-foot-high (50 m) waterfall and see it from the other side.

EMANUEL VIGELAND MAUSOLEUM

OSLO

Brother to the more-celebrated Gustav Vigeland, whose eccentric sculptures occupy a prominent park in central Oslo, Emanuel Vigeland will be remembered through his own strange and enchanting artistic work.

The Emanuel Vigeland Museum serves double duty as a mausoleum designed and decorated by Vigeland himself. Visitors enter the building by stooping through a heavy, low iron door. Inside, a large, darkened, barrel-vaulted room is completely covered with paintings that show human life from conception to death in explicitly erotic scenes. The 8,611-square-foot (800 m²) fresco took Vigeland 20 years to finish.

Entering the mausoleum is a solemn, even haunting, experience. Even the quietest footstep echoes across the barrel-vaulted ceiling for up to 14 seconds. A flashlight is needed to reveal the room's dark, painted walls.

Vigeland began construction on the building in 1926 with the intention of later filling it with his paintings and sculptures. Only one wall and the ceiling of the barrel-vaulted room were to be covered by paintings; the rest would be left bare to showcase other works.

Tasked with designing his own mausoleum, Emanuel Vigeland created a room where sounds echo for up to 14 seconds.

When Vigeland decided that the museum would also serve as his mausoleum, he had the windows sealed with bricks, lending the entire building an eerie atmosphere. He completed the fresco, finding inspiration in the burial chambers of antiquity and drawing especially from the dramatic stories of creation and original sin from Christianity. Named *Vita* ("Life"), the fresco focuses on humanity's sexual instinct portrayed by naked bodies captured in an impressive array of intimate acts.

After Vigeland's death, his ashes were put to rest in an urn that sits above the main entrance. Now run by a private foundation, the museum was opened to the public in 1959, more than a decade after Vigeland's death.

Today, the museum is open for only a few hours each week, but it plays host to several concerts (sometimes involving didgeridoos) throughout the year.

Grimelundsveien 8, Oslo. Take the metro to Slemdal. Ⓝ 59.947256 Ⓔ 10.692641

SWEDEN

YTTERBY MINE

STOCKHOLM

39
Y
Yttrium
88.906

Army lieutenant and part-time chemist Carl Axel Arrhenius was excited when, in 1787, he came across a strange, heavy black rock in an old quarry near the Swedish village of Ytterby. Arrhenius named the newly discovered substance ("elements" were not yet recognized) "ytterbite" after the town. A plaque now marks the site.

Also present in the quarry was a crude mineral called yttria, which was the oxidized form of yttrium. Silver yttria contained four rare silvery white elements: ytterbium (now used in electrodes and lasers), terbium (used to make microprocessor chips), erbium (used for medical lasers), and yttrium (used to make phosphors for LEDs), making the site the single richest source of elemental discoveries in the world, and giving the town of Ytterby an outsize presence on the periodic table.

The mine is located in the middle of Ytterby, which is a 40-minute drive from downtown Stockholm. Ⓝ 59.428524 Ⓔ 18.334887

SWEDEN SOLAR SYSTEM

STOCKHOLM

Created at a scale of 1:20 million, this country-spanning model is the world's largest representation of the solar system. It is anchored by Stockholm's spherical Globe Arena building, which represents the sun. The inner planets, all appropriately scaled, are dotted around Stockholm and its suburbs.

Further north are Pluto—still part of the lineup, despite its 2006 reclassification as a dwarf planet—and fellow trans-Neptunian objects Ixion, Sedna, and Eris. A plaque in Sweden's northernmost city of Kiruna, 592 miles (950 km) away, marks the spot for "termination shock," the point at which solar wind slows down and causes changes in the magnetic field.

In 2011, vandals snatched Uranus from the town of Gävle, 100 miles (1,600 km) from Stockholm. But in October 2012, a new model of Uranus appeared a few miles south, in the village of Lövstabruk. (The planet's new location reflects its orbit position when closest to the sun, so the solar system model is still accurate.)
Globentorget 2, Johanneshov, Stockholm. To see the sun, get the subway to Gullmarsplan. To see all the planets, you'll need to plan a road trip.
Ⓝ 59.294167 Ⓔ 18.080816

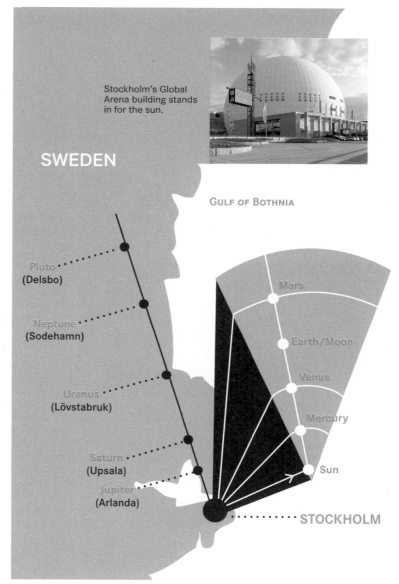

Stockholm's Global Arena building stands in for the sun.

SWEDEN

GULF OF BOTHNIA

Pluto
(Delsbo)

Neptune
(Sodehamn)

Uranus
(Lövstabruk)

Saturn
(Upsala)

Jupiter
(Arlanda)

Mars

Earth/Moon

Venus

Mercury

Sun

STOCKHOLM

The planets of Earth's solar system are laid out—to scale—throughout the country.

ALSO IN SWEDEN

UFO Memorial

Ängelholm · A memorial dedicated to a Swedish hockey player's supposed encounter with aliens, during which he says he received recipes for natural medical remedies.

Drottningholm Palace Theater

Drottningholm · The royal theater still uses 18th-century levers and pulleys to put on shows, in what they bill as the world's most genuine early opera experience.

Ven Island

Øresund Strat · Explore the site of the first modern observatory, built by Tycho Brahe.

Vasa Museum

Stockholm · The home of the hulking remains of the *Vasa*, a 17th-century warship that was meant to be the greatest vessel of her time—but sank in a matter of minutes on her maiden voyage.

SWEDEN

LUND UNIVERSITY NOSE COLLECTION

LUND, SCANIA

The hundred-strong nose collection at Lund University contains plaster casts of some notable Scandinavian snouts, including a cast of the metal prosthetic famed Danish astronomer Tycho Brahe wore after losing the bridge of his nose during a sword duel.

Lund University, Sandgatan 3, Lund. The noses are in the university's Museum of Student Life, a 10-minute walk from Lund's Central train station. ℕ 55.705673 Ⓔ 13.195374 ➤➤

➤➤ Post—World War I Facial Prosthetics

Soldiers fighting in World War I had to adapt to a new set of battle rules. Instead of fighting one another at close range, armies dug trenches and spent months living in them under appalling conditions, attempting to slowly destroy the enemy by launching gas grenades and firing machine guns.

A postwar soldier without and with his facial prosthetic.

It was machine guns, a recent addition to the warfare arsenal, that caught so many soldiers off guard. Unaccustomed to rapid-fire weapons, men would poke their heads out of the trench and be attacked by a hail of bullets, one of which would often hit them square in the face.

Injuries that resulted in facial disfigurement made reintegration into civilian life particularly difficult. In addition to bearing the mental scars of war, soldiers would return home looking like grotesque versions of themselves—a hole where an eye should be; a tongue lolling in the absence of a lower jaw; a ragged hole in the cheek that exposed a row of teeth. Mirrors were banned in convalescent hospitals to protect men from the devastation of seeing themselves.

Thousands of British soldiers with facial injuries underwent surgery at the skilled hands of Harold Gillies, a New Zealand–born surgeon who performed an early version of plastic surgery at Queen's Hospital in Sidcup, England. Gillies's groundbreaking reconstructive techniques gave many men the confidence to show themselves in public. For those whose faces could not be healed, there was another option: a mask.

English sculptor Francis Derwent Wood (at the Masks for Facial Disfigurement Department in the Third London General Hospital) and Pennsylvanian sculptor Anna Coleman Ladd (at the American Red Cross Studio for Portrait Masks in Paris) were the two most prominent mask makers, creating customized, hand-painted pieces of galvanized copper to conceal soldiers' wounds. Each mask took weeks to make. The process began with a plaster cast of an injured man's face and ended with the painstaking application of skin-colored paint, a glass eye, and hair or brushstrokes for eyebrows. A pair of wire eyeglasses or a ribbon held each mask against the face.

Plaster casts of injured faces.

Asia

The Middle East

IRAN · IRAQ · ISRAEL · PALESTINE · LEBANON · QATAR
SAUDI ARABIA · SYRIA · UNITED ARAB EMIRATES · YEMEN

South and Central Asia

AFGHANISTAN · BANGLADESH · BHUTAN
INDIA · KAZAKHSTAN · KYRGYZSTAN · PAKISTAN · SRI LANKA
TURKEY · TURKMENISTAN

East Asia

CHINA · HONG KONG · TAIWAN · JAPAN · NORTH KOREA
SOUTH KOREA

Southeast Asia

CAMBODIA · INDONESIA · LAOS · MALAYSIA · MYANMAR
PHILIPPINES · SINGAPORE · THAILAND · VIETNAM

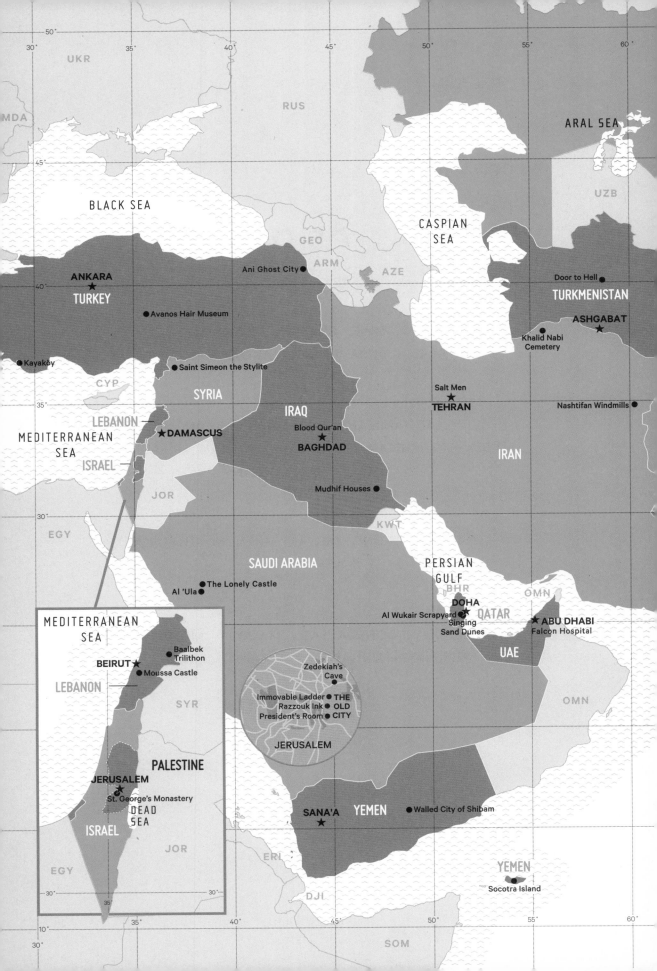

UKR

RUS

MDA

BLACK SEA

ANKARA ★

TURKEY

● Avanos Hair Museum

● Kayaköy

CYP

● Saint Simeon the Stylite

SYRIA

MEDITERRANEAN
SEA

LEBANON

★ **DAMASCUS**

ISRAEL

JOR

EGY

Ani Ghost City ●

GEO

ARM

AZE

CASPIAN
SEA

ARAL SEA

UZB

Door to Hell ●

TURKMENISTAN

ASHGABAT ★

Khalid Nabi
Cemetery ●

Salt Men ●

TEHRAN ★

Nashtifan Windmills ●

IRAQ

Blood Qur'an ★
BAGHDAD

IRAN

● Mudhif Houses

KWT

SAUDI ARABIA

Al 'Ula ● ●The Lonely Castle

PERSIAN
GULF

BHR

DOHA ★
Al Wukair Scrapyard ●★ QATAR
Singing
Sand Dunes ●

OMN

★ **ABU DHABI**
Falcon Hospital ●

UAE

OMN

MEDITERRANEAN
SEA

Baalbek
Trilithon ●
BEIRUT ★
● Moussa Castle

LEBANON

SYR

PALESTINE

JERUSALEM
★
St. George's Monastery

DEAD
SEA

ISRAEL

JOR

EGY

30°

30°

35°

Zedekiah's
Cave ●

Immovable Ladder ● **THE**
Razzouk Ink ● **OLD**
President's Room ● **CITY**

JERUSALEM

SANA'A ★

YEMEN

● Walled City of Shibam

YEMEN

● Socotra Island

ERI

DJI

SOM

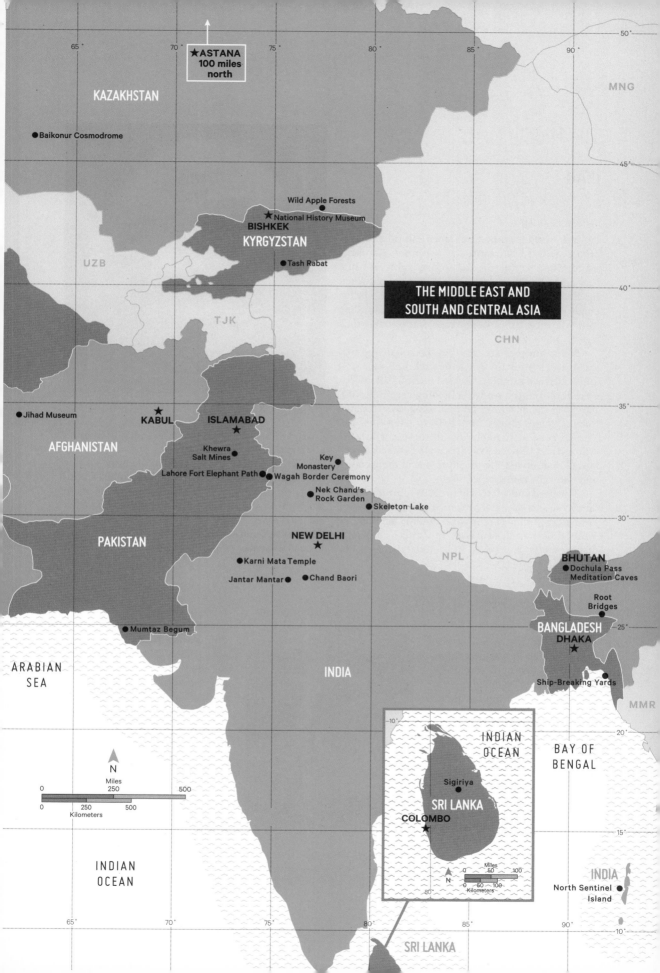

★ASTANA
100 miles
north

KAZAKHSTAN

●Baikonur Cosmodrome

Wild Apple Forests●
★ National History Museum
BISHKEK
KYRGYZSTAN

UZB

●Tash Rabat

TJK

THE MIDDLE EAST AND
SOUTH AND CENTRAL ASIA

CHN

MNG

50°

45°

40°

●Jihad Museum

★
KABUL

AFGHANISTAN

★ISLAMABAD
★

Khewra
Salt Mines●

Lahore Fort Elephant Path●●Wagah Border Ceremony

Key
Monastery●

Nek Chand's●
Rock Garden

●Skeleton Lake

PAKISTAN

NEW DELHI
★

●Karni Mata Temple

Jantar Mantar● ●Chand Baori

NPL

BHUTAN
●Dochula Pass
Meditation Caves

Root
Bridges●

35°

30°

25°

●Mumtaz Begum

ARABIAN
SEA

INDIA

BANGLADESH
DHAKA
★

●Ship-Breaking Yards

MMR

20°

N
Miles
0   250   500

0   250   500
Kilometers

INDIAN
OCEAN

INDIAN
OCEAN

BAY OF
BENGAL

10°

Sigiriya
●

SRI LANKA

COLOMBO
★

Miles
0   50   100

N   0 50 100
Kilometers

15°

INDIA

●North Sentinel
Island

10°

SRI LANKA

65°   70°   75°   80°   85°   90°

The Middle East

IRAN

SALT MEN OF CHEHRABAD

HAMZEHLU, ZANJAN

In 1994, workers at the Chehrabad salt mine uncovered a partial body buried in a tunnel. Naturally mummified by the salt, the corpse had long white hair and a beard, and was about 35 years old when he died, sometime around the 4th century. On his single preserved foot was a leather boot. Surrounding him were three iron knives, a rope, some pottery fragments, and a walnut.

Salt mummies are rare, but the surprising find was the first of many at Chehrabad. Between 1994 and 2010, six naturally preserved bodies—all male—were found in the mine. Following careful examination of the specimens, archaeologists estimated that the bodies ranged from 1,400 to 2,400 years old. The six men were likely all salt miners who were crushed and trapped when sections of the mine collapsed. Salt pulled the moisture out of their bodies and naturally mummified them.

The head and left foot of the 1994 mummy are on display at the National Museum of Iran in Tehran. Four of the other bodies were initially exhibited at Rakhtshuikhaneh Museum in Zanjan, but poor display methods resulted in bacterial damage. The

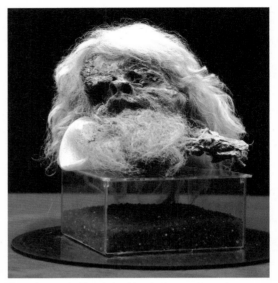

The salt-cured, silver-haired head of an ancient miner.

quartet of bodies is now on display in airtight cases at Zanjan Archaeology Museum.

The sixth mummy remains in the mine, as it is too fragile to be moved. In 2008, Chehrabad's mining permit was revoked, allowing archaeologists to study the area and piece together a portrait of the ancient miners' lives.

National Museum of Iran, 30 Tir Avenue, Emam Khomeini Avenue, Tehran. Ⓝ 35.687044 Ⓔ 51.414611

KHALID NABI CEMETERY

GOLESTAN

In the stunning green hills of northern Iran is a cemetery with headstones distinctly resembling male and female genitalia. Stone phalluses jut out from the ground at odd angles; lower to the ground are the rounded, clover-shaped stones. All told, the cemetery has a total of 600 monuments to sex organs sprawling across the vibrant landscape.

In a country known for its strict religious law, a cemetery dotted with penis-shaped headstones up to 6.5 feet (2 m) tall tends to stand out. Visitors can tell at once that the cemetery was created in a much different era. Due to the cemetery's proximity to the Turkmenistan border,

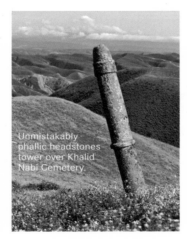

Unmistakably phallic headstones tower over Khalid Nabi Cemetery.

some experts suggest that phallus-worshipping peoples from Central Asia and India created it—although conclusive proof of this does not exist. There has been

little scholarship on the cemetery's origins within Iran because of a strong national embarrassment over the suggestive stones.

Aside from being home to hundreds of penis tombstones, the cemetery also contains the hilltop grave of Khalid Nabi, who was a Christian prophet from Yemen who died in the 4th century. His tomb—which has a conventional, non-anatomical shape—is a pilgrimage site for Turkmens, who place ribbons on his shrine. The combination of pilgrims and curious tourists makes for an odd mix of playful and pious visitors to Khalid Nabi.

The cemetery is about a 2-hour drive north of the city of Gonbad-e Kavus. Ⓝ 37.745472 Ⓔ 55.411236

Nashtifan Windmills

NASHTIFAN

Located on the arid and wind-swept plains of northeastern Iran, 30 miles (48.3 km) from the Afghan border, the small village of Nashtifan is keeping ancient traditions alive amid the winds of change. The town is home to some of the earliest windmills in the world, and the structures are still in use today.

Nashtifan is known for its uniquely powerful winds. The village's name is derived from words that translate to "storm's sting." Along the southern edge of town, a towering 65-foot-tall (19.8 m) earthen wall shelters residents from abrasive gales. The high wall houses over two dozen vertical axis windmills that date back to ancient Persian times. It's estimated the structures, made of clay, straw, and wood, are around 1,000 years old and were used for milling grain into flour.

During turbulent winter months, the handcrafted wooden blades whirl with a surprising velocity to power grindstones. With periodic repairs, these well-built turbines could continue to function for centuries, so long as there are caretakers willing to maintain them. For now, an amiable custodian named Ali Muhammad Etebari looks after the windmills on a volunteer basis. **The closest international airport is Mashhad. From there it's roughly a 4-hour drive to the windmills.** Ⓝ 34.431635 Ⓔ 60.174789

Also in Iran

Shah Cheragh

Shiraz · Mirrors and glass shards cover every inch of this beautiful mosque.

A thousand years later and still in use, these windmills are a source of local pride.

IRAQ

SADDAM HUSSEIN'S BLOOD QUR'AN

BAGHDAD

On his sixtieth birthday in 1997, Saddam Hussein requested a special gift: a Qur'an written in his own blood. In Islam, blood is considered *najis*, or ritually unclean. A Qur'an written in blood, therefore, is *haraam*—a sinful act of disrespect against the holy book. But none of this concerned Saddam. Over two years, Islamic calligrapher Abbas Shakir Joudi transcribed the holy book's 336,000 words using, if reports are to be believed, a total of 50 pints of blood collected from the dictator at regular donation sessions.

The 605-page "Blood Qur'an" went on display in 2000, housed behind glass in a marble building at the Umm al-Qura mosque complex in Baghdad. It made a fitting addition to the mosque, which was built in Saddam's honor to commemorate the tenth anniversary of the Gulf War.

Then came the 2003 invasion of Iraq and the fall of Saddam. As Baghdad burned and looting ensued, mosque leaders removed the Qur'an from display and hid it in a vault. Since Saddam's execution, the bloodied pages have been in an odd limbo—their very existence is forbidden within Islam, but it is also haraam for a Qur'an to be destroyed.

Umm al-Qura mosque, Baghdad. The Blood Qur'an is not on public display, but at least one journalist has convinced mosque clerics to bring a page out of storage for a quick look. Ⓝ 33.338273 Ⓔ 44.297161

MUDHIF HOUSES

MESOPOTAMIAN MARSHES

For thousands of years, *mudhifs*—large, arched communal huts made from reeds—have served as social and ceremonial hubs for the Marsh Arabs, or Madan, of southern Iraq. Weddings, dispute resolutions, religious celebrations, and community meetings all traditionally take place within the structures.

To build a mudhif, the Madan assemble 30-foot (9 m) lengths of reeds into bunches and bend them into arches. Rows of these arched columns form the basic structure of the house. Woven reed mats and latticed panels fill the gaps, forming the ceiling and walls. Each tribe's sheikh collects tributes from families in order to maintain the mudhif.

After the Gulf War in 1991, Saddam Hussein's regime drained the marshes as an act of revenge against those who had taken refuge there after participating in antigovernment uprisings, transforming the marshes into desert. With their food supply eliminated, about 100,000 Marsh Arabs fled, abandoning their traditional way of life.

With the 2003 defeat of Hussein came the removal of levees and the slow return of water to the marshes. However, droughts, new dams, and upstream irrigation projects have since reduced the levels again. A small number of Madan communities have returned to their old homes and rebuilt their mudhifs, but their ongoing survival in the marshes is far from assured.

The marshes are about 20 miles (32 km) northwest of Basra. Ⓝ 31.040000 Ⓔ 47.025000

Giant reed houses float atop the marshes of southern Iraq.

ISRAEL

THE IMMOVABLE LADDER AT THE CHURCH OF THE HOLY SEPULCHRE

JERUSALEM

Venerated as the site of the crucifixion, burial, and resurrection of Jesus, the Church of the Holy Sepulchre is perhaps the world's most sacred Christian pilgrimage site. It is also the location of a 150-year-old argument over a ladder.

Under an 1852 mandate, the care of the Church of the Holy Sepulchre is shared by no fewer than six Christian denominations: the Greek Orthodox, Armenian Apostolic, Roman Catholic, Coptic, Ethiopian Orthodox, and Syriac Orthodox churches. Upkeep of the Holy Sepulchre's edifice is carefully divided into sections. While some duties are shared, others belong exclusively to a particular sect. A set of complicated rules governs the transit rights through each section, and some remain hotly disputed. Arguments and fistfights over territory and boundaries are not uncommon.

One such area is a small section of the roof that is disputed between the Copts and Ethiopians. At any given time, at least one Coptic monk sits on a chair placed on a particular spot to express claim to the section. On a stifling summer day in 2002, a monk moved his chair eight inches over, to find shade. This was interpreted as a hostile act and violation of boundaries, and 11 were hospitalized after the fight that ensued.

The Church of the Holy Sepulchre's "immovable ladder" is a centuries-old symbol of this extreme territoriality. During the mid-1700s, a man belonging

Five wooden rungs have caused countless squabbles over the centuries.

to an unknown sect placed the ladder on a ledge against an exterior second-floor wall of the church. With the fear of inciting violence, no one has dared touch it since—except when it disappeared in 1997. A mischievous tourist allegedly plucked it from the ledge and hid it behind an altar, where it remained undiscovered for weeks. The ladder has since been put back into its "appropriate" spot.

Christian Quarter, Old City, Jerusalem. Ⓝ 31.778444 Ⓔ 35.229750

ZEDEKIAH'S CAVE/SOLOMON'S QUARRIES

JERUSALEM

Beneath the Muslim Quarter in Jerusalem's Old City is an underground quarry that goes by two names: Zedekiah's Cave and Solomon's Quarries. The names reflect the two main legends that surround this 750-foot-long (228.6 m) collection of caverns.

The first story is that King Zedekiah fled through the cave to escape from attacking Babylonians around 587 BCE. At the time, the legend goes, the cave extended all the way to Jericho—a distance of about 13 miles (21 km). The Babylonians chased Zedekiah to Jericho, capturing and blinding him. The dripping water in the cave is thus known as Zedekiah's Tears. The second story involves King Solomon, who is fabled to have used stones from the cave to build the First Temple in the 10th century BCE.

There is no archaeological evidence to support either premise. However, chisel markings on the walls suggest Zedekiah's Cave was one of the quarries that supplied limestone for King Herod's Second Temple and Temple Mount expansion. The stones of the Western Wall (also called the Wailing Wall)—Judaism's most sacred prayer site—may indeed have come from this cave.

Sultan Suleiman Street, near Damascus Gate, Jerusalem. Ⓝ 31.768967 Ⓔ 35.213878

PRESIDENT'S ROOM

JERUSALEM

President's Room

The Cenacle, room of the Last Supper

King David's Tomb

Just south of the Old City on Mount Zion is a building that houses two holy sites. On the ground floor is the Tomb of King David; upstairs is the Upper Room, or Cenacle—the location of the Last Supper.

Less conspicuous is the small domed chamber on the roof, known as the President's Room. From 1948 to 1967, when Jordan controlled East Jerusalem, Jews were prevented from visiting sacred places in the Old City, such as the Western Wall and the Mount of Olives. During this time, Mount Zion was one of the closest vantage points for viewing the forbidden sites. The Ministry of Religious Affairs established the President's Room so that Israel's first head of state, Chaim Weizmann, could keep watch over the Western Wall.

Weizmann never used the room. But his successor, President Yitzhak Ben-Zvi, did. Three times a year, he would ascend the stairs to the dome and look toward the Temple Mount.

Mount Zion, Jerusalem. The building is just southwest of the Zion Gate. Ⓝ 31.771639 Ⓔ 35.229014

ALSO IN ISRAEL

Museum on the Seam

Jerusalem · This collection of contemporary sociopolitical art sits on the border between East and West Jerusalem.

RAZZOUK INK

JERUSALEM

As the sole surviving pilgrimage-tattoo business, Razzouk Ink is a place where ancient artifacts meet contemporary machines and rich history intersects with modern technology.

Just inside the Jaffa Gate in Jerusalem's Old City, a big sign above a tiny shop reads TATTOO WITH HERITAGE SINCE 1300. For over 700 years the Razzouk family has been tattooing marks of faith. As Coptic Christians who settled in Jerusalem generations ago, the family had learned the craft in Egypt, where the devout wear similar inked inscriptions. Evidence of such tattoos dates back at least as far as the 8th century in Egypt and the 6th century in the Holy Land, where Procopius of Gaza wrote of tattooed Christians bearing designs of crosses and Christ's name. Early tattoos served as a way for indigenous Christians in the Middle East and Egypt to self-identify. Later, as the faithful came to the Holy Land on pilgrimage, the practice expanded to offer these travelers permanent evidence of their devotion and peregrination.

Razzouk Ink's stone walls and exposed beams lend antique character to the space. A museum-like case holds heirlooms, and an exhibition of pictures on the walls offers glimpses into the family's past. Pilgrims' accounts dating to the late 16th century show how purveyors such as the Razzouks must have tattooed back then, with sewing needles bound to the end of a wooden handle.

13 Greek Catholic Patriarchate St., Old City of Jerusalem Ⓝ 31.777236 Ⓔ 35.228100

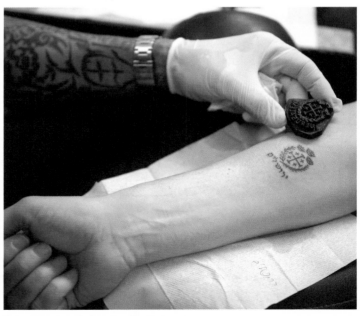

A permanent reminder of one's pilgrimage.

PALESTINE

ST. GEORGE'S MONASTERY

WEST BANK

Clinging to a cliff on the edge of the Wadi Qelt gorge is a Greek Orthodox monastery that's endured many centuries of turmoil and destruction.

The original monastery was founded in the 5th century by a group of cave-dwelling hermits. They chose the site because it was located next to the cave where the prophet Elijah is said to have been fed by ravens during the 9th century BCE.

A Persian invasion in the 7th century drove out the hermits and left the monastery in ruins. Around 500 years later, Crusaders rebuilt St. George's, only to be driven from the site following the Islamic re-conquest of Jerusalem.

The late 19th century saw St. George's restored once again. The monastery is now home to two churches, a small group of Greek Orthodox monks, and the tombs

Monks have been secluding themselves in this cliffside retreat since the 5th century.

of the five hermits who got the whole thing started.

A 15-minute drive from St. George's is the Mount of Temptation—so named because, according to the Bible, it is the place where Jesus was tempted by the devil. This mountain has its own cliffside monastery, established in the 6th century, which has only a single permanent resident. There is also a nearby collection of hermit caves—some of which are said to still occasionally be inhabited by ascetic monks. **Wadi Qelt, West Bank. St. George's is a 20-minute drive from Jerusalem, followed by a 15-minute hike or short camel ride. Ⓝ 31.844452 Ⓔ 35.414085**

LEBANON

MOUSSA CASTLE

BEITEDDINE, CHOUF

Moussa al Mamaari's desire to build a castle began in 1945 at the age of 14. Beaten by his teacher and mocked by his schoolyard sweetheart, Moussa vowed to rise above his impoverished origins and live in a castle of his own design.

It took decades of hard work, but Moussa got his wish. After dropping out of school to help his uncle restore old buildings, 20-year-old Moussa used his earnings to buy a plot of land in the Chouf mountains. There, unassisted, he lugged 6,500 stones into place; carved animals, geometric patterns, and plants into them; and slowly assembled the castle's walls.

The three-story Moussa Castle was built by one man over a lifetime.

Moussa Castle, with its medieval ramparts, turrets, moat, and drawbridge, opened to the public in 1967. Moussa filled the castle's three levels with thousands of artfully arranged weapons, mannequins in army uniforms, a two-headed taxidermy lamb, and a wax-figure re-creation of the Last Supper featuring an oddly maniacal-looking Jesus.

As for Moussa's childhood doubters, they receive their own tribute: One room is devoted to a classroom scene in which a wax-figure teacher, face contorted with rage, strikes a cowering student. **Moussa Castle is in the heart of the Chouf mountains between Deir Al Qamar and Beiteddine, south of Beirut. Ⓝ 33.700277 Ⓔ 35.583333**

It isn't clear how the Romans moved this 1,000-ton stone.

BAALBEK TRILITHON

BAALBEK, BAALBEK

In 15 BCE, when the city of Baalbek was a Roman settlement by the name of Heliopolis, Emperor Augustus ordered the construction of a grand and mighty temple. Dedicated to the god Jupiter, the temple was built on a layered base of massive foundation stones—so massive that it is unclear how they could have ever been maneuvered into place.

Just six columns are left of the Temple of Jupiter, but its base is mostly intact. On the western side are its three largest stone blocks, collectively known as a trilithon. Each block is 65 feet long, 10 feet wide, and 14 feet deep (20 m × 3 m × 4 m), and weighs in at around 800 tons. They are among the largest monoliths in history—but one mile away, in an ancient quarry, is an even bigger block, known as the Stone of the Pregnant Woman. This stone, weighing approximately 1,000 tons, is half-embedded in the ground, jutting up at an angle like a capsizing ship.

Given that even the largest modern transport equipment would struggle with such hefty stones, the presence of the trilithon and its even heavier quarry counterpart has perplexed archaeologists. A combination of winches, patience, and enormous manpower was surely involved in lifting the stones, but conspiracy theories abound. One hypothesis posits that beneath the temple is a launch pad that extraterrestrials used for their intergalactic spacecraft.

Baalbek is 53 miles (85.3 km) northeast of Beirut. Minibuses depart from the Cola intersection.
Ⓝ 34.006944 Ⓔ 36.203889

QATAR

AL WUKAIR SCRAPYARD

QATAR

As the crow flies, the scrapyard in Al Wukair is only 12 miles from the center of Doha, but it could hardly be more different. Leaving behind the city skyline and heading southwest, you encounter the kind of desert that is more akin to *Mad Max* than romantic rolling sand dunes.

It is here you will find the Al Wukair scrapyard, where thousands of unwanted vehicles languish in the desert. At any given time, the gigantic scrapyard accommodates around 20,000 cars, trucks, buses, excavators, bulldozers, cement mixers, and other construction equipment in varying states of neglect and disrepair. Many of the vehicles are burnt beyond recognition.

Walking through the endless rows of dust-covered scrap is an eerie post-apocalyptic experience.

But there is also tremendous beauty. Time and sand have worked their magic on these vehicles, and if you're willing to wander amid the wreckage with a camera in hand, you'll be rewarded with gorgeous photographs of industrial decay.

The scrapyard is run under the surveillance of the state police. Visits are not encouraged, but they are not actively discouraged either. At the entrance, there are several bungalow-like offices. It is required that you register yourself with officials upon arriving. An authorized escort accompanies visitors to the scrapyard.

The easiest way to reach the scrapyard is by heading west along Salwa Road, then turning south and following the signs for Mesaieed. Along the way, there is an exit for Al Wukair, which leads directly to the scrapyard. This route is longer than driving through the village of Al Wukair, but it bypasses all the traffic and countless speed bumps.
Ⓝ 25.115668 Ⓔ 51.468979

SINGING SAND DUNES

DOHA

When the air is dry and the wind picks up, a haunting moan rumbles across the sand dunes southwest of Doha. The area is one of dozens of places in the world where the sand sings and the dunes boom.

The sound, which can last for minutes and varies from a hum to a roar to a whistle, occurs when loose sand grains on the top layer of a dune cascade down its slope. The exact means by which this creates the noise are unknown, but researchers at Paris Diderot University have found that the size of the grain determines the pitch of the note.

You can amplify the boom by causing an avalanche—try running along the top of a dune or sliding down one on a homemade sled. Thick cardboard and large trays work well.

The dunes are 25 miles (40 km) southwest of Doha, the capital of Qatar. Avoid going if it has rained recently.
Ⓝ 25.038871 Ⓔ 51.405923

SAUDI ARABIA
THE LONELY CASTLE

MADA'IN SALEH

Among the dozens of ruins located in the archaeological playground of Mada'in Saleh, one stands alone. Literally. Carved into a massive boulder is the 1st-century Qasr al-Farid, or "the Lonely Castle."

The "castle" name is misleading, as the grand carving is actually a tomb that was built as part of the ancient Nabataean site of Mada'in Saleh. The Nabataean construction technique was to chisel tombs right out of the rock from the top down. Such is the case with Qasr al-Farid, although the monument appears to never have been completed. The incomplete portion, toward the base of the tomb, is a terrific window into the steps taken by the ancient carvers before the rougher work was polished away.

The castle is at the archaeological site of Mada'in Saleh, 311 miles (500.5 km) southeast of Petra. Ⓝ 26.790694 Ⓔ 37.952000

An unfinished tomb reveals ancient Nabatean carving methods.

The abandoned ruins of a 2,000-year-old village.

Al 'Ula

Al 'Ula

Walking through the narrow corridors of Al 'Ula in northwestern Saudi Arabia is like traversing a maze with ancient history at each turn. Once home to a bustling civilization, these 800 tightly packed mud-brick and stone houses—parts of which are more than 2,000 years old—are now abandoned ruins decaying in the desert sun.

The walled city of Al 'Ula was founded in the 6th century BCE, an oasis in the desert valley, with fertile soil and plenty of water. It was located along "Incense Road," the network of routes that facilitated the trading of spices, silk, and other luxury items through Arabia, Egypt, and India.

Though most of the original houses in the old town were rebuilt over the centuries, there are many remnants of traditional Arab architecture among the ruins. Reconstructed in the 13th century, old town Al 'Ula was inhabited until modern times. When the cramped space and poor infrastructure didn't meet 20th-century standards, residents abandoned Al 'Ula for a new town nearby, called Al-'Ula. The last family left town in 1983, and the last mosque service was held two years later.

Visitors to Al 'Ula often find themselves completely alone in the ruins of an ancient civilization, enjoying the silence and imagining what life in the winding streets would be like when there were hundreds of people living in the mud-brick huts. **Al 'Ula is a 90-minute flight from Riyadh, the Saudi Arabian capital. Ⓝ 26.624967 Ⓔ 37.915600**

SYRIA

Church of Saint Simeon the Stylite

Deir Semaan, Mount Simeon

The austerities of 5th-century monastic life were simply not strict enough for St. Simeon, who had a penchant for self-inflicted starvation, palm-frond girdles, and sleeping while standing up. After spending a decade at an Aleppo monastery, the ascetic St. Simeon proclaimed that God wanted him to be immobile. He climbed to the top of an abandoned pillar in the desert and remained there for the next 37 years, standing as much as possible, seldom eating, and tying his body upright to a pole to prevent himself from sleeping in a horizontal position.

Simeon died on the pillar in 459 CE, inspiring a rash of copy-cat "stylites"—from the Greek word *stylos*, meaning "pillar"—who spent their days preaching and praying atop columns. The Church of Saint Simeon was built in 491 CE on the site of the original stylite's pillar to honor his devotion. Only a small part of it remains—including the pillar,

Saint Simeon spent 37 years balanced on top of a 50-foot pillar. All that remains today is a stub.

whittled down to just a few feet from centuries of relic seekers who have carved off shards for themselves.

The church is a half-hour drive from Aleppo. Ⓝ 36.334166 Ⓔ 36.843888

UNITED ARAB EMIRATES

ABU DHABI FALCON HOSPITAL

ABU DHABI

Located conveniently close to the international airport, Abu Dhabi's Falcon Hospital is where birds of prey go for a talon trim or repair of a broken wing.

When the hospital opened in 1999, it was the first institution in the world to provide veterinary care exclusively for falcons. These days, the services have diversified—there is an on-site pet care center for domestic animals, as well as a shelter for abandoned creatures large and small. The focus, however, is still falcons, over 11,000 of whom come to the hospital each year from the UAE, Saudi Arabia, Qatar, Kuwait, and Bahrain.

Public tours of the Abu Dhabi Falcon Hospital began running in 2007. Visitors encounter an adorable yet confusing sight in the main clinic's waiting room: rows

The most highly prized falcons in the Middle East come for a talon trim or feather transplant.

of falcons sitting on Astroturf perches, some of them wearing hoods that cover their eyes and keep them calm.

In the clinic itself, you may observe a bird getting its talons clipped or getting broken feathers replaced. The room has drawers full of falcon feathers, which are sewed and glued to the injured birds' bodies. Many of the surgical tools used for procedures are modified or improvised versions of what doctors would use on human patients—tools meant for premature babies work well.

Falconry, originally practiced by ancient Bedouin hunters who relied on the birds of prey to gather them food, continues as a sport today. Now trained in a 21st-century way—with the help of drones and radio transmitters attached to their bodies—falcons remain highly prized in the Emirates. An annual falconry festival in Abu Dhabi lures hawks and their trainers from around the world for hunts, parties, and workshops in the desert.

Sweihan Road, Abu Dhabi. Turn right at the last gas station before Abu Dhabi Airport, and you'll end up at the Falcon Hospital. Tours run Sunday through Thursday and must be booked 24 hours in advance. Ⓝ 24.408265 Ⓔ 54.699379

YEMEN

SOCOTRA ISLAND

SOCOTRA

Describing the flora of Socotra Island is difficult, because it resembles nothing else on Earth. Take the dragon's blood tree, the island's signature specimen, named for its rich red sap. Its silhouette is best explained as that of an umbrella blown inside out. Then there is the desert rose tree—a delicate name that is no match for a bloated, gray trunk that sags beneath thin branches exploding with pink flowers.

These odd-looking organisms are just two of the hundreds of endemic flora and fauna species on this island off the coast of Yemen. The 78 x 28-mile (126 x 45 km) Yemeni territory's biodiversity is the result of long-term isolation—it has been at least 20 million years since the island broke off from Africa, the nearest land mass.

The construction of an airport in 1999 and the introduction of paved roads six years later have resulted in a moderate increase in tourism. That said, you will likely encounter deserted beaches, pristine sand dunes, empty volcanic caves, and shipwreck sites devoid of divers.

Flights to Socotra Island depart from Sana'a. Ⓝ 12.510000 Ⓔ 53.920000

Dragon's blood trees (left and center), and a bottle tree (right) are among the fantastical flora on Socotra Island.

Shibam's city of 16th-century mud high-rises was once a stopping point for traders along the spice route.

WALLED CITY OF SHIBAM

SHIBAM

Like those in New York's borough of Manhattan, the high-rises of Shibam were built on a rectangular grid of streets. Unlike Manhattan's, these "skyscrapers" are made of mud and date back to the 16th century, and the dusty streets are often overrun with goats.

Shibam, in the desert of central Yemen, is home to about 7,000 people. Located at the crossroads of Asia, Africa, and Europe, the small town was once a stopping point for traders traveling along the frankincense and spice routes.

The walled city of skyscrapers was built on a hill in the 1530s after a mighty flood destroyed much of

the existing settlement. Its huddled buildings, ranging from five to eight stories high, provided protection against the elements and deterred potential attackers. They continue to shelter the residents of Shibam.

The tower houses, however, are not immune to damage—fresh layers of mud must be applied to the walls regularly to replace sections eroded by wind and rain. A tropical storm in October 2008 brought another disastrous flood, causing some of the buildings to collapse.

Shibam is located 370 miles east of the Yemeni capital of Sana'a. Ⓝ 15.926938 Ⓔ 48.626669

South and Central Asia

AFGHANISTAN

A museum dedicated to overthrowing the Soviets, featuring a diorama of Russians being beaten with shovels.

JIHAD MUSEUM

HERĀT

The Jihad Museum was built in 2010 to pay tribute to the Afghan mujahideen, Islamic guerrillas who fought a bloody war against the Soviets after they invaded Afghanistan in 1979. During the decade-long conflict, the United States provided weapons and funding to the mujahideen to help them resist the USSR. Among the guerrillas fighting on the same side as the USA were Osama bin Laden and key members of the group that would later become Al Qaeda.

The conflict began in 1978, when Afghanistan's Communist party seized the country in a violent coup and began reforming some of its Islamic laws and traditions. In March 1979, armed Afghan rebels revolted in Herāt,

killing 100 Soviets who had entered Afghanistan to provide support to the new regime. The Afghan government responded by bombing the city, killing approximately 4,000 people. Uprisings spread across the country, compelling Soviet troops to invade in an attempt to quash rebellion. The ensuing struggle lasted almost a decade. Overpowered by the Afghan resistance, the USSR withdrew the last of its troops in 1989.

The museum's official aim—to educate future generations on Soviet-era jihadist resistance—suggests a restraint that is at odds with its presentation style. Scenes from Herāt are graphically depicted in the museum's central exhibit: a life-size diorama that shows green-tinged corpses

lying in bombed-out houses, women in burkas throwing rocks at troops, and bloodied Soviet soldiers slumped in a jeep, as rebels beat them to death with shovels. Surrounding these grim scenes is a 360-degree mural showing Islamist insurgents running through the streets as Russian helicopters bomb the city.

An array of war trophies adds to the sense of triumph—glass cases in the foyer hold Russian rifles, grenades, uniforms, and land mines, while the manicured garden surrounding the circular building contains Soviet tanks, cannons, and a helicopter.

Roodaki Highway, Herāt.
The museum is next to the
US consulate. Ⓝ 34.374166
Ⓔ 62.208888

BANGLADESH

CHITTAGONG SHIP–BREAKING YARDS

CHITTAGONG

The beaches of Chittagong are a massive graveyard for decommissioned ships and tankers. After plowing through the ocean for decades, their battered, rusting hulls sit fully exposed on the sand, waiting for workers in T-shirts, shorts, and flip-flops to tear them apart.

Chittagong is one of the largest ship-breaking yards in the world. Every year, 25,000 employees (down from an estimated 200,000) break down some 250 ships from around the world so their parts can be sold. Since the 1970s, vessels have been brought here to be deconstructed into heaps of steel, cables, generators, nuts, and bolts.

Bangladesh is the port of choice for two reasons: The labor is cheap and the safety standards are poor. Workers, many of whom are children, earn about a dollar a day for pulling apart the hulls by hand, often with little or no safety equipment. The workers inhale noxious fumes and are vulnerable to electrocution, falling debris, and explosions fueled by residual oil.

Organizations such as Greenpeace have been campaigning to impose stricter environmental and health standards at the Chittagong yards, or shift the work to more developed countries. However, mineral-poor Bangladesh is reliant on the steel it gets from the decommissioned ships. For the time being, Bangladesh's ship-breaking yards aren't going anywhere. **The yards, which run for miles along the coast, are not open to visitors, but the massive ships are visible from a distance. As you get closer, you will see bits of metal from the ships being offered for sale in the shops that line the street. Get a bus from Chittagong's railway terminal on Station Road. Ⓝ 22.442400 Ⓔ 91.732000**

Pulling apart colossal steel hulls by hand is one of the world's most dangerous jobs.

BHUTAN
DOCHULA PASS MEDITATION CAVES

DOCHULA

Perched at an altitude of 10,200 feet (3,109 m), Dochula Pass is one of Bhutan's most famous sites. A hidden surprise awaits the few who take the time to venture beyond this well-trodden place.

These curious visitors will discover the meditation caves tucked into the hills just above the pass. These tiny, open-faced caverns are built from stone and painted in colorful detail with Buddhist symbolism. The *druk*, or dragon—Bhutan's long-time national symbol—stretches over the cave entrances, bringing good luck and good tidings.

Meditation is a critical exercise for serious practitioners of Buddhism. To achieve enlightenment, monks and nuns will stay in one location, such as a meditation cave, for three days, three months, and three years. Each period is a trial for the next—if one can manage three days, one proceeds to three months. During this time, one may not speak with or cast eyes on another person. If you make it through three months, it's time to take on three years.

While you may not have three years to spare while visiting these caves, you'd be missing a hidden gem of Dochula Pass if you didn't take the extra ten-minute climb into the forest to explore these carefully constructed meditative retreats that nearly blend into the hills.

There is a café just up the hill from the memorial stupas of Dochula Pass. Head this way and continue upward, following paths built from oval-shaped stone steps. You'll soon encounter the meditation caves. ℕ 27.490086 Ⓔ 89.750300

INDIA
NEK CHAND'S ROCK GARDEN

CHANDIGARH, HARYANA/PUNJAB

When Nek Chand became a road inspector in 1951, the city of Chandigarh was in the midst of a dramatic reinvention. Small villages were being demolished to make way for streets, gardens, and sleek modern architecture. As construction continued through the 1950s, piles of debris—pottery fragments, bottles, glass, tiles, and rocks—littered the landscape.

Chand encountered these demolition sites and saw not junk, but potential. He began collecting materials from the scrap heaps and transporting them by bicycle to a forest gorge in Chandigarh's north. It was there that, in 1957, he began work on his own planned city: a sculpture garden, filled with thousands of human and animal figures, all made from recycled debris.

This one-man operation was, by necessity, a secret project. The forest area Chand chose for his garden was a government-designated no-build zone. Authorities remained unaware of Chand's ever-expanding sculpture park until 1975, when they were led there after Chand confided in the city's chief architect. They were astounded to discover 12 acres of statues, courtyards, man-made waterfalls, and pathways.

Despite the illegality of Chand's secret garden—and initial threats of demolition—state authorities allowed the project to expand, even providing Chand with a salary and a crew of 50 workers so he could devote himself to his creation full-time. In 1976, the rock garden opened to the public. It now sprawls over 30 acres, featuring parades of dancing women, gangs of monkeys, and a hillside animal stampede—all rendered in rock, glass, and pieces of colored tile.

Uttar Marg, Sector 1, Chandigarh. The Shatabdi Express—a 3.5-hour ride—runs from New Delhi to Chandigarh twice daily. Buses run from the Chandigarh railway station to Sector 1 in the northern part of the city. ℕ 30.760109 Ⓔ 76.801451

Stone sculptures, built secretly and illegally by an eccentric road inspector.

Skeleton Lake

ROOPKUND, UTTARAKHAND

When park ranger H. K. Madhwal discovered a lake ringed with thousands of human bones while walking in the Uttarakhand Himalayas in 1942, he raised a question that went unanswered for over 60 years: What killed the hundreds of people whose skeletons surrounded the lake? At first, the bones were thought to belong to Japanese soldiers who had stealthily crossed into India during World War II and perished in the high-altitude conditions. But carbon dating during the 1960s showed the estimated death date was wrong—very wrong. A broad range from the "12th to 15th century" was the best possible guess, but no cause of death could be established.

In 2004 the world got an answer to the mystery of Skeleton Lake. Radiocarbon testing at Oxford University narrowed the date of mass death to 850 CE, give or take 30 years. Analysis of skulls showed that, no matter their stature or position, all of the people died in a similar way: from blows to the head. The bodies had wounds only on their heads and shoulders, indicating the blows came from directly above.

After dismissing earlier theories—which included ritual suicide and attacking hordes—the scientists reached an unexpected conclusion: The travelers died from a severe hailstorm.

Hail is rarely lethal. But trapped in a valley without shelter and given no warning of the storm's severity, the 9th-century travelers could not escape the sudden barrage of tennis-ball-size spheres of ice.

Twelve hundred years after the storm, the green-tinged bones of hail victims still ring the lake, preserved alongside skulls and tattered shoes at an altitude of 16,500 feet (5,029 m).

The journey starts at Lohajung, a small pass where you can pick up essential supplies. Hire a guide and bring a mule or porter to help carry your belongings. The best time to go is in May or June, when there is no snow covering the lake and the skeletons are visible. You will need to be physically fit and prepared for high-altitude conditions. Ⓝ 30.262217 Ⓔ 79.731573

KARNI MATA RAT TEMPLE

DESHNOKE, RAJASTHAN

Twenty thousand rats scurry across the checkerboard floors of this temple, getting tangled in each other's tails and fighting for access to huge saucers of milk. Far from being regarded as vermin, the rats are venerated as the holy descendants of Karni Mata, who was worshipped as an incarnation of the Hindu goddess Durga during the 15th century.

The story of how Karni Mata's offspring took the form of rodents varies, but the most common version begins with her asking Yama, the god of death, to revive her drowned stepson. After first resisting, Yama gave in, promising that the boy and all of Karni Mata's male descendants would be reincarnated as rats.

Visitors to the temple are required to remove their shoes before walking inside. When shuffling among the droppings, spilled milk, and scrambling rodents, keep in mind that it is considered lucky for a rat to run across your foot. Another tip is to tread lightly. Temple rules state that if you accidentally step on one of the animals and kill it, you must

Visitors must remove their shoes before entering Karni Mata, a temple crawling with thousands of holy rats.

replace it with a rat made of solid gold. **National Highway 89, Deshnoke. Karni Mata is a 30-minute train ride from the larger town of Bikaner. Bring a pair of thick socks that you don't mind throwing out. Ⓝ 27.790556 Ⓔ 73.340833**

WAGAH BORDER CEREMONY

WAGAH, PUNJAB

The line separating India and Pakistan runs straight through the middle of Wagah, a Punjab village home to the only road joining the two countries that's open to international travelers. It is here that soldiers from each side conduct a precisely choreographed daily border-closing ceremony—a routine that combines flamboyant uniforms, fierce stares, and competitive high kicks.

Multiple wars and an ongoing dispute over Kashmir have fostered hostility between India and Pakistan—a hostility that finds its outlet in the aggressive dance moves of the border soldiers and the patriotic cheering of the daily crowds that come to watch them perform. The atmosphere is akin to a sporting match. Prior to the sunset ceremony, performers and visitors dance to traditional music pumped through loudspeakers on each side. As the anticipation builds, men with microphones rile up the flag-waving spectators in the bleachers.

The ceremony, performed since 1959, begins with a parade of soldiers on both sides of the border. Members of India's Border Security Force wear khaki uniforms with red fanned coxcombs on their turbans, while the Pakistan Rangers wear black uniforms with black coxcombs. In a region where the average male

height is 5.5 feet (168 cm), every soldier is over 6 feet tall (183 cm). They march in pairs with synchronized strides, their mustachioed faces dour through all the stomping and high-kicking.

At the border gates, the long-awaited confrontation occurs. The guards emerge from each side simultaneously while keeping a stern eye on one another. National flags are lowered—at exactly the same time, so that neither country can be accused of trying to "win"—after which an Indian and a Pakistani guard shake hands rigidly before retreating from the border with more stomps and high kicks. For the hundreds of Indian and Pakistani spectators, the ceremony is a source of national pride and a cathartic expression of built-up tension between the often-clashing countries.

Get a round-trip taxi from Amritsar—the driver will wait while you attend the ceremony. International visitors are entitled to VIP access and seating if they show their passports. Arrive early and travel light, because bags are not allowed. Ⓝ 31.604694 Ⓔ 74.572916

A dance of aggression, dominance, and coxcomb headgear.

KEY MONASTERY

DHAR LAMA CHUNG CHUNG

Resting snugly within India's Spiti Valley, the Key Monastery (or Ki, or Kye, or Kee) looks like a ramshackle temple of mysticism straight out of a fantasy novel, but this Buddhist training ground is actually the result of repeated attacks by Mongol hordes.

The exact age of the Key Monastery is not known, but it is believed to date back to at least the 11th century. The early structures erected by the original monks would have been built at a lower altitude than the current hilltop huts, which sit at 13,668 feet (4,166 m) above sea level. When Mongol barbarians attacked the monastery, reconstruction efforts then built upon what had gone before, eventually creating the erratic patchwork of rooms, tight hallways, and hidden courtyards that exists today.

Much of the monastery bears a distinct Chinese design influence, dating back to a period in the 14th century when the style came to the area. The interiors are also rich with historic murals and documents precious to the orders that have lived in the ever-changing monastery.

Today Key Monastery is still a fully functioning training ground for lamas, operated by a Gelug sect of Tibetan Buddhist monks. Around 250 monks reside on the site at any given time, training, farming, and generally keeping the thousand-year-old wonder alive. **Between May and October is the best time to visit, taking the roads from Manali via Kaza, as snow makes the (already bumpy and unfinished) route impassable. At the monastery, say yes to the butter tea. Ⓝ 32.297857 Ⓔ 78.011929**

CHAND BAORI STEPWELL

ABHANERI, RAJASTHAN

Built in the 9th century, but reminiscent of an M. C. Escher drawing, Chand Baori is among the largest and most elaborate of India's many stepwells. Locally known as *baori*, stepwells were tiered stone structures used to collect and store rainwater in arid climates. Often accented with arches, columns, sculptures, and geometric patterns, stepwells also served as village gathering places.

The four-sided Chand Baori is 13 stories tall, 100 feet (30.5 m) deep, and lined with 3,500 steps arranged in a spectacular zigzag pattern. Its intricate, multilevel facade overlooks a small pond of bright green water—a reminder of the magnificent structure's

3,500 Escher-esque steps lead down to a pool of stored rainwater.

practical purpose. Other beautiful stepwells worth visiting include Adalaj ki Vav and Rani ki Vav in Gujarat; Agrasen ki Baoli and Rajon ki Baoli in New Delhi; and Raniji ki Baori in Rajasthan. **Chand Baori is a 90-minute drive from Jaipur, on Jaipur-Agra Road. Ⓝ 27.007200 Ⓔ 76.606800**

NORTH SENTINEL ISLAND

ANDAMAN AND NICOBAR ISLANDS

The Stone Age hunter-gatherers who live on North Sentinel Island in the Andaman archipelago east of India may be the world's most isolated people—and they intend to stay that way, despite the increasing encroachment of the industrialized world.

From 1967 through the mid-1990s, Indian anthropologists embarked on periodic "contact expeditions" to North Sentinel Island. Approaching by boat, they attempted to coax out members of the tribe by depositing coconuts, machetes, candy, and, once, a tethered pig onto the beach. The Sentinelese almost always responded to these "gifts" by shooting

arrows, throwing stones, and shouting at the unwelcome visitors.

India discontinued its attempts at peaceful contact in 1997 and ruled that the islanders be left alone, but visits still occur—in 2006, a fishing boat drifted too close to the shore, and Sentinelese archers killed the two men on board. An Indian helicopter was sent to retrieve their bodies, but was also fired upon and could not land. In 2018, an evangelical missionary from the United States was killed while trying to approach the island. Efforts to recover his body were abandoned.

The hostility of the Sentinelese, whose population is estimated at between 100 and 200, is understandable considering the fates of tribes on other Andaman Islands. During British settlement, exposure to newly introduced diseases decimated tribal populations. More recently, roads constructed through forests on these other islands have allowed local sightseeing companies to offer "human safaris," during which travelers try to spot tribe members on their way to other island attractions.

India maintains a 2-mile buffer zone around North Sentinel Island. Access is strictly forbidden, but you can visit other islands in the archipelago. The main port of entry is Port Blair, where flights arrive from Chennai and Kolkata. Ⓝ 11.551782 Ⓔ 92.233350

The Sentinelese may be the most isolated tribe on Earth.

ALSO IN INDIA

New Lucky Restaurant

Ahmedabad · When Krishna Kutti realized there was a burial ground on the land he sought for his new restaurant, he simply incorporated the graves into the dining room decor.

Auroville

Bommayapalayam · This self-proclaimed experimental "city of the future," established by dubious spiritual leaders in 1968, centers around a golden geodesic dome named Mother Temple.

Hampi

Karnakaka · Known as "the last great Hindu Kingdom," Hampi flourished from the 14th to 16th centuries before being ransacked and abandoned. The surviving monuments evoke a grand riverside city.

JANTAR MANTAR ASTRONOMICAL INSTRUMENTS

JAIPUR, RAJASTHAN

Viewed without context, the 90-foot-tall (27 m) Samrat Yantra, or "supreme instrument," is a sand-colored stairway to nowhere. It sits in the middle of what looks like a skateboarder's half-pipe. The stately wooden doors at the base of its stairs and the small pagoda at its apex are merely mysterious. It is only when the Samrat Yantra's purpose is revealed that it becomes impressive: The Samrat Yantra is the world's largest sundial, built in the 18th century and accurate to within two seconds.

Jai Singh II, who became ruler of Amber (now Jaipur) at the tender age of 11, had a great enthusiasm for design, mathematics, and astronomy. Responsible for planning and building the city of Jaipur during the 1720s, Singh also took it upon himself to establish five observatories in North India. The largest, built between 1727 and 1734 in his hometown of Jaipur, is the best preserved.

The Jantar Mantar ("calculation instrument") of Jaipur is a collection of stone buildings used to determine local time, predict eclipses and monsoons, and track the movement of celestial objects. All of the instruments were built according to Hindu and Islamic astronomy and are impressively accurate due to their sizable dimensions.

Jantar Mantar is adjacent to City Palace near Tripolia Bazar. Ⓝ 26.924722 Ⓔ 75.824444

Indian ruler Jai Singh II's quest for scientific data resulted in the largest sundial in the world.

In one of the wettest places on Earth, bridges aren't built—they're grown.

ROOT BRIDGES OF CHERRAPUNJI

CHERRAPUNJI, MEGHALAYA

To cross the rivers and streams of the Cherrapunji forest, you must put your trust in a tree. There are no standard walkways to be found—instead, the tangled, twisting aerial roots of the rubber trees on the banks stretch across the water, forming a living, ever-growing bridge to the other side.

These organic bridges are the result of a little human guidance and a lot of patience. Members of the local Khasi tribe control their growth by first laying lengths of bamboo or betel nut tree across the water as a guide, then waiting for the roots of the rubber trees to follow along. As the roots grow, the Khasi add handrails made of vines and fill in gaps with mud and stones, creating a solid pathway. It takes up to 20 years for a bridge to become sturdy enough to cross, but once built, it continues to grow and strengthen for up to 500 years.

There are several root bridges in the Cherrapunji region. The most famous is the "Umshiang Double-Decker" root bridge at Nongriat. Its 60- and 80-foot (18 and 24 m) pathways, one atop the other, are made from the roots of the same tree.

The nearest city is Shillong. Bridges are accessible via jungle trek. Reaching the double-decker bridge requires a 6-mile (9.7 km) walk each way. Cherrapunji is one of the wettest places on Earth, so dress appropriately. Ⓝ 25.251513 Ⓔ 91.671963 ➤➤

➤➤ Other Notable Examples of Arbortecture

Arbortecture, or arborsculpture, is the art of shaping a living tree in order to create art or furniture. Using pruning, bending, and grafting, its practitioners spend years guiding each tree into a predetermined design. Arbortecture differs from topiary in that it shapes a tree's trunk or roots, rather than its foliage.

TREE CIRCUS AT GILROY GARDENS, GILROY, CALIFORNIA, USA
In 1947, arborsculpture pioneer Axel Erlandson established a roadside Tree Circus near Santa Cruz, California, to show off his grafted creations. Though the circus closed in 1963, some of his creations, such as a woven Basket Tree and the right-angled Four-Legged Giant, live on at Gilroy Gardens.

Gilroy Gardens

AUERWORLD PALACE, AUERSTEDT, GERMANY
Built by 300 volunteers in 1998, this willow dome is the centerpiece of a yearly summer music festival.

Each spring, a few dozen Auerworld supporters give the dome a ceremonial "haircut," trimming its wildest branches to maintain the mandala-influenced shape.

Auerworld Palace

KAZAKHSTAN

BAIKONUR COSMODROME

BAIKONUR

"Dear friends, known and unknown to me, my dear compatriots and all people of the world! Within minutes from now, a mighty Soviet rocket will boost my ship into the vastness of outer space. What I want to tell you is this. My whole life is now before me as a single breathtaking moment. I feel I can muster up my strength for successfully carrying out what is expected of me."

Those were the words of Yuri Gagarin on April 12, 1961, minutes before the cosmonaut lifted off in the *Vostok 1* spacecraft, becoming the first human to travel into space and enter orbit

around Earth. His journey began at the Baikonur Cosmodrome, the world's oldest and largest space launch facility, set in the desolate desert steppe of Kazakhstan.

The Soviet Union built the cosmodrome in 1955 as a secret missile testing site and space launch facility. Two years later, *Sputnik 1* launched from Baikonur, becoming the first man-made satellite in orbit and igniting the space race between the Soviets and Americans.

The cosmodrome is the world's most active spaceport, with a long list of historic launches. A month after *Sputnik 1*, a stray female dog named Laika hurtled into space aboard *Sputnik 2*, becoming the first animal to enter orbit and paving the way for human spaceflight. (Unfortunately, the canine pioneer's one-way suicide mission

was even shorter than planned—she died of heat exhaustion hours after launch, a detail that was only revealed in 2002.)

Before each launch, a Russian Orthodox priest clad in golden robes blessed the space-bound rocket, spraying holy water in the air and onto the faces of the assembled media.

The only way to see inside the cosmodrome and its space museum is on a guided tour. Since Baikonur is administered by Russia, tours leave from Moscow via a 3.5-hour chartered flight. For the most spectacular experience, time your visit for a launch—planned dates are available online. Be prepared for extreme weather, as Baikonur can reach –40°F (–40°C) in winter and 113°F (45°C) in summer. Ⓝ 45.965000 Ⓔ 63.305000

THE LAST WILD APPLE FORESTS

ALMATY

The common apple has its roots in one specific region of the world: *Malus sieversii*, its ancestor, grows wild in the Tian Shan mountains of Kazakhstan.

In the early 20th century, biologist Nikolai Vavilov first traced the apple genome back to a grove near Almaty, a small town whose wild apples closely resemble the Golden Delicious variety found at grocery stores today. Vavilov visited Almaty and was astounded to find apple trees growing wild, densely entangled and unevenly spaced, a phenomenon found nowhere else in the world.

Scientists believe the Tian Shan apple seeds were first transported out of Kazakhstan by birds and bears long before humans ever cultivated them. By the time humans did begin to grow and trade apples, the *Malus sieversii* had already taken root in Syria. The Romans dispersed the fruit even farther around the world. When modern genome sequencing projects affirmatively linked domestic apples to *Malus sieversii*, Almaty and its surrounding land were officially recognized as the origin of all apples.

The origin of all apples still flourishes.

Almaty means "father of apples," and the town touts its heritage proudly. A fountain in the center of town is apple-shaped, and vendors come out each week to sell their many varieties of domesticated apples at market.

The apple forests exist in patches along the Tian Shan mountain range. There are various protected sections in the Ile-Alatau National Park, but hiring a guide to take you there is recommended, as they are difficult to find. Ⓝ 43.092939 Ⓔ 77.056411

KYRGYZSTAN

TASH RABAT

AT-BASHI, NARYN

During the 15th century, the stone structure of Tash Rabat was a caravansary—a travelers' inn providing refuge for those journeying along the Silk Road. Protected by the high walls of the rectangular courtyard, human and animal travelers took shelter in its stalls to wash, rest, and prepare for the next leg of a long trip.

This desolate part of the trading route was particularly treacherous. Snow covers the ground for eight months of the year, and the area is subject to landslides, flooding, and earthquakes. The difficult conditions persist, which is why, for maximum safety and comfort, you should visit in summer and hire a local guide to drive you. For a fleeting insight into the Silk Road experience, camp in a yurt overnight at Tash Rabat.

Tash Rabat is a 6-hour drive south of the capital of Bishkek. Take precautions against altitude sickness—Tash Rabat is 11,500 feet (3,505 m) above sea level. Ⓝ 40.823150 Ⓔ 75.288766

The domed stone building of Tash Rabat gave weary Silk Road travelers a brief respite from the desert.

KYRGYZ NATIONAL HISTORY MUSEUM

BISHKEK

Kyrgyzstan's National History Museum is probably the only place in the world where you can find a ceiling mural of a naked Nazi in a horned helmet emerging from a wall of flames astride a demonic horse. It is definitely the only place you'll see a mural of Ronald Reagan in a skull mask, American flag T-shirt, and khaki cowboy hat riding a Pershing missile in front of a bunch of anti-nuclear demonstrators.

The museum, established in 1927, contains Kyrgyzstani cultural relics dating back to the Stone Age, such as armor, jewelry, coins, and weapons. The second and third floors became shrines to the legacy of the Soviet Union.

In Kyrgyzstan's time capsule of a museum, the Soviet Union still stands strong.

But images of communist heroes Lenin, Marx, and Engels inspiring the masses are gradually being replaced as the nation moves further away from its Soviet past. The outlandish murals, though, seem destined to stay.

Ala-Too Square, Bishkek. If you're feeling bold, flag down one of the overcrowded minibuses, locally known as *marshrutkas*. For a less stressful option, take a taxi. Ⓝ 42.876388 Ⓔ 74.603888

PAKISTAN

MUMTAZ BEGUM

KARACHI, SINDH

Resting in a shabby pavilion at Karachi Zoo is a creature by the name of Mumtaz Begum, a lounging fox with the head of a woman who can see the future and provide solid advice.

In reality, Mumtaz Begum is neither a fox nor a woman, but is actually played by performer Murad Ali, who inherited the role from his father. Each day he cakes his face with a thick layer of foundation, drawn-on eyebrows, and bright red lipstick. He then crawls into the box beneath Mumtaz's cage, jutting his head through the hole in the top to make it look as though his head is attached to the lounging fox carcass next to him. A shawl is placed around Ali's head to hide the connection, and then the visitors begin filtering in.

Ali's creature (known as a *kitsune*) is said to be able to see the future. Children and adults alike come to the zoo to ask Mumtaz Begum about everything from exam results to visa approvals. Ali gamely provides advice and peppers the interaction with references to his mysterious African origins.

Visitors coming to see Mumtaz Begum often leave small donations, cake, and juice, like

Pakistan's foremost prognosticating half-woman, half-fox can tell your fortune for a fee.

supplicants to a charlatan prophet. However, in true huckster fashion, anyone wishing to speak with the kitsune will need to pony up for a special ticket into the Mumtaz Mahal.

Karachi Zoo, Nishter Road and Sir Agha Khan III Road, Karachi. Buses, many of them alarmingly overcrowded, run along Nishter Road. Ⓝ 24.876228 Ⓔ 67.023203

KHEWRA SALT MINES

KHEWRA, PUNJAB

In 326 BCE, Alexander the Great and his army were making their way through present-day Pakistan on horseback when one of Alexander's steeds began voraciously licking the ground. When other horses joined in, soldiers dismounted to investigate, and discovered what is now the second-largest salt mine in the world.

Today, Khewra's 18-story-deep salt mine produces about 350,000 tons of pink Himalayan salt per year, a rate it is projected to maintain for the next 350 years. When salt leaves the mine, it is used for cooking and bathing. Within the mine, a visitors' section contains a mosque, a post office, and a "Palace of Mirrors," all made from salt bricks quarried from the 18 working levels below. The palace, with its illuminated floor of red, brown, and pink tiles, gives off a subterranean disco vibe.

The mine is a 2.5-hour drive south of Islamabad or 3-hour drive northwest of Lahore. Ⓝ 32.647938 Ⓔ 73.008394

ALSO IN PAKISTAN

Derawar Fort

Bahawalpur · In the Cholistan desert stands an enormous square medieval fortress with 98-foot-high (30 m) walls.

LAHORE FORT ELEPHANT PATH

LAHORE

Because it would be a shame to leave one's elephant parked outside the citadel, the magnificent Lahore Fort features a stepped entranceway crafted for an entire pachyderm parade.

As the Mughal Empire expanded across the Indian subcontinent in the 16th century, Lahore became an increasingly important stronghold. The city's fortress was built under the reign of Emperor Akbar between 1566 and 1605 and housed several Mughal and Sikh rulers over the following centuries.

The *Hathi Paer*, or elephant stairs, are part of the private entrance to the royal quarters and effectively allowed royalty to ascend all the way to the doorway before dismounting their enormous animals. In order to accommodate the lumbering creatures, the stairs were designed with wide treads but minimal height. (A balking elephant can really dampen the mood of a procession.) Although it's been centuries since a herd of jewel- and silk-laden elephants traveled several abreast along this sloping corridor, it was once certainly the most magnificent driveway in the world.

The elephant stairs pathway is located at the northwest corner of the Lahore Fort. Ⓝ 31.586606 Ⓔ 74.312300

Elephants in bas-relief line the base of the elephant path.

SRI LANKA
SIGIRIYA

MATALE, CENTRAL PROVINCE

When you've murdered your father and stolen your brother's crown, you need to find yourself a safe, vengeance-proof home. For King Kassapa I, who overthrew his father and buried him alive in the wall of an irrigation tank in 477 CE, that place was Sigiriya.

At the center of Sigiriya is the 650-foot-tall (198 m) hardened magma plug of an extinct volcano. Fearing an attack from his usurped brother Moggallana, Kassapa built a palace for himself on top of the rock and surrounded it with ramparts, fortifications, fountains, and gardens. A moat around the rock added an extra layer of protection.

Though Kassapa focused on keeping himself safe, he didn't skimp on the design details of Sigiriya. The stone stairway leading up the mountain is flanked by two huge lion paws carved into the rock. At the top of the 1,200 steps originally sat a lion's mouth, which people had to walk through to get to the palace. This explains Sigiriya's alternate name: Lion Rock.

Kassapa sequestered himself in his hilltop palace for 18 years before his greatest fear came to pass. Moggallana, having recruited an army in India, besieged Sigiriya and overpowered Kassapa's soldiers. Facing certain defeat, Kassapa turned his sword on himself and, with his death, granted Moggallana the kingship to which he had always been entitled. **For maximum comfort, go early to avoid the heat. From Colombo, get a bus to Dambulla and switch to a Sigiriya bus. The whole journey takes about 4 hours. Ⓝ 7.955154 Ⓔ 80.759803**

ALSO IN CENTRAL ASIA

ARMENIA
Khor Virap Monastery

Ararat · Visit a hilltop pilgrimage site that once held a saint in its dungeon for 13 years.

AZERBAIJAN
Naftalan Clinic

Naftalan · Bathe in crude oil at this spa devoted to petroleum-induced relaxation.

Mud Volcanoes

Baku · Gurgling volcanoes line the Caspian coast, occasionally erupting in flames.

GEORGIA
Stalin Museum

Gori · Established in 1957, this hometown tribute to the dictator is gradually changing its hagiographic presentation of Soviet history.

KAZAKHSTAN
Aral Sea

Thanks to aggressive irrigation practices, the Aral Sea has gone from being the fourth-largest lake in the world to a toxic desert strewn with rusty fishing vessels.

TURKEY
Cotton Castle

Pamukkale · Once a kind of Roman-era health spa, the spectacular rock formations below the ancient city of Hierapolis form a blindingly white natural cascading fountain.

At the top of Sigiriya, or "Lion Rock," is an ancient fort surrounded by gardens.

TURKEY

KAYAKÖY

KAYAKÖY, MUĞLA

Standing on a hill in the Kaya Valley region of Turkey are the deserted stone buildings of Kayaköy, a small town abandoned abruptly around 1922.

Once known as Levissi, the town was home to around 6,000 people, the vast majority of them Greek Orthodox. Then came World War I and the subsequent Greco-Turkish War, during which Turkey purged the Ottoman Greek people from its lands and Greece sent its Muslim residents to Turkey. In accordance with this population-swap policy, the inhabitants of Kayaköy were exiled to Greece. The Muslim inhabitants who arrived to take their place found Kayaköy's topography unsuited to their agricultural needs. They soon resettled to other parts of Turkey, leaving Kayaköy in the abandoned state that persists today.

There is much to explore, including hundreds of roofless homes and two eerie churches. At the top of the hill you'll get a glorious view down into the valley and out to the sea.

Kayaköy is 45 minutes south of Fethiye. Stay until after dark, when dramatic lights illuminate the crumbling corners of the village. Ⓝ 36.578922 Ⓔ 29.087051

The Greeks abandoned the Turkish town of Kayaköy.

ANI GHOST CITY

OCAKLI KÖYÜ

On the Turkish-Armenian border, scattered in the plains among the wildflowers, are the crumbling remains of a once mighty city. In the 11th century, Ani was home to over 100,000 people. Situated on a number of trade routes, the city became the capital of the Kingdom of Armenia, an independent state established in 961.

Ani was attacked by the Byzantines during the empire's takeover of the Armenian Kingdom in 1045. Two decades later, Seljuk Turkish invaders captured the city, murdered and enslaved its inhabitants, and sold the whole place to a Kurdish dynasty known as the Shaddadids.

The attacks continued in the 13th century, when the Mongols made two attempts—one thwarted, one successful—to capture the city. An earthquake in 1319 caused significant damage to Ani's many 11th-century churches. The city stumbled onward but was much smaller by the mid-17th century and completely abandoned by 1750.

Today Ani is a grand but ruined ghost town. Tensions between Turkey and Armenia have contributed to its neglect—it is an Armenian city but lies within Turkish borders, making conservation and restoration difficult. To visitors, Turkey omits all mentions of Armenia from descriptions of Ani's history and focuses on the city's Turkish and Muslim influences.

Ani is a 45-minute drive from the city of Kars. Bring snacks and water if you plan to spend the day exploring the ruins. Ⓝ 40.507636 Ⓔ 43.572831

Earthquakes, war, and vandalism have weakened the remaining structures of a once-bustling city.

AVANOS HAIR MUSEUM

AVANOS, NEVŞEHIR

Avanos, a small town in the Cappadocian region, has a millennia-long history of ceramics and pottery. But only one of its potters maintains a cave full of human hair.

In 1979, local potter Galip Korukcu was bidding farewell to a dear friend. When he asked for a memento, she snipped off a lock of her hair. Korukcu stuck the hair on one of the walls of his pottery shop, which is located in a cave. After hearing the story behind the wad of hair on the wall, pottery buyers began contributing their own locks.

The "hair museum" now crams an estimated 16,000 hair samples onto its walls. Visitors are invited to snip off a few strands of hair and attach them to a card with their contact details to add to the display. Pencils, paper, pins, and scissors are provided.

There is an added incentive for contributing to the collection: Twice a year, Korukcu asks a customer to choose 10 hair samples from the walls. The owners of the winning hair receive a free weeklong stay in the connected guest house and workshops with the master potter.

Firin Sokak 24, Avanos. Ⓝ 38.720612 Ⓔ 34.848448

TURKMENISTAN

DOOR TO HELL

DERWEZE, AHAL

When darkness falls, an orange glow illuminates the dusty plains outside of Derweze, a settlement of 350 in the middle of the Karakum Desert. The source of light is the "Door to Hell," a 200-foot-wide (61 m) crater that has been burning for over 45 years.

In 1971, Soviet geologists, looking for natural gas, accidentally burrowed into a huge cavern filled with methane, causing the ground to crumble and their drilling rig to collapse into the huge pit. With the pocket of gas punctured, poisonous fumes began leaking from the hole at an alarming rate. To avoid a potential environmental catastrophe, the geologists set the hole on fire. The crater has been burning ever since.

Following a visit to the Door to Hell in April 2010, Turkmen president Gurbanguly Berdimuhamedow recommended the hole be closed so the area's rich gas reserves could be tapped safely. Thus far the crater remains untouched, but with new pipelines and increased international interest in Turkmen gas reserves, the Door to Hell may not be open for much longer.

The crater is 160 miles (257.5 km) north of the capital city of Ashgabat, where you can hire a guide to drive you to the desert. Ⓝ 40.252777 Ⓔ 58.439444

The 200-foot-wide desert fire pit known as the Door to Hell has been blazing since 1971.

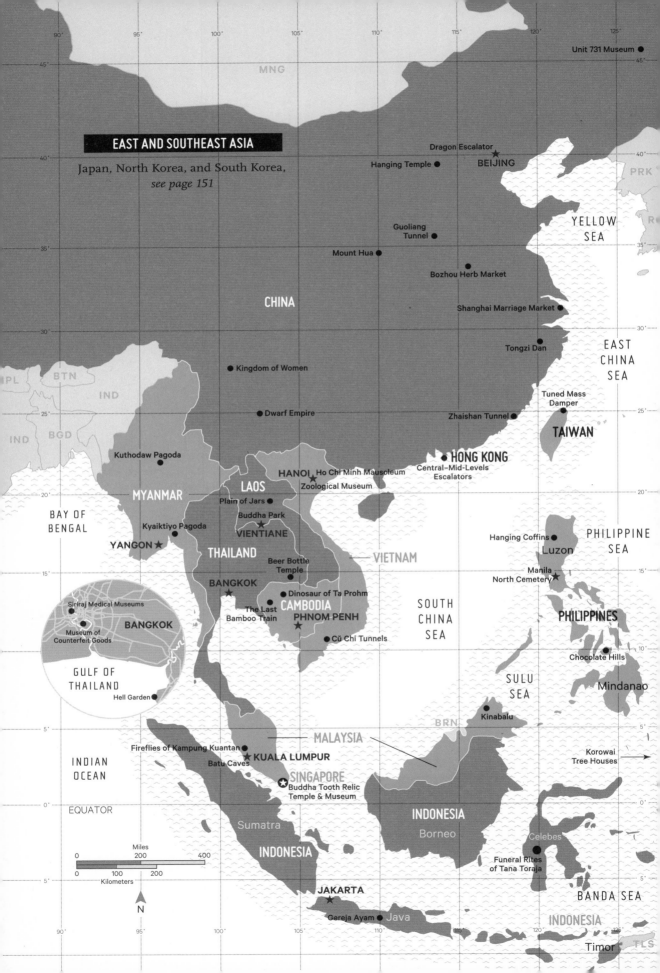

EAST AND SOUTHEAST ASIA

Japan, North Korea, and South Korea,
see page 151

MNG

Unit 731 Museum

Dragon Escalator ★
BEIJING
Hanging Temple ●

YELLOW SEA

PRK
R

Guoliang Tunnel ●

Mount Hua ●

Bozhou Herb Market ●

CHINA

Shanghai Marriage Market ●

EAST CHINA SEA

Tongzi Dan ●

NPL

BTN

IND

Kingdom of Women ●

Tuned Mass Damper ●

Zhaishan Tunnel ●

TAIWAN

IND

BGD

Dwarf Empire ●

Kuthodaw Pagoda ●

HONG KONG ●
Central–Mid-Levels Escalators

HANOI ●
Ho Chi Minh Mausoleum
Zoological Museum

MYANMAR

LAOS

BAY OF BENGAL

Plain of Jars ●
Buddha Park ★
VIENTIANE

Kyaiktiyo Pagoda ●

Hanging Coffins ●
Luzon

PHILIPPINE SEA

YANGON ★

THAILAND

Beer Bottle Temple ●

VIETNAM

Manila ★
North Cemetery

Siriraj Medical Museums ●

BANGKOK ★

Dinosaur of Ta Prohm ●

CAMBODIA

SOUTH CHINA SEA

PHILIPPINES

BANGKOK

Museum of Counterfeit Goods ●

The Last Bamboo Train ●

PHNOM PENH ★

Chocolate Hills ●

GULF OF THAILAND

Cu Chi Tunnels ●

SULU SEA

Mindanao

Hell Garden ●

INDIAN OCEAN

Fireflies of Kampung Kuantan ●
Batu Caves ●

MALAYSIA

BRN

Kinabalu ●

Korowai Tree Houses →

KUALA LUMPUR ★

SINGAPORE ★
Buddha Tooth Relic Temple & Museum

EQUATOR

Sumatra

INDONESIA

Borneo

INDONESIA

Celebes

Funeral Rites of Tana Toraja ●

Miles
0   200   400

Kilometers
0   100   200

BANDA SEA

JAKARTA ★

N

Gereja Ayam ●   Java

INDONESIA

Timor

TLS

Ascending the Longqing Gorge in serpentive style.

East Asia

CHINA
DRAGON ESCALATOR

BEIJING

Located roughly 53 miles (85.3 km) north of Beijing, Longqing Gorge is a pleasant change of pace from the crowded streets of China's capital. Home to the country's largest dam, the otherwise tranquil surroundings are slashed by a gigantic dragon escalator.

The bright yellow behemoth, which takes riders to the top of the dam, rises 846 feet (257.9 m). At the top you can have even more adventures, including a ride in a cable car or a boat trip on the artificial lake. If a boat ride sounds too slow for you, you can always bungee jump into the gorge.

You can ride the giant escalator as much as you want, but unfortunately, it only goes up. To come down, you can either take the stairs or pay a little extra to ride the toboggan.

Buses run from the Deshengmen bus terminal in Beijing to Gucheng, a short walk from the gorge. The journey takes around an hour. Ⓝ 40.456704 Ⓔ 115.974999

The temple of Hengshan clings to a cliff 246 feet off the ground in apparent defiance of gravity.

HANGING TEMPLE OF HENGSHAN

MOUNT HENG, SHANXI

Two hundred and forty-six feet (75 m) above ground, supported by a few thin wooden stilts driven into the cliff face, the Hanging Temple of Mount Heng dangles in apparent defiance of gravity. Built into the rock during the Northern Wei Dynasty (386–534 CE), the 40-room temple, connected by a dizzying maze of passageways, has survived the erosive effects of wind, sun, and snow due to its protected position.

Unusually, the temple caters to not one but three religions—elements of Confucianism, Taoism, and Buddhism are evident in its 78 statues and carvings. **Mount Heng is 2 hours southeast of Datong, where you can hire a driver to take you there and back.** **Ⓝ 39.673888 Ⓔ 113.735555**

UNIT 731 MUSEUM

HARBIN, HEILONGJIANG

Officially, Unit 731 was a lumber mill in Japanese-occupied Manchuria. Its staff received frequent deliveries of *maruta* ("logs"). Black smoke billowed from the chimney day and night.

Unit 731 was not, in fact, a lumber mill. It was a biological weapons research facility in which Japanese scientists conducted torturous, lethal experiments on humans—humans they referred to as "logs." Established in 1936, and operational during the second Sino-Japanese War through World War II, the facility functioned under the approval of Japan's Emperor Hirohito.

The atrocities conducted at Unit 731 consisted of in-house experimentation on predominantly Chinese and Russian individuals. In the laboratories, researchers studied blood loss by performing amputations. They infected patients with diseases such as gonorrhea and syphilis, and then conducted vivisections—removing the organs of living people—to observe their effects on the body. Patients received no anesthetic during these operations.

Planes dropped bombs filled with anthrax, smallpox, typhoid, cholera, and plague-infected fleas on Chinese villages, aiming to maximize death and devastation. On the ground, members of Unit 731 infected hungry children with lethal pathogens by giving them contaminated food and candy.

The secrecy surrounding Unit 731 has made it difficult to estimate how many people died in the experiments. Given the contamination of agricultural crops and water supplies that resulted from the aerial drops, the victims likely numbered in the hundreds of thousands.

In 1945, immediately following Japan's surrender, the members of Unit 731 fled the site, destroying as much evidence as they could on the way out.

Despite the destructive efforts of the departing researchers, portions of Unit 731—including the Frostbite Laboratory (where victims endured exposure to extreme cold), the Yellow Rat breeding room, and the incinerator that cremated the bodies—remain. The site opened as a museum in 1985, offering two floors of exhibits. Graphic photographs, medical instruments, and plaques re-create a period of history that Japan doesn't want to remember, and China will never forget. **23 Xinjiang Street, Pingfang, Harbin. Get a bus from Harbin train station. Ⓝ 45.608244 Ⓔ 126.639633**

MOUNT HUA

HUAYIN, SHAANXI

There are two ways to reach the 5,121-foot (1,561 m) North Peak of Mount Hua: an eight-minute cable car ride, or a four-hour climb up narrow staircases carved into the near-vertical cliffs. No matter which option you choose, there is more treachery to come.

The five peaks of Mount Hua, or Huashan, have long been a refuge for determined hermits unfazed by heights. Among the most notable of the hermits was the 10th-century Taoist priest Chen Tuan, who developed a martial art known as "water boxing" while living alone in a secluded high-altitude temple.

In addition to the exhausting stair climbs, the mountain offers hikers an extreme challenge in the form of its most terrifying path: Changkong Zhandao. This one-foot-wide (.3 m) walkway juts out from the cliffs connecting the north and south peaks. It is made of wooden planks—some wobbly, some rotting—that rest on iron bolts driven into the side of the mountain. There is a hand-rail of sorts—a thick chain pinned into the rock at intervals with large nails.

If you opt to walk the planks, your journey will begin with one of the most important purchases of your life: a safety harness, which vendors on the north peak have been renting since 2005. Technically, the harness is optional, but choosing not to attach yourself to the flimsy cable running above the path would be an act of folly.

At the end of Changkong Zhandao is a small shrine with a stunning view of the mountain range—the perfect place to pause and gather your courage before turning around and walking back down. Trains and buses run from Xi'an, 2 hours southwest. If the idea of scaling a cliff face in the dark sounds fun, climb the mountain at night so you can watch the sunrise from one of the peaks. Ⓝ 34.478861 Ⓔ 110.069525

A safety harness is recommended for the treacherous journey up Mount Hua.

华山奇景
HUASHAN SPECTACULAR SCENERY
长空栈道

BOZHOU MEDICINAL HERB MARKET

BOZHOU, ANHUI

Located at the juncture of two key railway lines, the dusty and rusty city of Bozhou is the capital of the Chinese medicinal herb industry. With a population of 3 million people, the city revolves around a massive 85-acre market, where some 6,000 traders come from every corner of southeast Asia to ply the ingredients found in traditional Chinese medicine.

Here you can find barrels of dried human placentas (for fainting sickness), dried fist-size stag beetles (for increased metabolism), dried flying lizards (also for metabolism), cockroaches (a topical anesthetic), crushed pearls to ingest with tea (for influenza), pencil-size millipedes bundled up and bound together in clumps (for a host of sicknesses), snakes (for arthritis), and a dozen different kinds of ants (for pretty much whatever ails you). Around every corner are hemp sacks overflowing with scorpions, seahorses, turtle shells, antlers, and every kind of root and flower imaginable.

Though the Bozhou herb market feels timeless—and there has indeed been an herbal market on the site for centuries—the market has recently undergone a boom as Westerners have increasingly adopted elements of traditional Chinese medicine. Today, the downtown is ringed with pharmaceutical factories and hotels for visiting traders.

Weiwu Avenue, Bozhou. The overnight train from Shanghai to Bozhou takes about 10 hours. Ⓝ 33.862205 Ⓔ 115.787453

GUOLIANG TUNNEL

GUOLIANG, HENAN

Located in the Taihang mountains, the village of Guoliang was once accessible only via a treacherous, 720-step cliffside staircase known as the "sky ladder." When its 350 inhabitants needed to stock up on food supplies or get medical attention, they would climb down these 720 steps, built without a railing during the Ming dynasty (1368–1644).

That arrangement changed in 1972, when villager Shen Mingxin and a dozen other men began to dig a road through one of the mountains. Using shovels, spikes, iron hammers, and dynamite that they purchased by selling livestock, the men spent the next five years hand-carving the tunnel out of rock almost 400 feet (122 m) up the mountain.

On May 1, 1977, the three-quarter-mile Guoliang Tunnel opened to traffic. Illuminated only by sunlight through its 30 rock-hewn windows and measuring under 20 feet (6 m) wide, it requires careful navigation by both vehicles and pedestrians. Drivers use their headlights and honk their horns at regular intervals to help prevent collisions.

From the city of Xinxiang, get a bus to Huixian, then switch to a Nanping-bound bus. The journey takes about 3 hours. Ⓝ 35.731287 Ⓔ 113.603825

Guoliang Tunnel, seen running along the mountain about halfway down the cliff, was dug by hand.

Urine-soaked "boy eggs" make for a pungent street food.

Tongzi Dan

DONGYANG, ZHEJIANG

Every spring, the streets of Dongyang fill up with egg vendors selling a popular seasonal product known as *tongzi dan*. Also called "boy eggs," the traditional delicacy is made by collecting the urine of young boys and using it to hard-boil eggs. After boiling in the steaming urine, the eggs are removed, their shells cracked, and they are placed back into the simmering urine to soak up the robust flavor.

Tongzi dan have been standard street fare in Dongyang in Zhejiang Province for hundreds of years. The smell of steaming urine wafting through the city heralds the arrival of spring. Residents attest to the appetizing taste and medicinal properties of the eggs, believing they increase blood flow and lower internal temperature. Despite these widely held beliefs, doctors in the area do not advocate the consumption of anything boiled in human waste.

The process of acquiring bulk quantities of fresh young urine is surprisingly simple: Local schools line their halls with plastic buckets, which boys under 10 use as toilets when the urge strikes. The receptacles are collected throughout the day and their contents poured into the cooking barrels. In order to keep the process as hygienic as possible, boys who are unwell are asked to refrain from using the buckets.

Dongyang, Zhejiang. Ⓝ 29.289634 Ⓔ 120.241561 ➻

➻ Other Asian Street-Food Eggs

BALUT, PHILIPPINES

Sold on the streets of the Philippines, *balut* is a boiled duck egg with a difference: It is fertilized. Crack the shell open, and snuggled against the yolk you'll see a veiny pink embryo in its fetal sac. Though recognizably avian, the fetus is too young to have developed a beak, claws, or feathers.

For 17 days, freshly fertilized duck eggs are stored in a warm place to allow the embryo to develop. Then, to prevent further maturation, vendors boil the eggs, killing the fetus and solidifying the yolk. Balut are served warm and salted, often as a snack to accompany beer.

CENTURY EGG, CHINA

The yolk of a century egg is forest-green and smells of sulfur-tinged ammonia. Its surrounding "white" is the color of rust, its texture similar to Jell-O.

The Chinese delicacy earns its pungent odor and distinctive look from a preservation process that raises its pH level. The recipe requires a duck or chicken egg to be encased in a mix of salt, clay, quicklime, ash, and rice hulls for several weeks. Eaten on their own or as a garnish for tofu dishes, century eggs are available at street stalls and in dim sum restaurants, and are often on the menu at birthdays and weddings.

ŌWAKUDANI BLACK EGGS, JAPAN

Ōwakudani, or the Great Boiling Valley, is a volcanic zone with hot springs, sulfur vents, and a great view of Mount Fuji. It is also the best place in Japan to eat a black-shelled, longevity-enhancing egg.

The eggs of Ōwakudani turn black when boiled in the thermal pools—the sulfur and minerals in the water react with the eggshell, causing the color change. A gondola lift brings cartons of eggs up the hill for boiling, after which they are sent back down to be sold in packs of five. Due to the presence of hydrogen sulfide and sulfur dioxide in the air, visitors are allowed only a brief look at the cooking process.

The acrobats at Circus World are among the world's best.

Shanghai Circus World

Zhabei Qu • It's a little out in the suburbs, but you can't miss the giant golden geodesic dome that hosts Circus World, where some of China's most skilled athletic performers and stunt divers combine a modern sensibility with traditional acrobatics, dance, and illusions.

1933 Slaughterhouse

Hongkou Qu • The rivers of cow's blood are long gone from this former slaughterhouse, leaving behind an eerie Escher-esque shell of a building, the last abattoir of its kind: British-designed, pre-communism-built, it is as haunting as it is beautiful.

Jewish Refugees Museum

Hongkou Qu • Shanghai (along with the Dominican Republic) was alone in taking in Jews fleeing Europe during World War II. The turbulent period is remembered at the Jewish Refugees Museum through artifacts, personal documents, photography, and archival copies of the *Shanghai Jewish Chronicle*.

M50 Art District

Putuo Qu • An artist's community dominates the district around an old textile mill on Moganshan Road, with studios open to the public, thought-provoking sculptures and murals, and more than a hundred galleries that show the work of both the well-established and cheeky newcomers.

Shanghai Postal Museum

Hongkou Qu • Now a protected landmark, the Postal Museum still functions as a regular post office, but inside its classic colonial architecture is a history of postal scouts, arctic expeditions, stamps made from silk and wood, and ancient oracle bones that marked the progress of far-off military battles.

Bund Tourist Tunnel

Huangpu Qu • This short underground railway connecting Pudong with the waterfront of the Bund offers a five-minute mind-bending train ride featuring seizure-inducing strobe lights and a psychedelic New Agey soundtrack.

Natural Wild Insect Kingdom

Pudong Xinqu • Billed as an educational science destination, the Insect Kingdom has a butterfly zone, a serpent and python area, beetles, and bugs of all kinds, and it might be the only museum with a Shrimp Appreciation Zone.

Oriental Pearl TV Tower

Pudong Xinqu • When it opened in 1994, this 1,535-foot (468 m) spire was the tallest structure in China, and although it lost that distinction a dozen years later, a trip to the glass-bottom observation deck at the top can still grab adrenaline junkies by the throat—and send acrophobes into a fetal position.

Jin Mao Tower Skywalk Harnessed Observation Deck

Pudong Xinqu • Take the elevator to the 88th floor of Jin Mao Tower, the third tallest building in China, skip the deck, and head for the glass-bottomed ledge called the Skywalk, where you can dangle from a harness over the street below.

Shanghai, circa 2020

Huangpu Qu • The entire city of Shanghai, as it hopes to look in the year 2020, has been scaled down to fit on the third floor of the Urban Planning Museum, over a thousand square feet (93 m²) of perfectly re-created Lego-like neighborhoods, soaring towers, and planned developments.

Shanghai Brush & Ink Museum

Huangpu Qu • The histories of Chinese ink, pens, and brushes are traced at this small museum, including examples of calligraphy from as early as the 4th century up

SHANGHAI MARRIAGE MARKET

SHANGHAI

During weekends, the walkways in the north part of People's Park are filled with middle-aged men and women affixing posters to the ground, the bushes, and lengths of string suspended at eye level. The posters advertise the glowing attributes of the goods they are offering: their marriage-ready sons and daughters.

Traditionally, Chinese marriages begin with parental matchmaking—before a potential couple meets, their parents will discuss the viability of the union, swapping information on looks, interests, and finances. In 21st-century Shanghai, the process can be difficult. Fast-paced lives, busy schedules, and a male-skewed sex ratio resulting from the country's former one-child policy all hinder parents who want to marry their children off before they hit the "crucial" age of 30.

The outdoor marriage bazaar draws hundreds of traders every week, each one clutching a piece of paper listing height, age, educational background, occupation, and spousal preferences. Some mothers and fathers bring a folding chair, settling in for a day of fielding offers from other matchmakers. The success rate is low—there are parents who have been coming every weekend for years—but, given the through colonial occupation, when wealthy merchants underwrote the artists who developed a new and vibrant Shanghai school of painting and calligraphic arts.

Yu Garden Zigzag Bridge

Huangpu Qu • Exquisite pavilions from the Ming Dynasty are all well and good, and Yu Garden certainly has those, but when you need to trick evil spirits, head to the zigzag bridge, because (if legend be believed) evil spirits can only travel in straight lines.

Ballroom Dancing at the Karl Marx and Friedrich Engels Statue

Huangpu Qu • In the French Concession at the northern end of the lush formal gardens of Fuxing Park, couples gather in the morning for some early-bird ballroom dancing under the watchful gaze of a 12-foot statue of Karl Marx and Friedrich Engels.

Waxworks Hall at the First National Congress of the Communist Party Museum

Huangpu Qu • In 1921, when the 13 original delegates of the Chinese Communist Party met for the first time, it was in this small residence in the French Concession. The meeting is re-created with wax figures in the tiny second-floor study.

Dajing Ge Pavilion

Huangpu Qu • The last remaining section of Shanghai's old city wall can be found here. Once more than 3 miles (5 km) of stone fortification, ramparts, and gates, the wall was torn down in 1912, leaving behind only 164 feet (50 m) on the Dajing Road.

Antique Music Box and Gallery

Pudong Xinqu • This collection of miniature and life-size mechanical wonders is as much automata as musical, with more than 200 animatronic birds in cages, twirling dancers, parlor scroll players, and a 1796 "musical device of reduced dimensions"—the world's oldest music box—made by Swiss watchmaker Antoine Favre-Salomon.

Animation & Comics Museum

Pudong Xinqu • Among the familiar Mickey, Jessica Rabbit, and Kung Fu Panda, this story temple to the animated arts celebrates the earliest forms of moving pictures, like sand painting and traditional Chinese shadow puppets.

Tian Zi Fang

DaPuQiao • The open-air Tian Zi Fang is a maze of galleries, trinket stalls, coffee shops, and makeshift craft studios, but it's the traditional Shikumen architecture of this alley bazaar that sets it apart from other stops along the tourist trail—and also what saved it from the redevelopment bulldozers.

Lu Hanbin Typewriter Museum

Yan'an Xi Lu • A temple to the writer's workhorse, this assemblage of more than 300 models and historical displays claims to be the third-largest typewriter museum in the world. With a small sitting area and no admission charge, it's like an internet cafe without the internet.

Propaganda Poster Art Centre

Xuhui • A collection of Chinese propaganda posters spanning a 30-year period from the start of the revolution in 1949 to the reforms of the late 1970s, including examples of idealized military victories and gauzy visions of the model communist life.

Longhua Revolutionary Martyrs Memorial and Cemetery

Xuhui Qu • Once a prison for the pre-Revolutionary Kuomintang party, this vast park of monumental statues and memorials is now a free museum of the Communist struggle, with preserved prison cells and 500 graves of Communist martyrs.

500 Arhats of Longhua Temple

Xuhui Qu • Inside one of the chambers of this Buddhist temple—the biggest and oldest in Shanghai, tracing back to the year 242—there are rows of foot-tall golden Buddhas.

Shanghai Astronomical Museum

Songjiang Qu • In 1900, on a quiet forested hilltop near the Basilica of Our Lady of Sheshan, Jesuit missionaries built China's first astronomical dome, where today there is a museum and public observatory. The original telescope is not only still working but is still one of the largest binocular refracting opticals in the country.

social stigma facing unmarried thirty-somethings, the marriage market has nevertheless thrived.
People's Square, Wusheng Road, Huangpu, Shanghai. Get the Metro to People's Square and walk west into the park.
Ⓝ 31.232229 Ⓔ 121.473163

A son-and-daughter meat market, run by mom and dad.

DWARF EMPIRE

KUNMING, YUNNAN

Before their performances in one of the twice-daily shows, the short-statured performers of Dwarf Empire theme park (also known as the "Kingdom of the Little People") get ready in miniature mushroom-shaped houses with crooked chimneys. Park visitors peer in through the undersized doorways as the actors apply makeup.

Established in 2009 by wealthy real-estate mogul Chen Mingjing, Dwarf Empire is an experience that invites ethical questions. Over a hundred performers, all under 4.5 feet (1.3 m) tall and recruited from around the country, live and work at the park—not in the mushroom houses, but in dormitories customized for their stature. Each day, the performers dance, sing, pose for photos, and sell refreshments to visitors, all while wearing fairy-tale costumes. When the audiences have left for the day, they sweep the park and stack the chairs before retiring to their dorm rooms.

China's treatment of its short-statured population is influenced by a widespread belief that disability is a punishment inflicted on those who have committed past-life sins. Many performers at the park have been ostracized from their families, denied health care or employment, and forced to live on the streets. For them, Dwarf Empire provides support, a steady paycheck, and the chance to display skills such as singing, kung fu, or break dancing.

That said, the park can feel exploitative. Presumably, many visitors are not there to appreciate the performers' skills, but to gawk at the novelty of being in an all-dwarf world. The resulting environment skates uncomfortably close to a human zoo.

Located next to the Butterfly Ecological Park, about 25 miles southwest of Kunming off the G56 Hangrui Expressway. Public buses run from the city center. N 24.850411 E 102.622266

ALSO IN CHINA

Beichuan Earthquake Memorial	Giant Mao Head	Kissing Dinosaurs	Jade Burial Suits	Hallstatt
Beichuan · Destroyed in the 2008 Sichuan earthquake, the city of Beichuan remains in ruins as a memorial to the thousands who died.	*Changsha* · This massive granite Mao-nument depicts the young Great Leader's windswept, youthful head.	*Erlian* · Two smooching apatosauruses tower over a road near the Mongolian border, their necks creating a tunnel for passing cars.	*Shijiazhuang* · Hebei Provincial Museum displays two intricate suits of armor made from thousands of jade tiles stitched together with gold thread.	*Luoyang* · China has built an exact replica of the Austrian village of Hallstatt to serve as a novelty high-end housing development.

KINGDOM OF WOMEN

LUGU LAKE, YUNNAN

Sitting astride the border of Yunnan and Sichuan provinces, and surrounded by mountains, Lugu Lake is a tranquil place. But most travelers don't come here for the scenery—they arrive tantalized by the promise of a "Kingdom of Women."

The Mosuo, a 50,000-strong Chinese ethnic group with a matriarchal social structure, live in villages around the lake. Women control their multigenerational households and maintain ownership of their homes and land. Children take their mother's surname, and inheritance is distributed through the female line.

An often misunderstood aspect of Mosuo culture is the practice of "walking marriage." When girls turn 13, they attend a coming-of-age ceremony and receive their own private bedroom. From this time onward, girls can receive male "visitors" at night. Visits take place by mutual agreement, and men arrive under cover of darkness, returning to their homes in the morning—the "walking" part of walking marriage. Women may have as many sexual partners as they wish, with no stigma attached, for as long as they desire.

When these trysts result in the birth of a child, the father plays no role in the child's upbringing, beyond occasional gifts and visits. Mothers are responsible for taking care of their children, with the assistance of family members in the household. A father never lives with his children—instead, he stays in the family home in which he grew up, helping to raise the children in that household.

In a country that famously prizes baby boys, the Mosuo are unique for valuing girls and striving for a balance of sexes in the home. If a household becomes skewed in either direction, its matriarch may adopt children of the appropriate sex. These children become equal members of the family.

Lugu Lake is a bumpy 6-hour bus ride from Lijiang, a former Silk Road town that has retained much of its ancient architecture. N 27.705719 E 100.775127

The Mosuo maintain one of the world's few matrilineal societies.

HONG KONG

CENTRAL—MID-LEVELS ESCALATORS

CENTRAL AND WESTERN DISTRICT

Residents of Hong Kong's affluent, elevated Mid-Levels commute to the city's main business district in a unique way: Instead of hopping on a train, tram, boat, or bus, they spend 20 minutes riding a series of hillside escalators.

Built in 1993 to ease road traffic, the Central–Mid-Levels Escalators comprise the world's longest outdoor covered escalator system. A total of 20 escalators, plus three moving walkways, snake along Cochrane Street and Shelley Street, linking the core urban area of Central to the residential Mid-Levels on Victoria Peak. The total distance covered is 2,600 feet (792 m), with a vertical climb of 443 feet (135 m).

Around 55,000 people use the escalator system every day. From 6 a.m. until 10 a.m., all escalators run downhill. Then it's switchover time—from 10:30 a.m. until

The world's longest outdoor covered escalator system transports commuters across Hong Kong's hilly terrain.

the midnight system shutdown, the mechanical walkways travel uphill. On the moving stairs, you'll glide past the vibrant shops and global restaurants of the bustling SoHo district.
Cochrane Street (between Queen's Road Central and Hollywood Road) and Shelley Street. Ⓝ 22.283664 Ⓔ 114.154833

ALSO IN HONG KONG

Chungking Mansions

Kowloon · Teeming with illegal goods and services, this towering maze of vice has some of the cheapest accommodation in the city.

TAIWAN

TUNED MASS DAMPER OF TAIPEI 101

TAIPEI

The view of the city from the 89th floor of Taipei 101, one of the world's tallest buildings at 1,667 feet (508 m), is spectacular. But turn your back to the urban panorama and you'll see something equally fascinating: a huge yellow sphere, suspended in the center of the building between floors 88 and 92.

The 728-ton globe is a tuned mass damper, a pendulum-like device designed to counter the effects of wind and seismic activity on high-rises. In strong wind, the upper levels of a skyscraper

will sway a few feet back and forth. The Taipei 101 damper, suspended from eight steel cables, provides a counterforce that offsets the movement and prevents people in the building from feeling unsteady. Given Taiwan's susceptibility to earthquakes—the city sits on the edge of two tectonic plates—the eye-catching damper is an essential architectural feature.
7 Hsin Yi Road, Section 5, Taipei. The building is a 15-minute walk from the Taipei City Hall subway station. You will ride to the observatory level in one of the world's fastest elevators, traveling at 38 miles (61 km) per hour. Ⓝ 25.033612 Ⓔ 121.564976

ALSO IN TAIWAN

Bei Tou Incinerator

Taipei · Dine in a revolving restaurant at the top of a waste incinerator's chimney.

An enormous pendulum helps keep Taiwan's tallest building from swaying in the wind.

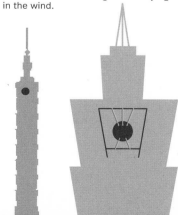

Step through the entrance (left) to this extensive underground tunnel and waterway system and you just might hear some Mozart.

ZHAISHAN TUNNEL

QUANZHOU

Following the Second Taiwan Strait Crisis in 1958, tensions between the People's Republic of China (PRC) and the Republic of China (ROC, now known more commonly as Taiwan) were at breaking point. The Taiwanese island of Kinmen, just a few miles from the mainland, had been shelled relentlessly during the four-week crisis, and a cold war lingered between the two factions.

Faced with potentially devastating artillery bombardments from the Chinese mainland, President Chiang Kai-shek of Taiwan ordered the construction of underground fortifications in the hard granite of Greater Kinmen, a strategically important island just off the coastline. This included the Zhaishan Tunnel, an incredible system of underground tunnels and waterways stretching for 2,592 feet (790 m).

The Zhaishan Tunnel was completed in 1966, comprising two interconnecting A-shaped tunnels. One was an underground waterway built to help protect boats from bombardment and allow for the safe unloading of cargo. It could shelter 42 small naval vessels.

As the cold war tensions gradually lessened, Taiwan went through a period of reform and social change. The Zhaishan Tunnel, meanwhile, was slowly filling up with sand. Maintaining the tunnel would have required money and manpower. With neither available, it was abandoned in 1986.

In the mid-1990s, an increased interest in preserving national historic sites and, more locally, remembering the men and women who fought for Kinmen turned the spotlight back onto the Zhaishan Tunnel. It was handed over to the Kinmen National Park to be restored and preserved. There was, however, one caveat: The Ministry of Defense would always be granted full access and control over the tunnel in times of war and for conducting military exercises.

Alongside its current role as tourist attraction, national historic site, and potential wartime boat shelter, the Zhaishan Tunnel also serves as a concert hall. The annual Tunnel Music Festival shows off the waterway's marvelous acoustics.

Jincheng Township, Kinmen County. Buses run from Jincheng station to the tunnel. Ⓝ 24.394329 ⒺＥ 118.320511

JAPAN, NORTH KOREA, AND SOUTH KOREA

JAPAN

MUSHIZUKA, SHRINE TO SLAIN INSECTS

TOKYO

In the garden of Kan'ei-ji Temple is a smooth, engraved boulder—a circa-1821 memorial honoring the slain victims of artist Masuyama Sessai. A man with a sizable conscience, Sessai ordered its construction himself, hoping it would console the spirits of those he killed. The slain were insects killed to serve as anatomy models for an illustrated scientific textbook, but erecting a stone shrine was the least the staunch Buddhist felt he could do. *Mushizuka*, the word engraved on the boulder, means "mound for insects."

1-14-11 Ueno Sakuragi, Taito-ku, Tokyo. Kan'ei-ji Temple is a 5-minute walk from Uguisadani Station's south exit. Ⓝ **35.721453** Ⓔ **139.774204**

ALSO IN TOKYO

Kabukicho Robot Restaurant

Tokyo · Neon strobe lights, techno music, drummer women in bikinis, and gyrating cyborgs are but a few notable features of this deeply confusing dining experience.

Nakagin Capsule Tower

Tokyo · This 13-story apartment building is crammed with 144 podlike residences straight out of a dystopian sci-fi novel.

Museum of Kites

Tokyo · This tiny museum hidden above a restaurant houses thousands of modern and traditional kites.

CITY GUIDE: More to Explore in Tokyo

Sogen-ji Kappa-dera Temple

Taitō-ku • This small temple is devoted to the folkloric and sort of cuddly "kappa," turtle-ish goblins who love cucumbers and are known to grab unsuspecting humans from bridges, wrestle with them, and occasionally drown them—so stop by, watch your back, and leave a cucumber.

Ghibli Museum

Shimorenjaku • Experience the wizardry of artist and filmmaker Hayao Miyazaki and explore his world: Play in the animation studio that fostered *My Neighbor Totoro* and *Spirited Away* and watch a short film from the master that can't be seen anywhere else.

Godzilla Head

Shinjuku • The King of the Monsters has stomped his way into a comfortable retirement age, but his giant scaly head still towers over the Toho theater complex, the studio behind the Godzilla franchise.

Pasona Tokyo Headquarters

Chiyoda-ku • Hidden in the Chiyoda district is a sky-scraping urban farm with hydroponic "fields" on the roof, exterior and interior walls, and a genuine rice paddy in the lobby.

Sanrio's indoor theme park boasts over 1 million visitors a year.

Alice in Wonderland Fantasy Dining

Ginza • A rabbit warren of storybook pages, a stack of books to make you feel like a caterpillar, hedges from the Queen's garden, and desserts shaped like the Cheshire cat—this Ginza restaurant will drop you squarely in the Bizarro World of Lewis Carroll.

The Giant Ghibli Clock

Higashishinbashi • Officially called the "NI-Tele Really BIG Clock," four or five times a day this wacked-out symphonic mega-machine spins, dances, whirs, and clanks, and as a side gig tells the time.

Nakagin Capsule Tower

Ginza • While the future of this groundbreaking experiment in modular living is uncertain, you can still experience the mostly unsuccessful—but undeniably thought-provoking—expression of micro-living, designed by famed Japanese "Metabolist" architect Kisho Kurokawa.

Roppongi Hills Garden Pond

Minato-ku • In 2003, these rejuvenated office towers, museums, shops, and hotels were a welcome upgrade for the city, but it's the underwater tenants who might be the most beguiling: little slips of silvery fish called *medaka*, direct descendants of those bred in space aboard the *Columbia* shuttle as part of a series of experiments in extraterrestrial reproduction.

Shakaden Reiyukai

Minato-ku • The temple headquarters of a 20th-century Buddhist off-shoot known as "Inner Trip Reiyukai" is a futuristic black pyramid, where welcoming monks provide free Japanese lessons, and 400 tons of drinking water are held in reserve—because you never know when you might need 400 tons of drinking water.

Gotokuji Temple

Gotokuji • As the birthplace of a kind of 17th-century meme, this Buddhist temple in the Setagaya district is overrun with thousands of porcelain and plastic *maneki-neko*, or good-luck cats, each raising one snowy-white paw to symbolize that this is a place of care and safety.

Lucky Dragon 5 Memorial

Koto Ward • A moving memorial to a little-known nuclear disaster, when the crew members of the trawler *Daigo Fukuryū Maru* (the "Lucky Dragon") unwittingly cruised into the warm and snowy nuclear fallout of a bomb test over a thousand times more powerful than Hiroshima.

Sengaku-ji

Minato-ku • The graves of the 47 Ronin—nearly deified figures of duty, commitment, and honor-bound revenge— are packed in tight rows at a temple befitting their legendary status.

PIGMENT

Higashishinagawa • Like something out of a painter's dreamscape, this Kengo Kuma–designed art-supply store connects ancient principles of Japanese design and ideology to modern aesthetics. Constructed almost entirely of bamboo, the store displays thousands of unmixed tints and hues with names like "Autumn Mystery" and "Luxury Twinkle."

Odaiba Statue of Liberty

Minato-ku • Originally installed as a temporary exhibit, this replica of Lady Liberty may not tower like the original, but it's still four stories tall and so popular, it's the site of dozens of daily photo ops.

Sanrio Puroland

Ochiai • If Hello Kitty was Mickey Mouse, this bright, loud, hyperactive world of pastels and super-morphed feline creatures would be her Disneyland. The blaring candy-colored cartoon world is broken only by the live stage shows and constant fireworks.

Anata No Warehouse

Kawasaki-ku • This faux-seedy video arcade looks like a maze of alleys straight out of a cyberpunk dystopia or a back alley of Hong Kong's Kowloon Walled City.

Ajinomoto MSG Factory Tour

Kawasaki-ku • Just south of Tokyo center at one of the world's largest monosodium glutamate factories, an albino panda named Aji-Kun welcomes you to the factory tour, some handmade MSG goes home with you, and a side-by-side taste test makes a pretty strong case for the often-maligned seasoning.

Aogashima Volcano Island

Aogashima Island • Although it's more than 200 miles (358 km) off the coast, the island of Aogashima—a volcano inside another volcano—is actually part of the city of Tokyo. A quick helicopter ride (or not-as-quick ferry) will take you to five square miles (8.75 km²) of remote and peaceful night skies, volcanic saunas, and the occasional fear-of-eruption twinge.

MEGURO PARASITOLOGICAL MUSEUM

TOKYO

A worm-infested dolphin stomach.

This small but memorable museum, established as a research collection in 1953, pays tribute to the tens of thousands of organisms that thrive at the expense of others. The first floor offers an overview of the parasite-host relationship and life cycle, while the second floor showcases 300 preserved specimens, including a dolphin stomach infected with roundworms, a heartworm-infested dog heart, and a turtle head whose eye sockets are now inhabited by leeches.

For reasons that mystify even museum staff, the institute has become a go-to date spot—young couples hold hands and gaze at the 29-foot (8.8 m) tapeworm extracted from the stomach of a man who ate infected trout.

4-1-1 Shimomeguro, Meguro, Tokyo. The museum is a 15-minute walk from Meguro station. Stop by the gift shop for a preserved parasite key ring. Ⓝ 35.631695 Ⓔ 139.706649 ➤➤

➤➤ Parasitic Worms and Their Effects on Humans

GUINEA WORMS (DRACUNCULIASIS)

You won't know you're infected with a guinea worm until a year after it has entered your body. That's when you'll notice a blister on your leg. Within three days, it will rupture, exposing what looks like a piece of white string. That is the guinea worm, and its journey through your body is far from over.

By the time they've spent a year in your connective tissue, where they grow up to 3 feet (1 m) long, guinea worms have generated and stored millions of eggs in their bodies. When a worm emerges from a burst blister, the temptation will be to yank it out. Bad plan. This can result in the worm breaking, causing the remainder of its body— and the eggs it holds—to putrefy and get stuck inside you.

When a worm pokes out of your leg, you must begin winding it around a stick to draw it out. Don't tug. Just wait for it to slither farther out, winding its body around the stick like cotton on a reel. The entire process may take months.

As the worm emerges, you will feel an intense burning sensation around the blister. The temptation to dunk your leg in the nearest pond or river will be overwhelming. Cruelly, this form of relief is how the life cycle of the guinea worm perpetuates: The worm will release larvae into the water. Water fleas eat the larvae. Humans drink water containing the larvae-infested fleas, and the entire process starts anew.

EYE WORM (LOIASIS)

An eye worm, or loa loa, is often asymptomatic, but when the worm makes its presence known, the effect is startling. Loa loa enter your body via the bite of a fly (commonly deer flies or mango flies) that is infected with larvae. Burrowing into your subcutaneous tissue and lungs, and traveling through your circulatory system, loa loa grow to more than 2.5 inches (6.35 cm) long and produce larvae that end up in your spinal fluid, urine, or mucus.

Though the most frequently observed symptom of loiasis is Calabar swellings (itchy red lumps, particularly on the forearms), the first sign of an infection may be a tickle in your eyeball. Loa loa can migrate through the subconjunctival tissues of your eye. In other words, you will be able to watch in a mirror as a worm wriggles beneath the top layer of your eyeball. The sensation will be painful, itchy, and unlike anything you've ever experienced.

TAPEWORMS (TAENIASIS)

Pause before eating pork or beef and check whether it's been

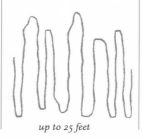
up to 25 feet

properly cooked. Raw or undercooked meat can harbor tapeworm larvae, which, once they reach your intestines, latch on and develop into adults of up to 25 feet (7.6 m) long.

Flat, ribbonlike tapeworms can live inside you for up to 18 years. Their bodies consist of 1,000 to 2,000 proglottids, or individual segments, giving them a ridged look. About 20 percent of the proglottids—the ones toward the rear end—are capable of producing eggs and behaving like individual worms. Segments that break off sometimes crawl out of the anus and down the thighs of their human host. Word is, it tickles.

Most people infected with tapeworms exhibit no symptoms, but a worm that has spent over a decade in your intestines may cause indigestion, abdominal pain, or weight loss. The first sign of an infection will likely be worm segments in the stool. They may be moving.

UENO ZOO ESCAPED ANIMAL DRILL

TOKYO

Every February, a papier-mâché rhino politely lunges at the staff of Ueno Zoo. The rhino, operated by a pair of zookeepers, is one of the fake creatures used in the zoo's annual Escaped Animal Drill.

Each year, the artificial rhino—along with staff members in furry monkey suits and one dressed as a bipedal tiger—attempts to storm the gates of the zoo and wreak havoc on the streets of Tokyo. Zookeepers band together to capture the "humanimals," encircling them with nets, loading tranquilizer guns, and tapping the ground with sticks. Some zoo staff even feign injury or play dead to heighten the authenticity of the scene. All participants carry out their duties earnestly, never cracking a smile.

The yearly drill is part of Ueno Zoo's emergency preparations for earthquakes and other natural disasters. It has become such a popular attraction in its own right that other Japanese zoos have started to copy the idea.

9-83, Ueno Park, Taito-ku, Tokyo. The Escaped Animal Drill is usually held between February 20–22, but contact the zoo directly to confirm. Ⓝ 35.714070 Ⓔ 139.774081

Once a year, Ueno Zoo's keepers dress as animals and try to break free.

THE WORLD'S LARGEST DRAIN

KASUKABE, SAITAMA

The G-Cans project, more officially known as the Metropolitan Area Outer Underground Discharge Channel, is a massive underground waterway and water storage area built to protect Tokyo from flooding during monsoons.

G-Cans opened in 2009 after 17 years of construction. With its 59 pillars, miles of tunnels, and 83-foot-high (25.3 m) ceilings, the vast space resembles an underground temple. Five 21-story concrete silos collect rainwater, preventing overflow of the city's rivers and waterways. The humongous drainage system can pump over 12,000 tons of water per minute—that's four and a half Olympic-size swimming pools.

There are daily free tours of the drainage system, but you will need to bring your own translator if you can't understand Japanese. This precaution ensures that, in the event of an emergency, you will be able to follow evacuation instructions.

Showa drainage pump, 720 Kamikanasaki, Kasukabe. Free tours of the drainage system are conducted daily in Japanese. The closest train station is Minami-Sakurai (a 40-minute walk), after a one-hour trip from central Tokyo. Ⓝ 35.997417 Ⓔ 139.811454

Tokyo's massive underground waterway is a temple to high-speed drainage.

FIREFLY SQUID OF TOYAMA BAY

TOYAMA BAY, ISHIKAWA AND TOYAMA

The firefly squid is a 3-inch-long (7.6 cm) cephalopod found in the waters surrounding Japan. Its standout feature—a series of photophores that make the squid glow a brilliant blue—is ordinarily concealed by the dark, 1,200-foot-deep (366 m) water it inhabits. But every year, from March to May, millions of firefly squid surface in Toyama to spawn and are swept ashore by the currents of the bowl-shaped bay.

This time of year is also prime fishing season. Nets trawl the predawn waters, hauling up piles of squirming, glowing creatures and turning boats into beacons. The beaches are bathed in a blue glow as the adult squid—who have a one-year life span—lay their eggs and prepare to die. The Japanese government regards the annual light show as a "special natural monument."

While the firefly squid are highly regarded for their magical visual effects, they are also prized for their tasty innards. After basking in the glow of the predawn bioluminescent bay, you can head to a sushi joint and feast on squid served raw, boiled, or turned into tempura.

If you'd like to learn more about the glowing squid before—or instead of—eating them, head to Toyama's Hotaruika Museum, which bills itself as the only firefly squid museum in the world. **Namerikawa fishing port, Toyama Bay. Sightseeing boats depart from the Namerikawa fishing port around 3 a.m. Ⓝ 36.788391 Ⓔ 137.367554**

The glowing bodies of bioluminescent squid light up the fishing port of Toyama Bay.

The monks of Shugendo mummified themselves while still alive by drinking tea made from poisonous sap.

SELF-MUMMIFYING MONKS OF SHUGENDO

MOUNT YUDONO, YAMAGATA

The monks of northern Japan who followed Shugendo, an ancient form of esoteric Buddhism, sought to achieve enlightenment through difficult, ritualistic physical and mental challenges. At least two dozen monks successfully enacted an extreme form of self-sacrifice: They brought about their own deaths by slow, excruciating self-mummification.

The entire process took about ten years. During the first of the three stages, the monks spent 1,000 days eating a strict diet of nuts and seeds, while taking part in a regimen of rigorous physical activity that stripped them of their body fat. (Due to its high water content and heat retention, fat accelerates decomposition.)

In stage two, the monks restricted their diet even further, consuming only bark, roots, and a tea made from the toxic sap of the urushi tree—a substance more conventionally used to lacquer wood. This caused vomiting, sweating, and excess urination, achieving the goal of bodily desiccation and insuring that any maggots attempting to feed on the post-mortem flesh would be poisoned.

Finally, a self-mummifying monk would lock himself in a stone tomb 10 feet (3 m) underground, where he would meditate and recite mantras while sitting in the lotus position. His only connection to the outside world was a bamboo air tube and a bell.

Each day, he rang the bell to indicate he was still alive. When the bell stopped ringing, the tube was removed and the tomb sealed.

Hundreds of monks attempted self-mummification; few were successful. A thousand days after the final ringing of the bell, when the tombs were opened, most bodies had decomposed. These monks were resealed in their tombs—respected for their endurance, but not worshipped.

The monk with the most outlandish story is Tetsumonkai, who, according to legend, killed multiple samurai and fell in love with a prostitute before joining the monastery. Newly devoted to a life of self-sacrifice, he castrated himself and then hand-delivered his carefully wrapped testicles to the lovelorn woman. In another incident he cut out his left eye in the hope that it would end the outbreak of ocular disease in Edo. Having decided he needed to leave his body to the world in order to bring salvation to mankind, Tetsumonkai entered his tomb in 1829. His mummy, still in the lotus position, is on display at Churenji temple at Mount Yudono.

From Tokyo, get the Shinkansen (bullet train) to Niigata and switch to an Inaho limited express, getting off at Tsuruoka. This will take about 4 hours. From there, get a Yudono-bound bus to Oami. Churenji and Dainichibo temples are within walking distance.
Ⓝ 38.531952 Ⓔ 139.985089 ➟

➤➤ Other Mummified Buddhist Monks

LUANG PHO DAENG

After dying during a seated meditation session, 79-year-old Buddhist monk Luang Pho Daeng was encased in a glass coffin at Wat Khunaram temple. That was in 1974. The monk has been on display ever since.

The passage of time has brought remarkably little damage to the body. Besides desiccation and a gradual brown mottling of the white skin, the only visible change is the loss of eyeballs—a development the temple monks dealt with by covering the empty eye sockets with sunglasses.

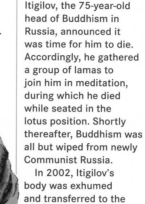

DASHI-DORZHO ITIGILOV

In 1927, Dashi-Dorzho Itigilov, the 75-year-old head of Buddhism in Russia, announced it was time for him to die. Accordingly, he gathered a group of lamas to join him in meditation, during which he died while seated in the lotus position. Shortly thereafter, Buddhism was all but wiped from newly Communist Russia.

In 2002, Itigilov's body was exhumed and transferred to the Ivolginsky Datsan, the most important Buddhist monastery in Russia. Itigilov's mummified remains are still there, sitting in the exact same lotus position as when he died in 1927. Itigilov is exhibited on major Buddhist holidays, during which pilgrims press their foreheads to the silk scarves that flow from Itigilov's hands out through a slot in the glass case.

The mummy of Luang Pho Daeng wears sunglasses so his desiccated eye sockets don't scare visitors.

THE MOUND OF EARS

MIMIZUKA, KYOTO

Tucked among houses on a narrow residential street in suburban Kyoto is a 30-foot-tall (9 m) grassy mound containing the ears and noses of tens of thousands of Koreans.

In 1592, Japanese military commander Hideyoshi Toyotomi led an invasion of Korea with the eventual goal of conquering China. Around 160,000 Japanese troops streamed into Korea with orders to murder indiscriminately.

The amount of respect and remuneration the soldiers received depended on their proving their kill tally. Traditionally, samurai took the severed heads of their victims as war trophies. But given the massive body count, soldiers began removing just the noses, or, less frequently, the ears, of the dead. Estimates vary wildly, but Japanese soldiers took body parts from perhaps as many as 150,000 Koreans. Reportedly, some of the victims were still alive when mutilated and survived the attacks.

Mimizuka literally translates to "mound of ears," despite the fact that it contains mostly noses. When the monument was dedicated in 1597, it was known as Hanazuka, or "mound of noses." The name changed decades later, when it was decided that the image Hanazuka evoked was simply too cruel. Severed ears were somehow more acceptable. **Shomen-dori, Higashiyama, Kyoto. Mimizuka is a short walk north of Shichijō station on the Keihan train line. Ⓝ 34.991389 Ⓔ 135.770278**

Peaceful though it appears, the Mound of Ears is a shocking testament to wartime brutality.

MARATHON MONKS OF MOUNT HIEI

HONSHU, SHIGA

The circumference of Earth is 24,901 miles (40,075 km). That's 2,000 miles (3,219 km) less than the distance the marathon monks of Mount Hiei must cover, on foot, over a period of seven years.

The marathon monks, or *gyoja*, belong to the Tendai sect of Buddhism, founded at the Enryaku-ji temple on Mount Hiei in the early 9th century. Seeking the status of living sainthood, they elect to undergo the rigors of *kaihogyo*, a seven-year program involving daily long-distance treks on little sleep and meager food rations.

Gyoja are allowed just one week of training before embarking on the first kaihogyo challenge: 100 consecutive days of 25-mile (40 km) walks. During this preparation week, other members of the sect clear the mountainous course of sharp rocks and sticks and weave 80 pairs of straw sandals for each monk to wear. Each pair lasts, at most, a few days.

The monks of Mount Hiei walk daily marathons, in straw sandals, for months on end.

For the next three and a half months, gyoja follow an unwavering daily routine. Each monk wakes at midnight and eats a small meal of rice or noodles before spending an hour in prayer. Then it is time to walk. Wearing a white robe, sandals, and a large rectangular straw hat, the monk follows a strict course, stopping at almost 300 stations along the way to pray and chant. He is forbidden from eating, drinking, or resting. Around his waist is a knife attached to a rope. In the event that the gyoja is unable to complete his kaihogyo, he is honor-bound to commit suicide using either implement.

When the monk completes his daily walk, he bathes, eats another small meal, and participates in chores and Tendai services before going to bed at 9 p.m. At midnight, he wakes and the entire process begins again.

The first three years of the kaihogyo follow the same pattern: one 100-day set of 25-mile (40 km) walks per year. In each of the fourth and fifth years, there are two 100-day sets. The greatest challenge comes on the 700th day, when, in an ordeal known as the *doiri*, the gyoja is denied food, water, and sleep for seven to nine days. The near-death experience, during which the gyoja sits upright in constant prayer, is intended to facilitate the death of the ego and the birth of a transcendent, interconnected being capable of leading others toward enlightenment. Originally, the doiri lasted 10 days, but the zero percent survival rate forced it to be shortened.

In the final two years, the distance of the walks increases—first to 37.5 miles (60 km), then to 52 miles (84 km). Approximately 50 monks have completed the seven-year kaihogyo since 1885. The most recent is Yūsai Sakai, one of three gyoja to have endured the process twice—from 1973 to 1980, and again, after a six-month break, from 1980 to 1987.

Enryaku-ji, the birthplace of Tendai Buddhism, is open to visitors. From Kyoto, take the Keihan Main Line to Demachiyanagi and transfer to an Eizan train to Yase-Hieizanguchi. From here the Eizan Cable Car will take you up Mount Hiei. Ⓝ 35.070556 Ⓔ 135.841111

ALSO IN JAPAN

The Tomb of Jesus

Shingo · According to a faithful few, Jesus died not on a cross on Calvary, but in northern Japan at the age of 106. A grave in the village of Shingo supposedly holds his remains.

Jigokudani Park

Yamanouchi · In the hot springs of snowy Hell's Valley is a spa for macaque monkeys.

WISTERIA TUNNEL

KITAKYUSHU, FUKUOKA

For most of the year, the wisteria tunnel at Kawachi Fuji Gardens is a latticed canopy overlaid with barren, twisting vines. But for a few weeks every spring, the tunnel is in magnificent bloom, its dangling flowers and sweet scent enveloping all those who walk its path.

The private garden is home to around 150 wisteria plants in shades of purple, pink, and white. Visit between late April and mid-May to see the wisteria in bloom—the exact dates vary each year.

2-2-48 Kawachi, Yahata-Higashi-ku, Kitakyushu. A JR train to Yahata Station followed by a bus to Kawachi Shogakko-Mae and a 15-minute walk will get you to the garden. Ⓝ 33.831580 Ⓔ 130.792692

A fragrant pastel passage that blooms for a few glorious weeks each year.

THE VILLAGE OF DOLLS

NAGORO, TOKUSHIMA

The school in the small village of Nagoro has a classroom full of silent students. Staring at an equally silent teacher, they sit motionless day after day, never calling out an answer or rustling their books.

The students and teacher are life-size dolls, created by Nagoro resident Ayano Tsukimi after the school closed due to low enrolment. Tsukimi, who was born in Nagoro, spent decades living in Osaka, Japan's third-largest city. When she returned to her childhood home, the population had dwindled from about 300 to 35 residents.

The dolls began as scarecrows, made to defend Tsukimi's veggie patch. But as Tsukimi created more and more, some in the likeness of deceased friends and relatives, she started placing them around the village in remembrance of the Nagoro residents she once knew. A doll in rain boots and wet-weather gear sits by a creek holding a fishing rod. A couple of elderly cloth people relax side by side on an outdoor bench, watching the world go by. Scenes like this are found all over the village.

Tsukimi estimates she has created 350 dolls, meaning they outnumber Nagoro's human residents 10 to 1.

To visit the village, board a JR train to Awa-Ikeda Station, followed by a bus to Kubo, then a bus to Nagoro. Ⓝ 34.043671 Ⓔ 133.802503

As residents die in the dwindling village of Nagoro, one local artisan is replacing them with cloth dolls.

BATTLESHIP ISLAND

HASHIMA, NAGASAKI

Hashima, an island off the coast of Nagasaki, is known by two nicknames: Gunkanjima ("Battleship Island") and Midori Nashi Shima ("Island Without Green"). The austere brutality conjured by these names is reflected in its appearance—Gunkanjima is a narrow lump of rock covered in the crumbling remains of a crowded concrete village.

The Mitsubishi company purchased the island in 1890 to establish a coal-mining facility for undersea reserves. In 1916, Japan's first concrete highrises sprung up on Gunkanjima—nine-story slabs of gray with cramped rooms and rows of identical balconies overlooking a claustrophobic courtyard. By 1959, over 5,000 coal miners and their families occupied these drab apartments, making the less-than-a-mile-long (1.2 km) island—with a population density of 216,264 people per square mile—the most overcrowded place on Earth.

Residents relied on the mainland for deliveries of food and, until 1957, water, but Gunkanjima was otherwise self-sufficient. Schools, playgrounds, cinemas, shops, a hospital, and even brothels operated in the tiny community. Steep concrete staircases that connected adjoining buildings were the only means of travel to ninth-floor apartments.

In January 1974, Mitsubishi officially closed its mining facility. All residents abandoned their homes for the mainland within two months, and Gunkanjima has been uninhabited ever since. Decades of typhoons, wind, rain, and seawater have caused massive degradation to the monolithic buildings. Wooden planks regularly fall from the disintegrating balcony railings, landing on the piles of crumbled concrete below. Contorted steel beams and rusted iron frames protrude from the walls. Hints of domesticity remain: a teacup; a tricycle; a television manufactured in the 1960s. The only sounds at what was once the world's most crowded place are the whipping wind and crashing waves.

Gunkanjima reopened to visitors in 2009, but official tours provide very limited access due to safety concerns. To actually explore the buildings, you would need to hop aboard an early-morning fishing boat for an unauthorized trip to the island. Official tour boats depart from Nagasaki Port and Tokiwa Terminal.
Ⓝ 32.627833 Ⓔ 129.738588

The lump of rock known as Battleship Island, now a ghost town in the middle of the sea, was once the most over-crowded place on Earth.

NORTH KOREA
KIJONG-DONG

DEMILITARIZED ZONE (DMZ)

Within the 2.5-mile-wide (4 km), 160-mile-long (250 km) heavily guarded demilitarized zone (DMZ) that separates North Korea and South Korea are two villages—one on each side of the border. Built in the 1950s, following the Korean War ceasefire, the northern village of Kijong-dong is a collection of well-appointed, multistory buildings that are home to 200 families. At least, that is the official story according to the North Korean government. In reality, Kijong-dong is an uninhabited propaganda village, built to convince South Koreans peering across the border of the North's economic success.

Viewed from a distance, Kijong-dong—or "Peace Village," as it's called by North Korea—is unremarkable, if a little drab. Look closer, however, and the trickery is revealed. Residential buildings have no glass in their windows. Electric lights—a rare luxury for rural Koreans—operate on an automatic timer. The only people in sight are maintenance workers, occasionally dispatched to sweep the streets to give the impression of ongoing activity.

One mile (1.6 km) from Kijong-dong, on the south side of the

The apparently populous, well-appointed village of Kijong-dong is not all that it seems.

border, is the village of Daeseong-dong. Its few hundred residents live in limbo—as residents of the DMZ, they are exempt from taxes and compulsory military service, but these perks come at the price of their freedom. An 11 p.m. curfew is strictly enforced, and relocating is prohibited.

The opposing villages offer a study in the often ridiculous one-upmanship between North and South Korea. In the 1980s, the South Korean government erected a 323-foot-tall (98 m) pole in Daeseong-dong and flew their country's flag. North Korea responded by building a 525-foot (160 m) flagpole—then the tallest

in the world—and raised an even larger flag over Kijong-dong.

Sixty endangered species roam around the DMZ, including the Amur leopard, Asiatic black bear, and red-crowned crane. They share their habitat with a million land mines. The Korea Tourism Organization offers nature tours of the DMZ, marketing it as "The Peace and Life Zone."

Entry to either village is not permitted, but both are visible from the UN-controlled Joint Security Area (JSA). You will be asked by the United Nations Command to sign a release form that includes a death waiver.
Ⓝ 37.941761 Ⓔ 126.653430 ➼

➼ Unplanned Nature Reserves

When humans leave, nature thrives. These "involuntary parks," to use a phrase coined by futurist Bruce Sterling, tend to be highly diverse ecosystems populated by species that thrive in the absence of agriculture and development.

EUROPEAN GREEN BELT
The Iron Curtain kept Europe segregated—physically and politically—for over 40 years. During this time, people fled the border regions, and the area between the Soviet-led East and NATO-aligned West became a corridor for wildlife. With the fall

of the Iron Curtain came increased development. A grassroots conservation initiative called the European Green Belt aims to preserve and protect the ecological network that runs from the top of Finland to Greece and Turkey.

CYPRUS GREEN LINE
The cease-fire border dividing the Greek and Turkish parts of Cyprus earned its "Green Line" nickname after a peace force commander used a green marker to draw a wobbly line across a map of the country in 1964. Now the zone is

verdant with flora and fauna, including the lapwing wading bird, the once-presumed-extinct Cyprus spiny mouse, and the mouflon—a wild sheep with large curled horns.

INTERNATIONAL FRIENDSHIP EXHIBITION

MOUNT MYOHYANG, NORTH PYONGAN

Like all national leaders, former North Korean rulers Kim Jong-il and Kim Il-sung received gifts from international politicians during their reigns. Uniquely, these presents are all on display in a 120-room museum that aims to showcase the world's undying love for the deceased despots.

Many of the gifts, which number around 100,000, are modest tokens of diplomatic etiquette—vases, ashtrays, books, and pens. At the other extreme is the big-ticket bounty received from Communists, terrorists, and despotic leaders keen to curry favor with North Korea. Animal trophies make a popular dictator-to-dictator gift: Fidel Castro handed over a crocodile-skin briefcase, while Nicolae Ceaușescu, the Romanian dictator who was

In order to see Kim Jong-il's bizarre collection of gifts from world leaders, visitors are required to wear white gloves.

overthrown and killed by his own people, proffered a bear head mounted on a red satin pillow.

Former Soviet ruler Joseph Stalin and Chinese chairman Mao Zedong each took a bigger-is-better approach, supplying a bulletproof limousine and armored train car, respectively. (Beyond their extravagance, these gifts show a certain

thoughtfulness and sensitivity, given both Kim Jong-il and Kim Il-sung's fear of flying.)

At the end of Madeleine Albright's diplomatic trip to the isolated country in 2000, the US secretary of state presented Kim Jong-il with a basketball signed by Michael Jordan. The ball joins a Sony Walkman, a Casio keyboard, an Apple computer, and a soccer ball signed by Brazilian soccer great Pelé.

Among all the creative gifts, the most perplexing comes from Nicaragua's Sandinista revolutionaries, who donated a stuffed, upright alligator holding a wooden tray of cocktail glasses. **Mount Myohyang is a 2-hour drive from Pyongyang, accessible only by guided tour. You will be provided with fabric covers to wear over your shoes so you do not sully the floors with filth. Photos are prohibited. Ⓝ 40.008831 Ⓔ 126.226469**

SOUTH KOREA

THIRD TUNNEL OF AGGRESSION

PANMUNJOM, GYEONGGI

The demilitarized zone (DMZ) separating North and South Korea is a 160-mile-long (257 km), 2.5-mile-wide (4 km) strip of land where attempts—or perceived attempts—to cross the border result in being shot to death. Two million soldiers patrol the tension-filled buffer, making it the most heavily guarded border on Earth. A stealth ground invasion would be impossible—which is why, following the 1953 ceasefire that ended the Korean War, North Korea began secretly digging tunnels.

South Korea discovered the Third Tunnel of Aggression, named for it being the third one found and its apparent intended function as a conduit for military invasion, in 1978. It is one of at least a dozen rumored tunnels beneath the DMZ, four of which have been found so far. Designed to provide the route for an attack on Seoul, the Third Tunnel measures just over a mile (1.6 km) long, ending 27 miles (43.5 km) from the South Korean city.

At first, North Korea denied digging the 6.5 × 6 foot (2 × 1.8 m) tunnel. Then the story changed: The tunnel was a North Korean coal mine, a statement officials backed up by hurriedly painting its walls black. Refusing to believe the creative explanation, South Korea took control of the passage and blocked the border line with concrete barriers. The tunnel is now open to visitors, who can get within a few feet of the barbed wire and machine guns waiting on the other side of the concrete.

Self-appointed "tunnel hunters" in South Korea continue to search for secret passageways in the DMZ, believing the country won't be safe from a North Korean invasion until all are found. Motivated by deep distrust of their neighbors to the north, some have spent decades poring over maps, searching the ground for clues, and financing fruitless drilling operations. **Tours to the tunnel, located near Panmunjom, in the DMZ, leave from Lotte Hotel in Seoul. Ⓝ 37.956000 Ⓔ 126.677000**

JEJU GLASS CASTLE THEME PARK

JEJU

"Museum" may be a more apt term for the Jeju Glass Castle theme park, since everything is made from shaped glass and is not really conducive to the roughhousing usually associated with a raucous funfair.

The delicate collection is divided into indoor and outdoor sections, each with its own variety of glasswork. The exterior garden features such amazingly naturalistic installations as a glass waterfall, glass flower beds, and a lake made of mirrors, with fish constructed out of used soju bottles. The indoor sights include a towering green glass beanstalk in the center of the exhibition hall, a room full of mirrors where visitors can get lost in their infinite reflections, and even a glass bookshelf with glass books.

Most of the over 200 pieces at the castle are not to be touched, but there are some interactive bits, such as a set of glass drums you can bang on if you dare.

3135-1 Jeoji-ri, Hangyeong-myeon 462, Nokchabunjae-ro, Hangyeong-myeon, Jeju-si. For an extra fee, you can make your own glass creations. Ⓝ 33.314810 Ⓔ 126.273490

ALSO IN SOUTH KOREA

Trick Eye Museum

Seoul · Become part of an optical illusion at this collection of trompe l'oeil paintings that seem to interact with visitors.

The park brims with delicate and unexpected glasswork.

JEJU MERMAIDS

SEOGWIPO, JEJU ISLAND

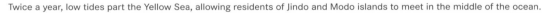

In the historically patriarchal society in South Korea, some of the small fishing islands off the south coast have flipped the script thanks to the *haenyeo*. Otherwise known as the Jeju Mermaids, these pioneering (and largely elderly) fisherwomen have become the heads of their family units.

The practice began around the 18th century as a way of getting around the high taxes male fishermen had to pay on their meager hauls of shellfish, octopus, and abalone. At the time, women were not taxed at all, so women began exploiting this loophole by taking over fishing duties from their husbands.

As this role reversal persisted, islands such as Jeju saw family units flip power structures almost completely as the women freedived for sea life in the icy East China Sea. The tradition has survived for hundreds of years, but now the haenyeo are in danger of disappearing as more and more of their daughters are choosing life in bigger mainland cities. Today the majority of haenyeo are women over the age of 50 who still go out and dive as deep as 30 feet (10 meters) to collect their family's main source of income.

Jeju is a beautiful volcanic island the haenyeo call home and is right next to the volcanic crater of Seongsan Ilchulbong. There are also several statues and art pieces in the area devoted to the haenyeo. Ⓝ 33.403554 Ⓔ 126.889611

JINDO-MODO LAND BRIDGE

JINDO, SOUTH JEOLLA

Moses may have parted the Red Sea to rescue the Israelites, but, according to South Korean legend, he is not the only one to achieve such a feat.

Twice a year, the waters separating Jindo and Modo islands recede, creating a causeway almost 2 miles (3 km) long and 120 feet (36.6 m) wide. The traditional explanation for this phenomenon is that a pack of tigers once attacked Jindo, sending its residents fleeing. Everyone escaped except an elderly woman, who prayed to the sea god to split the waters so she could pass safely to Modo. Her wish was granted.

Today, visitors to Jindo and Modo can relive the elderly woman's crossing of the Yellow Sea—once at the beginning of May and once in mid-June. On each of these days, visitors and tourists from each island pull on waterproof boots and walk to the middle of the causeway to meet one another and celebrate. Festivities are fleeting—the land bridge only lasts for about an hour.

Hoedong-ri, Jindo. Jindo is 6 hours from Seoul by bus. From the Jindo bus terminal, get a local bus toward Hoedong-ri. Ⓝ 34.407158 Ⓔ 126.361349

Twice a year, low tides part the Yellow Sea, allowing residents of Jindo and Modo islands to meet in the middle of the ocean.

Southeast Asia

CAMBODIA
DINOSAUR OF TA PROHM

ANGKOR, SIEM REAP

A carving on the wall of the Ta Prohm temple, built in the late 1100s, bears more than a passing resemblance to a stegosaurus. Since a 1997 guidebook first pointed out the strange carving, creationists have held up the Ta Prohm dinosaur as demonstrable "proof" that humans and dinosaurs once coexisted in Cambodia. There is even a replica of the carving on display at the Creation Evidence Museum in Glen Rose, Texas.

While the carved animal does seem to have a row of plates along its spine, it hardly makes a compelling argument for revising the prehistoric timeline. The bas-relief could just as easily be a depiction of a rhino or chameleon, with the "plates" forming a stylized version of foliage.

What's a stegosaurus doing on the wall of a 12th-century temple?

Angkor Thom, Siem Reap. Siem Reap is 6 hours from Phnom Penh by bus, leaving from Sisowath Quay.
Ⓝ 13.435000 Ⓔ 103.889167

THE LAST BAMBOO TRAIN

BATTAMBANG, BATTAMBANG

"Train" is a misleading word for this vehicle. "Queen-size wooden bed frame speeding along a rickety mine track" is a more accurate description. Known to locals as "norries," the bamboo trains of Battambang consist of two axles with welded-on wheels, topped by a 6 × 10-foot (1.8 × 3 m) bamboo platform. A noisy, sputtering engine, stripped from a motorbike or a piece of farming equipment, powers a drive belt that spins the rear axle.

To ride the norry, you simply sit on the platform—observing that your "engineer" is more than likely a child in flip-flops and taking note of the lack of safety rails or means by which to secure yourself—and hope that your balance doesn't betray you as you hurtle along the twisting rails at 30 miles (48 km) per hour.

Bamboo trains, an unregulated, improvised form of public transport, emerged after decades of Khmer Rouge rule decimated

Improvised transport for a region deprived of rail lines.

Cambodia's rail network. Using the poorly maintained tracks the French built in the 1930s, locals began running self-built norries made from scavenged spare parts. A network of norry routes sprang up, allowing people to travel and transport produce or animals.

Now that the restoration of national railway lines is underway, almost all of the norry routes have ceased operation. The sole remaining section runs from the outskirts of Battambang to a small village with a brick factory. Multiple bamboo trains thunder along the tracks in both directions. When two norries meet head-on, the passengers on the vehicle with the smaller load hop off, dismantle the train, and reassemble it once the other train has passed. Reassembly takes about a minute—one of the norry's many charms is that its parts are simply stacked on top of one another, with no nuts or bolts holding them in place. It's a detail you'll remember when you feel the wheels leave the tracks.

Rides on the bamboo train cost about $8 and can be booked through hotels in Battambang.
Ⓝ 13.068816 Ⓔ 103.202205

ALSO IN CAMBODIA

Pangolin Rehabilitation Center

Phnom Penh · This sanctuary, established in 2012, is dedicated to improving the plight of the rare scaly anteater.

Effigies of the dead linger among the living in Tana Toraja.

INDONESIA
Funeral Rites of Tana Toraja

TANA TORAJA, SOUTH SULAWESI

To the 650,000 Torajans of South Sulawesi, death is not the abrupt cessation of life, but a many-stage process that begins with a stopped heart and ends—sometimes years later—with burial.

The funeral rites of Tana Toraja consist of elaborate, multiday public ceremonies involving animal sacrifices, feasting, gift-giving, music, and a procession of village members toward the burial site. Wooden effigies, or *tau tau*, of the deceased are created and placed by the graves.

The cost of organizing and staging these rituals is so high that families may spend weeks, months, or even years raising the necessary funds. During this time, the wrapped corpse, chemically preserved with formalin, is kept in the ancestral home and referred to as "sick" or "asleep" rather than "deceased."

When there is enough money to purchase a herd of sacrificial buffalo, the festivities begin. On the first day, guests form a procession and offer gifts of food, drink, and sacrificial cattle or pigs to the family. The presents are given on an implicit quid pro quo basis—each gift is registered and announced to the crowd, allowing the villagers to keep track of debts paid or incurred.

Later in the week, an animal handler ushers each buffalo to the center of a ring and fastens a rope from its nose ring to a bamboo pole driven into the ground. As adults, children, and family pets look on, the handler raises a machete and drives it into the animal's neck, creating a gaping wound that gushes blood. The buffalo thrashes and writhes, blood spraying more forcefully with each movement. Gradually, the buffalo loses strength and collapses, dying in a pool of blood and mud. This process is repeated with dozens of buffalo and pigs.

According to Toraja belief, animal sacrifices prevent the spirit of the dead from lingering and bringing bad luck—souls travel to the afterlife on the backs of the buffalo. Back in the earthly realm, raw meat from the animal carcasses is distributed to guests, with high-ranking and wealthy people receiving the best cuts.

A week after the sacrificial ceremonies, the entire village parades the coffin to a burial site. The final resting place for a Torajan is usually a carved hole on the side of a cliff. Popular burial cliffs have rows of caskets guarded by tau tau.

The bus from Makassar, capital of South Sulawesi, takes 8 to 10 hours to reach Tana Toraja. Peak funeral season is after the harvest, from July to October. Ceremonies are public, and it is customary for visitors to bring a small gift for the family, such as coffee or cigarettes. Ⓢ 3.075300 Ⓔ 119.742604

GEREJA AYAM

MAGELANG

Should you be trekking through the thick forests of Magelang, Indonesia, try not to be too alarmed if you stumble upon a giant building shaped like a chicken. Known as Gereja Ayam, or the Chicken Church, this massive chapel is an unexpected sight both whimsical and fowl.

While the locals have dubbed it the Chicken Church (and it's easy to see why), the name is a bit of a misnomer since the visionary behind the avian-esque chapel actually intended for it to look like a dove. The architect, Daniel Alamsjah, received a holy vision that inspired him to create the dove-shaped church. He picked a forested hill near Magelang to build his pious tribute and created possibly the most bird-like building in the world, complete with giant, squawking head and ornate decorative tail feathers.

The church opened its doors (or spread its wings, so to speak) in the 1990s. It welcomed worshippers of all religions, and also offered charitable services to the local community. Unfortunately, the project was suspended in 2000 when funding ran out.

Gereja Ayam was vacated and left to the forest, where it continued to rot over time, becoming a bit more ghoulish with each year.

These days, the church has been cleaned up and turned into a proper tourist attraction. Local artists have covered the inner walls with vibrant murals, and there's a small café inside the chicken's rear end that sells traditional, tasty treats. You can even climb up to the top of the bird's head for amazing 360-degree views.

The church is open every day and is often part of the nearby Borobudur Temple tour. There is a small entry fee. ⑤ 7.605706 ⑥ 110.180483

The Chicken Church attracts worshippers of all faiths, locals and tourists alike.

A single notched pole allows access to these elevated houses, strong enough to support up to a dozen people and several animals.

KOROWAI TREE HOUSES

PAPUA

In the thick, remote rain forest of southeastern Papua, the Korowai people lived in total isolation up until the 1970s. Many members of the indigenous tribe still maintain a traditional lifestyle, which centers around a fascinating cultural trait: They are the architects of fantastic treetop homes built as high as 115 feet above the ground.

These unique dwellings protect families from the swarming mosquitoes below, as well as from troublesome neighbors and evil spirits. The tree houses are constructed in clearings with large banyan or wanbom trees selected as the main pole. Most huts are typically between 25 and 40 feet high, though some are built more than 100 feet (30.5 m) above the ground, reached by a single notched pole that serves as a ladder.

The floor is constructed first, then the walls and a sago palm tree roof are added, bound together with raffia. The flooring must be quite strong, as the tree houses often accommodate as many as a dozen people. Whole family groups, along with pets and other domestic animals, may live together in one treetop abode. The larger homes have separate living spaces for the men and women of the family, as well as firepits, and sometimes stairs.

The Korowai inhabit the region that stretches between the villages of Mabul and Yanirumah in the southeastern part of the province of West Papua, near the border of Papua New Guinea. An increasing number of Korowai people are moving away from the traditional nomadic lifestyle into the newly settled villages. To visit the tree houses, you will have to venture a few hours away from the settlements into the jungle. ⓢ 6.593292 ⓔ 140.163599

ALSO IN INDONESIA

Tanah Lot

Bali · The boat-shaped rock island of Tanah Lot holds a temple protected by a holy snake.

LAOS

XIENG KHUAN BUDDHA PARK

VIENTIANE

Though its weathered stone statues of gods, humans, animals, and demons look hundreds of years old, this sculpture park is the 1958 creation of Bunleua Sulilat, an eccentric priest-shaman whose mystic religious philosophies integrated Hinduism and Buddhism.

Sulilat claimed to have acquired his beliefs after falling into a cave and meeting a Hindu-practicing Vietnamese hermit named Keoku. There are 200 concrete sculptures at Xieng Khuan, all built by Sulilat and a few of his followers. They include a 400-foot-long (122 m) reclining Buddha, a Shiva whose eight arms are full of weaponry, and a three-story "Hell, Earth, and Heaven" pumpkin with a demon head inexplicably grafted onto one side. Visitors can enter through the demon's mouth and climb to the pumpkin's upper floors.

Following the Communist revolution in 1975, Sulilat fled across the Mekong River to the Thai city of Nong Khai, where he soon built another Buddha park called Sala Keoku. His mummified body is stored on the third floor of its pavilion.

Thanon Tha Deua, Vientiane. The Buddha Park is a half hour east of Vientiane. A bus from the city's morning market will take you to Friendship Bridge. From there, transfer to a minibus that will bump along uneven gravel roads to the park. Ⓝ 17.912289 ⓔ 102.765397

A "spirit city" filled with hundreds of Hindu and Buddhist statues.

400 FEET

350

300

250

200

150

100

50

0 FEET

SPRING TEMPLE BUDDHA
Mount Yao, Henan, China

BUILT: 2008
HEIGHT: 420 feet (128 m)
TOTAL HEIGHT*: 502 feet
(153 m)

The copper Spring Temple
Buddha stands on a 66-
foot (20 m) lotus throne,
which itself is on an
82-foot (25 m) pedestal.
Visitors are welcome to
hug the statue's toes,
which are all taller than
your standard adult
human.

*with pedestal

LAYKYUN SEKKYA
Khatakan Taaung, Myanmar

BUILT: 2008
HEIGHT: 381 feet (116 m)
TOTAL HEIGHT: 427 feet (130 m)

Laykyun Sekkya's golden-
robed Buddha took 12 years
to build. It stands on a 44-foot
(13.5 m) throne, behind
a reclining Buddha that's
equally huge and equally
golden. Both statues gaze
toward the gilded stupa of
the Aung Sakkya Pagoda.

USHIKU DAIBUTSU
Ushiku, Japan

BUILT: 1993
HEIGHT: 393 feet (120 m)
TOTAL HEIGHT: 394 feet
(120 m)

Standing on a lotus
throne atop a pedestal,
the Ushiku Daibutsu is a
bronze standing Buddha
with a four-level museum
inside. New Age music, low
lighting, and incense induce
a state of calm on your way
up to the observation deck,
located in the Buddha's
chest, and with observation
windows built into its chest.

GIANT BUDDHA STATUES OF ASIA

Buddhism is big in South and Southeast Asia—you can tell from the size of the statues. Colossal sculpted Buddhas smile serenely across the region, their imposing forms extending hundreds of feet high. Some are seated, others stand, but all serve as beacons from afar, guiding visitors to the temples and sacred sites they invariably guard.

The five statues drawn to scale below, placed next to a suddenly puny-looking Statue of Liberty, are the biggest Buddhas of them all.

THE GREAT BUDDHA OF THAILAND
Ang Thong, Thailand

BUILT: 2008
TOTAL HEIGHT: 302 feet (92 m)

Thailand's golden seated Great Buddha, the tallest in the country and 18 years in the making, was built on top of a single-story museum. It's made of cement, with a layer of gold painted on top. Stop by the temple's Buddhist hell garden to see sculpted sinners being sawn in half or forced through a meat grinder.

LESHAN GIANT BUDDHA
Sichuan, China

BUILT: 803 CE
HEIGHT: 223 feet (68 m)
TOTAL HEIGHT: 233 feet (71 m)

The Leshan Giant Buddha is carved out of a stone cliff on a tributary of the Yangtze River. Created in the Tang dynasty (7th–10th century), it is the tallest pre-modern statue on Earth.

STATUE OF LIBERTY
New York, USA

BUILT: 1886
HEIGHT: 151 feet (46 m)
TOTAL HEIGHT: 325 feet (93 m)

The thousands of mysterious stone jars left on Xieng Khouang plain may have once stored human remains.

PLAIN OF JARS

PHONSAVAN, XIENG KHOUANG

Ranging from 3 to 10 feet (1 to 3 m) tall, the stone jars scattered across the 500-square-mile (1,295 km²) Xieng Khouang plain in the Laos highlands date back to the Iron Age: 500 BCE to 500 CE. Though their exact purpose is unknown, archaeological surveys of the area during the 1930s uncovered charred human remains, suggesting the jars once functioned as funeral urns. A cave with two man-made holes in its ceiling likely served as a central crematorium.

Many of the vessels bear the marks of more recent history: During the Vietnam War, conflict spilled into Laos, and the Plain of Jars became a valued strategic location. From 1964 to 1973, the U.S. engaged in a "Secret War" in Laos, dropping millions of bombs over the area. Evidence of that war persists in the form of smashed jars, craters, and signs warning of unexploded ordnances (UXOs)—30 percent of bombs dropped did not detonate, and many remain hidden in the landscape despite the consistent work of UXO removal crews.

Phonsavan, at the center of the plain, is an 11-hour bus ride or half-hour flight from the capital of Vientiane. When exploring the sites, pay careful attention to markers left by the Mines Advisory Group, which indicate the areas that have been swept for UXOs. Ⓝ 19.430011 Ⓔ 103.185559

MALAYSIA
SYNCHRONIZED FIREFLIES OF KAMPUNG KUANTAN

KUALA SELANGOR, SELANGOR

As mating rituals go, this one is pretty magical: At night, in the mangrove trees on the banks of the Selangor River, thousands of male fireflies (*Pteroptyx tener*) gather and flash in unison to attract the females of the species. (Scientists haven't determined the exact biological reason for the year-round synchronized flashing, but it's definitely part of courtship.) Viewed from a long-boat, the fireflies look like tiny, twinkling string lights.

The Kampung fireflies were once much more numerous, but river pollution and development in their habitat have resulted in declining numbers over the last 10 years. That said, the silence of the boat ride, the pitch darkness, and the surviving clusters of bio-luminescent bugs still create an atmosphere of awe.

The Kampung Kuantan Firefly Park is a 45-minute drive from Kuala Lumpur. Boats run from around 8 to 11 p.m. Ⓝ 3.360616 Ⓔ 101.301090

ALSO IN MALAYSIA

Cat Museum

Kuching · Walk through the giant cat mouth to enter a feline-focused world of kitschy figurines, dusty taxidermy, and a mummified furball from ancient Egypt.

BATU CAVES

KUALA LUMPUR

The walk to Batu Caves begins at the feet of Murugan, the Hindu god of war and victory. A 140-foot (43 m) gold-painted statue of the deity greets visitors as they prepare to climb the 272 steps that lead to the trio of large caves. The trip up the stairs would be less challenging if not for the hordes of long-tailed macaque monkeys scampering back and forth.

Long familiar to locals, the caves became more widely known after American naturalist William Hornaday "discovered" them in 1878. When Tamil businessman K. Thamboosamy Pillay visited the caves in 1890, he noticed that the entrance was shaped like the Vel—the spear held by Murugan. This resemblance inspired Pillay to install Hindu shrines and statues in the main Temple Cave, turning the location into a sacred site.

Every year since 1892, Hindus have celebrated the Thaipusam festival at the caves. Occurring in late January or early February, the festival commemorates the occasion when Parvati, goddess of power, gave Murugan his Vel. Participants go on a pilgrimage to the caves, encumbered by a *kavadi*, or "burden." Types of kavadi range from simple tasks (such as carrying a brass jug of milk on their head) to more extreme forms, like piercing their cheeks, chest, back, or tongue with skewers. Dressed in yellow, red, and orange, the kavadi bearers embark on an eight-hour, nine-mile (14 km) procession to the caves, culminating in a climb up the stairs to make offerings to Murugan.

The KTM Komuter train from Kuala Lumpur to Batu Caves takes 25 minutes. Thaipusam occurs on the full moon during the Tamil calendar month of Tai (mid-January to mid-February). Ⓝ 3.237400 Ⓔ 101.683906

During the Hindu festival of Thaipusam, pilgrims whose bodies have been pierced with skewers flock to the Batu caves. The more pain, the more spiritual reward.

PITCHER PLANTS OF KINABALU

SABAH, MALAYSIAN BORNEO

Deep in the forests of Kinabalu Park lurks *Nepenthes rajah*, the largest of the pitcher plants. These carnivorous wonders feature liquid-filled cavities that lure and trap insects.

Ants are the preferred cuisine of the Nepenthes rajah, but the plant has been known to trap much larger organisms in its urn-shaped cavity. Rats, frogs, lizards, and birds have all been found in the rajah's sticky grasp.

Rat-eating plants make for fascinating botanical fodder, but Nepenthes rajah's considerable dimensions—the largest recorded urn can hold almost a gallon of water—probably aren't meant for trapping rodents. In 2011, a team of scientists led by Monash University carnivorous plant expert Dr. Charles Clarke published their findings after studying interactions between tree shrews, rats, and the Nepenthes rajah. The report's main discovery was that these small mammals habitually landed on the rim of the plant and fed on nectar from the lid while defecating into the urn. Nepenthes rajah was their food source and toilet rolled into one. And that suited Nepenthes rajah just fine—the rat and tree shrew feces provided the plant with valuable nitrogen.

As a result of this study, it is now thought that Nepenthes rajah may have evolved to accommodate the feeding and excreting behavior of the tree shrews. The distance from the front rim of the pitcher to the nectar on the lid corresponds to the length of the tree shrew's body, allowing it to dine and defecate simultaneously in comfort. It's a heartwarming example of a plant and animal living in perfect harmony.

The *Nepenthes rajah* plant can swallow small birds, lizards, and frogs.

Kinabalu Park, Sabah. A bus from Padang Merdeka Bus Terminal in the Sabah capital of Kota Kinabalu reaches the park entrance in about 2 hours.
Ⓝ 6.005837 Ⓔ 116.543310

MYANMAR

PAGODA OF THE WORLD'S LARGEST BOOK

MANDALAY

The world's largest book, finished in 1868, is not an oversized bundle of paper pages, but a collection of 729 marble tablets. Each 5-foot-tall (1.5 m) tablet is inscribed with 160 to 200 lines from the Tipitaka—the sacred text of Theravāda Buddhism. The tablets are housed in 728 domed white shrines, arranged in rows around a central 188-foot-high (55 m) golden pagoda. The entire construction is known as Kuthodaw Pagoda.

King Mindon Min, who founded Mandalay in 1857, began the project in 1860. He intended to create a book that would last for five millennia after the Buddha. If Kuthodaw Pagoda remains intact for the next 2,500 years, his wish will be fulfilled.
62nd Street, Mandalay. You can find the pagoda at the bottom of the South East stairway of Mandalay Hill.
Ⓝ 22.004181 Ⓔ 96.113050

KYAIKTIYO BALANCING PAGODA

MOUNT KYAIKTIYO, MON STATE

At 24 feet (7 m) tall, it may not be the biggest pagoda around—but it is certainly one of the most eye-catching. Kyaiktiyo Pagoda sits atop a huge golden boulder balanced precariously on the edge of a cliff. The boulder, painted gold by Buddhist devotees, sits on a natural rock platform but looks as though it's about to tumble down Mount Kyaiktiyo any second.

According to the legend, a Buddhist hermit was given one strand of hair from the Buddha himself, which he then gave to the king. For his gift, the king offered the hermit a stone shaped like his head, and used his magical powers to pull the boulder from the ocean. The king then built the small pagoda atop the rock to enshrine the Buddha's hair for eternity.

Pilgrims have flocked to the site for centuries. The hike uphill takes about 30 minutes—if you are unable or unwilling to make the climb, four porters will carry you in a bamboo sedan chair.

From Yangon, get a bus to Kinpun and then board the open truck that takes you up the mountain to base camp. Be prepared to surrender your personal space—trucks don't depart until they are crammed with people. Ⓝ 17.483583 Ⓔ 97.098428

The precariously perched golden rock defies gravity and inspires pilgrimages.

A golden temple is surrounded by hundreds of stone tablets that together comprise the world's largest book.

The dead outnumber the living ten to one.

PHILIPPINES

MANILA NORTH CEMETERY

MANILA

The largest graveyard in Manila is overflowing with the dead. And the living. Every day, 70 to 80 burials take place in the 133-acre cemetery and while its dead residents number approximately one million, there are an additional 10,000 residents who are very much alive.

Impoverished people who cannot afford more conventional housing often end up living inside their family mausoleums, sleeping on the stone tombs that house their dead relatives. To earn money, adults clean and repair tombs, while children carry coffins at funerals and collect scrap metal to sell. Some enterprising residents maintain shops in the uninhabited mausoleums, selling snacks, candles, and prepaid phone cards. A karaoke machine in one crypt allows visitors to sing pop hits for five pesos per tune.

If a family fails to pay the rent on a deceased relative's tomb for five years, the person's remains are excavated so a new burial can take place. As a result, the unclaimed skulls and bones pile up in the narrow aisles between graves, occasionally becoming toys for the children playing in the cemetery.

Many of the residents were born between tombs and have spent their entire lives in the cemetery. Though living among the dead seems extreme, it is a free, quiet, and—for many—safer alternative to living in Manila's slums.

A. Bonifacio Avenue, Manila. The graveyard is in the Santa Cruz district, next to the Chinese cemetery. Ⓝ 14.631476 Ⓔ 120.989104

CHOCOLATE HILLS

CARMEN, BOHOL

Chocolate may not be the first image that comes to mind when viewing the field of 1,268 conical, grass-covered hills on Bohol Island. But there is a logical explanation for the name: Each summer, the grass turns brown. Squint and, with a bit of imagination, you'll see a plain of giant chocolate kisses.

The hills, which range from 100 to 400 feet (30 to 122 m) tall, are limestone deposits from an ancient coral reef.

Get a Carmen-bound bus from the Dao terminal in Tagbilaran and ask the driver to let you off at Chocolate Hills, about 2.5 miles (4 km) before the Carmen stop. Ⓝ 9.916667 Ⓔ 124.166667

In Sagada, the dead are suspended from cliff faces for easier access to heaven.

HANGING COFFINS

SAGADA, MOUNTAIN PROVINCE

For 2,000 years, the people of Igorot Sagada have laid their dead to rest by jamming their bodies into compact wooden coffins and hoisting them up onto brackets driven into the side of a cliff. The practice protects the dead from floods and animals, and, according to Sagada beliefs, allows for easier passage to heaven.

Rows of pine caskets, some hundreds of years old, hang from the high bluffs of Echo Valley in Sagada. The Igorots embrace and actively prepare for death—elders, if physically able, carve their own coffins.

In summer, it will take you about 6 hours to travel from Manila to Sagada by bus or private car. During the rainy season, travel time doubles and roads are sometimes closed due to landslides. Ⓝ 17.083333 Ⓔ 120.900000

ALSO IN THE PHILIPPINES

Fire Mummies

Kabayan · In the mountain caves of Kabayan, tucked into hollow logs, are centuries-old mummies preserved via a smoking process.

Waterfalls Restaurant

San Pablo · Dine barefoot in a shallow river, seated beside a waterfall.

People were once mystified by the thousands of Hershey's Kiss–shaped hills on Bohol Island.

An exquisite (and expensive) temple honors an alleged tooth of the Buddha.

SINGAPORE

BUDDHA TOOTH RELIC TEMPLE AND MUSEUM

SINGAPORE

Completed in 2007, this $62 million dollar complex was built to honor a tooth fragment.

That may seem like quite an investment for such a small bit of dental debris, but the tooth purportedly belonged to one of the most famous religious figures in history—the Buddha.

The temple claims that the tooth is a relic recovered in 1980 from a collapsed stupa in Myanmar, but little additional information is provided. Experts have called into question the authenticity of the tooth, saying that it is most likely "the tooth of a cow or water buffalo, but definitely not a human."

The tooth is also only available for viewing during certain hours, but even if you miss the chance to see it, the temple itself is very impressive. It features multiple floors of Buddha statues, nagas (the popular Southeast Asian dragon-snakes that guard sacred relics), and impressive ceremonial venues in which to pray and meditate.

288 South Bridge Road, Singapore. There are a handful of other Asian temples that claim to have Buddha tooth relics, including the Temple of the Sacred Tooth Relic in Kandy, Sri Lanka. Ⓝ 1.281519 Ⓔ 103.844297

THAILAND

SIRIRAJ MEDICAL MUSEUMS

BANGKOK

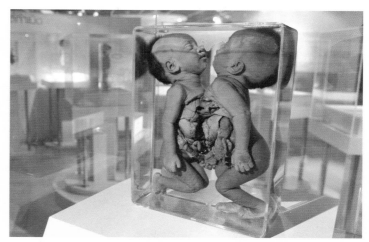

Conjoined twins are one of many confounding specimens in the Siriraj collections.

A walk through these six museums is a journey into the myriad ways human beings may be horrifically injured, killed, or deformed.

In the pathological museum, you'll see a collection of preserved fetuses and babies with congenital abnormalities, including a baby with cyclopia—a birth defect that results in a face with one malformed central eye, no nose, and no mouth. Next door, the parasitology collection displays multiple human scrota engorged to the size of basketballs due to elephantiasis.

The most gruesome specimens are displayed in the Songkran Niyomsane Forensic Medicine Museum. Shredded limbs that were recovered from car accidents float in glass jars. Photographs of decapitated train crash victims adorn the walls. Inside the cabinets, murder weapons sit beside the body parts they pierced. At the center of the room, standing in an upright glass case that resembles a phone booth, is Si Quey. The serial killer and cannibal died at the end of a noose after being sentenced for murdering and eating children during the 1950s. His black, shriveled body leans against one of the cabinet walls.

Adulaydejvigrom Building, Siriraj Hospital, 2 Phrannok Road, Bangkok. The museum is part of the Siriraj Hospital complex. Hop aboard the Chao Phraya Express boat at Bangkok's Central Pier and get off at Thonburi Railway. While the Pathology Museum is somewhat easy to find, it is worth tracking down the beautiful Congdon Anatomical Museum. Photography is not permitted. Ⓝ 13.757925 Ⓔ 100.485847

ALSO IN THAILAND

Goddess Tuptim Shrine

Bangkok · A shrine to a fertility goddess, complete with a forest of phallic offerings, has cropped up in a hotel parking lot.

Wat Samphran

Bangkok · An enormous dragon spirals around the exterior walls of this 17-story cylindrical temple.

MUSEUM OF COUNTERFEIT GOODS

BANGKOK

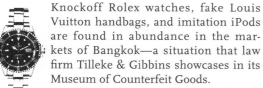

Knockoff Rolex watches, fake Louis Vuitton handbags, and imitation iPods are found in abundance in the markets of Bangkok—a situation that law firm Tilleke & Gibbins showcases in its Museum of Counterfeit Goods.

After raiding merchants on behalf of clients and seizing their forged goods to use as evidence, the firm ended up with rooms crammed full of counterfeit merchandise. In 1989, Tilleke & Gibbins began displaying 400 fakes in museum exhibits with the goal of educating the public on intellectual property infringement.

The museum's stash of illegal items now numbers over 4,000. T-shirts, perfumes, jewelry, cell phone batteries, and prescription drugs sit alongside their genuine counterparts, the differences often barely noticeable. Accompanying guides examine the societal impacts of counterfeiting—the operations support child labor, human trafficking, and the drug trade, among other ills. Consumer health and safety are also shown to be at risk, due to forged medications, car parts, and baby food that don't meet acceptable standards.

One of the more surprising aspects of the museum is the banality of some of its exhibits—apparently there is a market for counterfeit ballpoint pens, toothpaste, and stationery, along with the usual designer-label luxury goods.

Supalai Grand Tower, 26th floor, 1011 Rama 3 Road, Bangkok. Buses from the Khlong Toei MRT station stop right outside. Make an appointment at least 24 hours in advance. Ⓝ 13.683684 Ⓔ 100.548534

BEER BOTTLE TEMPLE

KHUN HAN, SISAKET

Every day, the monks of Wat Pa Maha Chedi Kaew Buddhist temple wake up surrounded by empty beer bottles. The order forbids intoxicants, but empty Heineken and Chang bottles are ever present, as they form the walls of the monks' living quarters.

Monks in Sisaket began collecting empty bottles in 1984 to promote recycling and keep the area litter-free. They amassed so many bottles of beer that they began to use them as building blocks for a temple and, eventually, a whole complex.

The main temple is comprised of approximately 1.5 million green Heineken and brown Chang bottles, set in rows. Inside are mosaics, created using pebbles and bottle caps. With the temple built, the monks moved on to new challenges: constructing a crematorium, prayer rooms, water tower, visitor restrooms, and residences—all made from beer bottles.

Construction is ongoing—the more empty bottles people bring, the more the monks will build.
Khun Han, Sisaket. Khun Han is a small village about an hour south of Sisaket. Ⓝ 14.618447 Ⓔ 104.418411

If you're thirsty for spiritual guidance, visit the temple of a million beers.

WANG SAEN SUK HELL GARDEN

CHON BURI

There's an awful lot of blood at Wang Saen Suk. It spurts from the mouths of the people being stabbed, gushes from the torso of the man being sawn in half, and bursts from the abdomen of the woman whose baby is being torn out of her body.

These gruesome scenes are sculptures created to decorate a Buddhist "hell garden" that shows what happens to those who behave badly. Buddhists believe in 16 main hells—eight hot and eight cold, all stacked on top of each other. Each hell is dedicated to specific offenses, and the punishments are tailored to the crimes.

The garden's focal point is an emaciated couple—one male, one female, each 30 feet (9 m) tall with ribs protruding, eyes bulging from exaggerated hollows, and tongues stretched to reach their hips. Surrounding the pair are figures in distress from an array of violent attacks. Some are having their heads eaten by dogs, while others stand helplessly as stern-faced men strip the skin from their bodies, revealing the glistening red layers beneath.

The garden is a popular destination for family day trips.

A depiction of the torture inflicted on those unfortunate souls who end up in Buddhist hell.

Sai 2, Soi 19, Saen Suk, Chon Buri. The garden is about 2 hours southeast of Bangkok. Ⓝ 13.297022 Ⓔ 100.910107

VIETNAM

HO CHI MINH MAUSOLEUM

HANOI

There is always a long line to see the embalmed body of Ho Chi Minh, who died in 1969, during the Vietnam War, at age 79. However, once you're inside the mausoleum, modeled after Stalin's final resting place in Moscow, things move quickly. Guards dressed in white enforce a policy of silence, shuffling people through the dimly lit space and allowing visitors only the briefest of pauses at the glass casket.

The hushed atmosphere reflects the national reverence for "Uncle Ho," who stood at this very spot in 1945 to read his declaration of independence, establishing the Democratic Republic of Vietnam. The need to pay tribute to the Communist revolutionary was strong enough to overpower his own wishes—in his will, Ho Chi Minh requested cremation.

Điên Biên, Ba Dinh District, Hanoi. More than 10 bus lines stop along the streets surrounding the mausoleum. Avoid visiting in October and November, when Ho Chi Minh's body travels to Russia for its annual makeover. Ⓝ 21.036667 Ⓔ 105.834722

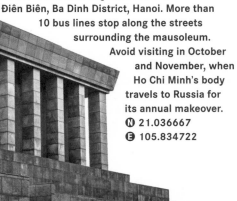

ZOOLOGICAL MUSEUM AT VIETNAM NATIONAL UNIVERSITY

HANOI, HOÀN KIÊM

Tucked away atop a staircase and only open by appointment, the zoological museum at Vietnam National University, Hanoi, is a charming, scruffy display of French Colonial taxidermy and preserved animal specimens.

The museum is divided into three rooms: Mammals, Reptiles and Fish, and Birds. The mammal room contains a parade of beasts—big cats, deer, bears, monkeys, and a baby elephant—all frozen midstride in an apparent stampede toward the door. A Komodo dragon, inflated blowfish, and jars full of snakes occupy the Reptiles and Fish room. For moth-eaten charm, you can't beat the Birds room, where eyeless, dust-covered gulls, owls, and pelicans stand or lie in rows, untouched since the early 20th century.

19 Le Thanh Tong, Hanoi. Look for the elephant skeleton at the top of the stairs. It marks the entrance to the little-known museum. Ⓝ 21.020579 Ⓔ 105.858346

CU CHI TUNNELS

HO CHI MINH CITY

Beneath the suburban Cu Chi district of Ho Chi Minh City is a network of tunnels that served as a home, air raid shelter, weapon storage facility, and supply route for the Viet Cong during the Vietnam War. For years, thousands of people effectively lived underground, only emerging after dark to gather supplies. It was a grim existence—the air was stale, the food and water scarce, and malaria spread fast through the claustrophobic, insect- and vermin-infested passages.

Construction on the tunnels began in the 1940s, as Vietnam fought to gain its independence from France. By the 1960s, the network stretched to over 100 miles (161 km). Tiny tunnel entrances, concealed beneath leaves on the jungle floor, required bodily contortion to squeeze into. To guard against enemy infiltration, the tunnel maintainers incorporated traps, such as dead-end passages and revolving floor panels that sent enemies tumbling into pits

of sharpened bamboo. Should a foe make it past these snares and into the underground city, the Viet Cong might respond with a handful of scorpions or a well-aimed snake to the face.

Large sections of the tunnels are gone, having collapsed or been destroyed, but a preserved section, enlarged to fit larger tourist bodies, is open to the public. Visits end with the seemingly inappropriate opportunity to fire AK-47s and M-16s at a shooting range.

Tour buses and public buses make the 90-minute trip from Ho Chi Minh City. Ⓝ 11.143511 Ⓔ 106.464471

ALSO IN VIETNAM

Dragon Bridge

Da Nang · Cross a 6-lane bridge, opened in 2013, dominated by a giant yellow steel dragon that breathes fireballs.

Cao Dai Holy See

Tay Ninh · The lavish temple of the Cao Dai movement is a brightly colored, dragon-infested beauty.

The underground Cu Chi Tunnels, used during the Vietnam War, are a nightmare for the claustrophobic.

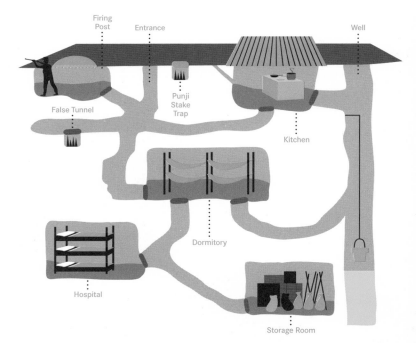

Firing Post · Entrance · Well · False Tunnel · Punji Stake Trap · Kitchen · Dormitory · Hospital · Storage Room

Africa

North Africa

EGYPT · LIBYA · MAURITANIA · MOROCCO · SUDAN · TUNISIA

West Africa

BENIN · BURKINA FASO · CAMEROON · GABON · GHANA · MALI
NIGER · SENEGAL · TOGO

Central Africa

CENTRAL AFRICAN REPUBLIC · CHAD
DEMOCRATIC REPUBLIC OF THE CONGO · REPUBLIC OF THE CONGO

East Africa

ETHIOPIA · SOUTH SUDAN · KENYA · TANZANIA · RWANDA · SOMALIA

Southern Africa

BOTSWANA · MALAWI · MOZAMBIQUE · NAMIBIA · SOUTH AFRICA
SWAZILAND · ZAMBIA · ZIMBABWE

Islands of the Indian
and South Atlantic Oceans

MADAGASCAR · SEYCHELLES · SAINT HELENA, ASCENSION,
AND TRISTAN DA CUNHA

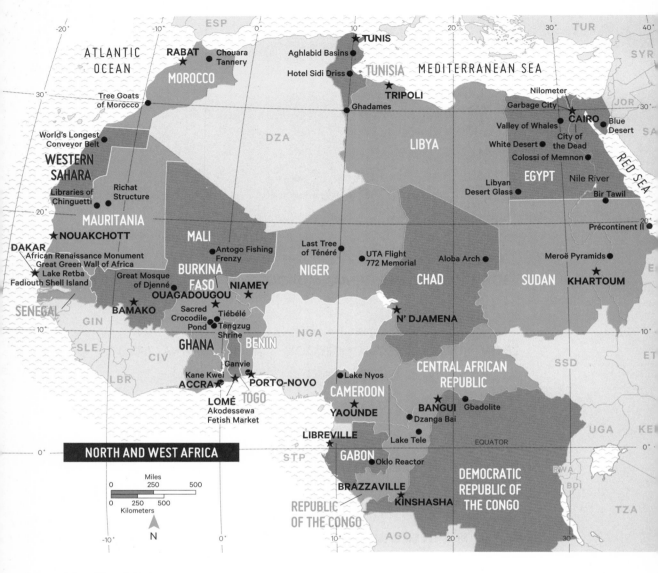

ATLANTIC OCEAN

MOROCCO

RABAT ★ Chouara Tannery

Tree Goats of Morocco

World's Longest Conveyor Belt

WESTERN SAHARA

Libraries of Chinguetti

Richat Structure

MAURITANIA

★ NOUAKCHOTT

DAKAR
African Renaissance Monument
Great Green Wall of Africa
★ Lake Retba
Fadiouth Shell Island

SENEGAL

GIN

MALI

BAMAKO ★

Great Mosque of Djenné

OUAGADOUGOU ★

BURKINA FASO

Antogo Fishing Frenzy

Sacred Crocodile Pond Tiébélé
Tengzug
Shrine

GHANA

BENIN

Kane Kwei
ACCRA ★
Ganvie
LOMÉ ★
Akodessewa Fetish Market

TOGO

★ PORTO-NOVO

SLE

CIV

LBR

TUNIS ★

Aghlabid Basins

Hotel Sidi Driss

TUNISIA

TRIPOLI ★

Ghadames

DZA

LIBYA

MEDITERRANEAN SEA

Nilometer
Garbage City
Valley of Whales
White Desert ●
CAIRO ★
Blue Desert

City of the Dead

Colossi of Memnon ●

EGYPT Nile River

Libyan Desert Glass ●

Bir Tawil

RED SEA

Précontinent II

Last Tree of Ténéré

UTA Flight 772 Memorial

Aloba Arch ●

Meroë Pyramids ●

NIGER

NIAMEY ★

NGA

CHAD

SUDAN KHARTOUM ★

N' DJAMENA ★

SSD

CENTRAL AFRICAN REPUBLIC

Lake Nyos ●

CAMEROON

YAOUNDE ★

BANGUI ★ Gbadolite ●
Dzanga Bai ●

ET

LIBREVILLE ★

Lake Tele ●

EQUATOR

UGA KE

GABON

BRAZZAVILLE ★

Oklo Reactor ●

STP

DEMOCRATIC REPUBLIC OF THE CONGO

RWA

BDI

REPUBLIC OF THE CONGO

KINSHASHA ★

TZA

AGO

NORTH AND WEST AFRICA

Miles
0   250   500

0   250   500
Kilometers

N

North Africa

EGYPT

NILOMETER

CAIRO

Long before the Aswan Dam was constructed to manage the flooding of the Nile, ancient Egyptians invented the nilometer to predict the river's behavior. The nilometer on Rhoda (or Rawda) Island in Cairo is an octagonal marble column held in place by a wooden beam that spans the width of an ornately carved well. The massive central column has markings on it that indicated where the river's water level was at any given time. This information was used to determine what conditions the future held: drought, which would mean famine; desirable, which would mean just enough overflow to leave good soil for farming; or flood, which could be catastrophic.

Although the behavior of the Nile could mean life or death for common people, only priests and rulers were allowed to monitor the nilometer. This is why many nilometers were built near temples, where priests would be able to access the mysterious instrument and appear prescient when they correctly predicted the river's behavior.

The Rhoda nilometer is on the southern tip of the island. The three tunnels that once let water into the stilling well have been filled in, allowing visitors to walk all the way down to the bottom. Ⓝ 30.007043 Ⓔ 31.224967

An essential instrument, housed in an elegant well.

BLUE DESERT

SINAI DESERT

When Egypt and Israel signed a peace treaty in 1979, Belgian artist Jean Verame wanted to mark the occasion. Verame journeyed to Egypt's Sinai Desert, near the resort town of Dahab, and created a "line of peace" by painting a stretch of its boulders bright blue. After decades under an unrelenting sun, the rocks are now less vibrant—but still a startling, cartoonish contrast to the beige and gray of the desert.

Verame went through official channels in order to paint the desert. Egypt's then-president Anwar Sadat signed off on the artwork, and the United Nations donated 10 tons of blue paint. The project was completed in 1981, the same year Sadat was assassinated by Islamic fundamentalists who were incensed by his signing of the peace treaty.

Hallawi plateau, between Dahab and St. Catherine.
Ⓝ 28.639722 Ⓔ 34.560833

WHITE DESERT

FARAFRA, WESTERN DESERT

Resembling giant mushrooms, atomic-bomb clouds, and, in one case, a chicken, the limestone rock formations of the White Desert are the heavily eroded remains of a former seabed. During the Cretaceous period, this portion of the desert was underwater, and chalk deposits from the skeletons of marine invertebrates accumulated on the ocean floor. To make a 100-million-year-old story short, the sea dried up, erosion dug odd shapes into the seabed, and now the White Desert is full of weirdly evocative rocks.

The best way to experience the bleached landscape of the White Desert is to camp overnight. As the sun sets and rises, the light on the rock formations changes, and their shadows morph. In the silence you may hear the soft patter of a fennec's paws. These adorably large-eared nocturnal foxes are native to the Sahara.

The White Desert is 25 miles (40 km) north of Farafra, an oasis where you can take a dip in a hot spring.
Ⓝ 27.098254 Ⓔ 27.985839

The eroded remains of a former seabed take the form of clouds, mushrooms, and a chicken.

CITY OF THE DEAD

CAIRO

The narrow, unpaved streets of the City of the Dead wind between densely packed, sand-colored buildings with weathered walls. These buildings are the tombs, family mausoleums, and intricately decorated funerary complexes of a 4-mile-long (6.4 km) Islamic necropolis that lies beneath Mokattam Hills.

The City of the Dead does not just house the deceased. Around half a million people also live there, sleeping, eating, and hanging their laundry in the centuries-old tombs. Many of the residents are families of those laid to rest, but the City of the Dead has also become a refuge for those forced out of Cairo's ever more crowded and expensive urban center. The northern cemetery, with its magnificent Qaitbay mosque, is more populated than the southern cemetery.

The necropolis lies under Mokattam Hills in southeast Cairo. Ⓝ 30.021667 Ⓔ 30.303333

The deceased and the living inhabit the same tombs.

Tens of thousands of freelance garbage collectors, known as Zabbaleen, pick up Cairo's trash.

GARBAGE CITY

CAIRO

The south end of the Manshiyat Naser ward in Cairo is better known as Garbage City. This is where the Zabbaleen—"garbage people" in Arabic—bring Cairo's household waste. The narrow streets, blocky apartments, and crammed courtyards are all piled with huge bags of trash waiting to be sorted.

Though the population of greater Cairo exceeds 17 million, the city has no municipal waste collection program. Instead it has tens of thousands of freelance Zabbaleen. For decades, this group has sustained itself by hauling household trash to Manshiyat Naser by truck and donkey-drawn cart. There, they recycle, reuse, and sell the refuse. Plastics and metals are carefully separated by color and composition, then sold as scrap. Pigs eat the organic waste.

This state of affairs shifted in 2003, when the Egyptian government attempted to downgrade the role of the Zabbaleen by bringing in corporations to handle waste disposal. The experiment was not a success—whereas the Zabbaleen received a fee for collecting trash from individual residences, the new system required people to take their garbage to communal collection points on the street. The corporate collectors were less efficient than the Zabbaleen and recycled a much lower percentage of refuse—about 20 percent compared with the Zabbaleen's 80

percent. Further threats to the Zabbaleen's livelihood came in 2009, when the government slaughtered Garbage City's hundreds of thousands of pigs, citing concerns over swine flu.

The vast majority of the Zabbaleen are Coptic Christians, a religious group that has suffered persecution and sectarian violence in the 90 percent Muslim country. Marginalized in the settlement of Garbage City, the Zabbaleen literally carved out safe spaces for themselves in the form of seven churches built into the limestone cliffs of Mokattam.

Garbage City lies at the base of Mokattam Hills in Manshiyat Naser, a ward in southeast Cairo.
Ⓝ 30.036230 Ⓔ 31.278252

ALSO IN EGYPT

Temple of Abu Simbel

Abu Simbel · Completed in 1244 BCE, this temple features four colossal statues of Rameses II, the reigning pharaoh at the time.

Muzawaka Tombs

Dakhla Oasis · See piles of mummies from Egypt's Roman era in these open rock-cut tombs.

Desert Breath

Near Hurghada · Completed in 1997, this giant land-art spiral of dots in the sand is slowly being consumed by the desert.

Aquarium Grotto Garden

Zamalek • The Fish Garden, just south of the aquarium on Gezira Island, is a rare spot of green within the chaotic city. Once part of a much larger estate, it was restored in 2000 but still feels like an old curio cabinet—if you can imagine a cabinet packed with sea life.

Agricultural Museum

Dokki • A threadbare remnant of Cairo's colonial past, many of the more decayed buildings in this sprawling complex are closed—but the areas still open are magical, with turn-of-the-20th-century taxidermy and out-of-the-past dioramas depicting early Egyptian daily life.

Lehnert and Landrock Bookshop

Qasar an Nile • The small gallery of glass-plate photographs by the two founders of this nearly 100-year-old shop will bring you back to a historic North Africa that is quickly fading from view.

Oud Shopping on Mohammad Ali Street

Abd el-Aziz • The Khan al-Khalili market is certainly a place to buy an oud, but if you're in need of the real thing from the hands of a true master, the place to go is Mohammad Ali Street, where more than a dozen shops specialize in professional-grade instruments. There they are made, repaired, and shipped to sellers all over the world.

Bayt al-Suhaymi

Qism el-Gamaleya • Once a family home of 17th- and 18th-century Ottoman merchants, this quiet and dignified museum is notable for its streamlined furnishings, elaborately carved wooden screens, and an elegant and organic design of wood and stone that would make Frank Lloyd Wright green with envy.

Bab Zuweila Gate

el-Darb el-Ahmar • One of the three remaining gates of the Old City walls, Bab Zuweila is an elegant example of Fatimid architecture, an Ottoman-era mash-up of Eastern and Western design. Climb up one of its two minarets via a tight spiral staircase.

Mosque of Ibn Tulun

Tolon • This 9th-century mosque in Islamic Cairo is thought to be the oldest in the city, and it's undoubtedly the biggest, with its long double arcades, pointed arches, mesmerizing scrollwork, and a perfectly square fountain holding down the center.

National Military Museum

Qism el-Khalifa • Part of the sprawling citadel on the western edge of the city, the National Military Museum is in a one-time harem palace of Muhammad Ali Pasha al-Mas'ud ibn Agha. The long history of the Egyptian military is captured through ancient armaments, dioramas of famous battles, and more recent materiel like a MiG-21 fighter jet.

The Monastery of Saint Samaan the Tanner

Qism el-Khalifa • The massive cave church of Cairo's Zabbaleen community is carved into a mountain, seats 5,000 parishioners, is named for a 10th-century tanner, and is one of the largest Christian churches in the Middle East.

Umm Kulthum Museum

el-Manial • To call Umm Kulthum's following cultlike is an understatement. Perhaps the most famous Arabic singer of the 20th century, the museum filled with all things "Umm" creates a portrait both expansive (her vast recording career and collection of gowns) and intimate (rhinestone reading glasses and little black day book).

Coptic Museum

Misr al-Qadimah • In 1908, a rare collection of early Coptic art—paintings, stone reliefs, tapestries, and metalwork of the early Christian era—found a home at this singular museum where thousands of artifacts are surrounded by intricate and ornate wood carvings.

Babylon Fortress

Misr al-Qadimah • Built by Trajan, a soldier-emperor who expanded the reach of Rome as far as it would ever get, this first-century CE fortress is still striking, with its striped stone walls and two millennia of history. Recent digs revealed ancient river walls that have held back the Nile for more than 2,000 years.

Cairo Geniza of Ben Ezra Synagogue

Misr al-Qadimah • The synagogue, built in the year 882 CE, is believed to be located on the spot where little baby Moses was scooped out of the Nile. Almost 2,000 years later, more than a quarter million fragments of medieval Jewish texts were found in the basement.

Hanging Church

Misr al-Qadimah • In Arabic, this 7th-century Coptic church is known as al-Muallaqah, or "the suspended," as it literally hangs over an ancient Roman gate. Its roof mimics Noah's Ark and 29 steps lead to its intricately carved doors.

Darb 1718

el-Fustat • In 2008, Egyptian multihyphenate Moataz Nasr created this vibrant and eclectic art and culture space for young performers and artists, with galleries, an independent film series, and workshops.

The Umm Kulthum Museum celebrates the singer known as "the fourth pyramid."

The fossilized skeletons of enormous footed whales litter the Egyptian desert.

VALLEY OF THE WHALES

WADI AL-HITAN, AL FAYYUM GOVERNORATE

There was a time when whales walked the earth. During the Eocene epoch —56 to 33.9 million years ago—there lived a whale suborder known as the Archaeoceti. The five families within this subgroup all bore a characteristic that set them apart from modern-day whales: limbs equipped with feet and toes.

During the Eocene epoch, these footed whales lived underwater. Though they didn't use their feet to walk, they are a missing evolutionary link between contemporary whales and their terrestrial ancestors.

In 1902, the first fossilized Archaeoceti skeletons were discovered in Egypt's Western Desert southwest of Cairo. After decades of fieldwork, the bones of 1,000 animals have been identified. Today, hundreds of partial skeletons remain in the sand, on display to any visitor who is willing to make the long trek through the desert. Wadi Al-Hitan has been a protected area since 1989, and even though it is remotely located and under strict management, the fossils are still vulnerable to visitors. In 2007, Egyptian authorities accused a group of Belgian diplomats of ignoring signs and steering two four-wheel-drives onto one of the whale skeletons. Cars are now prohibited from entering the site.

The site is 93 miles (150 km) southwest of Cairo. There are no paved roads; 4-wheel-drive vehicles are your best option. Leave your car at the entrance to avoid running over any 40-million-year-old whales. N 29.270833 E 30.043889

COLOSSI OF MEMNON

AL BAIRAT, LUXOR

The twin colossi (which no longer resemble twins) have loomed over the Theban Necropolis since 1350 BCE, battered for millennia by scorching desert sun and sporadic Nile floods.

These ancient Egyptian statues each depict the pharaoh Amenhotep III, who ruled during the 18th dynasty. They once flanked the entrance to his lost mortuary temple, which at its height was the most lavish temple in all of Egypt. Their faded side panels depict Hapi, god of the nearby Nile.

Though floods reduced the temple to no more than looted ruins, these statues have withstood many natural disasters. In 27 BCE, an earthquake shattered the northern colossus, collapsing its top and cracking its lower half. The damaged statue did more than merely survive the catastrophe: After the earthquake, it found its voice.

At dawn, when the first ray of desert sun spilled over the baked horizon, the shattered statue would sing. Its tune was more powerful than pleasant, a fleeting, otherworldly song that evoked the divine. By 20 BCE, esteemed tourists from around the Greco-Roman world were trekking across the desert to partake in the sunrise acoustic spectacle. Some say the sound resembled striking brass, while others compared it to the snap of a breaking lyre string.

The unearthly song is how these ancient Egyptian statues wound up with a name borrowed from ancient Greece. According to Greek mythology, Memnon, a mortal son of Eos, the goddess of dawn, was slain by Achilles. Supposedly, the eerie wail echoing from the cracked colossus's chasm was Eos crying to his mother each morning. (Modern scientists believe early-morning heat caused dew trapped within the statue's crack to evaporate, creating a series of vibrations that echoed through the thin desert air.)

Well-intentioned Romans silenced the song sometime between 196 and 199 CE. After visiting the storied statues and failing to hear their ephemeral sounds, Emperor Septimius Severus, reportedly attempting to gain favor with the oracular monument, had the fractured statue repaired. His reconstructions, in addition to disfiguring the statue so the fixtures no longer looked like identical twins, robbed the colossus of its famous voice and rendered its song a lost acoustic wonder of the ancient world.

The colossi make for a good stop on the way to or from the Valley of the Kings, which is a 15-minute drive away. N 25.720636 E 32.610445

The twin colossi once emitted
a mysterious, fleeting tune.

LIBYAN DESERT GLASS

Among the treasures buried in Tutankhamun's tomb was a jeweled, collar-style necklace featuring a scarab carved from pale yellow glass—a glass whose origin is still the subject of scientific debate.

Known as Libyan Desert glass or Great Sand Sea glass, the 98 percent silica substance is scattered in chunks all over the dunes of southwest Egypt. The glass formed

naturally under intense heat, but the exact manner of its formation is unclear. The prevailing belief is that a meteorite hit this part of the desert approximately 29 million years ago, superheating the sand and causing it to form glass when it cooled. Another possibility is that the meteorite exploded in the atmosphere, radiating heat strong enough to create the glass. **The glass is scattered around the desert along the Libya-Egypt border. If you'd rather not forage in the sand, go to the Egyptian Museum in Cairo to see Tutankhamun's necklace. N 30.047778 E 31.233333**

BIR TAWIL

On the border of Egypt and Sudan is Bir Tawil, an 800-square-mile (2,072 km²) patch of land that neither country wants to claim.

A border dispute in which each nation wants to foist land onto the other is unusual, but Bir Tawil is unappealing for good reason. There are two versions of the Egypt-Sudan border: the political boundary, a straight line established in 1899, and the more wobbly administrative boundary, established in 1902. The 1899 boundary gives the Hala'ib Triangle, a prized coastal section of land, to Egypt. The 1902 boundary puts the Hala'ib Triangle in Sudan. If either claims Bir Tawil,

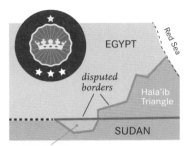

BIR TAWIL

it loses the much more valuable Hala'ib Triangle. And so explains the current stalemate.

Bir Tawil is one of the few bits of land on Earth to be regarded as *terra nullius*: unclaimed territory. There have, however, been attempts to establish sovereignty. In 2014, an American man named Jeremiah Heaton flew to Egypt and journeyed 14 hours through the desert in order to plant a flag in the sand of Bir Tawil. His mission was to claim the region for himself so that his seven-year-old daughter's dream of being a princess could be realized.

Heaton, who received criticism in the British and American press for reviving white colonialism in Africa, has thus far been unable to gain international recognition for his "Kingdom of North Sudan." **Bir Tawil is 150 miles (241 km) south of Aswan, a city on the Nile that was founded in ancient Egypt as the frontier town of Swenett. N 21.881890 E 33.705139 ➺➺**

➺➺ Terra Nullius and the Discovery Doctrine

Terra nullius is a loaded phrase. Meaning "nobody's land" in Latin, it also refers to the belief, popular among European colonizers, that uncultivated land may be claimed in the name of the Crown despite the presence of indigenous inhabitants.

English settlers used the terra nullius justification to establish penal colonies in Australia during the 18th century, despite the presence of Aboriginal Australians. It wasn't until 1992 that the Australian High Court officially rejected the concept that the country was terra nullius when the English arrived. Following this decision, Aboriginal Australians have been able to make claims on land for traditional use.

Closely related to the concept of terra nullius is the Discovery Doctrine, a set of principles that guided Europe's colonization of the Americas and informed westward expansion in the United States. Its origins are in the papal bulls of the 15th century. Pope Alexander VI's 1493 papal bull, for example, declared that any land in the Americas not inhabited by Christians was available to be "discovered." A Christian nation could claim sovereignty over such land, wrote Alexander VI, and explorers were to rid non-Christian inhabitants of their "barbarous" ways by instructing them in the Catholic faith.

LIBYA

GHADAMES

GHADAMES, TRIPOLITANIA

The old part of this ancient Roman oasis town, which is now uninhabited, is a maze of connected multistory mud-and-straw homes that are clustered together as though huddling against the swirling sands of the desert. Covered walkways between the buildings allowed people to socialize without having to face the heat of the Sahara. The passages linking the top floors of homes were used by women, while the ground-floor walkways were mostly used by men. Small ventilation holes in the walls let air flow through the alleys.

Some of the white walls of the homes are adorned with traditional designs of the Tuareg, the nomadic Berber people in the area. Red triangles, diamonds, moons, and suns are painted around windows and doors, above arches, and along stairs.

The newer part of Ghadames, which, unlike the old town, is equipped with electricity and running water, is still inhabited by around 11,000 people and welcomes overnight visitors.

There are flights from the Libyan capital of Tripoli to Ghadames, which is right near the spot where Libya meets the borders of Algeria and Tunisia.
Ⓝ 30.131764 Ⓔ 9.495050

The oasis town of Ghadames is known as the Pearl of the Desert.

The "Eye of the Sahara," a mysterious geological bull's-eye, is visible from space.

MAURITANIA

RICHAT STRUCTURE

OUADANE, ADRAR

The next time you're flying over Mauritania—or passing above Africa aboard the International Space Station—have a glance out the window and see if you can spot the Richat Structure. It shouldn't be too difficult: The circular feature is 30 miles (48.3 km) wide.

Also known as the Eye of the Sahara, the Richat Structure has somewhat mysterious origins. When the structure was first spotted from space in the mid-1960s, it was thought to be an impact crater from a meteorite. Geologists now believe that the massive bull's-eye is a heavily eroded mishmash of sedimentary, igneous, and metamorphic rock layers. After being pushed up into a dome formation—say, from a hot air balloon—these layers eroded at different rates, causing the striking pattern of concentric circles.

Though best viewed from above, the structure can be seen from the ground via a four-wheel-drive trip from the Mauritanian capital of Atar. Ⓝ 21.211111 Ⓦ 11.672220

LIBRARIES OF CHINGUETTI

CHINGUETTI, ADRAR

Soon after its founding around the 12th century, the desert mud-brick village of Chinguetti became a hub for trade, culture, and scholarship. Located on a trans-Saharan caravan route, it catered to the desert's nomadic population, receiving visitors with open arms—and open books.

The libraries of Chinguetti, owned and maintained by village families, contained a wealth of medieval Arabic manuscripts on science, mathematics, law, and Islam. Scholars, pilgrims, and the holy and venerated came to the village to pore over these handwritten leather-bound tomes and exchange ideas with one another.

Now desertification is causing the Sahara to encroach on Chinguetti. A few thousand people still live there, but sand is being pushed into the alleyways and swept against the walls. The libraries remain, but the dry air and swirling sand are a threat to the centuries-old manuscripts. Despite the risks, the families who own the texts prefer to keep them in place, and will display them for visitors using gloved hands.

Cars to Chinguetti leave from the nearby town of Atar. Chinguetti itself has few facilities, but some locals offer homestays, meals, and camel tours of the surrounding dunes. Ⓝ 20.243244 Ⓦ 8.836276

Chinguetti's holy manuscripts have been gathering dust since the Middle Ages.

MOROCCO
Chouara Tannery

FEZ, FEZ MEKNES

Wedged among the ancient buildings and serpentine passageways of Fez's Old Medina is a grid of stone wells, each filled with colored liquid. This is Chouara, an 11th-century tannery that still operates as it did a thousand years ago.

Animal hides are brought here to be preserved, dyed, and turned into the handbags, jackets, and wallets sold in the surrounding souks.

The process begins with the raw skins being soaked in a mixture of cow urine, pigeon feces, quicklime, salt, and water. This loosens the hair from the hides and makes them softer. After a few days of steeping in this concoction, the skins are hauled out and hung from balconies to dry. Then comes the dyeing. Tannery workers plunge the skins into the colored wells, leaving them for a few more days to absorb the hue.

Visitors are welcome to observe the tannery in action and are even given a gift upon arrival: a small sprig of mint to hold under the nose when the smell becomes too much.

Fes El Bali, Fez. The tannery is in the old, walled part of the city El Medina, which also contains the University of al-Qarawiyyin. Established in the year 859, it is the oldest existing university in the world. Ⓝ 34.066361 Ⓦ 4.970973

The colorful wells of Morocco's leather tanning industry date back to the 11th century.

TREE GOATS OF MOROCCO

TAMRI, SOUS-MASSA

Morocco's argan trees are infested with hordes of fruit-hungry goats, who hop up into the branches to pick out the fruit. Grown almost exclusively in the Sous Valley in southwestern Morocco, the rare and protected argan trees (*Argania spinosa*) produce an annual fruit crop—these delicious morsels attract delightful tree-climbing goats.

Argan trees are not the most aesthetically pleasing plant in the world, with their rough, thorny bark and crooked branches. But their forests still tend to attract admirers, thanks to the odd sight of the hoofed animals perching on impossibly precarious limbs high in the treetops to enjoy their seasonal feast. The spectacle is far from just a single ambitious goat climbing a single tree—the goats tend to swarm into the trees. As many as a dozen goats can be seen munching away in the branches of a tree at any one time.

Local farmers condone and even cultivate this bizarre feeding practice, keeping the goats away from the trees while the fruit matures and releasing them at the right time. There is also a secondary benefit to the goats' habits: After the animals finish eating the fruit off the tree, they expel the valuable nuts found inside the fruit. The nuts are pressed to create highly sought-after argan oil, one of the most lucrative plant oils in the world.

This memorable rural scene mostly happens in June when the argan fruit ripen. Ⓝ 30.682889 Ⓦ 9.834806

The argan tree is a jungle gym for nimble, hungry goats.

WORLD'S LONGEST CONVEYOR BELT SYSTEM

BOU CRAA, WESTERN SAHARA

The world's longest conveyor belt may not be as grand as the Great Wall of China, but it can be seen just as easily from outer space. The record-setting conveyance system is a winding chain of interlinked belts transporting phosphate from the mines in Bou Craa, Western Sahara, to the harbor town of El-Aaiún on the Atlantic Ocean, where it is shipped worldwide. The majority of the world's phosphate is mined in Bou Craa, the core of the so-called useful triangle in the Moroccan-controlled Western Sahara territory.

All told, the phosphate's leisurely journey covers a distance of 61 miles (98 km) from one end of the belt system to another. As the rocky ore makes its way across the landscape, strong desert winds blow the lighter particles of white powder off the belt, creating a bold ivory streak along the length of the transport system.

The start of the conveyor belt can be seen at the phosphate mines in Bou Craa. Ⓝ 26.188373 Ⓦ 12.696186

A conveyor belt so enormous it can be seen from space.

SUDAN

PYRAMIDS OF MEROË

MEROË, RIVER NILE

There are more pyramids in the northern region of the Sudanese desert than there are in all of Egypt. During Egypt's 25th dynasty—760 until 656 BCE—Meroë, now located in Sudan, was the capital of the Kingdom of Kush, ruled by Nubian kings who had conquered Egypt. The city, nestled against the Nile, contained a necropolis for royal burials.

As in Egypt, Nubian kings and queens were buried with gold, jewelry, pottery, and, occasionally, pets. Some royals were mummified, while others had their remains burned or buried whole. A sandstone pyramid, steeper and more narrow than the Egyptian variety, was built over each tomb.

In all, there were about 220 pyramids at Meroë. They remained relatively intact until the 1830s, when Italian treasure hunter Giuseppe Ferlini smashed the tops off 40 pyramids while searching for gold and jewels.

Meroë is a 3-hour drive north of Khartoum. A camel ride will get you to the pyramids. Bring water. Ⓝ 16.938333 Ⓔ 33.749167

Though less famous than their counterparts to the north, there are more pyramids in Meröe, Sudan, than in all of Egypt.

Explore the remains of Jacques Cousteau's underwater village.

PRÉCONTINENT II

PORT SUDAN, RED SEA

Précontinent II is the last remnant of a series of three French underwater habitats built between 1962 and 1965. Developed by oceanography pioneer Jacques-Yves Cousteau, the underwater "village" tested the ability of humans to live underwater without interruption for extended periods of time, at increasing depths.

Précontinent I, a 16-foot (5 m) steel cylinder fixed 30 feet below the surface, was the first underwater habitat. It launched in the Balearic Sea off the coast of Marseille in September 1962. Two aquanauts, Albert Falco and Claude Wesly, lived among infrared lamps used as heaters, a record player, a radio, three telephones, a video surveillance system, a library, a TV, and a bed. In the bottom of the habitat was an airlock that allowed the two men to access the ocean, where they studied fish and took measurements for underwater topographical maps. A year later, two Précontinent II habitats launched about 22 miles (35.4 km) northeast, off the coast of Port Sudan. The first, named the Starfish House, lasted for four

weeks and housed a group of oceanographers, as well as Simone Melchior Cousteau and the parrot Claude, who was supposed to warn the aquanauts of possible hazards in the air. The second habitat, Deep Cabin, was installed at a depth of 89 feet (27.1 m). Other structures included a tool shed and an air-filled hangar containing the Hydrojet Saucer DS-2, a two-person submarine equipped with three movable outside lamps, two cameras, a radio, a tape recorder, and a movable grappler. Précontinent III launched in September 1965 at 328 feet (100 m) below the surface, off the coast of Nice.

Of all the habitats, only part of Précontinent II remains underwater, and it has become the site of many diving tours from Sudan and Egypt. At the anchor place of the habitat you can find the remains of the toolshed, crusted with coral growth, and the fish cages, covered with sponges. A few meters deeper are the shark cages, covered with coral and crustaceans, and the hangar.

Shaab Rumi is a dive site 30 miles (48.3 km) from Port Sudan. Diving tours can be booked through various websites. Shaab Rumi is also a prime spot for shark sightings. Ⓝ 19.938736 Ⓔ 37.418697

TUNISIA

HOTEL SIDI DRISS

MATMATA, GABES

The small Berber village of Matmata is dotted with "troglodyte homes": traditional cave houses carved out of rock. Though the homes were created centuries ago, one of them, Sidi Driss, has a much more modern claim to fame: It was used as Luke Skywalker's childhood abode in *Star Wars: A New Hope* and *Attack of the Clones*.

The cave is now a hotel for *Star Wars* fans. For around $20 per night, you can live like a Jedi knight. It's not luxurious—the rooms are windowless, the beds are cots, and the occasional offensive odor travels on the wind—but it's certainly unique.

Shared taxis and buses leave from Gabes, 25 miles (40.2 km) away.
Ⓝ 33.545687 Ⓔ 9.968319

ALSO IN TUNISIA

Lac de Gafsa	Dougga
Gafsa · After appearing overnight in 2014, this mysterious desert lake has been dubbed both miraculous and possibly carcinogenic.	Twenty temples, an amphitheater, and a chariot-racing circle are among the highlights at this well-preserved ancient Roman town.

Sleep like a Skywalker at Sidi Driss.

AGHLABID BASINS

KAIROUAN, KAIROUAN

Kairouan is in a semi-arid region prone to drought, without any nearby rivers or natural water sources. In the 9th century, a solution took the form of the majestic Aghlabid Basins, a huge and highly advanced work of engineering.

The Aghlabid Basins are composed of two connected cisterns which together form an open-air reservoir, fed by a 36-mile (60 km) aqueduct that sources water from the hills beyond town. Water flows into the smaller pool, which serves as a sort of filter, collecting stray sediments before the water is transferred to the larger basin, which is an impressive 16 feet (5 m) deep and 420 feet (128 m) in diameter.

The basins would fill up with rainwater, which would be used for washing or emergency hydration, though contamination always posed a major risk. Still, the system was a remarkably sophisticated engineering feat for its time and is considered the largest hydraulic installation of the period. Originally, there were at least 15; just two remain today. Though the pools appear circular, they are in fact 17-sided polygons.

Ave Ibn El Aghlab, Kairouan. You can visit the basins, but don't drink the water. While in Kairouan, swing by the Great Mosque. Established in the seventh century, it's one of the oldest places for Islamic worship in North Africa. Ⓝ 35.686541 Ⓔ 10.095583

Cisterns so well built they still stand 1,000 years later.

Built on stilts in the middle of a lake, the village of Ganvie is home to about 30,000 people.

West Africa

BENIN

GANVIE

GANVIE, ATLANTIQUE

In the 17th and 18th centuries, a portion of present-day Benin was known as the Kingdom of Dahomey. Established by the Fon people, a West African ethnic group, Dahomey became a major part of the Atlantic slave trade following the arrival of the Portuguese.

Fon hunters worked with Portuguese slave traders, traveling around the region hunting for people to sell. One of the ethnic groups they targeted was the Tofinu, who lived in what is now central Benin.

Knowing that the Fon's religious beliefs prevented them from venturing into bodies of water, the Tofinu fled their homes and established Ganvie, a community of bamboo huts built atop stilts on Lake Nakoué. Having provided protection for the Fon during the slave-trading days, Ganvie lived on, and has adapted to the demands of the 21st century. Motorized boats zigzag around its 3,000 buildings, which include a school, post office, church, bank, and mosque. About 30,000 people live in the village, traveling between huts by canoe and earning an income by fishing. **Ganvie is on the northern edge of Lake Nakoué, north of the coastal city of Cotonou, about 4 hours from Porto Novo. Ⓝ 6.466667 Ⓔ 2.416667**

BURKINA FASO

TIÉBÉLÉ

TIÉBÉLÉ, NAHOURI

The walls of the mud-brick homes at this village near the Ghanaian border are also canvases for cultural expression. Women of the Kassena ethnic group, who have lived in the region since the 15th century, work together to decorate the huts with geometric patterns, people, and animals. They use mud, chalk, and tar to paint, then cover the designs with a protective layer of varnish made from the boiled pods of a locust bean tree. **Tiébélé is 19 miles (30 km) east of the city of Pô, where you can hire a driver to take you to the village. Ⓝ 11.095982 Ⓦ 0.965493**

The women of Tiébélé turn every village wall into a geometric mural.

CAMEROON

LAKE NYOS

MENCHUM, NORTHWEST REGION

Lake Nyos killed over 1,700 people in a single night, but its victims did not drown. None were even in the lake—many died in their beds, in homes up to 15 miles (24 km) from shore.

The bizarre disaster began with the buildup of carbon dioxide in the lake, which sits in the crater of a dormant volcano. Gas rose from an underground magma chamber and dissolved into Lake Nyos, slowly creating a highly pressurized bottom layer saturated with carbon dioxide.

On the evening of August 21, 1986, just after 9 p.m., the lake erupted. A huge cloud of carbon dioxide burst from the water, smothering local villages and asphyxiating the people and animals within them. Those who survived spent hours unconscious

Seventeen hundred people died when these calm waters burped a bubble of carbon dioxide.

from oxygen deprivation. They awoke surrounded by bodies, with no indication of what had happened.

Since this catastrophic event, French scientists have implemented a degassing program at Lake Nyos. In 2001, they installed a pipe that runs to the bottom of the lake and allows the gas to escape at a regular, safe rate. Two more pipes were added in 2011. A solar-powered alarm system monitors carbon dioxide levels—should the lake explode again, there will at least be some warning.

The lake, part of the Oku Volcanic Field, is about 200 miles (322 km) northwest of Yaoundé.
Ⓝ 6.438087 Ⓔ 10.297916

GABON

OKLO REACTOR

MOUNANA, HAUT-OGOOUÉ

At a University of Chicago athletic field on December 2, 1942, a crowd of excited physicists gathered to watch their nuclear reactor, CP-1, go critical. At the time, they

believed they were witnessing the world's first self-sustaining fission chain reaction. In fact, CP-1 was the world's second uranium fission nuclear reactor—the first went critical approximately 1.7 billion years earlier, underneath the ground in the Oklo region of Gabon. This one was entirely natural.

Oklo's land is rich in uranium deposits, which France mined for decades beginning in the 1950s. In 1972, a routine analysis of samples from the Oklo mine showed an unusually low amount of uranium-235, one of the three isotopes found in naturally occurring uranium deposits. Ordinarily, the deposits consist of about 0.72 percent uranium-235. The Oklo sample contained only 0.717 percent uranium-235—not a huge difference, but enough to alert scientists to the fact that something unusual had taken place: a natural nuclear chain reaction.

The chain reaction began in the Precambrian era, when groundwater flowed through cracks in the ore and made contact with uranium-235. Generally, in a nuclear reactor, uranium-235 absorbs a stray neutron, causing its nucleus to split—fission—and releasing energy, radiation, and free neutrons. These neutrons then get absorbed into more uranium-235 atoms, causing more nuclei to split, which releases yet more neutrons in a chain reaction. Water acts as a neutron moderator, slowing down the fast-moving free neutrons that are released from split nuclei and giving them a better chance at further fission.

In all, 15 reactor zones have been discovered in the area surrounding Oklo's uranium mine. The waste products of one reactor remain, located on a slope of a mining pit and encased in a concrete block to prevent them from sliding into the trench.

The uranium mine, now closed, is on the N3 road in Mounana, a town of around 12,000 in the Oklo region.
Ⓢ 1.394444 Ⓔ 13.160833

GHANA

SACRED CROCODILE POND

PAGA, UPPER EAST REGION

The world's most docile crocodiles prowl the waters of this Paga pond. According to local tradition, each of the calm crocs represents the soul of a Paga villager. It is therefore forbidden to harm or disrespect the animals. But you're welcome to sit on their backs and pose for photos.

To see the crocs up close, you'll need to pay a guide, who will fetch a live chicken and whistle to summon the toothy beasts. Once your designated crocodile has sated itself with the flesh of the fowl and entered a state of post-gorging lethargy, you can pet its tail and sit astride its body. The crocodiles are not inclined to attack—probably because they are so well fed with fowl—but you are advised to steer clear of their snouts. Just in case.

The pond is on the Burkina Faso border, 25 miles (40.2 km) northwest of Bolgatanga. Ⓝ 10.98147 Ⓦ 1.115642

KANE KWEI CARPENTRY WORKSHOP

ACCRA, GREATER ACCRA REGION

A Ghanaian teacher was once buried in a ballpoint pen. A singer was laid to rest inside a microphone, while a laborer was interred in a hammer. These "fantasy coffins," built in the shape of items representing the deceased's occupation, passions, or aspirations, were made by craftsmen at Kane Kwei Carpentry Workshop.

The studio was established in the 1950s by Seth Kane Kwei, a member of the Ga ethnic group of coastal Ghana. The Ga believe that when someone dies, they move on to another life and continue to exert influence on their living descendants. Family members make sure to honor them and secure their from-the-grave goodwill by staging elaborate funerals involving hundreds of guests and a procession. The centerpiece is the casket, which is custom-designed to please the recently deceased.

Since Seth Kane Kwei's own death in 1992, the workshop has continued with a group of dedicated craftsmen who gladly sculpt coffins in the shape of lobsters, robots, sandals, high-top sneakers, bananas, and garbage trucks.

Should you be interested in a fantasy coffin, you can order one to be shipped anywhere in the world. You don't have to have your funeral in mind: In addition to their practical usage, fantasy caskets have acquired international value as works of art. A coffin in the shape of a Nike sneaker is on display at the Brooklyn Museum in New York, while an eagle-shaped casket holds court at the British Museum. Both are the work of Paa Joe, who was trained by Seth Kane Kwei.

Teshie First Junction, Accra. The open-air Kane Kwei workshop is on a dirt road, between a barbershop and a clothing store. Ⓝ 5.579425 Ⓦ 0.108690

With Kane Kwei coffins, death need not be a dull affair.

TENGZUG SHRINE

BOLGATANGA, UPPER EAST REGION

In order to visit the Tengzug Shrine, you must show respect by removing your shirt. Then you may walk the path toward the sacred site smeared with the blood of recently sacrificed animals.

The shrine is a place of worship for the Talensi, a northern Ghanaian ethnic group. Animal sacrifice is a significant part of the group's way of life, and as you wander through the village of Tengzug, you'll notice that the mud homes have adjoining shrines covered in chicken blood, feathers, and partial carcasses.

The main Tengzug Shrine at Tongo Hills is a cave atop a pile of boulders. Visitors—male and female alike—must enter topless. Inside are piles of feathers, sacrificial implements, and assorted remains of animals loved and lost. If the sights and smells of the cave are a bit too much to bear, turn your head to the hills—the vista is lovely.

Tengzug is 10 miles (16 km) southeast of Bolgatanga. Ⓝ 10.718635 Ⓦ 0.799999

MALI

ANTOGO FISHING FRENZY

BAMBA, GAO

Fishing is illegal at Lake Antogo, except for one day each year during the dry season. On that day, thousands of men—women are prohibited from participating—surround the small lake, wait for the gunshot that signals "go," and rush into the water, clamoring to seize catfish with their bare hands. After about 15 minutes of splashing and struggling, the fishermen emerge triumphant and mud-covered, carrying reed baskets full of gasping fish. The lake is all but empty and the ritual is over for another year.

The Antogo fishing frenzy is a tradition of the Dogon, a Malian ethnic group who inhabit the country's central Mopti region. Before desertification reduced the lake to its current small size, fishing happened year-round. Now that the fish are few, the annual ritual offers a fair chance for Dogon from all over Mopti to come together and capture food from the lake. Each year, men from multiple villages converge on the lake, located in a desert where temperatures often exceed 120°F (48.8°C).

The captured fish are presented to an elder from the nearby village of Bamba, who distributes them fairly among the gathered throng. The entire ritual, from catching the fish to sharing them, promotes unity among the disparate Dogon villages.

Lake Antogo is 120 miles (193 km) from Timbuktu. The date of the fishing frenzy is determined by village elders and changes every year, but generally occurs during May. Ⓝ 17.033644 Ⓦ 1.399999

Once a year, the world's wildest fishing trip takes place at Lake Antogo.

GREAT MOSQUE OF DJENNÉ

DJENNÉ, MOPTI

In 2014, the UNESCO Heritage–listed Great Mosque of Djenné sustained damage to its walls and had to be repaired by members of the community. But that wasn't cause for concern—it happens every year.

Like hundreds of other buildings in Djenné, the Great Mosque is made of mud. It was built in 1907, but the town's packed-mud architectural style dates back to at least the 14th century. To create the buildings, masons form mud and straw into bricks, allow them to dry in the sun, and stack them to form walls. A layer of mud plastered on top provides a smooth surface and better stability.

Though the buildings are sturdy and often sprawling—the Great Mosque can hold 3,000 worshippers—they are still vulnerable to the elements. Rain, humidity, and temperature changes cause cracks and erosion in the walls. Djenné's mud masons regularly band together and repair the mosque to keep it from falling apart.

Djenné is an 8-hour drive from Bamako, including a ferry ride across the Bani River.
Ⓝ 13.905278 Ⓦ 4.555556

Djenné's adobe architecture includes the largest mud-brick building in the world.

NIGER
LAST TREE OF TÉNÉRÉ

TÉNÉRÉ

For decades, a single, solitary acacia tree stood alone amid the vast desert of Ténéré. It was the only tree for over 250 miles in any direction, and became a landmark for travelers crossing northeast Niger.

In 1939, French military commander Michel Lesourd visited the tree to observe the construction of a well beside it. The acacia's roots were found to extend to a depth of 115 feet (35 m), where the water table began. "One must see the Tree to believe its existence," wrote Lesourd. He described it as a "living lighthouse."

roots, 10 stories deep

The last tree of Ténéré was destroyed in 1973 not by the harsh desert or the unforgiving weather, but by one human: a truck driver. The allegedly drunk Libyan man plowed into the acacia and snapped its trunk.

A simple sculpture of a tree, made from old pipes, fuel barrels, and auto parts, has since replaced the real one in the baked and barren Ténéré region of the Sahara. The fallen acacia still exists: It was taken to the National Museum of Niger in Niamey, where it stands in a fenced-in structure, safe from rogue drivers.

Around 150 miles (241 km) east of Agadez. Ⓝ **16.984709** Ⓔ **8.053214**

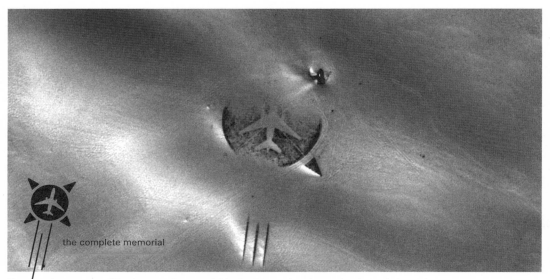

the complete memorial

Few people get to see this plane crash memorial in the Sahara, but those who do won't forget it.

UTA FLIGHT 772 MEMORIAL

TÉNÉRÉ

On September 19, 1989, UTA Flight 772 to Paris had been in the air for just over 45 minutes when a suitcase bomb placed in the hold by Libyan terrorists exploded, destroying the plane and killing all 170 people on board. Wreckage rained onto the Ténéré region of the Sahara desert in Niger, hundreds of miles from the nearest town.

Eighteen years after the crash, relatives of the victims journeyed to

Ténéré to build a memorial. When they arrived, the parts of the plane not removed by crash investigators were still lying strewn in the sand.

Joined by 140 locals from Agadez, the nearest city, the group spent six weeks living in the desert and constructing a monument to the downed plane. Using dark stones trucked in from a site 44 miles (71 km) away, the workers built a 200-foot-wide (61 m) circle and filled it in to create a life-size silhouette of a DC-10 in the sand. A ring of 170 broken mirrors—each one representing someone aboard

the flight—lies at the perimeter of the circle. The crashed plane's starboard wing, brought in from its landing place 10 miles away, stands upright at the northern point of the circle. On the wing is a list of the 170 passengers' names.

The memorial is a tough, multiday drive from the nearest settlement, but it is visible to planes flying above it. Sand is gradually burying the rocks and mirrors.

Ténéré is a region in the south central Sahara, about 262 miles (421.6 km) east of Agadez. Ⓝ **16.864930** Ⓔ **11.953712**

SENEGAL
GREAT GREEN WALL OF AFRICA

SENEGAL TO DJIBOUTI

It began as a wonderfully simple solution to the threat of the expanding Sahara desert: Plant a long green belt of trees across the entire width of the African continent to hold back the encroaching sands. Dubbed the "Great Green Wall," the plan was to grow a giant drought-resistant forest about 5,000 miles long and 10 miles wide across the southern edge of the desert, from Senegal in the east to Djibouti in the west.

The Sahara is currently the second-largest desert in size, smaller only than Antarctica. However, unlike its frozen relative, the Sahara is actually expanding at an alarming rate, threatening the farmlands in the Sahara and Sahel regions, where food security is an increasing problem.

The ambitious Great Green Wall initiative was started in 2007 by 11 African countries in order to fight drought, climate change, and creeping desertification. The name stuck, but the plan itself has evolved considerably over the last decade.

Rather than planting a narrow corridor of trees at the edge of the desert, the vision is now more of a mosaic of green landscapes across the entire arid region. The idea is to bring the land back to life by encouraging the natural regeneration of the drylands with a more grassroots approach that encourages green harvesting practices. The new goal is to regreen 100 million hectares of land by 2030. It's about 15 percent underway so far and still growing. The initiative now includes 21 African countries surrounding the Sahara on all sides.

Conservation and greening efforts in Senegal and Niger have seen the most success so far, restoring 9 million hectares of land.
Ⓝ 14.917014 Ⓦ 5.965215

AFRICAN RENAISSANCE MONUMENT

DAKAR

Atop one of the twin hills in the Mamelles district stands a mighty—and mightily confusing—monument. Sixteen stories tall, the bronze African Renaissance Monument is more than one and a half times the height of the Statue of Liberty. It depicts, in Soviet socialist realist style, a man with a bare, ripped torso holding an infant aloft in one arm and guiding a near-naked woman with the other.

In 2006, then-president of Senegal Abdoulaye Wade began planning a massive hilltop monument that would represent the country's emergence from centuries of slavery and colonialism. To build a budget-friendly monument, he turned to Mansudae Overseas Projects, a division of North Korea's government-run propaganda art factory. The company specializes in constructing huge, Soviet-style statues for cash-strapped nations.

The African Renaissance Monument was inaugurated in 2010 to mark the 50th anniversary of Senegal's independence from France. Unable to afford the $27 million price tag, Wade paid North Korea in the form of state-owned land in Senegal.

When the monument was unveiled, Wade was nearing the end of a 12-year presidency marred by alleged corruption, vote rigging, and self-serving changes to the constitution. His claim that intellectual property laws entitled him to 35 percent of revenue from tourism at the monument was met with understandable ire from fed-up Senegalese.

Despite this controversy, the monument still stands, surrounded by half-built houses and piles of litter.
Avenue Cheikh Anta Diop, Dakar.
Ⓝ 14.722094 Ⓦ 17.494981

One and a half times taller than the Statue of Liberty, Senegal's much-derided sculpture was designed by a North Korean propaganda art factory.

LAKE RETBA

RUFISQUE

The pink waters and white shores of Lake Retba are deceptively inviting. Also known as Lac Rose, the lake has a salt content that rivals that of the Dead Sea, which makes for increased human buoyancy, as well as a busy salt industry. Salt collectors arrive daily, covered in shea butter to protect their skin from the harsh salinity, and spend up to seven hours a day collecting the precious mineral from the lake bed.

The lake's pink tint comes from the salt-loving microorganism *Dunaliella salina*, combined with a high mineral concentration. The water constantly changes hues, with the most stunning pink shade appearing during the dry season. In windy weather, and during the short wet season, the lake is less strikingly pink, due to the rain, which dilutes the salinity. With the salt levels reaching upward of 40 percent, Lake Retba can sometimes take a more sinister shade, appearing blood red, a much less comforting place for your imagination to go when gazing at the surreal view.

The lake is 25 miles (40.2 km) north of Senegal's capital, Dakar. In a car, the ride will take you less than one hour. Ⓝ 14.838894 Ⓦ 17.234137

ALSO IN SENEGAL

Fadiouth Shell Island

All the walls on this island are made with shells, as is the cemetery.

Salt collectors strike the riverbed to loosen the salt (above), which then dries in the baking sun (below).

The savvy voodoo shopper's source for chimp paws, desiccated cobras, and dog heads.

TOGO

AKODESSEWA FETISH MARKET

LOMÉ, MARITIME

When troubled by illness, relationship problems, or financial woes, voodoo practitioners in the West African nation of Togo go to the fetish market of Akodessewa. Located in the capital city of Lomé, the market has a row of tables piled high with dog heads, elephant feet, chimpanzee paws, desiccated cobras, and gorilla skulls. These are all fetishes, or talismans: objects infused with the power of the divine that are used to heal and protect.

Togo and neighboring Benin are where voodoo, known locally as vodun, began. Today, about half of Togo's population continues to hold indigenous animist beliefs. The fetish market, which is suffused with the smell of decaying flesh, is a sort of al fresco pharmacy, the perfect place to stock up on ingredients for rituals.

Tourists are welcome to peruse the offerings and visit one of the traditional healers in the huts behind the tables. During one of these consultations, the voodoo priest or priestess will ask you to describe your ailment, then consult with the gods to determine your prescription. Animal parts are ground up with herbs and held to a fire, which produces a black powder. Traditionally, a healer will make three cuts on your chest or back and rub the powder into the wounds. Tourists of a squeamish persuasion can opt to buy a wooden doll or just apply the powder to unbroken skin.

There are no set prices for the remedies; healers toss cowry shells to ask the gods what you ought to pay. If the price seems exorbitant, you are welcome to say so. The healer will keep consulting with the gods until you reach a mutually agreeable fee.

The market is located in the suburb of Akodessewa, just east of Lomé's airport. Ⓝ 6.137778 Ⓔ 1.212500

Central Africa

CENTRAL AFRICAN REPUBLIC

Dzanga Bai

BAYANGA, SANGHA-MBAÉRÉ

Ordinarily, forest elephants are an elusive bunch. Smaller than African bush elephants and now numbering under 100,000 due to poachers and deforestation, they travel the forests of the Congo Basin in small groups.

There is one place, however, where you can see 100 forest elephants at once: Dzanga Bai, a protected clearing surrounded by the dense forests of Dzanga-Ndoki National Park. Each day, the elephants stroll into the reserve along with forest buffalo, antelope, and wild boar. An elevated platform allows visitors to watch the wildlife on parade from a prime vantage point.

It's a peaceful scene, but not long ago, Dzanga Bai was a site of terrible violence. In May 2013, poachers stormed the clearing and killed 26 elephants for their ivory. The fight against poaching is ongoing and fraught with complications—forest elephant tusks are denser than the tusks of bush elephants, making them more valuable.

Following the 2013 poaching attack, Dzanga Bai reopened to visitors in July 2014. Researchers there are conducting an Elephant Listening Project, which studies how elephants use low-frequency sounds to communicate. They have been listening to the Dzanga Bai elephants since 1990.

The clearing is northwest of the village of Bayanga. It is accessible via a 40-minute walk through dense forest, on a path created by elephants. Ⓝ 2.950584 Ⓔ 16.367569

CHAD

Aloba Arch

ENNEDI

In the Ennedi Plateau of northeast Chad is a sight rarely seen outside China and the southwest United States: a natural arch of monumental proportions.

A natural arch is exposed stone in which a hole forms due to erosion or lava flow, leaving a frame shape in the rock. Aloba Arch spans about 250 feet (76 m), which is impressive in itself. But what really ups the drama factor is its height: The arch is located around 394 feet (120 m) above ground, which is about as high as the top floor on a 32-story building.

Natural arches spanning over 200 feet are rare. Of the 19 catalogued by the Natural Arch and Bridge Society, nine are in China and nine are in the Colorado Plateau in the southwest US, leaving Aloba as an awe-inspiring anomaly. (The Ennedi Plateau contains many other natural stone arches, but most of them have spans in the tens of feet, rather than the hundreds.)

The arches are several days' drive northeast from Chad's capital of N'Djamena. A 4-wheel drive is a must for navigating the sands. Ⓝ 16.742404 Ⓔ 22.239354

A towering arch in the middle of the desert.

DEMOCRATIC REPUBLIC OF THE CONGO

GBADOLITE

GBADOLITE, NORD-UBANGI

In the early 1960s, Gbadolite was a small village of mud huts. Then came Mobutu. When Mobutu Sese Seko seized control of the Democratic Republic of the Congo in 1965, it was the beginning of a 32-year reign characterized by tyranny against the people and extravagance for himself.

During his three decades of ruling the nation—which he renamed Zaire in 1971—Mobutu shut down trade unions, tortured dissidents, staged public executions, and embezzled billions of dollars. One of the self-serving ways in which he spent this money was to establish and maintain a luxury residence just outside of Gbadolite that became known as the "Versailles of the Jungle."

Powered by a hydroelectric dam Mobutu established in 1989, the newly lavish area featured well-appointed homes, schools, and hospitals, a five-star hotel, a Coca-Cola factory, and three large palaces, one of which was made up of Chinese pagodas. The Gbadolite airport had a VIP terminal accented with gold, and its runway could accommodate pricey supersonic Concordes. Foreign dignitaries would land here, then be whisked off in Mercedes-Benzes to Mobutu's private palace. There, they could take a dip in one of two swimming pools, lounge on Rococo furniture, and dine on sumptuous meals featuring gourmet ingredients flown in from France.

Following Mobutu's ousting in 1997, and his death from prostate cancer just a few months later, the shiniest parts of Gbadolite began to dull. Mobutu's hundreds of staff, including chauffeurs, chefs, and servants, stopped going to his palace. Weeds popped up among the marble, stained glass, and gold.

The gates to Mobutu's palace are still intact, but many of its roofs are gone. The steel beams of the ceiling form an eerie skeleton of the once-lavish residence. A makeshift school operates out of one of the Gbadolite buildings.

Some wall murals of Mobutu, which depict him in his signature leopard-skin hat, have survived the decay. His smiling face lives on in Gbadolite. **Gbadolite is about 8 miles (13 km) south of the Central African Republic border. Former workers at Mobutu's mansions, as well as their children, give tours of the ruins. ⓝ 4.283333 ⓔ 21.016667**

The grandeur of the "Versailles of the Jungle" has faded since the death of its despotic founder, Mobutu Sese Seko.

REPUBLIC OF THE CONGO

LAKE TELE

LIKOUALA

The Mokèlé-mbèmbé, Congo's version of the Loch Ness Monster, is said to live in this lake. Surrounded by tropical swampland, the circular body of water is difficult to access, which only fuels the stories about the apatosaurus-like creature supposedly lurking in its depths.

The cryptid has inspired many a proof-seeking pilgrimage to the lake, including a spate of trips during the 1980s led by American, British, Dutch, Japanese, and Congolese adventurers. In 1981, American engineer Herman Regusters claimed to have seen Mokèlé-mbèmbé during a two-week expedition, but none of the local men accompanying him claimed the same.

Monster-seeking trips like these have yielded little beyond muffled sound recordings and blurry photos of distant, indiscernible objects. But for cryptozoologists, the lack of resolution just fueled more imaginative speculation about what Mokèlé-mbèmbé gets up to over there. **Lake Tele is in the deepest jungle, surrounded by swamp forests teeming with gorillas, elephants, and swarms of bees. You'll need to fly into Impfondo (where you'll need to get a lake-visit permit from the ministry office), then drive to Matoko, get a riverboat to Mboua, and trek for 30 miles (48 km) through the wilderness. ⓝ 1.346967 ⓔ 17.154360**

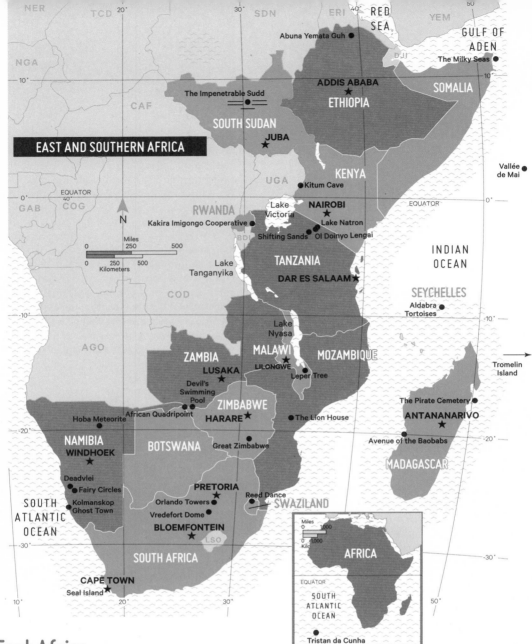

EAST AND SOUTHERN AFRICA

East Africa

ETHIOPIA

ABUNA YEMATA GUH

EAST TIGRAY

If church can be an enlightening experience, imagine going to worship in a rough-hewn painted cave atop a towering sandstone pinnacle, only reachable via a daredevil climb with 650-foot (198 m) drops on all sides.

At Abuna Yemata Guh, this risky and thrilling experience is common practice for a few dedicated priests. The monolithic place of worship is reachable only by a 45-minute ascent on foot. There are cliff faces to scale, rickety bridges to cross, and narrow ledges to traverse. After navigating the valley beneath the church, you must ascend the half-mile-high sandstone pinnacle, searching for rare footholds to avoid the long drop. Adding to the general sense of dread,

the route passes by an open-air tomb filled with the skeletal remains of deceased priests (although it's said that none of the priests died from falling off the cliff).

If the intense climb and the gorgeous view of the valley below aren't enough to take your breath away, the interior of the church surely will. The cave's ceiling is covered with two beautiful frescoes, featuring intricate patterns, religious imagery, and the faces of nine of the twelve apostles of Christ. The church also contains an Orthodox Bible with vibrant, colorful sheets made of goatskin. Abuna Yemata Guh is so sacred that some Ethiopian parents even risk bringing their babies all the way to the top of the cliff to have them baptized in the church.

The church is a 3.5-hour drive from Axum. The last portion of the climb, a vertical rock wall, must be done barefoot. Hire a guide to show you where to put your feet—and whatever you do, don't look down.
Ⓝ 13.915330 Ⓔ 39.345254

An Ethiopian Orthodox priest stands on a narrow path leading to the church.

The White Nile's vast swamps are the largest in the world.

SOUTH SUDAN

THE IMPENETRABLE SUDD

JONGLEI

The particularly squelchy chunk of land surrounding much of the White Nile in South Sudan constitutes the world's largest swamp. The Sudd, which means "barrier" in Arabic, is a vast wetland clogged with papyrus plants, water grasses, hyacinths, and other plants that have clumped together to create a giant impenetrable stew of greenery.

The size of the Sudd varies according to whether it's wet or dry season, but can expand to cover 50,000 square miles (130,000 km²)—roughly the size of the entire state of Louisiana. Swamp villages have been built on some of the floating islands of vegetation, which can measure up to 18 miles (29 km) across. Hippos and crocodiles hang out in the Sudd's shallower waters, while over 400 species of birds pay a visit during migration season.

The entire Sudd acts as a massive sponge, soaking up rainfall and the water that flows in from Lake Victoria in neighboring Uganda. This is bad news for boats, which carry saws to cut through the reeds and grasses. There have been proposals to establish channels that would cut a swath through the Sudd and allow for easier passage, but such an operation would disturb the ecosystem and displace its human inhabitants.

The Sudd spreads across the states of Unity and Jonglei. ⓔ 8.380439 Ⓝ 31.712002

KENYA

KITUM CAVE

MOUNT ELGON NATIONAL PARK, WESTERN PROVINCE

Kitum Cave extends 600 feet (183 m) into an extinct volcano. Its salty walls are lined with scratches, troughs, and pits, which resemble the scars left by miners searching for diamonds or gold. But these carvings are not the work of humans—Kitum Cave's walls have been carved out by elephants.

The pachyderms that roam the park eat a diet of forest vegetation low in sodium. To get their fill of salt, elephants come to the cave and run their tusks along the walls, dislodging chunks of rock that they then crush and lick.

Elephants aren't the only animals that hang out in Kitum Cave. Buffaloes, antelope, leopards, and hyenas are known to prowl its depths. Beyond the inherent danger these animals pose, you'll need to be especially careful of the Egyptian fruit bats that swarm around the entrance. It is thought that the bats are vectors for Marburg hemorrhagic fever, an Ebola-like virus that two cave visitors contracted during the 1980s.

Salt-licking elephants are constantly renovating the interiors of Kitum Cave.

Both travelers—a 15-year-old boy and a 56-year-old man—died within days from the disease, for which there is still no vaccine. **The road to Mount Elgon leads from the town of Kitale, where you can get a minibus to the park. Ⓝ 1.133333 ⓔ 34.583333**

ALSO IN KENYA

Maasai Ostrich Farm

Kajiado · Take a ride on the world's largest bird—then eat its meat.

Gedi Ruins

Malindi · The remains of a mysterious town sit here surrounded by a tropical forest overlooking the Indian Ocean.

Marafa Depression

Malindi · The intricate folds and ridges of these sandstone gorges reveal whites, pinks, oranges, and reds in the rock.

TANZANIA

Ol Doinyo Lengai

NGOBRON, ARUSHA

Along the East African Ridge in Tanzania emerges a volcano known as Ol Doinyo Lengai, translating from Maasai to "Mountain of God." The name signifies a volcano of immense power, but the defining feature of Ol Doinyo Lengai is something else entirely.

This volcano, sprouting over 10,000 feet (3,000 m) tall in the middle of a plain, is the only active volcano on Earth that spews carbonatite lava instead of silica. This unusual composition makes the lava relatively cool: a mere 950°F (510°C), almost half the temperature of silicate magma. Carbonatite lava is black and gray in daylight, instead of the quintessential glowing red. When the lava solidifies, it turns white. And since the magma can cool and harden in seconds, it sometimes shatters in midair and sprinkles down the slopes.

Aside from the unusual temperature and color, the lava that pours from Ol Doinyo Lengai is also known to be the least viscous—or most "runny"—in the world. The lava doesn't just spout from the main crater either; little peaks on the volcano's surface called "hornitos" spew it too.

Ol Doinyo Lengai is a favorite among scientists, as it's easier to study than its dangerously hot counterparts—in fact, it's sometimes called a "toy volcano."

You can climb to the crater of the volcano with a guide. But it's a very taxing hike best reserved for the fit and well-equipped. Treks begin around midnight and return 9 to 10 hours later. ⑤ 2.763494 ⓔ 35.914419

A volcano frosted in white lava.

Flamingos are among the few animals that can handle the hot, intensely salty waters of Lake Natron.

LAKE NATRON

MONDULI, ARUSHA

Visit Lake Natron during breeding season and you'll find it teeming with flamingos, their cotton-candy feathers contrasting beautifully with the mountainous backdrop. But this is no paradise: with a pH of 10.5—compared with ammonia's 11.6—the lake is caustic enough to burn your skin.

The shallow lake's high alkalinity comes from sodium compounds—primarily sodium carbonate—that flow in from the mountains. The water is hot—up to 140°F (60°C)—and often tinted a rusty red due to the presence of pigmented cyanobacteria.

The harsh conditions keep most animals away, but those that stay provide spectacular sights. A 2-million-strong flock of flamingos comes to Lake Natron every year to feed on algae and breed. The caustic conditions are perfect for warding off predators seeking to disturb the flamingo nests. On the downside, Lake Natron is the only regular breeding ground for East Africa's lesser flamingos, meaning that any environmental threats to the lake would have a severe impact on the species.

Proposed lakeside power plants and sodium carbonate processing facilities have thus far not materialized—a state of affairs that conservationists would like to preserve indefinitely.

Camp at the edge of the lake to see the flamingos at sunrise. Ⓢ 2.416667 Ⓔ 36.045844

320-foot-long dune

moves 50 feet
per year

MAGNETIC SHIFTING SANDS

OLDUVAI GORGE, GREAT RIFT VALLEY

As the site of many early hominid fossil finds, Olduvai Gorge on the eastern Serengeti plains is widely regarded as the cradle of humanity. Right near it is a majestic pile of ash.

The ash—which originates from the Ol Doinyo Lengai volcano—is arranged in a crescent-shaped dune measuring roughly 320 feet (98 m) long. Wind is constantly reshaping the mound as its iron-rich magnetic grains fight back by clinging together. The result is a dune that is creeping across the desert at a rate of up to 50 feet (15 m) per year. **Ngorongoro Conservation Area. To appreciate the magnetic nature of the ash, grab a handful from the dune and throw it in the air—the grains find their way back to each other. ⓢ 2.920776 ⓔ 35.390521**

RWANDA

KAKIRA IMIGONGO COOPERATIVE

NYAKARAMBI, EASTERN PROVINCE

In Rwanda, cattle are an important status symbol, and if you venture into the rural areas of the country you're almost guaranteed to cross paths with these gloriously long-horned denizens. You may even find yourself scraping their dung off your shoes. Or, if you're lucky, hanging it on your living room wall.

For generations, cow dung has been used in an art form found only in Rwanda: imigongo. Legend has it that in the 18th century, Prince Kakira was the first to use cow manure, mixed with ash and clay for color, to decorate the interior walls of his house. The practice took the name of the prince's domain, Gisaka-Imigongo.

Today's geometric imigongo artwork is generally painted on portable wooden panels, plates, or wall hangings. The vibrant paintings are composed of colors sourced from natural materials—rusty red from the natural soil, white from kaolin, or white clay, and black from the ashes of banana peels. There are several options for visiting an imigongo workshop, the most well-known being the Kakira Imigongo Cooperative. Here you can watch artisans making imigongo and even step inside a dizzying hut decorated in traditional imigongo style. Continue down the road between Nyakarambi and Rusumo and you'll see several other workshops along the way. **Nyakarambi is about 3.5 hours southeast of Kigali, the country's capital, by bus. ⓢ 2.271910 ⓔ 30.696677**

Contemporary works in a medium that is many hundreds of years old.

SOMALIA
THE MILKY SEAS
INDIAN OCEAN

Tales of a great glowing ocean had long been told within the sailing world, but the scientific community largely ignored them. Such bioluminescence, they reasoned, would require an impossible concentration of bacteria. Then a scientist named Steven Miller decided to do some further investigating.

In 2005, while searching for recorded accounts of the phenomenon, he came across the 1995 logs of the SS *Lima*, which recorded crossing milky seas 150 nautical miles east of Somalia. While it was presumed that no area would be large enough, or have a sustained enough glow, to be captured by satellite imagery, Miller, with the help of Steve Haddock, acquired archival data from the US Defense Meteorological Satellite Program for the night that the *Lima* recorded the phenomenon. When they overlaid it with the coordinates recorded by the *Lima*, they suddenly saw it shining up at them: a huge bright area off the horn of Africa.

Believed to be caused by the bioluminescent bacteria *Vibrio harveyi*, the glowing area is over 6,000 square miles (15,400 sq km) and is visible by satellite. It remains unclear how such a large congregation of bacteria can exist.

A few hundred milky seas have been documented around the world since 1915. The majority

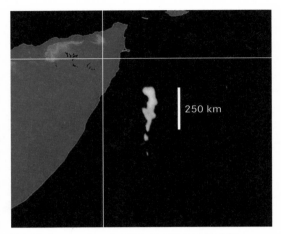
Satellite images show the bioluminescence at night.

show up in the northwest Indian Ocean, usually during summer monsoons. Their glow can last between several hours and several days at a time and is only visible at night.

If you have the time, budget, and inclination to venture 150 nautical miles from Somalia into the Indian Ocean, in the dark, you may encounter the milky seas. You'll just have to hope the stars—as well as enough bacteria—align. Ⓝ 11.350908 Ⓔ 51.240151

Southern Africa

BOTSWANA
AFRICAN QUADRIPOINT
CHOBE

There are several places where the borders of three nations touch, but the rare confluence of four nations in one spot exists only in Africa. Here, the corners of Zambia, Zimbabwe, Botswana, and Namibia meet.

The African quadripoint sits in the middle of the Zambezi River, which cuts between the countries. Quibblers point out that this is not a true quadripoint but instead a pair of tri-points separated by thin strips of real estate. Regardless, the jurisdictional headache of having four countries so close to one another has resulted in some conflict.

At one point the ferry that carried people across the river from Namibia to Botswana became a point of contention, with both countries laying claim to the transport. A small fight broke out. Thank goodness it never escalated into a quadriputal war.

Zambezi River, an hour's drive west of Victoria Falls. The Kazungula Ferry travels across the river from Zambia to Botswana, via Zimbabwe. It's a 3-minute trip. Ⓢ 17.791100 Ⓔ 25.263334

MALAWI

LEPER TREE

LIWONDE, MACHINGA

 A baobab tree in quiet, wildlife-filled Liwonde National Park bears a small, hand-painted sign that challenges the apparent tranquility of the scene. The sign reads: "The grave for people who suffered from leprosy in the past."

In Malawi, as in many places around the world, people with leprosy—also known as Hansen's disease—have often been ostracized. Though the bacteria that causes the disfiguring disease can be contagious, most of the world's population is naturally immune. Nevertheless, in accordance with traditional Malawian religious beliefs, those who died of leprosy could not be buried, as it was believed that they would contaminate the earth.

A look inside the Liwonde baobab, which is cut open on one side, shows the legacy of such beliefs. At the bottom of the hollow trunk lies a tangle of human skeletons. These were people with leprosy who got tied together and bundled into the baobab. It is unclear whether they were alive or dead at the time of their internment.

Liwonde National Park. The most prominent landmark in the park's southern area is Chinguni Hill, which is encircled by a dirt road. The tree is near the base of the hill.
Ⓢ 15.030231 Ⓔ 35.247495

MOZAMBIQUE

THE LION HOUSE

GORONGOSA

The Gorongosa region, located in Great Rift Valley, is one of the most biodiverse places on the planet, and efforts have been made to preserve its unique ecosystems since the 1920s. By 1940, the wildlife reserve was such a popular attraction that a safari lodge was built to accommodate tourists. Unfortunately, it was built right on a floodplain.

Every year during the rainy season, the Mussicadzi River would flood the building, and the lodge was abandoned before it was even fully completed. Then the lions moved in. In 1960, Gorongosa became a national park, and today the magnificent animals and their unexpected residence are a famous landmark at the park.

For generations lions have been coming back to the house. Some of the estimated 50 to 70 living on the reserve use it as a safe haven for their cubs and a place to take shelter from the rain. The animals are known to ascend to the roof of the building for a better vantage point from which to watch their prey for hours, carefully choosing the best warthog or bushbuck to target for their next meal. And of course they use the house to do what cats do best: lie around and nap. Being at the top of the food chain has its advantages.

Fly from Maputo to Beira (a 3-hour drive from the park) or Johannesburg to Chimoio (a 2-hour drive). The park has well-appointed villas and bungalows for lodging, as well as a campsite for the more intrepid. Ⓢ 18.690239 Ⓔ 34.313589

It may look abandoned, but this shelter is a clubhouse of sorts for local lions.

A former diamond mining settlement is being claimed by the desert.

NAMIBIA

KOLMANSKOP GHOST TOWN

LÜDERITZ, KARAS

During the 1920s, the desert town of Kolmanskop was inhabited by hundreds of German diamond miners, and boasted a hospital, theater, casino, bowling alley, and gym. Now Kolmanskop is a ghost town, bleached by the sun and swallowed by sand.

The town was built in 1908 following the discovery of diamonds in the area. At the time, Kolmanskop lay in German South-West Africa, a colony that lasted from 1884 until 1915. Diamond mining operations were fast and furious until World War I caused a slump in sales. In 1926, a richer source of diamonds was discovered south of Kolmanskop. The town struggled on for a few more decades before being abandoned for good in 1954.

The homes of Kolmanskop still stand, but their floors are piled with sand, which has begun to reclaim the buildings. Some of the mini dunes are marked with wavy lines—the tracks of snakes that occasionally slither through the sandy halls.

Tours of Kolmanskop leave from the Insell Street boat yard in Luderitz, a harbor town 15 minutes away.
Ⓢ 26.705325 Ⓔ 15.229747

DEADVLEI

NAMIB-NAUKLUFT PARK

Rising from a parched white clay bed surrounded by soaring red sand dunes, the scorched branches of Deadvlei's trees twist toward a cloudless sky. Though long leafless and dead, the trees have yet to decompose—the climate is too dry to permit decay. Instead they stay stuck in the cracked and thirsty clay, skeletons of their former selves.

It's difficult to picture now, but a river once flowed into Deadvlei. Its trees bore flowers and provided canopies of shade. Then came the drought, around 1100 CE, which robbed the clay pan of its nourishing water. Sand dunes swept around the edges, blocking the flow of the river and leaving Deadvlei bare.

A few hardy plants have survived the stark terrain. Patches of salsola, a succulent saltbush, and nara, a bumpy-skinned melon, cling to the sand. Except for the camera-toting visitors, they're the only signs of life in a cinematic barren land.

Tours to Sossusvlei, the clay pan where Deadvlei lies, leave from Windhoek. It's a 6-hour drive and you can stay overnight. Ⓢ 24.760666 Ⓔ 15.293373

Seven-hundred-year-old scorched trees emerge from the salt pan, surrounded by the world's tallest sand dunes.

THE HOBA METEORITE

GROOTFONTEIN, OTJOZONDJUPA

The largest known meteorite on earth lies on a farm near the town of Grootfontein, on the very spot where it landed around 80,000 years ago. One of the reasons it remains in situ instead of sitting behind glass at a museum is its weight—at over 60 tons, it's roughly as heavy as a US Army tank.

Found in 1920, the Hoba meteorite measures 9.8 feet long by 9.8 feet (3 × 3 m) wide, and stands 3.3 feet high. Though a sizable crater must have been created at the moment of impact, 80,000 years' worth of erosion has erased the dent in the sand.

The meteorite's iron-and-nickel surfaces bear the marks of vandals—visitors to Hoba have been known to chisel off a piece of the space rock to keep as a souvenir. The addition of a surrounding amphitheater and anti-vandalism measures during the 1980s seem to have quelled attempts to take chips off the old block, but you can still touch the meteorite as much as you like.

The meteorite is about 16 miles (25.7 km) west of the town of Grootfontein, along the D2859 road. Ⓢ 19.588257 Ⓔ 17.933578

The largest meteorite ever found crashed to Earth around 80,000 years ago.

FAIRY CIRCLES OF SOUTHERN AFRICA

NAMIBIA DESERT

Small, circular patches of bare ground form like bald spots in the otherwise thick grass fields of Namibia and South Africa, in places where humans rarely set foot. The mysterious origin of these so-called fairy circles have bewildered scientists and spurred local legends about supernatural footprints and UFO landings.

Unlike their distant cousins, the crop circles of Britain, the fairy circles of southern Africa don't suddenly appear. In fact, they grow and shrink over time as though they were alive. Scientific research has established that the bald patches occur naturally; however, examination of the soil in and around the circles reveals no abnormal insects or parasites, no unusual fungi to speak of. In 2017 a study by a team of ecologists from Princeton University posited that the patterns are the result of termites establishing circular borders against competing colonies. Other experts have disputed this, however. All science has discovered conclusively thus far is that the fairy circles are not a hoax.

One spot to see the circles is NamibRand Nature Reserve, which is a 5-hour drive south of Windhoek, Namibia's capital.
Ⓢ 25.018789 Ⓔ 16.016973

Some call these mysterious circles the footprints of the gods.

caption tk

SOUTH AFRICA

ORLANDO TOWERS

JOHANNESBURG, GAUTENG

Orlando Towers, a self-professed "vertical adventure facility," caters to a niche crowd: those who have always wanted to bungee jump from a bridge suspended between the cooling towers of a decommissioned power plant, 33 stories from the ground.

From 1951 until 1998, the towers handled excess heat at Orlando Power Station. Following the closure of the plant, a rope access specialist named Bob Woods turned the towers into an extreme-sports venue. You can now rappel down the side of one of the towers, rock climb up the other, freefall off the rim into a giant net, or bungee off the bridge. Experienced BASE jumpers can even jump off one of the towers with no safety equipment, provided they bring their own parachute and sign a comprehensive legal waiver.

Since 2002 the once-drab towers have been covered in color. One is painted with murals representing South Africa, while the other is devoted to advertising space. The towers have become a landmark of Soweto, the ramshackle township that gained infamy as a ghetto for black residents of Johannesburg during apartheid.

Dynamo Street at Old Potch Road, Orlando, Soweto.
Ⓢ 26.253394 Ⓔ 27.927189

Bungee jumping from a brightly painted former cooling tower, anyone?

SEAL ISLAND

CAPE TOWN, WESTERN CAPE

In False Bay—named in the 17th century by disgruntled sailors who mistook it for the adjacent Table Bay—lies a rocky strip of land overrun with seals. Around 60,000 of the barking, flopping, waddling brown fur seals jostle for space on the half-mile-long, 164-foot-wide (50 m) Seal Island.

From a tour boat, the cacophonous jostling is a compelling sight, but it's the smell—a unique bouquet of rotting fish and excrement—that makes the boldest impression. This, however, is before you view the effects of the "ring of death," a circle of great white sharks that surrounds the island waiting for seals to enter the water. When they do, the sharks pursue them relentlessly.

It's not unusual to see a great white suddenly breach the surface and hurtle into the air with a seal struggling in its jaws. The sharks are fast, brutal, and capable of chomping seals to death in under a minute. It's a violent sight, and you can't help but root for the seal. Sometimes, after a thrilling battle, they get away.

To the sharks surrounding the island, the herds of noisy, smelly seals constitute a tasty buffet.

Seal Island cruises leave from Hout Bay. The trip takes about 45 minutes. Ⓢ 34.137241 Ⓔ 18.582491

VREDEFORT DOME

VREDEFORT, FREE STATE

When entering the small village of Vredefort, it may not immediately be apparent that you are at the center of one of the planet's most violent moments. Spread out from the town in a 185-mile (298 km) diameter are the rippling remnants of one of the oldest and largest known meteorite impact craters still visible on Earth. Though it can only be fully viewed from space, and 2 billion years of erosion have flattened the crater, the rippling rings around Vredefort speak to the sheer impact of the event when a 6-mile (9.7 km) wide mountain-size object slammed into the earth at over 25,000 miles per hour. **The crater is 75 miles (121 km) southwest of Johannesburg. Guided tours of the dome leave from Parys, a town north of the crater.** Ⓢ **26.997842** Ⓔ **27.360668**

The oldest impact crater can be seen from space.

SWAZILAND

THE REED DANCE

LUDZIDZINI, HHOHHO

For a week every year, in late August or early September, tens of thousands of girls and young women storm Ludzidzini, the village of the royal family in Swaziland. Barefoot, bare-breasted, and adorned with brightly colored skirts, beaded necklaces, and pompoms, they parade in front of the king and his family, singing and dancing while holding machetes.

This annual ritual, known as the Umhlanga, or Reed Dance, is held primarily to celebrate the chastity of the girls and young women involved. Virginity is a prerequisite for participation, which reflects both a traditional social value in Swaziland and modern concerns about HIV transmission in a country where one in four people is living with the virus.

The ceremony begins with the girls being separated into groups according to their age. Each group then journeys to a patch of wetlands, where the girls cut and bundle reeds using their machetes. Over the next few days, the girls travel back and forth from the wetlands to the queen mother's palace, bringing back bundles of reeds intended to patch holes in the fence surrounding the palace.

After a day of rest and preparation, the girls return to the palace in bright sashes, skirts, and jewelry. The king and his family sit and watch as wave after wave of girls sing and dance before them. The public are also invited to attend these two days of celebrations, though photography is prohibited.

In addition to promoting the traditional social values of female virginity and cooperative labor, the ceremony has a practical purpose: The King of Swaziland, Mswati III, has often used the parade to scout for wives. His thirteenth wife, Inkhosikati LaNkambule, and his fourteenth wife, Sindiswa Dlamini, were both plucked from Reed Dances. **Ludzidzini Royal Village is between the capital city of Mbabane and Manzini. Exact dates of the Reed Dance differ each year, as they depend on astrology.** Ⓢ **26.460652** Ⓔ **31.205313**

Young women prepare to dance for the Swazi king. One may become his next wife.

ZAMBIA

The Devil's Swimming Pool

LIVINGSTONE, SOUTHERN PROVINCE

The small pool with the terrifying view at the edge of Victoria Falls takes your average swimming experience and adds a hefty splash of anxiety. Situated just off tiny Livingstone Island on the Zambian side, the "Devil's Pool" invites you to sit at the top of the world's largest sheet of falling water in relative safety. A naturally formed rock barrier separates you from the roaring falls and keeps the pool's current weak, preventing you from getting swept over the edge and into the 355-foot-tall (108 m) waterfall.

Tour guides lead swims to the Devil's Pool from Livingstone Island when the water level is low enough—usually between August and January. ⓢ 17.924353 ⓔ 25.856810

Take a daring dip in the world's highest infinity pool, formed naturally by rocks.

The largest ancient structure south of the Sahara was part of a city that may have been home to 18,000 people.

ZIMBABWE

Great Zimbabwe

MASVINGO, MASVINGO PROVINCE

The soaring stone walls that snake around the perimeter of what was once Great Zimbabwe provide an indication of the grandeur of the former city. Built by the Bantu people between the 11th and 15th centuries, Great Zimbabwe consisted of three parts: an ellipse-shaped Great Enclosure with 36-foot (11 m) walls, a citadel on a hill, and a scattering of stone dwellings in a valley.

Before it became overpopulated in the 15th century, leading to its abandonment, Great Zimbabwe was a thriving medieval trading center. Archaeological excavations, which began during the early 20th century, have unearthed glass and porcelain from China and Persia, as well as gold and coins from Kilwa, an island off Tanzania.

Excavations also uncovered eight carved soapstone birds, known as the Zimbabwe Birds. These are now Zimbabwe's national symbol, appearing on its flag, coat of arms, and banknotes.

Great Zimbabwe is about 17 miles (27 km) south of Masvingo, which itself is a 4-hour drive from Harare, the capital of Zimbabwe. ⓢ 20.266667 ⓔ 30.933333

At sunrise the dark baobabs provide contrast with the shifting pastel hues of the sky.

Islands of the Indian and South Atlantic Oceans

MADAGASCAR

AVENUE OF THE BAOBABS

MORONDAVA, MENABE

Along a stretch of dirt road that leads from Morondava to Belon'i Tsiribihina stand rows of baobab trees, their stout trunks glowing and fading as the sun passes overhead. This is the Avenue of the Baobabs, one of the more striking spots for appreciating the *Adansonia grandidieri*, one of seven baobab species endemic to Madagascar.

Hundreds of years old and standing up to 98 feet tall (30 m), the baobabs look like trees that have been uprooted and replanted upside down. Their branches, which only sprout from the very top of the trunk, are adorned with flat clusters of leaves that catch the light at sunset. Dusk and dawn are the best times to visit.

The baobabs are 45 minutes north of Morondava. The roads are best navigated with 4-wheel drive.
Ⓢ 20.250763 Ⓔ 44.418343

....

THE PIRATE CEMETERY

ÎLE SAINTE-MARIE, ANALANJIROFO

Here lies a dastardly fellow.

During the 17th and 18th centuries, the small island of Île Sainte-Marie near the East Indies trade route was the off-season home of an estimated 1,000 pirates. Buccaneers from all over the world lived in wooden huts adorned with flags that signified which captain's "crew" they belonged to.

When pirates died, they were buried in a scenic, palm-shaded hilltop cemetery overlooking the water. Legend has it that the notorious William Kidd is buried in a large black tomb in the cemetery, sitting upright as punishment for his dastardly deeds. He was actually buried in England, but his legendary ship, the *Adventure Galley* (rediscovered in 2000), was left docked near the island, and his booty is said to be buried somewhere in the surrounding sea. The crumbling cemetery, its graves half-covered by tall, swaying grass, is open to the public.

The cemetery is on the small island of Île Sainte-Marie, 4 miles (6.4 km) off the coast of eastern Madagascar. Hour-long flights to the island depart from the capital city of Antananarivo. Ⓢ 16.894317 Ⓔ 49.905893

TROMELIN ISLAND

Located almost 300 miles (483 km) east of Madagascar and measuring just 1.1 by 0.43 miles, Tromelin Island would be a mere splat of sand if not for its incredible, awful history.

In 1761, the French cargo ship *l'Utile* set sail from Madagascar bound for Mauritius—then known as Île de France. On board were approximately 160 slaves the crew had picked up in Madagascar, with the intent of selling them in Mauritius. But while en route, the ship ran into a reef. The stern shattered, sending the ship, 20 of its crew, and about 70 of the slaves to a watery grave. Many of the slaves who drowned were trapped belowdecks, with the hatches closed and, in some cases, nailed down.

The survivors struggled to Tromelin Island, a sandy, treeless, wind-battered speck equipped with nothing but turtles, seabirds, and coral. Some of the sailors managed to dig a well, but the lack of food and shelter made conditions intolerable. By salvaging wreckage from the ship, the survivors were able to build a seaworthy raft—one big enough to hold all the gentlemen and sailors, but not the slaves.

The white survivors climbed aboard the raft and set sail for Mauritius, promising to return for the remaining slaves. The rescue party never came—the governor of Mauritius balked at the idea of risking white lives to save what he considered human cargo.

Fifteen years after *l'Utile* was shipwrecked, a French warship captain named Bernard de Tromelin visited the island that now bears his name. There he discovered seven of the slaves from *l'Utile*—all women—and one baby boy who had been born on the island. The group had spent the previous decade and a half surviving on turtles and shellfish and sheltering in houses made from coral bricks. Little is known about the ultimate fates of these survivors, but it seems they were granted the right to live as free people on Mauritius.

A team of French archaeologists performed excavations on Tromelin Island in 2006 and unearthed cooking utensils made from copper salvaged from *l'Utile*, as well as remnants of the coral houses and evidence of a communal oven. The island is now home to a weather station that monitors winds and detects cyclones in the Indian Ocean.

There are no commercial flights to the island, but if you can find your own plane, you can land on the island's unpaved airstrip. Another option is to get there by boat, though there are no harbors—you'll need to anchor offshore. Accommodation is nonexistent. Bring a tent, a sleeping bag, and a lot of provisions. Ⓢ 15.892222 Ⓔ 54.524720

SEYCHELLES
VALLÉE DE MAI

PRASLIN

Praslin, the second largest of the Seychelles's 155 islands, is a paradise of secluded beaches, dense vegetation, and clear, tranquil waters. At the center of the island is Vallée de Mai, a forest home to a most unusual palm tree: the coco de mer. Endemic only to the Seychelles, the coco de mer produces the largest seeds of any plant in the world. Weighing in at up to 60 pounds (27 kg), the seeds are also remarkable for their distinctive shape. Simply put, they resemble a curvaceous woman's buttocks.

In 1881, devoutly Christian British general Charles Gordon visited Praslin and caught sight

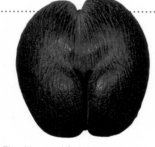

The 60-pound fruit known as the "love nut."

of the scandalous seeds. So struck was he by their sensuous shape that he concluded the coco de mer was the tree of knowledge, and its seed the original forbidden fruit. That meant, of course, that Vallé de Mai was the Garden of Eden.

Though that geography may sound a bit wonky, at least according to the Book of Genesis, the coco de mer seeds are a suggestive sight worthy of the "forbidden fruit" label. A mythology has arisen around the coco de mer's supposedly aphrodisiac properties—the seed is regarded as a fertility symbol, which accounts for its popularity as a souvenir. In 2012, the Seychelles government imposed a ban on the export of coco de mer kernels in response to illegal poaching and trade. To take a seed home, you'll need to buy it from a licensed dealer and obtain an export permit. Expect to pay top dollar.

If you're content to just see the seeds, a hike through Vallée de Mai will expose you to plenty. You may also spot black parrots, a threatened species found only on the island of Praslin.

The park is open between 8 a.m. and 5:30 p.m. Ⓢ 4.331692 Ⓔ 55.740093

An Aldabra giant tortoise out for a stroll.

THE TORTOISES OF ALDABRA

ALDABRA ATOLL, OUTER ISLANDS

The Galápagos Islands may be the land most associated with giant tortoises, but there's another archipelago where these terrapins roam: Aldabra, a group of four islands located 435 miles (700 km) east of Tanzania.

Aldabra is the second-largest coral atoll on Earth. Over 100,000 giant tortoises live there—the largest population in the world. The tortoises, which can grow to 550 pounds (250 kg), share their remote home with green sea turtles, coconut crabs, hammerhead sharks, manta rays, and oceanic flamingos. A small research station houses a few scientists, but that's the extent of the human presence.

The Seychelles Islands Foundation oversees access to Aldabra. Trips are limited to those engaged in nature tourism and education, and some areas of the atoll are off-limits. To get there, you need to get a chartered plane to Assumption Island followed by a chartered boat or private yacht to Aldabra. ⑤ 9.416681 ⑥ 46.416650

....

SAINT HELENA, ASCENSION, AND TRISTAN DA CUNHA

TRISTAN DA CUNHA

An 8-mile-wide (13 km) part of a British Overseas Territory, Tristan da Cunha is the most remote populated island in the world. The nearest mainland city, located 1,743 miles (2,805 km) east, is Cape Town in South Africa. The journey from Cape Town takes seven days by boat—traveling by air is not an option, as there is no airport on the island.

Every inhabitant of Tristan da Cunha—269, at last count—lives in the island's only settlement, Edinburgh of the Seven Seas. Established in the early 19th century, the village is located on the north coast and is home to 70 families, all of whom are farmers. Electricity is supplied by diesel generators. The island's lone road, a narrow, winding path, is flanked by bungalow-style cottages, potato patches, and roaming cows. The looming volcanic cliffs and low-lying mist create a secluded, hazy setting.

It's a peaceful, pared-back existence with few anxieties—unless the volcano erupts. Such was the case in 1961, when earthquakes, landslides, and an eruption from one of the north vents sent the entire population fleeing to England via Cape Town. (Put off by England's busy streets and savage winters, most returned two years later after getting the all-clear from geologists.)

Now that the volcano has calmed down, life on Tristan da Cunha is an exercise in patience and planning. There is a grocery store, but orders must be placed months in advance so the goods can be loaded onto scheduled fishing vessels and delivered. A hospital equipped with X-ray machines, a labor and delivery room, operating room, emergency room, and dental facilities takes care of most health concerns, but patients requiring more specialized treatment must be transported to South Africa or the UK.

Boats leave from Cape Town. You'll need to get approval from island administrators before visiting the island. ⑤ 37.105249 ⑩ 12.277684

Oceania

Australia
New Zealand
Pacific Islands

**FIJI · GUAM · MARSHALL ISLANDS · MICRONESIA
NAURU · PALAU · PAPUA NEW GUINEA · SAMOA
VANUATU**

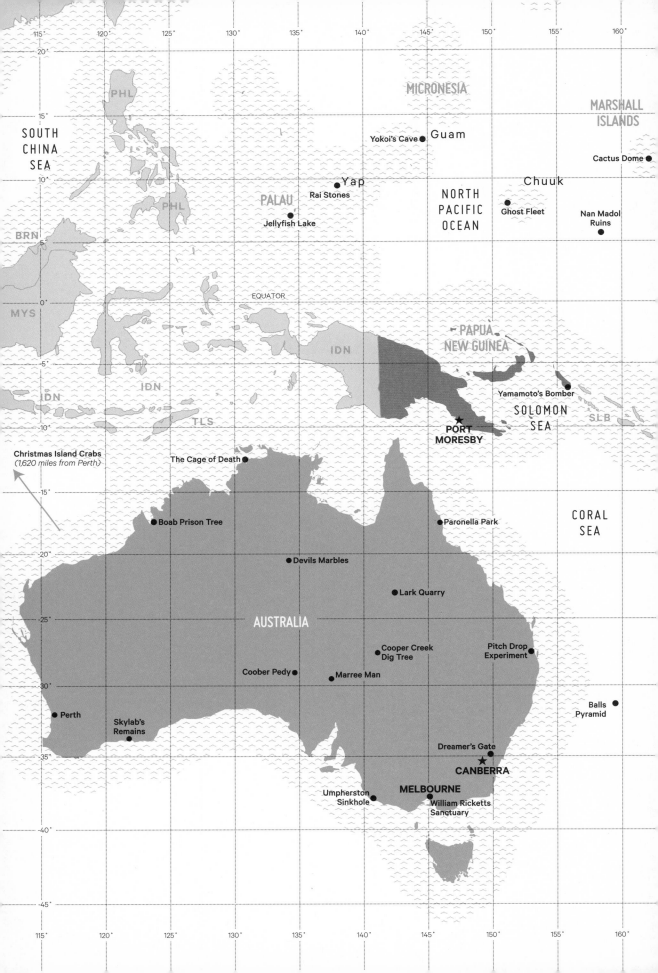

MICRONESIA

MARSHALL ISLANDS

SOUTH CHINA SEA

PHL

PHL

BRN

MYS

IDN

IDN

IDN

TLS

Yokoi's Cave ● ● Guam

● Yap

PALAU

Rai Stones ●

Jellyfish Lake ●

NORTH PACIFIC OCEAN

Chuuk

Ghost Fleet ●

Cactus Dome ●

Nan Madol Ruins ●

EQUATOR

PAPUA NEW GUINEA

Yamamoto's Bomber ●

SOLOMON SEA

SLB

★ PORT MORESBY

Christmas Island Crabs
(1,620 miles from Perth)

The Cage of Death ●

CORAL SEA

● Boab Prison Tree

● Paronella Park

● Devils Marbles

● Lark Quarry

AUSTRALIA

Cooper Creek ●
Dig Tree

Pitch Drop ●
Experiment

Coober Pedy ●

● Marree Man

● Perth

Skylab's ●
Remains

Balls ●
Pyramid

Dreamer's Gate ●

★ CANBERRA

MELBOURNE

Umpherston ●
Sinkhole

William Ricketts
Sanctuary

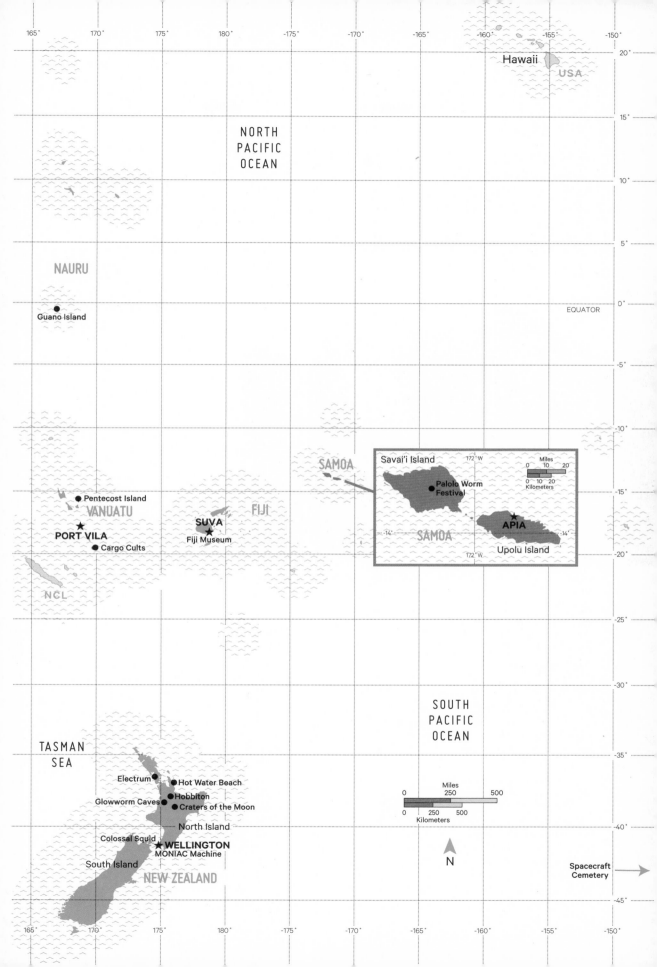

20°

15°

NORTH
PACIFIC
OCEAN

Hawaii   USA

10°

5°

NAURU

0°   EQUATOR

● Guano Island

-5°

-10°

SAMOA

Savai'i Island   172° W   Miles
                                              0   10   20

-15°   ● Palolo Worm
                                          Festival   0 10 20
                                                              Kilometers

● Pentecost Island
VANUATU   FIJI   -14°   -14°
                                                  SAMOA   ★ APIA
★ PORT VILA   SUVA ★
                      Fiji Museum   172° W   Upolu Island
● Cargo Cults

-20°

NCL

-25°

-30°

TASMAN
SEA   SOUTH
                                            PACIFIC
                                            OCEAN

-35°

Electrum ●   ● Hot Water Beach
Glowworm Caves ● ● Hobbiton   Miles
               ● Craters of the Moon   0   250   500
North Island
                                          0   250   500
Colossal Squid ●   Kilometers
★ WELLINGTON   -40°
MONIAC Machine

South Island   N

NEW ZEALAND   Spacecraft →
                                                         Cemetery

-45°

AUSTRALIA

DEVILS MARBLES

WAUCHOPE, NORTHERN TERRITORY

The Devils Marbles, known as Karlu Karlu ("big boulders") to local Aboriginal groups, consist of hundreds of rocks scattered across the landscape, varying in diameter from 1.5 feet to 20 feet (.45–6.1 m). Some boulders are stacked and appear to be balancing precariously atop one another.

According to Aboriginal mythology, the boulders are the work of Arrange, the Devil Man, who walked through the valley twisting a hair-string belt. As he twisted, clumps of hair fell to the ground, becoming rocks. On his walk back, Arrange spit on the ground, forming more boulders.

In 1953, a marble was transported to the town of Alice Springs and used in a memorial for John Flynn, the founder of a mobile medical service for the outback. The move was controversial—the site is sacred to local Aboriginal groups—and after more than 40 years, the rock was returned to its original location. In 2008, the government also gave back possession of the 7-square-mile (18 km²) Devils Marbles Conservation Area to its Aboriginal owners. Today, Aboriginal communities and local government work together to manage the site.

Aboriginal myths attribute Wauchope's split, stacked, and balanced boulders to the Devil Man.

Wauchope, Northern Territory. The Devils Marbles are a long journey on the Stuart Highway from either Alice Springs (242 miles/389.5 km to the south) or Darwin (679 miles/1,092.7 km to the north). The nearest small town is Wauchope, a good place to stock up on supplies. Out of respect for indigenous residents, don't climb the boulders. ⓢ 20.566667 ⓔ 134.266667

THE CAGE OF DEATH AT CROCOSAURUS COVE

DARWIN, NORTHERN TERRITORY

While you are being lowered into the first of Crocosaurus Cove's saltwater crocodile enclosures, you may notice claw marks on the walls of the acrylic cage in which you're standing. Try not to let that bother you—there are 1.5 inches (3.9 cm) of protective plastic between you and the 18-foot-long (5.5 m) lethal reptile.

Though this urban wildlife park houses other reptiles and fish, the main attraction is its collection of saltwater crocodiles. Compared with alligators, the saltwater species is larger, faster, and more inclined to attack humans. "Salties," as they're known locally, are abundant in the rivers and estuaries of northern Australia, where swimming is strongly discouraged.

Crocosaurus Cove offers a rare chance to splash alongside a saltie in the Cage of Death, a transparent cylinder that is lowered into the crocodile enclosures. Though no humans have ever been injured, the same can't be said for the crocodiles. In 2010, a crocodile named Burt—the 80-year-old star of *Crocodile Dundee*—lost a front tooth when he lunged at two football players who were taunting him from the cage.

In 2011, a cable holding the Cage of Death snapped and dropped a pair of tourists into the tank with an enormous crocodile named Choppa. Luckily for the couple, Choppa ignored them and they were quickly rescued.

58 Mitchell Street, Darwin. Thirteen bus routes stop along Mitchell Street. Unsurprisingly, you'll need to sign an indemnity release form before stepping into the Cage of Death. ⓢ 12.462333 ⓔ 130.839162 ➻

Despite the name, there's a good chance you'll survive the Cage of Death.

AUSTRALIA'S OTHER DEADLY CREATURES

Australia's reputation as a land of killer animals is a little unfair. Though the country is home to 21 of the world's most venomous snakes, marine creatures whose sting can kill within minutes, and five highly poisonous spiders, most of the continent's dangerous creatures do not seek out people to attack. A bite or sting usually happens in self-defense after a human unwittingly invades an animal's habitat. Here are some creatures—and one plant—to look out for during your travels.

1 BOX JELLYFISH
Named for the cubelike bell from which its lethal tentacles dangle, the box jellyfish is an almost-transparent creature that zooms silently through the water at up to 7 feet (2.1 m) per second. Its tentacles, the longest of which extend to 10 feet (3 m), have tiny harpoons that inject venom when touched. A kiss from a box jellyfish results in a temporary tattoo of its tentacles and up to 8 hours of agonizing pain. In rare cases, stings cause death by cardiac arrest.

TREATMENT: Pour vinegar over the tentacles to disable the stinging mechanism, then remove them from the skin using a towel or gloves.

2 STONEFISH
With their craggy, mottled complexion, stonefish camouflage themselves under mud or sand in the calm shallows of Australia's tropical waters. Accidentally step on one and up to 13 venom-filled dorsal spines will send neurotoxins coursing through your body, resulting in terrible pain, redness, swelling, muscle weakness, and short-term paralysis.

TREATMENT: Immerse limb in hot water to destroy the venom. In more serious cases, seek antivenom.

3 BLUE-RINGED OCTOPUS
This small, unassuming yellow cephalopod covered in neon-blue circles will not hesitate to bite when stepped on. The bite itself is painless, but the venom that spills forth from its salivary glands can cause muscular weakness, temporary paralysis, and respiratory failure. In many cases, the victim remains conscious and alert, but is unable to breathe or move.

TREATMENT: Utilize artificial respiration until the patient can breathe unassisted. There is no antivenom.

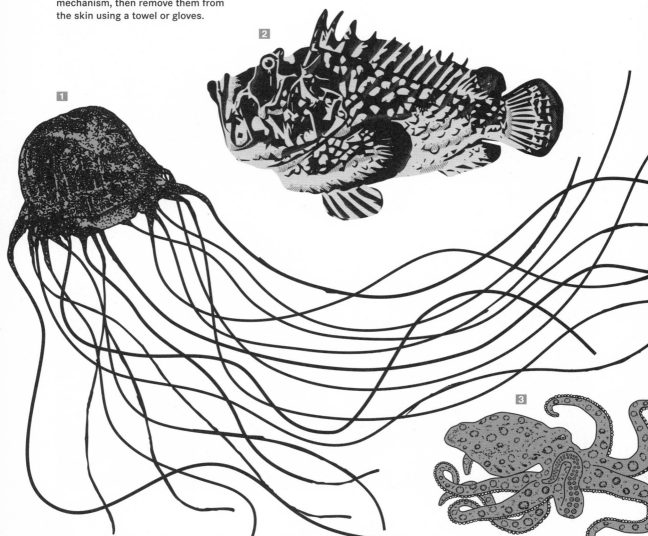

(AND ONE DEADLY PLANT)

4 COASTAL TAIPAN

Native to the seaside regions of northern and eastern Australia, the coastal taipan has a light brown, slim body, grows up to 6 feet (1.8 m) long, and has the largest fangs of any snake in the country at up to half an inch (1.3 cm) long. These sizable teeth can sink into your leg several times before you realize what's happened. When the taipan detects prey, it will freeze, raise its head off the ground, and then lunge and deliver multiple bites.

TREATMENT: The coastal taipan is the world's third-most-venomous snake, capable of disabling the human nervous system and issuing a one-two punch of neurotoxins that weaken the muscles, then prevent the blood from clotting. To avoid bleeding to death or suffering extensive muscle and kidney damage, bite victims should be administered antivenom as soon as possible.

5 SYDNEY FUNNEL-WEB SPIDER

Pervasive funnel-web phobia is an understandable phenomenon in Sydney. Unlike most other deadly animals, these 2-inch (5.1 cm), dark brown spiders have a habit of wandering into backyards, swimming pools, and homes. During summer, males leave their burrows and travel in search of a mate. A bite to an unsuspecting human causes symptoms within 30 minutes, including rapid heart rate, muscle spasms, sweating, tremors, and breathing difficulties. Though death is possible, there have been no recorded fatalities since the introduction of funnel-web antivenom in 1981.

TREATMENT: Get to a hospital for a dose of antivenom. In the meantime, apply pressure and a tight bandage, and immobilize the wound with a splint to prevent the poison from spreading.

6 GYMPIE GYMPIE

A 4-foot-tall (1.2 m) plant found in the forests of Queensland, the gympie gympie catches your eye with its large heart-shaped leaves, juicy pink berries, and dear little flowers. Then it stings you and all the prettiness is subsumed by a world of pain. Tiny hairs on the stalks and leaves dig into your flesh and break off, causing agony, swelling, and redness that can last for months.

TREATMENT: The strange but medically approved way to remove stinging hairs from the skin is to apply a hair-removal wax strip.

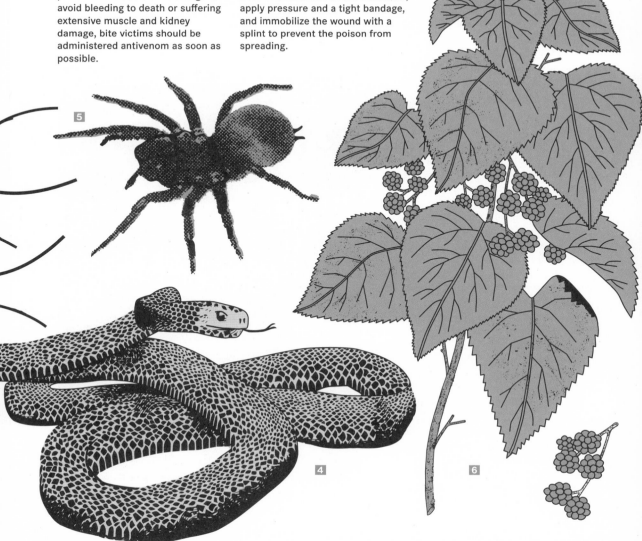

Pitch Drop Experiment

BRISBANE, QUEENSLAND

To view the experiment that the University of Queensland School of Mathematics and Physics boasts is "more exciting than watching grass grow," you'll need to go to the display cabinet in the school's foyer. There, beneath a glass dome, you will see a funnel filled with asphalt. It doesn't seem to be doing anything other than sitting there, but do not be deceived: You are looking at the world's longest running lab experiment in action.

The Pitch Drop Experiment began in 1927, the brainchild of UQ physics professor Thomas Parnell. His aim: to demonstrate that pitch—a term for thick, solid polymers such as asphalt—is not solid, but a very viscous liquid.

To prove this, Parnell poured a heated sample of pitch into a closed funnel and let it settle, a process that took three years. In 1930, he cut off the end of the funnel, allowing the pitch to flow freely. Which it did. Very, very slowly. The first drop fell into the beaker in 1938, with the second and third drops following in 1947 and 1954. With the installation of air-conditioning in the building came a reduction in the flow rate—the eighth drop, which hit the beaker in November 2000, took over 12 years to fall.

Just over 13 years later, in April 2014, the ninth drop fell. The next data point is expected to be collected in 2027.

Physics Annexe, University of Queensland, Brisbane. Get a bus from central Brisbane (Adelaide Street) to UQ (Chancellor's Place). ⓢ 27.497854 ⓔ 153.013286

Paronella Park

MENA CREEK, QUEENSLAND

Since childhood, baker José Paronella had dreamed of building a Moorish castle. In 1913, the adventurous 26-year-old left his village in Catalonia, Spain, and moved to tropical northern Australia. There, he eventually found wealth as a sugar cane farmer, and was finally able to pursue his dream.

In 1929, Paronella purchased a plot of rain forest in Queensland and began building his castle by hand, using sand, clay, old train tracks, gravel from the nearby creek, and wood taken from abandoned houses. By 1935, the structure had expanded to include a pool, café, cinema, and ballroom, as well as tennis courts and villa gardens with a grand staircase—all open to the public.

After Paronella's death in 1948, the building suffered decades of neglect, but thanks to conservation efforts, the castle is alive once again. Lush tropical plants have encroached upon and mingled with Paronella's hand-built stairs and fountains, making them look like they sprouted from their natural surroundings.

1671 Japoonvale Road, Mena Creek. Bus tours leave from Cairns, which is 75 miles (120.7 km) north of Mena Creek. ⓢ 17.671856 ⓔ 145.917067

Fountains and lush gardens are all part of a Catalan baker's self-built fantasyland.

Site of an ill-fated decision that sent two explorers to their doom.

COOPER CREEK DIG TREE

COOPER CREEK, QUEENSLAND

An old eucalyptus tree on the banks of the Bullah Bullah waterhole marks the spot where a cruel twist of fate claimed the lives of two ambitious explorers.

In 1860, Robert Burke and William Wills led an expedition from Melbourne, aiming to travel to the northernmost part of Australia—a journey of around 2,000 miles (3,218.7 km). At the time, the desert terrain, where summer temperatures exceeded 122°F (50°C), had yet to be explored by non-indigenous people. Burke and Wills, neither of whom had any exploration experience, brought 19 men, 23 horses, 26 camels, and 6 wagons stuffed with far more supplies than was sensible. (A Chinese gong and a large oak table were among the more impractical baggage.)

After spending months journeying to Cooper Creek, Burke and Wills split from the rest of the group to make the trip to the northern coast. Those left at Cooper Creek were instructed to wait three months while the two men completed their journey to and from the coast. After holding out for more than four months, the Cooper Creek group buried food and supplies beneath a tree marked with the date and departed.

Nine hours later, Burke and Wills arrived back at Cooper Creek. They encountered the "Dig Tree," excavated the cache of supplies, and found a letter from the other group. Too exhausted to try to catch up with them, the men rested for two days. Before heading south, they buried a letter of their own in the cache, detailing their plans, in case a rescue party should arrive. But Burke and Wills made a crucial mistake: They did not change the date or message originally etched into the tree. When members of their original group returned to Cooper Creek carrying extra supplies, they saw no evidence that Burke and Wills had been there, and departed.

With their rations running low and all their animals dead, Burke and Wills stayed alongside the creek, too weak to carry water. Knowing the end was near, Wills returned to the Dig Tree and buried his diary. An entry on June 21, written a week before he and Burke died, reads thus:

"We have done all we could . . . our deaths will rather be the result of the mismanagement of others than of any rash acts of our own. Had we been come to grief elsewhere, we could only have blamed ourselves, but here we are returned to Cooper Creek, where we had every reason to look for provisions and clothing, and yet we have to die of starvation."

The tree is 23 miles (37 km) northeast of Innamincka, a tiny outback town in South Australia near the Queensland border. Markings on the Dig Tree are still visible. ⓢ 27.617267 ⓔ 141.078583

LARK QUARRY DINOSAUR STAMPEDE

WINTON, QUEENSLAND

Ninety-five million years ago, a herd of 150 dinosaurs—made up of chicken-size coelurosaurs and ostrich-size ornithopods—ran for their lives when a predator (possibly a Muttaburrasaurus) lurched onto the scene with teeth and talons bared. The chaotic exodus, which probably took just a few seconds, left a tangle of fossilized tracks on the ground—tracks which now form the world's only evidence of a dinosaur stampede. A conservation building constructed over the tracks protects them from erosion and damage.

Lark Quarry Conservation Park, Winton-Jundah Road, Winton. ⓢ 23.016100 ⓔ 142.411400

The world's only evidence of a dinosaur stampede.

SKYLAB'S REMAINS

ESPERANCE, WESTERN AUSTRALIA

On July 12, 1979, the United States' unmanned space station, Skylab, began its reentry into the Earth's atmosphere. Things got a little precarious when the space station missed its intended reentry target. The debris was supposed to fall 810 miles south-southeast of Cape Town, South Africa, but a 4 percent calculation error caused the debris to land about 300 miles east of Perth in Western Australia.

For the people of Esperance (current population 10,421), it was quite the event. Strange lights and sonic booms were just the start. NASA officials soon arrived, and locals were incentivized to hand over any debris they found. Local government offices handed out plaques to debris discoverers, and the *San Francisco Examiner* caused a frenzy by offering $10,000 to the first person to arrive at their office with an authentic piece of Skylab. The office, naturally, was in the US, and potential winners had just 72 hours to get there. Stan Thornton, a 17-year-old from Esperance, managed to claim the prize.

At the same time, the local Esperance Museum started to build its collection. Today, it contains all kinds of Skylab artifacts that fell to Earth, including large titanium nitrogen spheres and oxygen tanks, fragments of metal and insulation foam, ruined circuit boards, a portion of the main hatch, and a storage freezer.

Just outside the museum entrance stands a model of Skylab on a pedestal. A nearby billboard states:

IN 1979, A SPACESHIP CRASHED OVER ESPERANCE. WE FINED THEM $400 FOR LITTERING. PAID IN FULL.

While in the Esperance region, check to see if nearby Lake Hillier is pink. Sometimes it has a striking bubblegum hue. Ⓢ 33.858962 Ⓔ 121.893997

BOAB PRISON TREE

DERBY, WESTERN AUSTRALIA

During the 1890s, police used the bulbous, 15-foot-wide (4.6 m) hollow trunk of this 1,500-year-old boab tree as a temporary jail for Aboriginal prisoners en route to sentencing in the small town of Derby. Long before its use as a makeshift prison, the tree was part of an indigenous legend. The story goes that the boab, once tall and prideful, learned humility when the spirits turned it upside down, causing its roots to grow into the sky.

Broome Highway, Derby. Skywest flights from Perth to Derby (a 3-hour trip) depart every weekday. A fence around the tree prevents visitors from going inside it, partly out of respect for its connections to Aboriginal beliefs, but also because snakes like to sleep in the trunk. Ⓢ 17.350738 Ⓔ 123.669919

Rabble-rousers used to be flung into the belly of the Boab Prison Tree.

UMPHERSTON SINKHOLE

MOUNT GAMBIER, SOUTH AUSTRALIA

Mount Gambier is a city built on a porous foundation of eroding limestone caves and craters. In 1864, a gentleman named James Umpherston purchased a plot of land containing a large sinkhole. This was no oversight.

Far from being perturbed by the sinkhole's presence, Umpherston decided to transform the pit into a sunken recreational garden open to all. By 1886, he had filled the hole with a variety of ferns and flowers. Visitors flocked to the new garden, entering it via a set of wooden stairs.

After Umpherston's death in 1900, the garden began to deteriorate into an overgrown garbage dump. By 1976, however, plans were afoot to restore its former beauty. Department of Woods and Forest staff uncovered Umpherston's terraces, and planted new flowers and shrubs. Just like its original incarnation, the garden was an instant hit.

Umpherston Sinkhole continues to flourish, with its blooming pink and lilac hydrangeas and its edges dripping with hanging ivy that conceals the caverns below. Its popularity has extended to local possums, who scamper among the plants at night in search of food.

Jubilee Highway E, Mount Gambier. Mount Gambier is a 6-hour bus ride from Melbourne's Southern Cross station, or just over 6 hours from Adelaide's Central station. ⑤ 37.835267 ⑥ 140.802465

Ivy drapes over the edge of a 150-year-old sinkhole garden.

MARREE MAN

MARREE, SOUTH AUSTRALIA

Charter pilot Trec Smith was flying over the outback toward Coober Pedy in June 1998 when he saw it: a 2.6-mile-tall (4.2 km) drawing of a naked indigenous man, his left arm raised and ready to launch a hunting stick toward unseen prey.

The perfectly proportioned figure, formed by wide lines dug 10 inches (25.4 cm) into the ground, seemed to have been freshly carved. But despite the planning, precision, and sheer boldness required to create it, no one came forward to claim authorship—and apparently no one witnessed its creation.

The situation only got stranger. Anonymous press releases appeared, ostensibly written from an American perspective. They used US units of measurement, referred to local places with awkwardly formal names, and referred to the Native American Great Serpent Mound in Ohio. In June 1999, a fax from the UK revealed that a message had been buried beneath the Marree Man's nose. Authorities surrendered to curiosity and dug it up to discover a plaque that was decorated with an American flag, the Olympic rings, and a quote about Aboriginal hunting from a 1936 book on outback Australia.

The created-by-Americans angle seemed to be a red herring planted by an audacious eccentric. That's where Bardius Goldberg, now considered the most likely culprit, comes in. Goldberg, an artist prone to provocation, had been making Aboriginal-style dot paintings near the desert town of Alice Springs when he got into a dispute with the traditional landowner, Herman Malbunka. Goldberg then allegedly used a borrowed GPS and a tractor to send a spiteful message to Malbunka.

Unfortunately, Goldberg died in 2002 (he developed septicemia after losing a tooth in a bar fight), meaning the mystery of the Marree Man has never been officially solved. Goldberg's other schemes included planting eucalyptus trees in the shape of a giant kangaroo and installing a magically disappearing Virgin Mary in the wall of a house. To those who knew him, he was the only possible culprit.

The lines of the Marree Man have faded due to erosion, but the figure is still visible from the air. Tours to the Marree area, incorporating scenic flights over the Marree Man and nearby Lake Eyre, depart from Adelaide. Ⓢ 29.437780 Ⓔ 137.468077

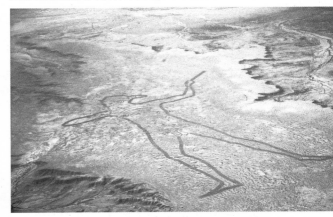

The 2.6-mile-tall figure seems to have been created as a hoax by an eccentric artist.

ALSO IN AUSTRALIA

Litchfield Termite Mounds

Adelaide River · What look like craggy tombstones are actually houses built by ants.

Sun Pictures

Broome · The world's oldest outdoor cinema still screens multiple films per night.

Tessellated Pavement

Eaglehawk Neck · This plateau came into being when pressure at the Earth's crust caused cracks to appear in the rock at perpendicular angles, creating a tiled effect.

Eden Killer Whale Museum

Eden · See the full skeleton of Old Tom, a 22-foot-long (6.7 m) orca who once cooperated with hunters to herd fellow whales.

Hamelin Pool Stromatolites

Gascoyne · The rock-like formations in the shallow waters of Hamelin Pool Marine Nature Reserve are not rocks, but stromatolites: living, growing organisms that provide a rare glimpse of life on Earth as it was 3.5 billion years ago.

Wunderkammer

Melbourne · This downtown boutique sells artifacts and ephemera from the natural and scientific worlds.

Lake Hillier

Middle Island · The water in this lake is a startling bubblegum pink.

Undara Lava Tubes

Mount Surprise · White cockroaches, caramel-colored pseudoscorpions, and eyeless silverfish are among the rare insects and arachnids that lurk in these caves.

Twelve Apostles

Port Campbell · The "apostles"—which now number seven—are actually a line of limestone stacks off the coast of a national park.

Regent Street Station

Sydney · Groups of mourners with picnics in tow once assembled at this Gothic station to get the train to Rookwood Cemetery.

Museum of Human Disease

Sydney · Australia's only publicly accessible pathology museum offers a grim reminder of the consequences of an unhealthy lifestyle.

Fort Denison

Sydney Harbor · This tiny island, once home to the occasional banished starving prisoner, is also known as "Pinchgut."

Burning Mountain

Wingen · Mount Wingen has been smoldering for thousands of years due to a slow-moving, burning coal seam.

Newnes Glowworm Tunnel

Wolgan Valley · The path of a onetime railway tunnel is now home to thousands of glowing insects.

COOBER PEDY

COOBER PEDY, SOUTH AUSTRALIA

For most residents of Coober Pedy, opening a window to get some fresh air isn't an option. More than half of the desert town's 2,000 people live underground in houses excavated from rock, a lifestyle that protects them from the extreme dry heat.

Coober Pedy was established in 1915, after a 14-year-old named Willie Hutchison discovered opals in the ground while looking for water. The town has since become the world's largest producer of the white gemstones, which form when silica and water combine in

Living underground helps Coober Pedy residents stay cool during the brutal summers.

the cracks of sedimentary or igneous rock.

The cave-dwelling residents attend underground church services, browse underground art galleries, and manage underground hotels. For those recreational pursuits that simply aren't possible in a cave, there are creative modifications—at the town golf course, players begin games during the cool of night, using glowing golf balls and carrying a small piece of turf for teeing off. **The town of Coober Pedy is 526 miles (846.5 km) northwest of Adelaide. Two-hour flights to Coober Pedy on the Rex airline depart from Adelaide. Alternatively, an overnight Greyhound bus takes 11 hours. To avoid the hottest temperatures, visit between April and November. ⓢ 29.013244 ⓔ 134.754482**

DREAMER'S GATE

COLLECTOR, NEW SOUTH WALES

Once a five-inn town that attracted gangs of bushrangers—19th-century, forest-dwelling outlaws who robbed coaches—Collector has since quieted down to a 300-person village with few visitors. At its entrance is a striking sight: a 23-foot-tall, 112-foot-long (7 m, 34 m), sand-colored, Gothic-style gate that seems to have emerged organically from its sun-bleached surroundings.

The Dreamer's Gate is the work of local artist Tony Phantastes, who created it as a tribute to his deceased father. Phantastes began building it in 1993, using a skeleton of galvanized wire and piping, and adding burlap, plaster, chicken wire, and cement. The five-panel design incorporates twisting tree limbs, hands,

a sleeping man's face, and circular windows that show the landscape beyond.

In 1999, the local council issued a stop-work order, claiming the gate was structurally unsound. Phantastes fought back, but lost his case and, in an attempt at compromise, added steel support beams to the back of the gate.

Dreamer's Gate remains intact but unfinished. As the years pass, it becomes a little more bleached and rusted, its broken archway symbolizing Phantastes's incomplete dream.

Church Street, Collector. Collector is a 2.5-hour drive from Sydney. The gate is opposite the Bushranger Hotel, once frequented by Ben Hall's legendary bushranger gang. A monument at the hotel pays tribute to Constable Samuel Nelson, shot dead by an outlaw on that spot in 1865. ⓢ 34.912724 ⓔ 149.436323

The massive, half-built roadside fence is a haunting, modern ruin.

AUSTRALIA'S BIG THINGS

In 1963, a 16-foot-tall (4.9 m) bagpipe-playing Scotsman appeared on the northeast corner of Scotty's Motel in Adelaide. The concrete structure, designed to attract travelers to the roadside lodge, sparked an Australian cultural phenomenon: the big thing.

Families road-tripping along the country's major highways during the 1980s and '90s encountered many an oversized object. Initially erected as promotional tools for stores, museums, and lodgings, the big things became attractions themselves. For many Australians, stopping for a photo in front of the Big Banana was a summer vacation ritual.

There are over a hundred big things in Australia, but most look a little run-down these days. Low-cost airline flights are gradually replacing the great Australian road trip. Low-tech big things now evoke feelings of nostalgia—and a bit of cultural cringe.

Big Mango
39 feet (11.9 m)
Bowen, QLD

Big Boxing Crocodile
26 feet (7.9 m)
Humpty Doo, NT

Big Banana (above)
16 feet (4.9 m)
Coffs Harbour, NSW

Golden Guitar
39 feet (11.9 m)
Tamworth, NSW

Big Merino
49 feet (15 m)
Goulburn, NSW

GIANT KOALA
46 feet (14 m)
Dadswells Bridge, VIC

BIG NED KELLY
20 feet (6.1 m)
Glenrowan, VIC

BIG PRAWN (above)
20 feet (6.1 m)
Ballina, NSW

BIG PINEAPPLE
53 feet (16.2 m)
Woombye, QLD

BIG GALAH
26 feet (7.9 m)
Kimba, SA

Long Beard and Earthly Mother.

WILLIAM RICKETTS SANCTUARY

MOUNT DANDENONG, VICTORIA

The winding mountain paths in the quiet forest of the William Ricketts Sanctuary are lined with 92 ceramic faces that seem to have grown straight out of the rocks and boulders. Each face depicts a real figure in the life of sculptor William Ricketts. Never trained in sculpture, Ricketts grew up surrounded by Aboriginal mythology, in which ancestral beings are believed to have created the land's natural features during what is called "the Dreamtime."

Ricketts began creating sculptures of indigenous elders, adults, and children in the 1930s—a time when Victorian government policy still removed Aboriginal children from their parents to be raised in white homes and institutions. In his depictions, Ricketts often sculpted white men wearing crowns made from bullets, with dead animals at their feet.

Ricketts continued sculpting until his death in 1993 at the age of 94. Opened to the public in the 1960s, his sanctuary offers a place for quiet reflection and the appreciation of nature.

1402-1404 Mt. Dandenong Tourist Rd., Mount Dandenong, Victoria. Get a train from Flinders Street Station in central Melbourne to the Croydon stop. From there, get a bus to the sanctuary. Ⓢ 37.832715 Ⓔ 145.355645

CHRISTMAS ISLAND CRABS

CHRISTMAS ISLAND

Christmas Island, an Australian territory in the Indian Ocean, is populated by 1,500 humans and 100 million crabs.

Each year at the start of the rainy season, the ground is transformed into a vast, churning red carpet as the island's crabs leave their forest burrows and scuttle to shore in order to spawn. During their perilous journey, which can take several weeks, the crabs must cross roads, withstand seaside winds, and steer clear of yellow crazy ants.

The ants, who, true to their name, move frantically and erratically when disturbed, were accidentally introduced to Christmas Island in the 1920s. Since then they have formed super-colonies, wreaking havoc on the island's ecosystem. Red crabs, despite their larger size and exoskeletons, are no match for the jets of lethal acid that swarms of ants spray into the crabs' eyes and mouths.

Concerned by the insects' destructive effect on red crabs and Christmas Island ecology as a whole, the Australian government has established a seven-member Crazy Ant Scientific Advisory Panel. Tactics thus far include a four-year research study into the ants' reliance on honeydew, and the 2009 release of 13 tons of insecticide on super-colony areas. Though this last measure made a significant dent in the crazy ant population, their numbers are rapidly increasing once again.

Christmas Island is a 4-hour flight from Perth. Exact timing of the migration varies depending on the weather and the phases of the moon, but the crabs tend to start moving in November. Ⓢ 10.447525 Ⓔ 105.690449

BALLS PYRAMID

LORD HOWE ISLAND GROUP

Darkness had fallen. Two scientists were perched 330 feet (100.6 m) above the sea on a shard of an old volcano shaped like a giant jagged dagger. It was in this precarious position that they laid eyes on the best possible surprise: a nest of 24 giant stick insects, each the size of a human hand.

David Priddel and Nicholas Carlile embarked on their 2001 trip to Balls Pyramid, a 1,844-foot-tall (562 m) volcanic remnant off the east coast of Australia, motivated by a shaky belief: that the stick insect, long thought to be extinct, was still alive. The insect, also referred to as a "tree lobster" on account of its unusually large size, once roamed the forests of nearby Lord Howe Island. Following the 1918 introduction of black rats to the island—which escaped from a supply ship that ran aground—the insects disappeared, and by 1930 were considered extinct.

Balls Pyramid, located 12 miles (19.31 km) south-east of Lord Howe, is hardly lush with vegetation—it is almost entirely rock, its near-vertical cliff faces inhospitable to fauna and off-limits to mountain climbers without government permission. But beneath a spindly shrub growing from a crack, there they were: two dozen tree lobsters.

No one could figure out how they got there. Were they carried by birds? Did their eggs float across on the sea? At first, the Australian government couldn't decide whether they ought to be moved. But in 2003, a team from the National Parks and Wildlife Service scaled the pyramid and collected two pairs of stick insects for breeding in captivity. One pair died shortly after, but the pair dispatched to the Melbourne Zoo—"Adam and Eve"—met with success, producing eggs that became the foundation of the zoo's now-thousands-strong tree lobster population.

373 miles (600.3 km) northeast of Sydney. Flights link Sydney and Lord Howe Island, taking just under 2 hours. ⑤ 31.754167 ⑥ 159.251667

Home of the tree lobster, long thought to be extinct.

NEW ZEALAND

ELECTRUM

MAKARAU

In addition to being one of New Zealand's wealthiest people, Alan Gibbs can control lightning. His farm is home to a sculpture called *Electrum*, the world's largest Tesla coil.

At night, from his balcony 50 feet (15.2 m) away, Gibbs can flip a switch that sends lightning streaming from the top of the 38-foot (11.6 m) structure, which consists of a simple sphere atop a column. When the coil discharges, it sends up to three million volts into the surrounding air. You can hear an earsplitting crackling sound and feel the hair on your head standing on end. That burning smell is ozone, the result of oxygen molecules breaking up and reforming in new combinations.

Electrum is the artistic realization of Gibbs's long-held fascination with lightning. To create the work, he commissioned late American artist Eric Orr, whose sculptures often combined fire, water, and a sense of high drama. With the help of electrical engineer Greg Leyh and teams in New Zealand and San Francisco, Orr designed and built a cylinder with a hollow sphere (known as a Faraday cage) on top. At its first test demonstration, during which a constant stream of lightning discharged amid a tremendous buzz, Leyh sat protected inside the sphere reading a book—an homage to the famous photograph of Tesla reading in a chair while a coil erupts into a lightning frenzy behind him.

At 38 feet tall, *Electrum* is the world's largest Tesla coil.

Since 1998, the sculpture has been part of Gibbs Farm—a large bayside property an hour north of Auckland with more than 20 large-scale, site-specific art installations. All have been commissioned by Gibbs, who collaborates with artists on the abstract-minimalist works. Adding to the unusual landscape are roaming zebras, giraffes, alpacas, emus, and goats. **Kaipara Coast Highway, Makarau. The farm is a private residence, but its art park is open to the public by appointment. ⑤ 36.616196 ⑤ 174.491259**

HOBBITON

MATAMATA, WAIKATO

In a hole in the ground there lived a hobbit—and that hole is located in Matamata, New Zealand. What remains of the Shire movie set for Peter Jackson's *Lord of the Rings* and *Hobbit* adaptations lies on Alexander Farm, a 2-square-mile (3.2 km²) family property with 13,000 sheep and 300 beef cattle.

Jackson selected the site after an aerial survey of the area revealed its hills were the perfect size and shape for the cozy hobbit houses described in J. R. R. Tolkien's books. To transform the farm into a Hollywood-approved

The little house where Frodo's perilous journey began.

Hobbiton, the film crew built a mill and bridge, brought in foliage, and wired individual fake leaves to a dead tree, all after enlisting the help of the New Zealand Army to pave an access road.

After taking a 1-hour guided tour of Hobbiton, visitors are welcome to bottle-feed a lamb or watch one being shorn. **501 Buckland Rd., Hinuera, Matamata. InterCity buses from Auckland take just over 3 hours to reach Matamata. A free shuttle runs from the bus stop to the farm. ⑤ 37.879794 ⑤ 175.650222 ➤➤**

➡ Epic Film Sets You Can Visit

BIG FISH

Private island in a river near Montgomery, Alabama, United States

The custom-built set of Spectre, the friendly, hidden town in this 2003 Tim Burton film, still sits on its little island. Years of weather have taken their toll, though—the buildings' rustic facades are wearing off to reveal the Styrofoam beneath.

The island is privately owned. Seek permission before visiting.

STAR WARS

Chott el Jerid, southwest of Nefta, Tunisia

The Lars Homestead, Luke Skywalker's Tatooine home, still stands in the Tunisian desert—thanks to some particularly dedicated fans. In 2010, *Star Wars* enthusiast Mark Dermul led a pilgrimage to the set location and was disheartened by the dilapidated state of the Skywalker igloo. After two years of fund-raising and negotiating with the Tunisian government,

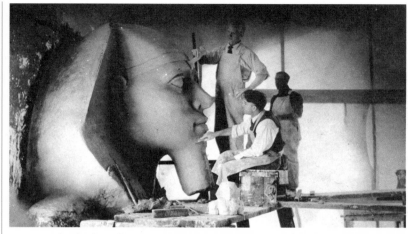

Artists sculpt the face of Rameses II for Cecil B. DeMille's *Ten Commandments*.

Dermul and five friends returned to the desert and repaired the plaster set piece.

THE TEN COMMANDMENTS

Guadalupe, California, United States

For almost 90 years, the Guadalupe-Nipomo Dunes hid a massive bounty of Egyptian "relics." A 720-foot-long (219 m) Egyptian palace, 21 five-ton (4,535.9 kg) sphinx sculptures, and four 30-foot (9.1 m) pharaoh statues were among the set pieces for Cecil B. DeMille's *The Ten Commandments*, which was filmed in 1923. Despite building the largest set ever built at the time, DeMille ordered his workers to dismantle and bury the set after its month-long filming had been completed. The movie set lay hidden until 1983, when Peter Brosnan, a filmmaker and DeMille fan, set out to unearth the secret bounty. Brosnan and his crew found the "Lost City" after following clues featured in DeMille's autobiography, but lacked the funding for a proper excavation. It was not until October 2012 that digging finally commenced. Pieces of the set are still buried in the desert, but several unearthed objects are on display at the Dunes Center on Guadalupe Street.

Dig your own hot tub before the tide comes in.

HOT WATER BEACH

COROMANDEL PENINSULA, WAIKATO

Arrive on Hot Water Beach at the right time, and you'll be able to dig yourself a custom-size hot tub out of the sand, with adjustable water temperature. A small section of the beach sits on an underground geothermal water trough. A little digging between low and high tide releases the warm water, making it possible to create a personal spa.

Finding the temperature sweet spot can be tricky—on some patches of the beach, the underground water is hot enough to scald. The best approach is to fill a bucket with cold seawater and pour it into the hot tub if you start sizzling.

If diligently dammed, hot tubs will last for about four hours before being claimed by the sea as high tide approaches.

Check tide times before you go. Aim to arrive about 2 hours before low tide to claim your spot, start digging, and get the most time out of your hot tub before it washes away. Ⓢ 36.886044 Ⓔ 175.822721

CRATERS OF THE MOON

TAUPO, WAIKATO

Swirling wisps of steam billow from the basins and bubbling mud pools at Craters of the Moon, a geothermal field encircled by a raised wooden walkway.

The area was not always a hotbed of activity. The craters began to appear in the 1950s, when the installation of a nearby geothermal power station caused a reduction in underground water pressure, allowing hot water to come to the surface and escape as steam.

Since then, the land has been in a state of constant shift. New craters—up to 65 feet (19.8 m) deep—form during hydrothermal eruptions, which occur about once per year. Blowholes emitting steam and gas pop up more frequently—so much so that the walkway needs regular rerouting to bypass new vents and avoid scalding visitors.

Karapiti Road. Wairakei. Taupo is a 5-hour bus ride or 45-minute flight from Auckland. The Craters of the Moon site is 3 miles (4.8 km) north of the city center. Ⓢ 38.646667 Ⓔ 176.103753

A geothermal field dotted with steamy blowholes.

Bioluminescent larvae light the way through Waitomo's softly glowing caverns.

WAITOMO GLOWWORM CAVES

WAITOMO, WAIKATO

The tour of Waitomo's glowworm caves ends with a silent boat ride in the dark beneath a dense scattering of blue-tinged stars. Or at least that's what it looks like. The dots of light on the ceiling are actually bioluminescent fungus gnats.

This remarkable sight greeted local Maori chief Tane Tinorau and English surveyor Fred Mace when they explored the Waitomo caves for the first time in 1887. Entering through a stream and paddling on a raft by candlelight, the two men were astonished to discover the beauty of the caves, which formed approximately 30 million years ago. Return visits yielded greater rewards—the pair found an entry point on land and, by 1889, were guiding visitors through the caves for a small fee.

The Waitomo caves contain magnificent natural limestone formations that resemble cathedrals, pipe organs, and twisted columns. But the main attraction is, of course, the glowworms. Found only in New Zealand, *Arachnocampa luminosa* emit a blue-green light during their 6-to-12-month larval stage. The bioluminescence occurs due to chemical reactions in the gnat's excretory organs, and, along with dangling feeding lines, helps attract prey to the silk webs where the larvae live. The hungrier a larva is, the more brightly it glows. These are the gnat's glory days—after emerging from the pupal stage mouthless, they will die of starvation within 100 hours, devoting their short adult life to mating and, if female, laying about a hundred or so eggs.

39 Waitomo Caves Road, Waikato. The nearest major city is Hamilton, a 1-hour drive or bus ride away. Photography is forbidden in the caves and you are asked to remain silent when near the glowworms.
⑤ 38.250961 ⑥ 175.170983

ALSO IN NEW ZEALAND

Baldwin Street

Dunedin · The steepest residential street in the world came about by accident: When drawing up road plans for Dunedin in the mid-1800s, London-based city planners used a standard grid system without regard for the local terrain.

Hundertwasser Public Toilets

Kawakawa · The crooked tiles, glass-bottle windows, and bursts of clashing color that make up the walls of this public restroom are the hallmarks of artist Friedensreich Hundertwasser.

Lake Tekapo

Mackenzie Basin · This turquoise lake gets its hue from "rock flour," a powder made when glaciers grind rocks as they go.

Te Wairoa Buried Village

Rotorua · In 1886 this settlement was buried beneath the ash of an erupting volcano.

Rere Rock Slide

Rere · The Wharekopae River cascades down this slanted 200-foot (61 m) rock formation, allowing daring visitors to slide down it on tires and boogie boards.

Mrs. Chippy Monument

Wellington · A bronze kitty adorns the grave of Harry McNeish, the polar explorer who brought the original feline to the Antarctic.

Bridge to Nowhere

Whanganui · A 130-foot-long (39.6 m) concrete bridge stretches across the untamed greenery of Mangapurua Gorge. Then it simply stops.

MONIAC MACHINE

WELLINGTON

The 1949 creation of crocodile-hunter-turned-economist Bill Phillips, the MONIAC (Monetary National Income Analogue Computer) demonstrates the workings of a national economy, using flowing water to represent the movement of money.

The 6-foot-7-inch (2 m) machine, which Phillips made from spare parts in his landlady's garage while studying at the London School of Economics, consists of transparent plastic tanks mounted on a wooden board, connected by plastic tubing. Each tank

How does a national economy work? Let the MONIAC explain with flowing water.

represents a part of the economy, such as imports, health, and education. Adjusting the flow rate of the water causes changes throughout the system, providing a clear simulation of the system-wide effects of spending, saving, investment, interest rates, and taxation.

Phillips created the MONIAC to demonstrate the ways in which small changes can have complex and far-reaching results within a national economy. Fourteen of the hulking, noisy machines were built following the initial prototype, but the advent of computers during the 1950s soon rendered them obsolete. One of the few surviving models is on display at the Reserve Bank Museum. Another can be found at the Science Museum in London.

Reserve Bank Museum, 2 The Terrace, Wellington. The museum is a 10-minute walk from the Wellington train station.
Ⓢ 41.278997 Ⓔ 174.775217

COLOSSAL SQUID AT TE PAPA MUSEUM

WELLINGTON

Like its cousin the giant squid, the colossal squid is an elusive beast. The cephalopod, which is shorter but heavier than the giant squid, lives 3,000 feet (914.4 m) deep in the pitch-black waters surrounding Antarctica. The first clues of its existence came in the form of two sucker-covered arms found in the belly of a sperm whale. Until 2007, only three complete colossal squid specimens had ever been captured and recorded.

In February of that year, the New Zealand fishing vessel *San Aspiring* was hunting near the Ross Ice Shelf when something tugged at the boat's fishing line with an unusually strong force. The crew pulled up the line to discover a 1,000-pound (453.6 kg) colossal squid attached to an Antarctic toothfish, refusing to let go. Knowing they'd happened upon something special, the fishermen hauled the squid on board, froze it, and plotted a swift course back to New Zealand.

Wellington's Te Papa Museum—a national institution focused on art, history, the natural world, and Maori culture—gratefully received the specimen, preserving it in formalin while staff pondered what to do with it. The squid remained in frozen storage for over a year until scientists thawed it in a specially built tank filled with ice and salt water. Webcams provided a live broadcast of the 60-hour thawing and examination process.

Though initially estimated at 30 feet (9.1 m) long, the squid measured 13 feet 9 inches (4.2 m)—a discrepancy attributed to postmortem tentacle

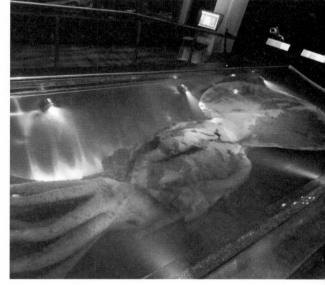

Dredged up from the Antarctic deep, Messie the squid has found a more cozy home.

shrinkage. Nicknamed "Messie," after its scientific name of *Mesonychoteuthis hamiltoni*, it is on display in a horizontal tank, forming the centerpiece of Te Papa's colossal squid exhibit. Here you will learn—or be reminded—that the squids have three hearts, rotating hooks on the ends of their tentacles, and an esophagus that passes through the center of a donut-shaped brain.

55 Cable Street, Wellington. Te Papa Museum is a 20-minute walk, or a 10-minute bus ride, from Wellington Station. Ⓢ 41.290502 Ⓔ 174.781737

SPACECRAFT CEMETERY

PACIFIC OCEAN

As ships, stations, and other satellites come crashing down to Earth, many end up making planet-fall at the same patch in the Pacific Ocean, around 3,000 miles east of New Zealand and 2,000 miles north of Antarctica. For years, these downed science vessels have simply sunk down to the bottom of the sea in a place now known as the Spacecraft Cemetery.

While a great deal of debris and smaller satellites burn up on reentry, larger items—including entire space stations—need to be disposed of in a way that keeps the hazardous materials out of public circulation. So into the dark depths of the ocean they go. Among the more than 250 craft that have been scuttled at the spot since 1971 are unmanned satellites, waste freighters carrying astronaut poop, and, possibly most remarkably, the entire decommissioned Russian space station, Mir.

Technically, the spacecraft cemetery surrounds the Oceanic Pole of Inaccessibility, which is the place on Earth farthest from any landmass. That makes it quite tricky to get to. (The ocean is also about 2.5 miles deep.) This is probably one place best appreciated from afar.
Ⓢ 43.579692 Ⓦ 142.720088

Deep in the Pacific Ocean lies the technology of outer space.

PACIFIC ISLANDS

FIJI

FIJI MUSEUM

SUVA, FIJI

Reverend Thomas Baker's grave mistake was allegedly attempting to remove a comb from the hair of a Fijian chief. For this breach of etiquette in 1867, the Methodist missionary became the last man to be killed, boiled, and eaten by villagers of Nabutautau. After that incident, Fiji abandoned cannibalism.

The sole of Baker's boiled shoe, as well as the specially designated "cannibal forks" used to eat him, are on display at the Fiji Museum. Other items in the collection, including a 44-foot-long (13.4 m), double-hulled war canoe and a copy of the 1874 deed of cession that granted the British ownership of the archipelago, piece together a portrait of Fiji's indigenous and colonial past.

At a tribal ceremony in 2003, Nabutautau villagers formally apologized to Reverend Baker's descendants for eating him, claiming the village had been experiencing "bad luck" since the event.
Thurston Gardens, Cakobau Road, Suva, Viti Levu.
Ⓢ 18.149635 Ⓔ 178.425746

A cannibal fork was used to keep human flesh from touching the lips of a chief.

GUAM
YOKOI'S CAVE

TALOFOFO

The accepted date for the end of World War II is August 14, 1945, even if Japan did not formally surrender until September 2. What some people don't know, however, is that for many Japanese soldiers the war ended much later.

An official count of 127 so-called holdouts or stragglers surrendered in various places in the Pacific Area between 1947 and 1974. This number does not include the many who died in their hiding places, only discovered decades later.

For these holdout soldiers, strong militaristic principles made surrender impossible: It was better to die or be captured than surrender. In some cases, they did not even know about the end of the war. Some of the holdouts continued fighting the American troops or, later, the police, while others just went into hiding. The stragglers believed it impossible to return to Japan, as they feared they would be treated as deserters and punished with the death penalty.

One of those stragglers was Shoichi Yokoi, a tailor by trade who was conscripted to the Japanese Army in 1941. Yokoi was part of the Japanese forces on Guam when the American troops under General Douglas MacArthur conquered the island in summer 1944. US forces advanced quickly, and while many Japanese soldiers were captured or killed, Yokoi and nine other men retreated deep into the jungle.

The men quickly realized that such a big group would be easily discovered. Seven of them left; what happened to them is unknown. The three remaining men, Yokoi included, split up to different hiding places in the area, but kept visiting each other. It took Yokoi three months to dig his "cave," not far from the Talofofo

A prisonlike underground cave served as a hiding place for Yokoi years after World War II.

Falls, about 7 feet (2.1 m) underground. Supported by large bamboo canes, the small underground room was about 3 feet high and 9 feet long, with a small hidden entrance and a second opening to provide an air supply. Inside, he hid all day and stored his few belongings. Yokoi only left his cave at night, lived on caught fish, frogs, snakes, or rats, and learned to use the unknown fruits and vegetables he found. Two of his biggest treasures were a self-made eel trap and a self-made loom, with which he made clothes from the fibers of hibiscus bark.

The three men heard around 1952 that the war was over. They were not sure if the information was true and feared for their lives if they were captured or surrendered, so they decided to stay in hiding. Around 1964, when Yokoi wanted to visit the other two men, he found them dead and buried them. He believed that they died of starvation. Other sources say they died in a flood.

Finally, in 1972, two local fishermen discovered Yokoi on the banks of the Talofofo River, and when, afraid for his life, he charged them, they captured him. He begged the two men to kill him. Instead they took him home, fed him his first real meal in 28 years, and brought him to the authorities. Two weeks later Yokoi returned to Japan and was welcomed as a hero. He felt differently. His famous words were: "It is with much embarrassment, but I have returned."

After Yokoi's death at age 82, the original cave was protected as a historical monument, but it collapsed. In its place, a replica of the cave was erected, along with a shrine and memorials for the last three Japanese stragglers. Some of Yokoi's belongings from his time in the cave can be seen in a museum at the entrance of the Talofofo Falls Resort Park. **From the entrance of the Talofofo Falls Resort Park, you can get a funicular to Talofofo Falls. From there it is about a quarter-mile walk to the cave. A monorail will also take you there. Ⓝ 13.322826 Ⓔ 144.736497**

MARSHALL ISLANDS

CACTUS DOME

ENEWETAK ATOLL

With its ring of verdant islands surrounding a deep sapphire lagoon, the Enewetak coral atoll was a beautiful place to launch the world's first hydrogen bomb. After capturing the atoll from Japan during World War II, the US evacuated the islands, exhumed their fallen soldiers to send home their remains for reburial, and conducted a series of nuclear tests.

Between 1948 and 1958, 43 weapons exploded over Enewetak. Among these was Ivy Mike, a hydrogen bomb 500 times bigger than Hiroshima's Little Boy. By the time testing ceased, the entire atoll was highly radioactive, its reefs and islands dotted with craters that each measured several hundred feet in diameter.

When evacuated residents began returning to Enewetak during the 1970s, the US government determined it ought to decontaminate the islands. In 1979, a military team arrived and gathered up contaminated soil and debris, mixing it with cement and piling the sludge into a 350-foot-wide (107 m) blast crater on Runit Island to the atoll's east. When the mound reached 25 feet (7.6 m) high, army engineers covered it with a saucer-shaped concrete cap. It was dubbed the Cactus Dome, after the Cactus bomb that caused the crater.

The US declared Enewetak safe for habitation in 1980. Currently, about 900 people live on the atoll, though none live on the Cactus Dome. A 2008 field survey of the dome noted that 219 of its 357 concrete panels contain defects such as cracks, chips, and vegetation taking root in joints.

You can charter a small plane from Majuro, the capital of the Marshall Islands, 90 minutes away. Prepare for a bumpy landing—the Enewetak airstrip is not well maintained. Ⓝ 11.552593 Ⓔ 162.333333

After decades of nuclear testing on the Marshall Islands, the US government covered its radioactive sins with a concrete dome.

FEDERATED STATES OF MICRONESIA

NAN MADOL RUINS

MICRONESIA

Off the coast of a remote Micronesian island lie the ruins of a once great city of man-made stone islands. These ruins represent the remains of megalithic

Man-made islets dot the coast of Pohnpei.

architecture on an unparalleled scale in Micronesia.

The construction of the nearly 100 artificial islets started around the 8th or 9th century, and the megalithic structures were built in the 13th to 17th century. This is about the same time as the stone construction of the Cathedral of Notre-Dame in Paris and Angkor Wat in Cambodia. The population of the city was probably more than 1,000. Most of the islets served as residential areas; however, some of them served a special purpose, such as food preparation, coconut oil production, or canoe construction.

There are no sources of fresh water or possibilities to grow food on Nan Madol, so all supplies had to be brought in from the mainland. According to local legend, the stones used in the construction of Nan Madol were flown to the location by means of black magic. Archaeologists have located several possible quarry sites on the main island, but the exact method of transportation of construction material has not been determined.

The complex of Nan Madol is constructed on a series of artificial islets in the shallow water next to the eastern shore of Pohnpei island. An island-hopper flight or sailboat will get you to Pohnpei. Ⓝ 6.840464 Ⓔ 158.331699

GHOST FLEET OF TRUK LAGOON

CHUUK

In February 1944, still reeling from the bombing of Pearl Harbor, the American military targeted a Japanese military base at what was then known as Truk Lagoon. Japan had converted the atoll into a major naval and logistical hub, building roads, trenches, and communications on its islands and stationing battleships, submarines, aircraft carriers, and other giant vessels in the waters.

The Project Hailstone mission began at sunrise on February 17. Five hundred aircraft departed from the nearby Marshall Islands, joining submarines and surface ships in the attack on Truk. Though the Japanese, fearing such a raid, had removed many of their larger ships from the area a week earlier, the damage was extensive.

Forty-seven ships and 270 aircraft were sent to the bottom of the lagoon. About 1,700 Japanese servicemen went down with them.

Truk's sunken vessels remain at the bottom of the lagoon, comprising the world's largest ship graveyard. Riddled with torpedo holes, the ships have released some of their contents. Gas masks, rotting shoes, unopened bottles of beer, and phonograph records drift silently along the coral-covered decks, a sobering reminder of the crew members' daily lives.

In addition to being the final resting place for warships, Truk Lagoon is a mass war grave for those who perished in the attack. In the 1980s, divers retrieved the remains of approximately 400 Japanese crew members. The bones were taken to a Japanese air base and cremated, and the ashes interred at the National Cemetery for the War Dead in Tokyo. The remains of the other 1,300 servicemen are scattered in the lagoon.

Truk's ghost fleet attracts much marine wildlife, including sharks, manta rays, and turtles, as well as scuba divers. The ships continue to rust and deteriorate, causing ecological concerns—the three tankers on the lagoon floor contain about 32,000 tons (29 million kg) of oil, or about three-quarters of the amount spilled during the Exxon-Valdez disaster. **Truk Lagoon is also known as Chuuk Lagoon. Flights from Guam arrive at Chuuk airport, on Weno Island in the middle of the lagoon. The trip takes about 90 minutes. Ⓝ 7.416667 Ⓔ 151.783333**

The world's largest ship graveyard holds 47 vessels, 270 aircraft, and the remains of about 1,300 Japanese servicemen.

RAI STONES

YAP

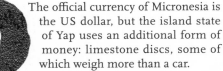

The official currency of Micronesia is the US dollar, but the island state of Yap uses an additional form of money: limestone discs, some of which weigh more than a car.

Hundreds of years ago, Yapese explorers journeyed 280 miles (450.6 km) west in bamboo canoes to the island of Palau, where they encountered limestone for the first time. After negotiations with the people of Palau, the Yapese established a quarry, using shell tools to carve disc-shaped stones they named *rai*.

The stones varied in diameter from a few inches to 12 feet (3.7 m) and weighed up to 8,000 pounds (3,628.7 kg). A hole punched into the center of each disc allowed the explorers to carry the larger stones on poles to their bamboo canoes. Keeping afloat during the long journey home was a more treacherous undertaking.

Back on Yap, rai became a sort of currency, customarily exchanged in marriages and as part of political deals and inheritances. The value of each rai depended on its size, but also on its provenance—if explorers died during the expedition to retrieve a stone, it acquired a higher value. A transaction did not require the physical exchange of a disc, merely the acknowledgment of a transfer of ownership. In fact, one of the rai stones in active circulation sits on the bottom of the Pacific Ocean, having tumbled from a canoe during a storm.

Rai stones were not immune to the vicissitudes of inflation. During the 1870s, an Irish American adventurer named David O'Keefe accompanied the Yapese to Palau, where he used imported tools to carve the limestone. His methods sped up the rai production process, with negative consequences: Yapese placed a lower value on O'Keefe's stones than on the discs carved with traditional shell tools, and the sudden abundance of rai brought down its overall worth.

The quarrying of rai stones ended at the beginning of the 19th century, but the Yapese still exchange discs to commemorate traditions. Many of the 6,500 remaining rai are displayed in rows at outdoor "banks"—jungle clearings and village centers. Theft is not a concern.

Sites include the villages of Maaq and Gachpar on Tomil Island. United Airlines flights to Yap depart from Guam twice a week. ⓝ 9.533333 ⓔ 138.116667

NAURU

GUANO ISLAND

YAREN DISTRICT

For a few decades after Nauru became an independent nation in 1968, its 6,000 citizens were among the richest, per capita, in the world. After being controlled since the 1840s by Germany, Britain, Australia, New Zealand, and Japan, Nauru's residents could finally call the 8-square-mile (20.7 km²) island their own—and they imported pricey sports cars to drive on its lone paved road.

Nauru's mighty wealth came from a single source: seabird droppings. Mining companies seized upon the phosphate from the fossilized guano—also known as bird (or bat) poop—in 1908, excavating and exporting it in abundance. The Nauru government, foreseeing the exhaustion of the phosphate supply, established a fund to channel mining money into investments that would secure the country's financial future. Unfortunately, much of the spending—such as the strange and disastrous investment in a four-hour-long 1993 West End musical about Leonardo da Vinci, which closed within five weeks of its premiere—yielded losses. When the phosphate ran out at the beginning of the 21st century, Nauru was left with no money, no natural resources, and no plan for recovery.

The plundered island's sudden poverty led to political instability, the 2005 grounding of its national airline, and unconventional approaches to generating income. Since 2001, Nauru has accepted financial aid from Australia in exchange for hosting a controversial refugee detention center. Asylum seekers hoping to gain entry to Australia by boat are diverted to Nauru and detained for months while the Australian government evaluates their requests for refuge.

Today, 9,500 Nauruans consume food and water imported from Australia—mining has left most of the land rocky, barren, and unsuitable for vegetation. Power shortages are common. Abandoned four-wheel drives and rusted equipment occupy the foreground of the island's coastal landscape, providing a constant reminder of its prosperous past and uncertain future.

The Republic of Nauru's Our Airline flies weekly to Nauru from Brisbane, Australia, and Nadi, Fiji. ⓢ 0.530083 ⓔ 166.931906

They don't sting.

PALAU
JELLYFISH LAKE

EIL MALK

A million golden jellyfish spend their days softly pulsating in a small saltwater lake in eastern Palau, their sting too soft to harm snorkelers. About 12,000 years ago, water from the Pacific Ocean flowed into the lake basin, bringing jellyfish with it—but not their predators. As a result, the golden jellyfish multiplied without restraint.

The jellyfish do have one nemesis—anemones that lurk in the shadows of the water, ready to eat them. The jellyfishes' consistent daily migration pattern helps keep them safe and well fed. They spend 14 hours bobbing up and down to gather plankton before chasing the sun from the east to the west, crisscrossing the lake each day.

Snorkeling is welcome in the lake, and the cnidarians will do you no harm as they waft past your body. The only danger is 50 feet (15.2 m) below the surface, where high levels of ammonia and phosphate may cause fatal skin poisoning. For this reason, scuba diving is prohibited.

East Eil Malk, also known as Mecherchar, is one of the southern islands in the Palau archipelago. United Airlines flies to Palau via Guam. Getting to the lake involves a boat trip and a steep 10-minute hike. Do not wear sunscreen, as it pollutes the lake. Ⓝ 7.161111 Ⓔ 134.376111

PAPUA NEW GUINEA
YAMAMOTO'S BOMBER

BUIN, BOUGAINVILLE

Deep in the jungle north of Buin lies the wreckage of a particularly significant Japanese World War II bomber. At dawn on April 18, 1943, this plane departed from Rabaul, Papua New Guinea. Onboard was Admiral Isoroku Yamamoto, commander in chief of the Japanese navy's Combined Fleet and mastermind of the attack on Pearl Harbor. Yamamoto was headed for a tour of Japanese front lines in the South Pacific, intending to boost morale following a series of setbacks in the area.

What the admiral and his officers didn't know was that US naval intelligence had intercepted and deciphered his travel plans. With Yamamoto's full itinerary in front of them, the US seized the opportunity to ambush his flight.

The mission—Operation Vengeance—was audacious. Eighteen American planes took off from Guadalcanal in the Solomon Islands, flying hundreds of miles north toward the Japanese bomber. The distance was so great that they had to carry extra fuel tanks. Flying over Bougainville, the Americans spotted Yamamoto's bomber and its escort planes. A dogfight ensued, and the admiral's bomber, spewing smoke, crash-landed in the jungle.

The Japanese rescue party who retrieved the admiral the next day said they found Yamamoto's body sitting upright under a tree, after he was thrown from the bomber. Rescue leader Lieutenant Hamasuna claimed that Yamamoto's white-gloved hand was still clutching the hilt of his sword—a detail that preserved the dignity of the admiral in the eyes of the dispirited Japanese.

To get to the wreckage, you'll need to take a 3-hour road trip from Arawa to Buin. Then it's an hour-long jungle trek to the crash site. Traveling with a local guide is recommended. Ⓢ 6.785649 Ⓔ 155.646687

ALSO IN THE PACIFIC ISLANDS

Great Pacific Garbage Patch

A Texas-size swirl of plastic fragments floats in the middle of the Pacific Ocean, made up of trash swept from the West Coast beaches of the US.

Alofaaga Blowholes

Savai'i, Samoa · Ocean water barrels through a series of coastal lava tubes, eventually blasting skyward through holes at the surface.

The wreckage of Admiral Yamamoto's World War II plane is still in the Bougainville jungle.

An Island to Oneself

New Zealander Tom Neale spent 16 years alone on Suwarrow, a tiny Cook Islands coral atoll 200 miles (321.9 km) from the nearest populated land. He rode out violent storms in a rickety shack, ate a diet heavy on fish and coconuts, and wore nothing but a loincloth on his sun-browned body. It is the remarkable tale of a shipwrecked survivor, but for one detail: Neale did not land on Suwarrow by accident. He traveled there willingly and enthusiastically, determined to live a simple, solitary existence on an island he could call his own.

Neale was 50 when he first set sail for Suwarrow in 1952. Formerly a shopkeeper in Tahiti, he became fixated on the atoll after meeting author Robert Frisbie and hearing him tell of its charms.

To prepare for his stay, which he planned to last for years, Neale stockpiled supplies—bewildering store owners in the small town of Rarotonga by buying their entire inventories of flour, sugar, kerosene, and coffee beans. As word spread about his grand plans, villagers offered gifts and even companionship—Neale politely refused the advances of several women who asked to accompany him, suspecting it would not be long before he came to resent their company. He did, however, choose two nonhuman companions: a cat named Mrs. Thievery (named for her favorite hobby) and her kitten, Mr. Tom-Tom.

Armed with tins of food, tools, seeds, and a motley collection of paperbacks, Neale arrived on Suwarrow following 6 days at sea.

Gradually, Neale adapted to island life. He spent the daylight hours building, cleaning, and tending to his garden, and ushered in nightfall sitting on a wooden box at the beach, watching the sun set while drinking a cup of tea. The necessary task of killing six pigs who threatened to destroy the garden brought Neale much anguish—after spearing the first one and hearing its screams, he wrote in his journal of feeling "melancholy," and decided to bury the animal rather than eat it.

Ten months after his arrival on Suwarrow, Neale received his first visitors: two couples on a yacht, who were astonished by the orderliness of his abode. (Not one to lower his living standards just because he lived in a shack, Neale boiled his sheets weekly, kept a tidy room, and ate using a tablecloth.) When the visitors departed a few days later, leaving well wishes and a half-drunk bottle of rum, Neale took up a new project: reconstructing a destroyed pier that had embarrassed him with its unsightliness.

For the next six months, he spent up to five hours every day dragging, rolling, and carrying large pieces of coral to the water's edge, setting them on the pier's fragmentary foundations. When the task was finally complete, Neale had single-handedly built a neat, structurally sound jetty using just rocks, patience, and hard work. The next day, a violent storm attacked

A latter-day Robinson Crusoe with his catch.

the island, completely destroying the newly finished pier.

Thus began a run of bad luck. Having exhausted his tobacco supply, Neale experienced torturous cravings. He dreamed nightly of soothing cigarettes, chocolate, beef, and fat, juicy duck. But the mental torture would soon pale in comparison to physical pain. Casting an anchor on the beach one day, Neale felt a searing stab along his spine and suddenly every movement was agony. Taking 4 hours to make the short journey back to his shack, Neale laid down in his bed and, between bouts of unconsciousness, hoped for a miracle.

Incredibly, a miracle indeed arrived—in the form of Peb and Bob, two American yachtsmen who stopped in at the atoll while en route to Samoa. (Peb, as it turned out, was more formally known as James Rockefeller Jr., a member of one of the US's wealthiest dynasties.) The duo, astonished to find Neale on the "uninhabited" island, fed him, massaged his back, and, having returned

him to good health, departed with a pledge that they would send a ship to fetch him. Two weeks later, the promised ship arrived, plucking Neale from his two-year life of island solitude and ferrying him back to Rarotonga.

Neale did not take kindly to being back among the comforts of civilization. Clocks were an intrusive nuisance, cars moved noisily and too fast, and trousers compared unfavorably with the comfort of a cotton loincloth. All Neale wanted was to return to Suwarrow, but the government forbade it. Disconsolate, he took a job in a warehouse. Six years passed before a friend with a 30-foot (9.1 m) boat offered to take Neale back to his beloved atoll.

Neale's second stay on Suwarrow lasted two and a half years, ending only when the increasing presence of pearl divers at the atoll began to wear on his patience. A 3-year break in Rarotonga allowed him to pen his island memoir, An Island to Oneself, before returning for his final, decade-long stay on Suwarrow. In 1977, stricken with stomach cancer, he was transported to Rarotonga, where he died at the age of 75.

Despite his long stretches of solitary living, Neale claimed he never felt lonely. The few times he wished someone was with him, he wrote in An Island to Oneself, were "not because I wanted company but just because all this beauty seemed too perfect to keep to myself."

SAMOA
PALOLO WORM FESTIVAL

SAVAI'I

For most of the year, Samoan palolo worms live in burrows in coral reefs, feeding on algae and slowly growing to their adult length of 12 inches (30.5 cm). But over one or two nights every October, they rise to the surface and engage in a swarming, floating, mucus-laden reproductive frenzy.

During the annual swarm, the worms release their tail segments (known as the epitoke), which contain sperm or eggs. The segments float up to the surface, where the casings dissolve, releasing their contents and triggering the reproduction process.

At that time, local families grab their nets and cheesecloth and wade into the water, hoping to scoop up a delicious serving of headless worms. Palolo, which has a salty, fishy flavor, is a delicacy in Samoa. Between filling buckets with the wriggling creatures, locals pop handfuls of the worm segments into their mouths completely raw. The next day, palolo on toast is the meal of choice.

Safotu, Fagasa, Asau, Papa, and Puleia reefs surrounding the island of Savai'i. The timing of the palolo swarm differs every year. Locals have developed a range of methods for predicting the date using the lunar calendar. Aim to be there 7 days after the full moon in October. Ⓢ 13.613956 Ⓔ 172.420349

VANUATU
PENTECOST ISLAND LAND DIVING

PENTECOST ISLAND

There is a boy, perhaps 7 years old, standing on a platform that juts out from a tower made of tree trunks and vines. He is about 30 feet (9.1 m) above the ground. He is naked except for a traditional penis sheath attached to a vine belt. Tied around his ankles are the frayed ends of liana vines. Women in grass skirts cheer and dance. The boy closes his eyes, places his hands in the prayer position, and jumps. A loud snapping noise accompanies the moment when his body brushes the ground before bouncing back up and dangling from the vine. The boy is shaken, but smiles as two men help him stand. That was the warm-up. The really exciting stuff is yet to come.

The males of southern Pentecost Island have been land diving for centuries. The ritual, in which boys and men leap from increasingly higher levels of a 100-foot-tall (30.5 m) tower using vines as bungee cords, is performed to ensure a good yam harvest. Over the years, it has acquired additional meaning: Boys who make the leap following circumcision at age 7 or 8 gain acceptance into manhood.

Land diving takes place every year on Saturdays between April and June. At this time of year, the liana vines are at their most elastic. After spending up to 5 weeks building the tower, a team of men tills the soil at the landing patch to soften it. A village elder selects two vines for each jumper, taking height and weight into consideration. The margin for error is slim—too short and the jumper may swing back and hit the tower; too long and he may die or become paralyzed after hitting the ground hard.

The ritual begins with the youngest jumpers, who leap from the lowest platform. To ensure a plentiful yam harvest, the diver's shoulders must make contact with the ground. (As a safety precaution, the head is tucked against the chest during the dive.) The loud snap heard on impact is not—in most cases—a human spine breaking, but the snapping of the support beam.

As the age and experience of the divers increase, so does the level from which they leap. The ritual reaches a climax when a man dives from the very top. When he lands, villagers erupt into applause and cheers, surround him, and hoist him into the air. **Pentecost Island is a 50-minute flight from Port Vila. The land diving site is a 5-minute walk from the airport. Ⓢ 15.717317 Ⓔ 168.179243**

Using vines as bungee cords, a man bravely leaps off a 100-foot tower to ensure a good yam harvest.

Tanna's true believers await an unconventional messiah.

CARGO CULTS OF TANNA

TANNA

On a small island in the southern part of the Vanuatu archipelago, devoted believers await the second coming of an American deity who will bring divine gifts in the form of TVs, refrigerators, and Coca-Cola. They are members of a cargo cult: an anthropological label for a tribal society that engages in religious practices designed to attract goods—or "cargo"—from more technologically advanced cultures.

Cargo cults rose to prominence during World War II, when hundreds of thousands of American and Japanese soldiers flooded into the islands of the Pacific region, bringing items that reflected material wealth and industrialization. Encountering mass-produced goods such as candy and radios—and having no concept of manufacturing processes—some island residents believed the goods were divinely created.

When the war ended and the soldiers went home, the cargo disappeared. Cult members believed that goods were dispatched to them by the gods, but intercepted by Westerners. They responded by setting up mock airstrips, airports, and offices, hoping to attract the cargo deliveries back to their rightful destination in Tanna.

Most of the cargo cults disappeared in the decades after the war, but the John Frum Movement lives on in Tanna. Cult members worship Frum, a messiah with mutable characteristics. To some, he is white. To others, black. For most, he is American, likely based on a soldier who brought cargo to Vanuatu during World War II: "John from America."

Though Frum's appearance varies, his mission is consistent: to shake off the restrictions of colonial rule and restore the independence and cultural freedom of the Tanna people. Cult followers believe Frum will return on February 15—an annual holiday known as John Frum Day—of an unspecified year, bearing food, household appliances, vehicles, and medicine.

Celebrations on John Frum Day have a distinctly American feel. Men in jeans with "USA" painted in red on their bare chests perform military drills, holding sticks of wood shaped like rifles. Above them, the American flag flies high from a bamboo pole.

The John Frum Movement coexists with other cargo cults in Tanna: The Tom Navy faction holds a US naval officer as its figurehead, and the Prince Philip Movement regards the Duke of Edinburgh as a pale-skinned mountain spirit and eagerly awaits his messianic arrival.

Sulphur Bay, Tanna. Flights to Tanna via Port Vila depart from Australia, New Zealand, Fiji, and New Caledonia. The center of the John Frum Movement is at Sulphur Bay. ⓢ 19.515486 ⓔ 169.456501

Canada

Western Canada

**ALBERTA · BRITISH COLUMBIA · MANITOBA
NORTHWEST TERRITORIES · NUNAVUT
SASKATCHEWAN · YUKON**

Eastern Canada

**NEWFOUNDLAND AND LABRADOR · NOVA SCOTIA
ONTARIO · PRINCE EDWARD ISLAND · QUEBEC**

WESTERN CANADA

ALBERTA

THE BANFF MERMAN

BANFF

In a glass case at the Banff Indian Trading Post is what looks like the back end of a snapper fish glued to the top half of a desiccated monkey. Tufts of white hair cling to the 3-foot-long creature's emaciated form. One webbed hand reaches out as if in the throes of death. A maniacal grin bares two rows of tiny, pointed teeth. This is the Banff Merman.

The legend of the merman began with Norman K. Luxton, founder of the Indian Trading Post and the Sign of the Goat Curio Shop. Accounts of its origin differ, but following a failed

Meet the mysterious 100-year-old man-beast of the Seven Seas.

around-the-world expedition, Luxton either caught, bought, or built the hokey-looking man-beast around 1915. The Banff Merman has been entertaining and provoking the imaginations of visitors ever since.

Indian Trading Post, 101 Cave Avenue Banff, 80 miles (129 km) west of Calgary. Next door is the Buffalo Nations Luxton Museum, also founded by Norman Luxton.
Ⓝ 51.171971 Ⓦ 115.571960

THE BADLANDS GUARDIAN

MEDICINE HAT

Gaze out of an airplane window above the Badlands east of Medicine Hat, and a Native American chief will stare back at you. Over millennia, erosion and weathering fashioned the rocky terrain into the shape of a human head, complete with feathered headdress. A gas well and road leading down from the chief's ear resemble earphones.

An Australian woman going by the name of "supergranny" discovered the head on Google Earth in 2006. Following a naming competition, the 820 × 740-foot (250 × 225.5 m) figure came to be known as the Badlands Guardian. (Rejected monikers include Space Face, Chief Bleeding Ear, The Listening Rock, Jolly Rocker, and Pod God.)

The Badlands Guardian is an example of pareidolia, the phenomenon of an overactive imagination perceiving recognizable shapes in ambiguous stimuli. Faces, particularly religious ones, are frequently found in inanimate objects—the Virgin Mary has appeared, among other places, in a grilled cheese sandwich, on an expressway underpass, and in a pile of chocolate drippings at a California candy factory.

The Badlands Guardian is east of Medicine Hat, a few miles south of Many Islands Lake. It is only visible by air—there is no public access to the site. Ⓝ 50.010600 Ⓦ 110.115900

Seen from the air, the folds of Medicine Hat's badlands resemble a man listening to music.

ALSO IN ALBERTA

Sunnyslope Sandstone Shelter

Linden · Standing alone on a prairie, this shelter likely dates to the early 20th century, when homesteaders built it to protect themselves from the elements.

WORLD'S LARGEST BEAVER DAM

IMPROVEMENT DISTRICT NO. 24

The beavers in Canada's Wood Buffalo National Park have been hard at work for decades, and their tree-chomping labor has paid off. The furry architects have created the largest beaver dam in the world.

The dam, which is about a half mile (.8 km) long, is so massive it even shows up on satellite images. It remained hidden within the Alberta wilderness until 2007, when a researcher spotted it while looking at Google Earth. The beavers are currently building new dams nearby which, when joined with the main structure, could add over 300 more feet (91.4 m) to its length.

Beavers are one of the few species capable of creating structures that are significant enough to be seen from space. The toothy critters are remarkable environmental engineers. Their dams reroute streams and even alter entire ecosystems. These creations, which are built to last, are barriers that form ponds, which act like defensive moats to protect the beavers from predators like wolves and bears.

It's likely the beavers began working on the Alberta dam sometime in the 1970s, making it a multi-generational architectural endeavor. The hodgepodge of mud, branches, stones, and twigs is cloaked in a layer of grass, meaning it's been there for a while. The dam stretches across a remote wetland area, which provides the creatures with both plenty of fresh water and bountiful building materials.

The isolated beaver dam is difficult to reach for any curious human. American researcher Rob Mark was the first to set foot on it, in 2014. His journey took nine days. Since the dam can be seen from space, you could always fly over it—in a standard plane or aboard a spacecraft. Ⓝ 58.271474 Ⓦ 112.252228

BRITISH COLUMBIA
SAM KEE BUILDING

VANCOUVER

In 1913, Chang Toy, owner of the Sam Kee import-export company, sold most of his corner property at Carrall and Pender Streets to the city of Vancouver for the construction and widening of West Pender Street. It was not the most amicable transaction—Toy was not adequately compensated, and the city's disrespect toward him was part of the anti-Chinese sentiment that prevailed at the time.

Left with only a strip of land to call his own, Toy recruited architects Bryan and Gillam to create a two-story building with attention-grabbing dimensions. The 6-foot-wide (1.8 m) structure is Guinness-certified as the narrowest commercial building in the world.

8 West Pender Street, Vancouver. The building is a 7-minute walk from the Stadium-Chinatown stop on the SkyTrain. Ⓝ 49.280416 Ⓦ 123.104715

The world's narrowest commercial building is 6 feet (1.8 m) wide.

FREE SPIRIT SPHERES

VANCOUVER ISLAND

Spending a night in a Free Spirit Sphere is a moving experience. The three spheres, named Eve, Eryn, and Melody, which are about 10 feet (3 m) in diameter, are suspended from trees in a forest just north of Qualicum Beach. When a breeze sweeps through the forest, the spheres sway gently.

Each 1,100-pound (500 kg) cedar-and-spruce orb is equipped with a bed, dining table, storage space, built-in speakers, and big circular windows for gazing at the forest. Spiral staircases and drawbridges provide access to the spheres, which hang about a story above the ground. Toilets and showers are located in separate buildings nearby—Eryn's outhouse is shaped like a mushroom.

The spheres are the work of Tom Chudleigh, who operates the property with his wife, Rosey. Chudleigh used shipbuilding techniques to handcraft them, aiming to create natural-looking, nutlike globes that blend in with the environment.

420 Horne Lake Road, Qualicum Beach. The ferry from Vancouver (Horseshoe Bay) to Nanaimo takes an hour and 40 minutes. Then it's an hour's drive north to the spheres. Ⓝ 49.378834 Ⓦ 124.616330

Sleep inside a softly bouncing wooden ball suspended from the trees.

THE TREE ON THE LAKE

PORT RENFREW

Living up to its name, Fairy Lake is in a remote and unspoiled landscape near the town of Port Renfrew. Sticking up out of the lake's stillness is a partially submerged log. Clinging to that log for dear life is a tiny, living Douglas fir tree. Tourists, boaters, and hikers come seeking it as a unique window into nature and rebirth.

The "bonsai" tree has attracted more than a few photographers to capture its struggle of endurance, including a winner of the National History Museum of London's Wildlife Photographer of the Year award. **Fairy Lake is on Vancouver Island, about 5 miles east of the town of Port Renfrew. Take Parkinson Road to the turnoff for Deering Road, follow that to the end, and turn right onto Pacific Marine Road. Follow that all the way to the lake. The little tree will be on your right, about a quarter mile past the turnoff for Fairy Lake Recreation Site. Ⓝ 48.589443 Ⓦ 124.349811**

Bonsai-like serenity and a symbol of survival.

VANCOUVER POLICE MUSEUM

VANCOUVER

Housed in a former coroner's court and morgue, this museum displays confiscated weapons, counterfeit currency, police uniforms, and other artifacts that chronicle Canadian crime-busting.

The best stories are found in the morgue, which retains its original autopsy bays, sinks, and drawers. It was here that a coroner conducted the autopsy of Errol Flynn after he died suddenly at the age of fifty in 1959. (The hard-living actor's recorded afflictions at the time of death included myocardial infarction, coronary thrombosis, coronary atherosclerosis, cirrhosis of the liver, and diverticulosis of the colon.)

A wry sense of humor pervades the museum's exhibits. A sign on one of the morgue drawers reads, PLEASE DO NOT OPEN MORGUE DRAWERS (DO YOU *KNOW* WHAT USED TO BE IN HERE??). The answer is not just "bodies"—one presiding coroner only used 17 of the 18 drawers, preferring to keep the last one as a beer fridge.

240 East Cordova Street. Get a bus along Hastings Street and walk a block north to Cordova. Ⓝ 49.282269 Ⓦ 123.099345

LARGE ZENITH TELESCOPE

MAPLE RIDGE

At the Liquid Mirror Observatory, scientists use a 20-foot-wide (6 m) spinning puddle of mercury to survey distant galaxies. While most reflecting telescopes incorporate solid mirrors made of glass and aluminum, the Large Zenith Telescope uses a pool of spinning mercury to reflect starlight. The centrifugal force of the rotating mercury forms a parabola, the shape required for the mirror to reflect light.

The main advantage of liquid metal telescopes is that they are significantly cheaper to build than their solid-mirror counterparts. The downside is that they can only point in a fixed direction—tilting the telescope would cause the reflective liquid to spill out. The Large Zenith Telescope is so named because it points toward the zenith—in other words, straight up.

mercury surface

rotating mirror platform

Malcolm Knapp Research Forest, Maple Ridge. The observatory is 43 miles east of Vancouver. It's a 45-minute hike from the gate.
Ⓝ 49.287579 Ⓦ 122.572759

ALSO IN BRITISH COLUMBIA

Boswell Embalming Bottle House

Boswell · When he retired after 35 years in the funeral business, David H. Brown took half a million empty embalming fluid bottles and turned them into a two-story home.

Enchanted Forest

Revelstoke · In this old-world forest fairy-tale land, over 350 folk figurines are hidden among 800-year-old cedars.

Kitsault

Kitsault · The shiny new settlement that popped up in 1979 to cater to employees of a molybdenum mine lasted just 18 months before a molybdenum price crash rendered it a ghost town.

Hundreds of mineral-rich pools make up the giant painter's palette that is Spotted Lake.

SPOTTED LAKE

OSOYOOS

From fall to spring, there is nothing unusual about Spotted Lake. But come summer, the reason for its name becomes clear: Evaporation reveals hundreds of round pools on the bottom of the lake. Each spot is colored according to the type and concentration of its minerals.

Spotted Lake has long been a sacred site for the First Nations of the Okanagan Valley. Its minerals, which include magnesium sulfate, calcium, sodium sulphates, and traces of silver and titanium, are believed to have healing properties. During World War I, these therapeutic minerals were harvested and used to make ammunition.

Okanagan Highway 3, 6 miles (9.7 km) west of Osoyoos. The lake is on private land, but you can catch a glimpse from the highway.
Ⓝ 49.078018 Ⓦ 119.567502

MANITOBA
Narcisse Snake Orgy

NARCISSE

For a few weeks every year, Interlake teems with piles of writhing serpents. In late April, tens of thousands of harmless red-sided garter snakes emerge from their winter dens, slithering over one another in search of a mate—the largest single concentration of garter snakes in the world. A multiweek reproductive frenzy follows, during which tangled balls of snakes constantly form and disperse. It's hypnotic and unexpectedly audible—on a dry day you'll hear the animals' scales rubbing together. **The dens are 4 miles north of Narcisse along Highway 17. To see the snakes mating, visit in late April or early May. Ⓝ 50.734526 Ⓦ 97.530355**

There's a whole lot of hissing going on.

NORTHWEST TERRITORIES
Diavik Diamond Mine

LAC DE GRAS

An open-pit spiral on an island in a lake, the Diavik Diamond Mine is spectacular when viewed from above. But the most remarkable thing about it is its remoteness. The mine is located 190 miles (306 km) north of the nearest town, Yellowknife in the Northwest Territories.

There is only one road to Diavik, and it can be used for just nine weeks each year. The reason? It's made of ice. At the end of every December, when the lakes and ponds have frozen, workers begin a six-week around-the-clock construction process to open the mine for resupply in late January. The route closes for the season by early April—the exact date depends on the thinness of the ice. Trucks bringing fuel, cement, explosives, and construction materials must stick to the road's 16-mph (25 kph) speed limit.

The mine began producing diamonds in 2003, and was soon yielding 3,500 pounds (1,588 kg) of the precious stones per year. In 2012, Diavik completed the transition from open-pit to underground mining in order to reach more diamonds. Its employees arrive and depart from a private airstrip. Heated access tunnels and readily available baked goods make the −20°F (−29°C) nights more bearable. **Lac de Gras, North Slave Region, Northwest Territories. Ⓝ 64.496100 Ⓦ 110.664280**

Pingo Canadian Landmark

TUKTOYAKTUK

The arctic region of Tuktoyaktuk is stippled with pingos, dome-shaped hills with a core of ice and a layer of soil on top. Just west of the Tuktoyaktuk, at Pingo Canadian Landmark, there are eight of these arctic landforms—including Ibyuk, the highest pingo in Canada at 160 feet tall and 984 feet wide (49 m by 300 m). The eight domes, which resemble baby volcanoes, are among the 1,350 pingos of the Tuktoyaktuk peninsula.

Three miles west of Tuktoyaktuk. Pingo Canadian Landmark is best accessed by boat from Tuktoyaktuk. Depending on the time of year, you may encounter caribou, grizzly bears, or snow geese among the pingos. Ⓝ 69.399722 Ⓦ 133.079722

   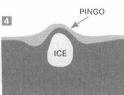

Pingos are hills made of earth-covered ice. Above, the formation of a pingo.

■ water-saturated sand ■ permafrost ■ soil   ice

NUNAVUT

HAUGHTON IMPACT CRATER

DEVON ISLAND

Twenty-three million years ago a meteorite slammed into Devon Island, the largest uninhabited island on the planet. The impact melted the surrounding rocks, leaving a 12-mile-wide (19 km) scar on the High Arctic landscape. Due to the cold, dry climate, which encases the ground in permafrost for most of the year, the island's geological conditions have remained stable ever since. They also bear a striking resemblance to a place humans have yet to visit: Mars.

The idea of using the Devon Island crater as a test lab for Mars exploration experiments originated with SETI Institute planetary scientist Dr. Pascal Lee in 1997. Working with NASA, Lee and a team of about 30 researchers make yearly summer visits to the spot, conducting field tests to help plan for Mars expeditions. Using a tent city as base camp, the scientists drive all-terrain vehicles to simulate rovers, operate automated drills to look for water, take

---

➼ Spaceship Architecture of the Arctic

In the Canadian Arctic, high winds, subfreezing temperatures, and the difficulty of transporting building materials create challenging design constraints for Nunavut's architects.

IQALUIT AIRPORT

Known locally as the Yellow Submarine, Iqaluit's airport terminal is built to withstand the elements. The airport's extra-long runway has been used by Airbus to cold-weather test new aircraft designs.

ST. JUDE'S CATHEDRAL

The "Igloo Cathedral" is the seat of the Anglican Diocese of the Arctic. It opened in 2012, replacing an earlier church that was destroyed by arson.

walks in spacesuit prototypes, and conduct mapping tests using robots.

Though human colonization of Mars is still but a distant dream, the Haughton crater experiments offer a practical look into how we might get there. New research projects take place every summer, and the eventual goal is for Red Planet–bound astronauts to use the crater as a training ground before blasting off to Mars for real.

Baffin Bay. ℕ 75.198235 𝕎 89.851182 ➤➤

The Mars-like landscape of Haughton crater is an ideal place to train astronauts for a Red Planet mission.

NAKASUK SCHOOL
The students who go to school in this two-story fiberglass structure don't much miss having windows during Iqaluit's dark winter months.

IGLOOLIK RESEARCH STATION
Built by the Canadian government in the 1970s, this UFO-shaped building holds research labs and offices for scientists from the Department of Environment.

NORTH WARNING SYSTEM SITE
The North Warning System is a 3,000-mile-long (4,828 km) string of early-warning radars in Cambridge Bay, built by the US and Canadian governments in the 1980s, to detect Soviet attacks across the North Pole.

GRAVES OF BEECHEY ISLAND

RESOLUTE

Standing eerily on Beechey Island, a peninsula off Devon Island in the Canadian Arctic, are four gravestones belonging to three members of an ill-fated expedition to the Northwest Passage and one of the men who went looking for them.

In 1845, Sir John Franklin led an expedition to find the Northwest Passage, a direct route from the Atlantic Ocean to the Pacific Ocean across the Arctic, on two ships that were called "unstoppable" at the time. They were stopped.

The expedition visited Beechey Island for a winter encampment. There, three members of the 129-person crew—John Torrington, William Braine, and John Hartnell—were buried near the shore on an otherwise desolate plain. The rest of the crew abandoned the ships later, when they got stuck in the ice near King William Island. Evidence has been found to suggest they resorted to cannibalism before perishing.

How the three young men died is still unknown despite the fact that their bodies stayed well preserved—essentially mummified—in the frozen arctic ground. Lead poisoning from their canned food was a leading theory, though it is disputed now. The men's remains were exhumed in the 1980s, examined, and reburied. In recent years, both of Franklin's ships have been discovered, and it is hoped that more mysteries can be solved.

The fourth grave marker at the site is that of Thomas Morgan. A member of one of the many expeditions launched to find out what had happened to Franklin and his crew, he died of scurvy in 1854. **Charter a plane from Edmonton to the very remote and very cold Arctic hamlet of Resolute. From there, climb aboard a well-stocked ship to journey to Beechey Island. And you come back now, you hear?**
Ⓝ **74.714333** Ⓦ **91.825265**

The markers stand out on the otherwise barren shoreline.

SASKATCHEWAN
TWISTED TREES

ALTICANE

Trembling aspen trees, so named for the way their leaves shiver in the wind, normally grow tall, straight, and thin. But a grove of aspens near Alticane, known as the "Twisted Trees," has somehow ended up as a crooked and mangled mess.

The oddest thing about the Twisted Trees is the fact that they are surrounded by perfectly normal trembling aspens. No one knows exactly when this patch of forest became so gnarled, but theories abound. One local legend tells of an alien who emerged from a UFO to urinate into the soil, causing it to become contaminated and warp the trees.

The scientific explanation is more mundane. Quaking aspen grow in vast groves consisting of what appear to be individual trees, but which are actually "clones" of an original source. The entire stand is a single giant organism.

At some point, a genetic mutation seems to have affected how these trees grow—and not just one tree, but the entire clonal forest. Though the mutation may have made some creepy, gnarly trees, it apparently wasn't devastating enough to impede their growth.

The Twisted Trees are about 3 miles southwest of Alticane.
Ⓝ 52.900372 Ⓦ 107.479533

The crooked aspens of Alticane twist their trunks for unknown reasons.

URANIUM CITY

URANIUM CITY

Candu High School, named after a type of nuclear reactor invented in Canada, is located on Uranium Road, not far from Fission Avenue in Uranium City.

As evidenced by its nomenclature, this northern Saskatchewan settlement is—was—a small town sustained by a single industry. Established in 1952 to house workers from the nearby Beaverlodge uranium mines, Uranium City flourished in the 1950s, '60s, and '70s thanks to British and American nuclear weapons programs.

Then, without warning, the town's reason for being vanished. On December 3, 1981, Eldorado Nuclear Limited announced that the uranium mine would be shutting in six months. At the time, about 4,000 people lived in the city—almost all of them dependent on income from the mine. Candu High, freshly built, had just opened to students. But with jobs wiped out, a mass exodus soon took place by air and ice road—the only ways in and out of the city.

Today Uranium City is not quite a ghost town, though its abandoned, crumbling buildings make it look like one. Around 70 people still live in the settlement, enduring Januaries with an average daily low of –25.4°F (–31.9°C). There is no hospital, but among the ruins there is a school, a bar, a hotel, and a general store that receives grocery deliveries by air once a week.

Uranium City is a 4-hour flight from Saskatoon, departing twice weekly.
Ⓝ 59.569326 Ⓦ 108.610521

The closure of Beaverlodge Mine caused the collapse of Uranium City.

YUKON
WATSON LAKE SIGN POST FOREST

WATSON LAKE

The Sign Post Forest beside the Alaska Highway is a trail of wooden poles covered top to bottom with directional signs and license plates. Visitors are encouraged to show their hometown pride by adding a sign to the collection. (If you forget to bring one, the visitor's center provides the materials for you to make one.)

Sign Post Forest dates back to 1942, when an injured American soldier from Illinois was tasked with building signs in the area. Pining for home, he snuck in a sign that read "Danville, Illinois: 2,835 miles." The one-time lark soon became a popular tradition—there are now more than 72,000 signs in the forest. It is the pride of Watson Lake, which has a population of 800, or approximately 90 signs for each resident.

Mile 635, Alaska Highway, Watson Lake. The Sign Post Forest makes for a great pit stop during a road trip along the Alaska Highway—the 1,387-mile-long (2,232 km) road that links Alaska with the Yukon and British Columbia. Ⓝ 60.063716 Ⓦ 128.713954

Walk by 72,000 roadside signs—not on a highway, but in the woods.

SOURTOE COCKTAIL

DAWSON CITY

The story of the Sourtoe begins with Captain Dick. A former cowboy, truck driver, and professional wolf-poisoner, Dick Stevenson was poking around a cabin on the outskirts of town in 1973 when he found a jar. Inside, preserved in alcohol, was a human toe.

Known for his madcap schemes—he boasts of having organized "the first nude beauty contest north of the 60th parallel"—the captain wondered how he could best make use of the toe. Then, after several drinks, it hit him: *cocktail garnish.*

The Sourtoe, served every night at the Downtown Hotel since 1973, is Stevenson's enduring creation. Originally it adhered to a strict two-ingredient recipe: champagne plus pickled toe. Over the years, however, the Sourtoe rules have relaxed. Any liquid, alcoholic or otherwise, may now be used, but drinkers must abide by the official chant: "Drink it fast or drink it slow, but the lips have got to touch the toe." Those who accomplish the wince-inducing task receive a certificate of membership in the Sourtoe Cocktail Club.

The same toe is used for every drink—alcohol keeps it sterile—but incidents of accidental swallowing have resulted in a succession of donated toes. The first toe, amputated from a man's frostbitten foot in the 1920s, went down the throat of an intoxicated miner in 1980. Toe number two, donated when its owner developed an inoperable corn, went missing a short time later. A baseball player swallowed toe number three, another frostbite casualty, in 1983. Five toes have since been donated to the bar, the most recent arriving in a jar with a message: "Don't wear open-toed sandals while mowing the lawn."

Sourdough Saloon at the Downtown Hotel, 1026 Second Avenue, Dawson City. Ⓝ 64.062336 Ⓦ 139.433435

Does a square mile of sand just south of the Arctic Circle count as a desert?

CARCROSS DESERT

CARCROSS

A one-square-mile desert, surrounded by snow-covered mountains, in a Canadian territory just south of the Arctic Circle. How can this be?

The answer is that Carcross Desert, though certainly sandy and often referred to as the world's smallest desert, is actually the remains of a glacier. The Carcross area—"Carcross" is derived from "Caribou Crossing"—was once a glacial lake. Over thousands of years, as the glaciers retreated, the water level lowered, leaving behind the layer of silt that once formed the bottom of the lake. This silt, shaped by the wind into dunes, became the world's most adorable desert.

Summer activities at Carcross Desert include off-roading and sandboarding. In winter, when the sand is covered in snow, bring your skis and snowboards. **The teeny desert is an hour's drive south of Whitehorse. Ⓝ 60.187222 Ⓦ 134.694722**

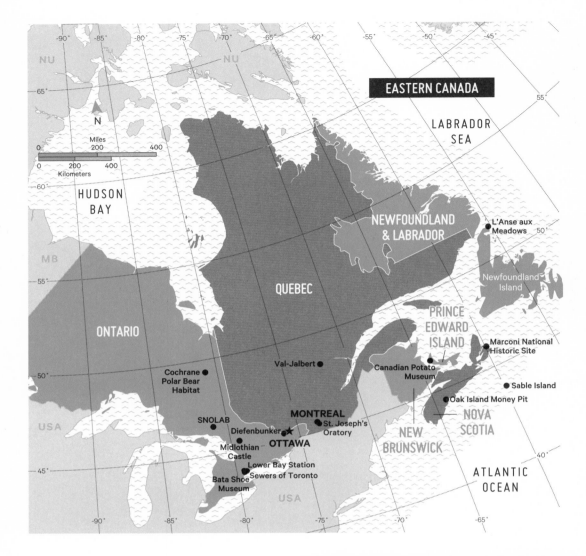

LABRADOR
SEA

NU

NU

65°

55°

HUDSON
BAY

60°

MB

55°

NEWFOUNDLAND
& LABRADOR

• L'Anse aux
Meadows   50°

Newfoundland
Island

QUEBEC

ONTARIO

PRINCE
EDWARD
ISLAND

Marconi National
Historic Site

50°

Cochrane ●
Polar Bear
Habitat

Val-Jalbert ●

Canadian Potato
Museum

● Sable Island

SNOLAB
●

Midlothian
Castle

Diefenbunker ★

MONTREAL
● St. Joseph's
Oratory

OTTAWA

● Oak Island Money Pit

NOVA
SCOTIA

40°

USA

NEW
BRUNSWICK

45°

Lower Bay Station
Sewers of Toronto

Bata Shoe
Museum

USA

ATLANTIC
OCEAN

Miles
200   400

0   200   400
Kilometers

N

NEWFOUNDLAND AND LABRADOR

L'ANSE AUX MEADOWS

ISLANDS BAY

The Vikings were here five hundred years before Columbus stumbled upon the Americas. Around 1000 CE, Vikings landed at the northernmost tip of Newfoundland and built a village. Known as L'Anse aux Meadows, it is the only recognized Viking settlement in North America.

The site was discovered in 1960 by Norwegian explorer Helge Ingstad. Excavations during the 1960s unearthed the remains of eight buildings as well as a stone oil lamp, a bone knitting needle, a sharpening stone, and traces of butternuts, a species of walnut that does not grow at the latitude of the settlement. This indicates that the Vikings must have traveled farther south to find sustenance.

The wood-and-sod buildings, which resemble Viking dwellings in Greenland, have been

The reconstructed lodgings of North America's first European visitors.

reconstructed to give visitors an idea of what L'Anse aux Meadows looked like. Site guides dressed as Vikings add to the immersive experience.

L'Anse aux Meadows is on the tip of the northern peninsula, 270 miles (434.5 km) north of Deer Lake. The site is open from June through September.

Ⓝ 51.598918 Ⓦ 55.530883

NOVA SCOTIA

OAK ISLAND MONEY PIT

OAK ISLAND

A deep, booby-trapped pit on uninhabited Oak Island may hold the Holy Grail. The Ark of the Covenant could also be down there, along with Marie Antoinette's jewels, Blackbeard's pirate treasure, and documents confirming that Francis Bacon was the actual author of Shakespeare's entire oeuvre.

Speculation over the contents of the pit began in 1795, when three men wandering on the eastern side of the island discovered a circular dent in the ground and a pulley attached to the branch of an adjacent tree. The men decided the scene was odd enough to grab shovels and start digging. After allegedly encountering layers of logs at 10 feet, 20 feet, and 30 feet down (3, 6, 9 m), the trio abandoned their burrowing.

This is where the story starts to get wilder and less plausible. Subsequent excavations by excited treasure hunters apparently uncovered pick marks, coconut fibers, and a stone inscribed with mysterious symbols. Once the dig reached below 90 feet (27.4 m), the pit began to flood. Attempts to bail out the water resulted in more water rushing in. "A trap," the diggers thought, "a clever trap to protect the treasure."

By 1861, people were literally dying to find the mythical riches that lay below. A boiler burst while workers were draining the hole with a steam-powered pump, mortally scalding one man and causing the pit to collapse again.

Despite over 200 years of treasure hunting—including corporate-funded remote-camera–enhanced

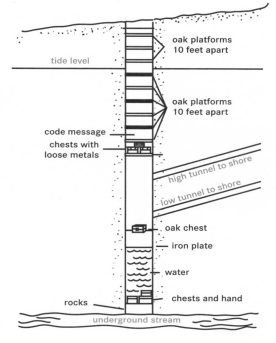

excavations during the 1960s—no one has solved the mystery of the Oak Island money pit. The likely story—that the pit is a naturally occurring sinkhole devoid of precious relics—goes ignored by the treasure hunters who still scour the island. A popular History Channel show features a new generation of hunters. So far, no treasure.

The island is privately owned and usually off-limits, but the Friends of Oak Island Society runs occasional tours. Ⓝ 44.512740 Ⓦ 64.288570

MARCONI NATIONAL HISTORIC SITE

GLACE BAY

On this spot in December 1902, Guglielmo Marconi ushered in the wireless age. Using a transmission station surrounded by four 200-foot (61 m) wooden aerials, Marconi sent and received Morse code signals to a station in Cornwall, England. It was the first official transatlantic use of electromagnetic waves for sending a message and the beginning of wireless global communication.

Marconi's station, constructed at Table Head in the small former coal mining town of Glace Bay, was relocated in 1905, but the tower foundations and a visitor's center remain. Now preserved as the Marconi National Historic Site, the center has a model of the original station as well as photos, artifacts, and information on the history of radio.

15 Timmerman Street, Table Head. The site is a half-hour drive from Sydney or a one-hour drive from Louisbourg. Ⓝ 46.210969 Ⓦ 59.952542

The age of radio began in Nova Scotia.

SABLE ISLAND

HALIFAX

Sable Island, which is made entirely of sand, is a low-lying stretch of land in the middle of the ocean. Due to its low profile and isolated location, the oversize sandbar has caused so many shipwrecks, it's earned the moniker "Graveyard of the Atlantic."

Sable Island is also known for its population of some 550 wild horses (descendants of the Shetland pony). A small number of scientists live on the island to study its flora and fauna, but it is otherwise empty of human activity. **Sable Island is 186 miles southeast of Halifax, Nova Scotia. Travelers to the island must register with Parks Canada. Ⓝ 43.934509 Ⓦ 59.914032**

A place where wild horses roam free.

ONTARIO

DIEFENBUNKER

OTTAWA

From 1962 to 1994, a small white shed in the middle of a field would have become national headquarters of the Canadian government in the event of a nuclear attack.

The "Diefenbunker," named after Prime Minister John Diefenbaker, is the largest of about 50 shelters built across Canada during the Cold War. Beneath the unassuming shed, a steel-lined blast tunnel leads to a four-story bunker with enough supplies to house over 500 government and military personnel for one month.

If a bomb were to hit, the prime minister and his staff would head directly to the bunker—alone, for there was no room to accommodate family members. Once there, they would strip and shower in the decontamination chamber, then make their way to the dormitory-style quarters. The spartan rooms contained bunk beds and little else—though the prime minister was granted the relative luxury of a single bed and a private shower.

Several of the bunker's design features aimed to lessen cabin fever and claustrophobia. In addition to the overall color scheme of serene blues and greens, the bunker's support columns were painted with vertical black stripes to make the ceiling appear higher. The floors of the narrow hallways have horizontal stripes, in an effort to make them look wider. Had these measures proved inadequate to keep someone calm and rational, a confinement cell was available.

As with other shelters built during the height of Cold War paranoia, the Diefenbunker was never used for its intended purpose. Decommissioned in 1994, the facility now operates as a Cold War museum. Stops on the tour include the situation room, a broadcast studio built to keep the surviving members of the nation informed, and a vault that would have kept the Bank of Canada's gold reserves safe from nuclear destruction. **3911 Carp Road, Carp. The bunker is 20 miles (32 km) west of Ottawa. Ⓝ 45.351819 Ⓦ 76.044741**

A nuclear shelter built for Canada's prime minister is now a Cold War museum.

Hundreds of screaming concrete heads adorn an eccentric teacher's 300-acre farm.

MIDLOTHIAN CASTLE

BURK'S FALLS

Peter Camani has a suggestion for where your earthly remains should go when you die. "Why settle for a small underground plot in the suburbs," he writes on his website, "when you have the option of joining a vibrant creation that fills the landscape?"

The "vibrant creation" to which he refers is a forest of 18-foot-tall (5.5 m) screaming heads made from cement and—if his idea catches on—the ashes of deceased humans.

The field of screams is just one component of Camani's ongoing art project on a 300-acre (1.2 km²) former farm just outside the village of Burk's Falls, Ontario. It all began in 1989, when the former high school art teacher began building a house on the property. Dubbed Midlothian Castle, the dwelling features a screaming head for a turret and a dragon for a

chimney that appears to breathe smoke whenever the fireplace is blazing.

In 1995, Camani began installing sculptures on the land surrounding the house. There are now over 100 scattered through the landscape—84 screaming heads, giant half-buried hands, trees with ghoulish faces, and the Four Horsemen of the Apocalypse.

Inspired by the Druids, Camani intends to continue sculpting screaming heads for as long as he's able. Though he has long envisioned a forest of sculptures made from cement and human ash—with the names and bios of the deceased written on them—he has only created one so far.

Midlothian Ridge, RR #1, Burk's Falls. The castle is around 150 miles (240 km) north of Toronto.
Ⓝ 45.595141 Ⓦ 79.537376

LOWER BAY STATION

TORONTO

Toronto's ghost subway platform has been empty since 1966.

Beneath Toronto's Bay subway station on the Bloor-Danforth line is Lower Bay, a platform that only operated for six months before it was boarded up.

The ghost platform opened in 1966 and was intended to ease commutes by joining the Bloor-Danforth and Yonge-University lines. In theory, passengers would benefit from being able to reach their destination without having to change trains. In practice, confusion reigned: Trains to the same destination arrived on different platforms, and travelers weren't sure whether to wait at the upper or lower location.

Since its closure in September 1966, Lower Bay has been used for film and television productions, often playing the part of a New York subway station. Although not usually viewable, the subterranean platform is occasionally open for special city events such as the annual Doors Open Toronto.

Bay Station, 64 Bloor Street West. The entrance to Bay Station is on Bloor Street West at Bellair Street. ⓝ 43.669539 ⓦ 79.392154

BATA SHOE MUSEUM

TORONTO

This museum tells the story of thousands of years of humanity, one pair of shoes at a time. The collection—which includes such historically resonant items as ancient Egyptian sandals, samurai shoes made of bear fur, and Queen Victoria's ballroom slippers—is drawn from the shoe stash of founder Sonja Bata, who began stocking up on fascinating footwear in the 1940s. The collection now numbers over 13,000 items, 1,000 of which are on display at any one time in the museum's permanent and rotating exhibits.

Around 13,000 shoes from all over history live in a shoebox-shaped building.

The shoes on show don't just illustrate fashion trends—they reflect their cultural contexts. A common theme is restriction and impediment, as seen in tiny lotus shoes once worn by Chinese women with bound feet as well as an impossibly tall velvet-covered platform shoe donned in the days of 16th-century Italy.

The museum has also managed to nab some cast-off footwear from 20th-century celebrities. See Robert Redford's cowboy boots, Elvis's blue patent-leather loafers, and an ankle boot that once cradled the foot of John Lennon.

327 Bloor Street West, Toronto. ⓝ 43.667278 ⓦ 79.400139

ALSO IN ONTARIO

Cheltenham Badlands

Caledon · These rolling red rock hills, striped with green from groundwater-induced oxidation, contrast impressively with the bright blue sky above.

Flowerpot Island

Ontario · On this delightful little island in Lake Huron, two rock pillars shaped like flowerpots guard the shore.

Ottawa Jail Hostel

Ottawa · Once a prison known for its inhumane practices, Ottawa Jail now welcomes backpackers for overnight stays in the cells.

Monkey's Paw

Toronto · This bookstore has the world's first "Biblio-Mat," a random-book vending machine.

SEWERS OF TORONTO

TORONTO

It takes a truly special set of circumstances to turn sewage into a landmark.

Toronto's subterranean waste management network is something to behold. The sewage tunnels in Toronto are wide and high-ceilinged, so large and well-built that they look more like soggy subway tunnels than rivers of waste water.

Daring adventurers and mischievous youths have long ventured into the city's underground passageways to map their trajectory and examine their current state. Each leg of the sewer system has its own story. The Garrison Creek Sewer running underneath the west end of the city, smoothly beveled and circular like a pneumatic tube, was once, as the name suggests, a creek. In the late 19th century, after that creek became full of human waste, the city wisely thought it best to just go ahead and bury the whole thing.

The tunnels are an unsung marvel of public engineering. That's to be expected from Toronto, a city that takes its sewage seriously, as evidenced by the ornate water treatment plant nicknamed the "Palace of Purification." That's where all the contents of these majestic tunnels end up—including explorers, if they follow the path long enough.

Entrances to the tunnels are exactly where you'd expect them to be—manholes, maintenance shafts, spillways, and water treatment offshoots—but entry is not technically allowed. Visiting is therefore difficult and granted only with permission and guidance by a public works employee.
Ⓝ 43.646747 Ⓦ 79.408311

Toronto's well-maintained sewer system is over 100 years old.

Take a relaxing dip with an 800-pound polar bear.

COCHRANE POLAR BEAR HABITAT

COCHRANE

A playful 800-pound (363 kg) bear named Ganuk is currently the sole ursine resident of Cochrane Polar Bear Habitat. The world's only facility dedicated solely to captive polar bears, Cochrane aims to care for rescued and zoo-raised bears while educating people on their conservation. Ganuk has free rein over the habitat's three expansive outdoor enclosures. His hobbies include digging, getting his head stuck in barrels, and eating bucket-size popsicles made with watermelon pieces, tuna, peanut butter, and marshmallows.

Ganuk can also literally paint you a picture of life in the habitat—zookeepers occasionally put nontoxic paint and canvases in front of him with the aim of encouraging his creative side. The resulting paw-printed abstract art is sold in the gift shop for up to $200 apiece.

Visit Ganuk between May and September and you'll be able to swim alongside him. There is a wading pool for humans right next to the bear pool, separated only by a 2-inch (5 cm) wall of glass.
1 Drury Park Road, Cochrane, a 7-hour drive from Toronto.
Ⓝ 49.057664 Ⓦ 81.023397

SNOLAB

LIVELY

A mile and a half underground, beneath the Creighton nickel mine, a team of astrophysicists is trying to solve the mysteries of the universe. They work at SNOLAB, a laboratory devoted to searching for neutrinos—neutral subatomic particles—and dark matter. The laboratory needs to be so far underground in order to shield the sensitive detection systems from interference caused by cosmic radiation.

The site is best known for its Sudbury Neutrino Observatory (SNO), an experiment that ran from 1999 to 2006, which used a 40-foot-wide (12 m) vessel filled with heavy water (water containing a large amount of the hydrogen isotope deuterium) to detect neutrinos produced by fusion reactions in the sun.

SNO's successor, SNO+, is currently being prepared. The new experiment will continue the search for neutrinos using a tweaked version of the existing equipment.

Creighton Mine, 1039 Regional Road 24, Lively, Greater Sudbury. Due to contamination concerns, anyone entering SNOLAB must shower and change on-site. Ⓝ 46.473285 Ⓦ 81.186683

Over a mile underground, a sphere studded with 9,600 photomultiplier tubes helps search for solar neutrinos.

PRINCE EDWARD ISLAND

CANADIAN POTATO MUSEUM

O'LEARY

The excitement starts building when you see the 14-foot (4.3 m) fiberglass potato mounted on a pole like a popsicle. The oversize tuber entices you through the doors of a museum that is charmingly effusive about potatoes—particularly the Prince Edward Island potatoes harvested from the surrounding fields.

Journey through time and across continents at the Potato Interpretive Center, where you can trace the spud's origins in South America and follow it as it's introduced to Europe and North America. Take a moment to reflect solemnly on the diseased potato exhibit, where potatoes are displayed in miniature coffins beside signs detailing their causes of death. View a rust-encrusted 19th-century thresher in the Antique Farming Equipment room.

The museum experience ends, naturally, with a visit to the café, where french fries, potato cinnamon buns, and potato fudge compete for the chance to dazzle your taste buds.

For a potato-themed vacation, visit O'Leary in the last week of July for the annual Potato Blossom Festival, which celebrates the plants in full bloom. Events include a farmer's banquet, fireworks, and the Miss Potato Blossom pageant.

1 Dewar Lane, O'Leary. The museum is a 45-minute drive from Summerside northwest along Highway 2. It's open from mid-May to mid-October. Ⓝ 46.703346 Ⓦ 64.234841

Get ready to have your potato-related assumptions challenged.

QUEBEC
Val-Jalbert Ghost Town

CHAMBORD

The village of Val-Jalbert, 150 miles (241 km) north of Quebec City, sprang up in 1901 when forestry entrepreneur Damase Jalbert built a pulp mill powered by Ouiatchouan River waterfalls. Though Jalbert died just three years later, the mill survived a resulting cash shortage and began to thrive under new owners. From 1909 until the early 1920s, Val-Jalbert was a prosperous community. Its 200 mill workers and their families had access to a school, a general store, a butcher's shop, and a post office, all enhanced with electricity.

In 1927, however, the reduced global demand for pulp forced the mill to close. At that time, Val-Jalbert had a population of 950. By the early '30s, it was a ghost town. Transformed into a tourist attraction in 1960, the village is now open for exploration from June to October. Many of the buildings have been restored while others sag inward, their wooden walls splintered and ridden with holes. Much of the mill's original machinery is intact, and the old butcher's shop now houses an old-timey photo booth.

Before you leave, head to the river's viewing platform for a close look at Ouiatchouan Falls, which, at 236 feet (72 m), is taller than Niagara Falls.
Highway 169, between Chambord and the city of Roberval. Ⓝ 48.444660 Ⓦ 72.164561

The Crutches of St. Joseph's Oratory

MONTREAL

Soaring 506 feet (154 m) above street level, the top of St. Joseph's dome is the highest point in Montreal. The hilltop basilica is the legacy of Brother André, a Quebecois man known for providing hope and reassurance to the sick. In 1904, he approached the city's archbishop for permission to build a shrine to St. Joseph, of whom he was a lifelong devotee. Afflicted by poor health since childhood, Brother André often prayed to St. Joseph and encouraged others to trust in his healing powers.

The archbishop couldn't fund Brother André's dream, but he did allow him to build a modest chapel. It wasn't until 1924 that construction on St. Joseph's Oratory basilica finally began. It was completed in 1967, 30 years after Brother André's death at the age of 91.

People who are unwell or chronically ill, or have limited mobility flock to St. Joseph's for comfort and the chance to be miraculously healed. The oratory's Chapel of Brother André is lined with rows of crutches left by those who claim to have been given the ability to walk again. (Brother André himself was canonized as a saint in 2010, so miracles are credited to his intercession as well as St. Joseph's.)

The heart of Brother André sits inside a glass box in the oratory's reliquary. In March 1973, a band of brazen thieves stole the organ. Over a year later, following an anonymous tip-off, the heart was recovered from a locker in the basement of an apartment building. It is now back on display.
3800 Chemin Queen Mary, Montreal. The oratory is near the Côte-des-Neiges metro station. There are 283 steps from the street to the basilica—pilgrims traditionally climb the middle set of 99 steps on their knees. Ⓝ 45.492171 Ⓦ 73.616943

NEW BRUNSWICK

World's Largest Axe

Nackawic · Built in 1991, this massive tool has a 50-foot (15.2 m) handle and a time capsule embedded in its axe head.

Old Sow Whirlpool

Deer Island Point · Legend says this tidal turbulence got its odd name because it sounds like a grunting pig.

Walking aids left by those who say Brother André restored their paralyzed legs.

USA

West Coast
CALIFORNIA · OREGON · WASHINGTON

Four Corners and the Southwest
ARIZONA · COLORADO · NEVADA · NEW MEXICO · TEXAS · UTAH

Great Plains
**IDAHO · KANSAS · MONTANA · NEBRASKA · NORTH DAKOTA
OKLAHOMA · SOUTH DAKOTA · WYOMING**

The Midwest
**ILLINOIS · INDIANA · IOWA · MICHIGAN · MINNESOTA · MISSOURI
OHIO · WISCONSIN**

The Southeast
**ALABAMA · ARKANSAS · FLORIDA · GEORGIA · KENTUCKY
LOUISIANA · MISSISSIPPI · NORTH CAROLINA · SOUTH CAROLINA
TENNESSEE · VIRGINIA**

The Mid-Atlantic
**DELAWARE · MARYLAND · NEW JERSEY · NEW YORK
PENNSYLVANIA · WASHINGTON, DC · WEST VIRGINIA**

New England
**CONNECTICUT · MAINE · MASSACHUSETTS · NEW HAMPSHIRE
RHODE ISLAND · VERMONT**

Alaska and Hawaii

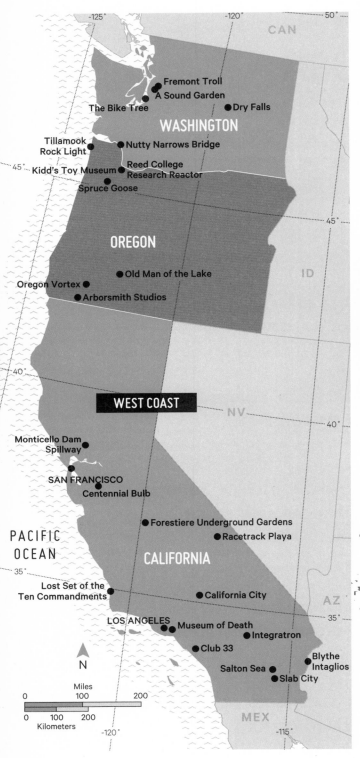

CALIFORNIA
BLYTHE INTAGLIOS

BLYTHE

Similar to the Nazca Lines of Peru, the Blythe Intaglios are huge designs scraped into the ground of the desert. There are six figures in three locations, the largest of which is a 171-foot-tall (52 m) human. The intaglios were likely carved by Mojave and Quechan Indians, but the date of their creation is unknown.

15 miles (24 km) north of Blythe on Route 95.
Ⓝ 33.800333 Ⓦ 114.531883

The Native Americans who scratched enormous human figures into the Colorado Desert could only have imagined what they'd look like from the air.

171 feet / 52 meters

SALTON SEA

BOMBAY BEACH

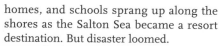

In the 1950s and '60s, Bombay Beach was a thriving seaside resort. Guests swam, water-skied, and golfed during the day, then headed to the North Shore Yacht Club to party into the night.

Now Bombay Beach is a bleached, rusted, abandoned wasteland. The water smells of salt, petrol, and rotting fish. The shores, once lined with sunbathers, are covered in green sludge and desiccated fish carcasses. It's an apocalyptic landscape.

To understand how this place turned from paradise to purgatory, you need to know the story of the Salton Sea. Bombay Beach lies not on the Pacific Coast, but in the middle of the Colorado Desert. In 1905, the Colorado River swelled, breached its levees, and flooded the desert valley known as the Salton Sink. The water flowed for two years, creating a 15 × 35-mile (24 × 56 km) lake dubbed the Salton Sea.

Though the creation of this inland sea—the largest lake in California—was an accident, it initially appeared to be a happy one. Birds flocked to the area, and fish thrived in the Salton Sea. Developers seized upon the rare setting and branded it the "Salton Riviera," a "miracle in the desert." Hotels, yacht clubs, homes, and schools sprang up along the shores as the Salton Sea became a resort destination. But disaster loomed.

By the late 1970s, the ecosystem was deteriorating rapidly. With no drainage outlet, almost zero yearly rainfall, and runoff flowing in from nearby farms, the sea was polluted with pesticides and turned saltier than the Pacific Ocean. Thousands of dead fish washed up on the beach and rotted. As they decomposed, the sand became coated in a layer of fragmented fish skeletons.

Hints of the Salton Sea's heyday still litter the shores. Boarded-up motels, rusting boat frames, and cracked concrete swimming pools covered in graffiti are a few of the sights that remain. People do live here—Bombay Beach is home to around 250 residents, who travel the barren landscape by golf cart and must drive 40 miles (65 km) to stock up on groceries.

There is one part of the Salton Sea that's shiny and new: The North Shore Beach and Yacht Club, long abandoned, was refurbished in 2010 and opened as a community center.

Start your visit at the Salton Sea History Museum, located at 72–120 South Lincoln Street, Mecca. Ⓝ 33.253533 Ⓦ 115.710179

Sun-bleached and over-salinated, Salton Sea turned from a resort paradise to a wasteland.

Sculptor Joe Holiday's woolly mammoth made of discarded truck tires.

SLAB CITY

NILAND

Known to its residents as "the last free place," Slab City is an isolated, off-the-grid desert community of squatters. Established in the early 1960s on a former United States Marine Corps training base, it is home to a motley mix of artists, travelers, retirees, and snowbirds. There is no local government, no running water, and no waste disposal system. Power comes from solar panels and generators; nightfall brings pitch darkness to the rattlesnake- and scorpion-infested sands.

Slab City's population varies from a few dozen in summer—when temperatures reach up to 120°F (49°C)—to a few thousand during the cooler months. Residents live in RVs, vans, tents, and converted school buses, and venture to a hot spring to bathe. Saturday is talent show night at The Range, an outdoor concert venue with mismatched chairs and tattered couches for seating.

Within Slab City is a sub-community, East Jesus. The art collective grew from an old shipping container in 2006, and centers on an inhabitable, ever-growing artwork made entirely from salvaged materials. Among them are a giant mammoth made from blown-out car tires, a television graveyard, and a swing set made from the rusted skeletons of old car seats. According to its custodians, East Jesus is "populated by an ever-rotating cast of artists, builders, writers, musicians, freethinkers, merry pranksters, wandering messiahs, the dispossessed, the damned." **East Beal Road, Niland. Drive toward Niland on Highway 111, turn east on the street by United Grocery. Travel for 3.5 miles (5.6 km). Salvation Mountain marks the entrance. ℕ 33.258889 𝕎 115.466389**

THE INTEGRATRON

LANDERS

Whisper while standing in the middle of the Integratron—an acoustically perfect dome in the desert—and you'll feel as though your brain is talking to you via vibrations. The sensation is odd and otherworldly, much like the Integratron itself.

In 1952, former flight mechanic George Van Tassel had a life-changing experience. As he described in his 1955 book, *I Rode a Flying Saucer*, Van Tassel was sleeping when a creature from the planet Venus woke him up, took him aboard his spacecraft, and telepathically divulged the secret of eternal youth. Rejuvenation of the human body, according to the Venusian visitor, required the construction of a domed structure, built without metal and featuring time-travel and anti-gravity devices.

Armed with these instructions, Van Tassel set about building what he called the Integratron in the desert north of Joshua Tree National Park. The location was important—Van Tassel believed the spot contained powerful geomagnetic forces that, when harnessed by the Integratron, could recharge human cells.

Using wood, fiberglass, concrete, and glass—but no nails, screws, or any other metal—he constructed a two-story, 38-foot-high (12 m) domed building. The time-travel device and anti-gravity chamber, however, never came to be—Van Tassel died of a heart attack in 1978 with his creation incomplete.

These days the Integratron offers "sound baths," which involve up to 30 people lying in a circle on the floor, feet pointed outward, while a facilitator strikes a series of quartz bowls that resonate at different frequencies. (According to the owners, this experience provides relaxation and "sonic healing" by causing parts of the body to vibrate individually.)

The dome is so acoustically sensitive that soft sniffs and shuffles are clearly audible to people on the opposite side of the room. It is for this reason that sound baths begin with a stern warning: no snoring. **2477 Belfield Boulevard, Landers. The Integratron is a 2.5-hour drive east of Los Angeles. Public sound baths happen twice a month. You can also book a private bath. ℕ 34.293490 𝕎 116.403976**

Experience a quartz-crystal sound bath inside a desert dome.

CITY GUIDE: More to Explore in Los Angeles

California Institute of Abnormalarts

North Hollywood • This venue for out-there music, B movies, and self-described freak shows, and home to a 100-year-old mummified clown is owned by an ex-mortician with a love for curios.

The Abandoned Old Zoo

Griffith Park • Built in 1912 and abandoned in the 1960s, the old Griffith Park Zoo is open to the public for hikes, picnics, or a climb inside an abandoned monkey cage.

Ennis House

Los Feliz • A monumental architectural wonder constructed mainly from concrete blocks, this Frank Lloyd Wright marvel has the cracks and pockmarks from nearly a century of seismic shifts, but its Mayan-inspired design hasn't suffered one bit.

Philosophical Research Society

Los Feliz • This library and bookstore has been at the center of altered consciousness, obscure religious philosophies, and study of the occult since its founding in 1934, but the library is a closed-stacks collection, so please don't try to take anything home. Karma (or maybe the librarian) will put a stop to that.

Snow White Cottages

Los Feliz • These eight storybook cottages, just a few blocks from Walt Disney's original studio, very likely inspired one of the most famous animated films of all time—and decades later ended up on the other side of the cinematic spectrum as a location in the David Lynch thriller *Mulholland Drive*.

Museum of Broken Relationships

Hollywood • The brainchild of two Croatian artists, this offshoot of the Zagreb original provides the brokenhearted with a place for their love affair leftovers, all donated anonymously, inviting the viewer to impose their own personal stories of heartache on the parade of seemingly mundane objects.

Holyland Exhibition

Los Feliz • The rumored real-life inspiration for Indiana Jones, Antonia Futterer was the last of a breed, an LA original. Tour his personal collection of Middle Eastern arts and artifacts, all packed into a hillside Los Feliz bungalow.

The Echo Park and Mar Vista Time Travel Marts

Venice • It can feel like an endless trek from Echo Park to Mar Vista, but these eccentric fronts for a nonprofit writing workshop have all the necessary supplies for your trips across town or to the fourth dimension, including dinosaur eggs, canned woolly mammoth chunks, Fresh-Start clones, and robot toupees.

Bob Baker Marionette Theater

Westlake • A Los Angeles Historic-Cultural Monument and the oldest children's theater in LA, with thousands of marionettes (not creepy ones!) and free cups of ice cream—who doesn't love a great puppet show?

Velveteria

Chinatown • With thousands of trippy, retro, black-lit beauties, the art of velvet painting is not only kept alive with this collection but elevated to a high (in more ways than one) art.

Skeletons in the Closet

LA County • This coroner's office gift shop will set you up with a supply of toe tags, body bags, and chalk-outline beach towels. The healthy mail-order business for their not-so-healthy-sounding catalog helps to raise money for safe teen driving programs.

The Bradbury Building

Downtown • Story goes that the architect of this deco downtown building—made famous in movies like *Blade Runner* and *The Artist*—wouldn't start building until he got an otherworldly OK from his dead brother.

Clifton's Cafeteria

Downtown • This psychedelic, kitschy, giant-redwood-themed cafeteria has been operating for 80 years under the gaze of stuffed bears and mounted moose heads, and is guided by its own golden rule: "Dine Free Unless Delighted."

Velaslavasay Panorama

University Park • This theater and exhibition hall—with a lush backyard garden—provides patrons with a 360-degree experience of wonder, a kind of early virtual reality from a time before the movies, when worlds on spinning painted backdrops could transport a crowd to dreamy and exotic landscapes.

Museum of Jurassic Technology

Culver City • For nearly 30 years, this stalwart curiosity cabinet has provided LA with hushed, soft-focus exhibits that are confounding, freakishly mind-bending, and challenge your sense of what is real.

The Wende Museum

Culver City • Housed in an inconspicuous business park is a stash of secret Eastern bloc spy equipment and Cold War–era paraphernalia—and a big stretch of the original Berlin Wall.

Saydel, Inc.

Huntington Park • Never pay retail for your religious effigies, magic soaps, and New Age candles—the folks at Saydel bring them to customers at low-low wholesale prices.

Old Town Music Hall

El Segundo • The Mighty Wurlitzer has been pumping out the tunes at downtown El Segundo's historic home for silent and classic movies since 1968 under the tickling fingers of in-house historian and organ master Bill Field.

The Bradbury Building's soaring interior.

CLUB 33

ANAHEIM

In Disneyland's New Orleans Square, near the exit to the Pirates of the Caribbean ride, is an unmarked gray-green door. The door itself is unmarked, but to its right is a sign bearing the number 33 and a brass speaker box with a panel that hides a buzzer. This is the entrance to Club 33, an ultra-exclusive, exorbitantly priced club that Walt Disney himself established so he would have a swanky place to entertain his VIP guests.

You must be a Club 33 member—or the guest of one—to dine at the secret restaurant. Membership requires an initiation fee of $25,000, followed by annual dues of $10,000. Membership is capped at 500 people, and it can take many years to be accepted.

Lunch or dinner at Club 33 is by reservation only. After being buzzed in through the secret door, you'll be ushered to the second-floor dining room, passing one of only two bars in Disneyland that serves alcohol. The restaurant is decked out in 19th-century New Orleans style, complete with a wooden elevator and wicker toilets. Food and drinks are served by doting staff wearing special Club 33 uniforms. They are only too happy to point out the original *Fantasia* animation cels on the walls and the table from the set of the first *Mary Poppins*.

Walt Disney never saw Club 33—he died in December 1966, five months before it opened. Evidence of his personal touch, however, remains. In one corner of the former trophy room is an animatronic California turkey vulture. Disney envisioned the bird surprising guests by talking to them via a voice actor hidden in a booth. Microphones were installed around the room for this purpose—but have since been disconnected.

33 Royal Street, Disneyland, Anaheim. To gain access, befriend a Club 33 member, book your meal months in advance, and prepare to spend a great deal of money on food, drinks, and exclusive souvenirs.
Ⓝ 33.810987 Ⓦ 117.921459

Jim Jones

Aileen Wuornos

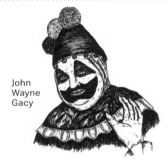

John Wayne Gacy

MUSEUM OF DEATH

HOLLYWOOD

Your visit to the Museum of Death begins with a test: Look at the photo on the wall next to the front desk. It shows a man's freshly mutilated body, parts of it scattered across a road following a truck crash. If the picture makes you feel queasy, this place is not for you.

Established in 1995, the museum is a graphic, shocking ode to the myriad ways humans shuffle off this mortal coil. A 45-minute self-guided tour through the small building takes in capital punishment, cult suicides, traffic accidents, and serial murders. A display of body bags, coffins, and mortician instruments serves as a reminder that death is the great equalizer.

Standout exhibits include the Heaven's Gate room, a re-creation of the scene that greeted investigators when they entered a San Diego mansion in March 1997 to discover 39 cult members had committed suicide. Each person had swallowed a lethal dose of phenobarbital and—covered in a purple shroud and wearing brand-new Nikes—lay neatly on a bunk bed, believing their souls would be transported to a higher realm on an alien spacecraft. In the museum version, mannequins lie on a set of beds taken from the actual house. They wear shrouds and Nikes removed from two of the deceased followers.

High-profile cases such as the Black Dahlia, Charles Manson, and John Wayne Gacy receive name checks, but the more fascinating artifacts belong to lesser-known murderers. On one wall is a set of photos taken by a couple. The images depict the duo grinning at the camera as they hold up the shredded body parts of the man they have just killed. This being the days before digital cameras, the twosome were caught and charged with murder after they took the incriminating film to a lab to be developed.

6031 Hollywood Blvd., Hollywood. Take the metro to Hollywood and Vine and walk 2 blocks east.
Ⓝ 34.101943 Ⓦ 118.321201

LOST SET OF THE TEN COMMANDMENTS

GUADALUPE

In 1983, a trio of film buffs journeyed to the Colorado Desert in search of the lost Egyptian City of the Pharoah. Using clues gleaned from Cecil B. DeMille's posthumously published autobiography, they found sphinxes and pharaohs buried in dunes.

DeMille had the inside scoop because he built the City of the Pharaoh himself—it was the film set for his 1923 silent epic, *The Ten Commandments*. Built by a team of 1,500, the lavish set featured a 720-foot-long (220 m) palace decorated with hieroglyphs, four giant statues of Pharaoh Ramses II, and a grand path lined with 21 five-ton sphinxes.

When the shoot ended—behind schedule and over budget—DeMille was left with a massive array of plaster buildings and props. Transporting it all back to Hollywood would have been exorbitantly expensive, but the director didn't want to leave the set in the desert where low-budget filmmakers could use it to shoot their own epics. So DeMille dismantled the sets and buried the parts in the sand.

The City of the Pharaoh lay beneath the dunes for 60 years. When filmmaker Peter Brosnan read a cryptic joke in DeMille's autobiography about hapless archaeologists finding a lost Egyptian city in the dunes, he spearheaded the search for the Lost City of Cecil B. DeMille.

Brosnan and archaeologists surveyed the site using ground-penetrating radar. Lack of funds and permit issues with the county thwarted plans for a full-scale dig. The set pieces made of plaster crumbled into dust when unearthed.

Much of DeMille's city remains buried, but a few excavated artifacts—including a makeup compact, film reel tin, and re-created sphinx head—are on display at the nearby Dunes Center.

Guadalupe-Nipomo Dunes Center, 1065 Guadalupe Street, Guadalupe. Ⓝ 34.972501 Ⓦ 120.572081

CALIFORNIA CITY

CALIFORNIA CITY

From the air, this collection of streets resembles a printed circuit board. Its carefully planned cul-de-sacs and concentric curved roads are neat and densely packed. The layout of streets and services is logical, aesthetically pleasing, and well suited to a major city. There's just one thing missing: people.

In 1958, a former sociology professor named Nat Mendelsohn purchased 80,000 acres of land in the Mojave Desert. His grand plan was to create California's next great metropolis—a thriving, car-centric city to rival Los Angeles. Mendelsohn mapped out a network of roads, named them all, and oversaw their construction. The streets surrounded Central Park, an 80-acre green expanse incorporating an artificial lake. Model homes and architectural drawings showed an enticing, affordable version of the American Dream.

By January 1959, there were 65 homes in California City. But this influx of people did not herald a mass migration. As the years went by, a trickle of families established homes, but for the most part, the carefully laid out streets remained quiet and empty. Mendelsohn bailed on his planned city in 1969, selling it to a Denver-based sugar and mining company.

By 1990, the population was hovering at just over 6,000. In 2000, it was 8,388. As a major metropolis, California City was a total failure. But its cracked, sand-sloughed streets now attract off-road adventurers, who enjoy careening around the curves of the uninhabited roads on motorcycles and ATVs. The current population is around 14,000, but the houses are spaced oddly. Some blocks are crammed with homes, while others have just one dwelling. All are buffered by a swath of streets that carve up the vacant lots.

Located about 100 miles (160 km) north of Los Angeles, between State Route 14 and US Route 395. Ⓝ 35.125801 Ⓦ 117.985903

Miles of streets without houses are the only remains of a midcentury ghost city that was planned to rival Los Angeles.

RACETRACK PLAYA

DEATH VALLEY

When no one is looking, the rocks on the dry lake bed of Racetrack Playa move. Some have only traveled a few inches. Others have journeyed half a mile. All of them leave telltale trails—some straight; some curved; others erratic and jerky, as if the rock changed its mind along the way.

Until December 2013, no one had ever witnessed the rocks in motion, but plenty had offered theories to explain their movement. A 2010 NASA study concluded that melted snow had streamed from the surrounding mountains and flooded the playa. At night, according to NASA, the water froze around the bottom of the rocks, creating an "ice collar." Over the next month, more water from the mountains arrived, creating a slippery surface and allowing the ice-collared rocks to float on the playa. Fierce winds of up to 90 miles (145 km) per hour sent the stones skidding across the plain.

It was a sound enough theory, but actual evidence was hard to come by. No one is allowed on the playa when it is wet, as their footprints would scar the ground, and research must be noninvasive—meaning rocks can't be disturbed and cameras must be hidden in the landscape.

Then in 2013, paleobiologist Richard Norris and his cousin, research engineer James Norris,

The mysterious heavy stones of Death Valley can sail up to a half mile across the desert.

happened to be in the right place at the right time. In front of their eyes, wind pushed an ice floe across the playa, causing one of the rocks to slide along the slick surface of the lake bed. From their position on the mountainside next to the playa, the Norris cousins began taking photos. The evidence was in and the mystery solved.

Death Valley National Park. The 26-mile (42 km) road to Racetrack Valley begins near Ubehebe Crater. You'll need a 4-wheel drive vehicle. Once you reach the playa, cars are prohibited. Bring sturdy shoes.
Ⓝ 36.681069 Ⓦ 117.560258

FORESTIERE UNDERGROUND GARDENS

FRESNO

It took a few minutes of digging in the dirt before Baldassare Forestiere realized he'd made a terrible mistake. It was 1906, and the Sicilian had arrived in California to fulfill his dream of becoming a fruit merchant. But when he stuck a shovel into the 80 acres of Fresno land he'd just purchased, Forestiere was dismayed to discover hardpan, a layer of soil too dense for planting fruit trees.

The hard soil wasn't the only problem. Fresno sweltered in summer, regularly reaching 115°F (46°C). Determined to make the best of the situation, Forestiere came up with a new plan: He took the gardens underground, where the soil is softer and the air is cooler.

Fruit trees stay cool in a subterranean villa built over four decades.

It was a long-term project. Over 40 years, and without ever putting his design plans on paper, Forestiere hand-carved catacombs, arches, and alcoves beneath the hardpan surface. He planted oranges, lemons, grapefruit, mulberries, and grapevines. Skylights and open-air courtyards provided enough light to keep the fruits growing without overheating them.

Forestiere didn't just keep his gardens underground—he lived there himself. Taking advantage of the cooler subterranean temperatures (which range from 10 to 30 degrees lower than the surface), he installed a bedroom, bath, kitchen, and dining area.

The Forestiere Underground Gardens only stopped expanding when Forestiere died in 1946. Today the gardens are open to visitors. You can sample fruit from some of the trees Forestiere planted—it's for sale on site.

5021 West Shaw Avenue, Fresno. The gardens are a 10-minute drive north of downtown Fresno, right off Highway 99. Tours are available. They are closed during winter.
Ⓝ 36.807573 Ⓦ 119.881809

Golden Fire Hydrant

The Castro • The gilded hydrant below Dolores Park earned its special status when it continued to function in the aftermath of the 1906 quake, saving much of the Mission District from burning down.

Head of the Goddess of Progress

Civic Center • The Goddess of Progress was a 20-foot-tall (6 m) bronze statue on top of San Francisco's original city hall. Her head, which weighs 700 pounds (317.5 kg), sits in the current city hall.

McElroy Octagon House

Cow Hollow • This eight-sided powder-blue home is a rare remnant of the octagon-shaped house-building craze of the second half of the 19th century.

Vaillancourt Fountain

Embarcadero • Despite drawing criticism since its installation in the early 1970s, this artistic jumble of angular cement pipes stands strong.

Westin St. Francis

Financial District • Since 1938, this hotel has provided a coin-washing service to prevent guests' hands being sullied by filthy cash.

Corporate Goddess Sculptures

Financial District • A dozen eerily faceless, draped figures loom over pedestrians from 23 floors up at 580 California Street.

Musée Mécanique

Fisherman's Wharf • Save up your quarters for the world's largest privately owned collection of coin-operated arcade machines.

California Academy of Sciences Herpetology Department

Golden Gate Park • The academy's 300,000-strong collection of jarred reptile specimens was amassed over 160 years. Viewing is by appointment only.

Drawn Stone

Golden Gate Park • A huge crack in the ground outside the de Young Museum was put there on purpose by the wry English artist Andy Goldsworthy.

Buena Vista Park Tombstones

Haight-Ashbury • Broken Gold Rush–era gravestones line the gutters of this park's paths.

Secret Tiled Staircase

Inner Sunset • The 163 colorful steps in this staircase form a vibrant mosaic that leads you to a smashing view of the city.

Sutro Egyptian Collection

Lakeshore • The antiquities housed here include two intact mummies, three mummified heads, and a mummified hand.

San Francisco Columbarium

Lone Mountain • This beautifully restored neoclassical atrium offers thousands of alcoves for burial urns. Reservations accepted.

Wave Organ

The Marina • Waves make music by crashing against the 20 pipes of this lovingly built organ, located on a jetty in the San Francisco Bay.

Palace of Fine Arts

Marina District • A piece of San Francisco's 1915 Panama-Pacific International Exposition survives in the form of this collection of classically inspired buildings, gardens, and fountains.

Institute of Illegal Images

Mission District • Mark McCloud's collection of acid blotters, amassed since the 1960s, makes for a uniquely hallucinogenic art gallery.

Good Vibrations Antique Vibrator Museum

Nob Hill • Shake things up with a visit to an institution that honors the Victorian cure for "female hysteria."

Seward Street Slides

Noe Valley • Two steep concrete slides provide thrills for the bold, provided you wear sturdy pants and bring your own cardboard sled. Adults must be accompanied by a child.

Cayuga Park

Outer Mission • Beside some elevated train tracks is a park filled with mystical wooden sculptures.

Yoda Fountain

The Presidio • A bronze version of the big-eared Jedi master stands atop a fountain at the Letterman Digital Arts Center.

Prelinger Library

SoMa • Organized in a unique way that encourages browsing, this privately funded library is open to all for "research, reading, inspiration, and reuse."

CENTENNIAL BULB

LIVERMORE

In June 2015, Fire Station 6 hosted a raging party in honor of a lightbulb. The occasion was a special one: It marked one million hours of service for the bulb, which hangs on a cable from the station's ceiling.

The Centennial Light Bulb, as it is affectionately known, was first screwed into a socket in 1901. Having shone ever since, with just a few minor interruptions, it has been Guinness-certified as the longest-lasting bulb in the world. Initially installed in the fire department's hose cart house, and moved to other fire stations over the years, it found a permanent home at Fire Station 6 in 1976.

There is no obvious reason why this light has lasted so long. Though its output has dimmed from 60 watts to four, the Centennial Bulb's glow has stayed steady for decades and shows no signs of weakening. There are no plans for what to do with it should the dark day of its demise arrive. **Fire Station 6, 4550 East Avenue, Livermore. You are welcome to visit the bulb if there are firefighters around to let you in. Ring the bell at the rear of the station and see if anyone answers. Ⓝ 37.680283 Ⓦ 121.739524**

MONTICELLO DAM SPILLWAY

LAKE BERRYESSA, NAPA COUNTY

Following heavy rain, a giant plughole funnels water from Lake Berryessa. The 72-foot-wide (22 m) Morning Glory spillway, known locally as the "glory hole," clears excess water to prevent spilling over the Monticello Dam.

While other drain-style spillways exist, the one at the Monticello Dam is the largest. Water rushes through the tapered pipe, under the dam wall, at up to 362,000 gallons (1,370,319 L) per second, traveling 700 feet (213 m) to exit at the south side of the canyon beneath the dam.

The deadly force of the flow means swimming near the dam is prohibited—in 1997, a UC Davis graduate student who defied the ban drowned after being pulled into the hole. During the dry season bikers and skateboarders have been known to circumvent the barbed-wire fences and use the exit section of the spillway as an unauthorized skate park.

The spillway is best seen during the rainy season when the lake is full. Ⓝ 38.512201 Ⓦ 122.104748

ALSO IN CALIFORNIA

Winchester Mystery House

San Jose · The former home of Sarah Winchester, heir to the Winchester gun fortune, contains doors to nowhere, stairs that stop suddenly, and secret passages.

Dymaxion Chronofile

Stanford · From journals to blueprints to dry-cleaning bills, Buckminster Fuller documented his life in staggering detail. The full archive is stored at Stanford University Library.

Lake Berryessa's massive overflow drain is known affectionately as the "glory hole."

OREGON

REED COLLEGE RESEARCH REACTOR

PORTLAND

Strictly speaking, a liberal arts college does not *need* a nuclear reactor. But there are myriad benefits of having one. Undergraduates at Reed College—an institution that does not offer a nuclear engineering program—have been using an on-campus reactor to conduct experiments for their theses since 1968.

Students using the reactor come from all majors, including English, philosophy, history, and art. Before being allowed to press any buttons, they must attend a year of seminars on nuclear safety and undergo a seven-hour exam administered by the Nuclear Regulatory Commission.

The water-cooled Reed reactor, which sits at the bottom of a 25-foot-deep (7.6 m) tank, generates neutrons that can be used to make materials radioactive. This allows students to determine the quantity of elements in a sample by measuring the amount of radiation they emit—a process known as neutron activation analysis. For instance, a history major might irradiate a pottery shard in order to find out where it came from by analyzing its trace elements.

Used only for university and community research, Reed's reactor is considered a "zero-risk facility" as far as public safety is concerned. That said, the student operators are still careful—following a federal inspection, they removed the family of rubber ducks from the surface of the tank water on the grounds that they violated safety protocols.

Reed College, 3203 SE Woodstock Boulevard, Portland. The reactor is on the east side of campus. Visits and tours must be arranged at least a week in advance. ℕ 45.480571 𝕎 122.630091

Undergraduate students operate this nuclear reactor—after going through stringent training, of course.

KIDD'S TOY MUSEUM

PORTLAND

It would be easy to miss the entrance to Kidd's Toy Museum but for the sign on the blue warehouse door. Beneath the official opening hours is a phrase hinting at the low-key nature of the place: "Other hours by appointment or chance."

Behind the blue door is a hoard of toy planes, trains, cars, mechanical banks, dolls, and collectible characters, dating mostly from 1869 to 1939. Every item is from the personal collection of Frank Kidd, who has spent a lifetime acquiring thousands of vintage toys.

Among the charming glimpses into childhoods past are some toys that evoke abhorrent beliefs of earlier eras—you'll see racist images and phrases on some of the mechanical banks, and the collection of golliwog dolls harks back to minstrelsy.

1301 SE Grand Avenue, Portland. The museum is in an unmarked building one block north of the Southeast Grand/Hawthorne light rail stop. Look for the blue door on the west side of the street, south of Southeast Main Street. ℕ 45.513518 𝕎 122.661068

THE SPRUCE GOOSE

MCMINNVILLE

"I put the sweat of my life into this thing. I have my reputation all rolled up in it and I have stated several times that if it's a failure I'll probably leave this country and never come back. And I mean it."

So said Howard Hughes in 1947 at a Senate War Investigating Committee hearing. "This thing" was the H-4 Hercules plane, also known as the Spruce Goose, the Flying Boat, and the Flying Lumberyard among Hughes's many detractors.

By the time he appeared before the Senate committee, Hughes had been working on the Spruce Goose for five agonizing years. In 1942, the US War Department needed to move troops and tanks across the ocean to the World War II front. German U-boats were attacking Allied ships in the Atlantic, so a plane big enough to house a tank was needed for the job. Hughes, the notoriously eccentric owner of the Hughes Aircraft Company, won the development contract.

There were several logistical issues right off the bat. Metals were needed for the war effort, so the plane was to be made of wood. It was to be the largest aircraft ever constructed—five stories tall, with a wingspan longer than a football field. The original contract called for three planes to be made within two years, but Hughes worked so slowly that the war was over well before he finished building one.

By 1947, Hughes had spent $23 million in government funds and the H-4 still hadn't flown. Skeptics questioned whether it was capable of flight at all. In August came the Senate hearings, during which Hughes curtly refuted allegations of misappropriating federal dollars. In November, determined to prove his critics wrong, Hughes invited members of the press aboard the flying boat for nonflight "taxi tests" in Long Beach harbor.

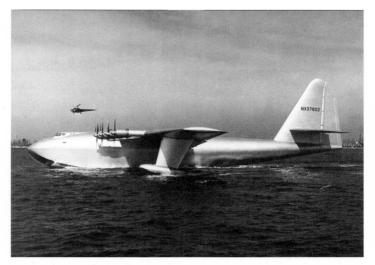

Howard Hughes's hefty wooden plane aimed high but flew for a total of one minute.

During the third test, however, the H-4 lifted off and flew 70 feet (21 m) for about a mile. That one-minute flight was the first and last time the Spruce Goose flew.

Ever the optimist, Hughes kept the H-4 in a climate-controlled hangar for the next 29 years, attended by a full-time crew. After Hughes died in 1976, the Spruce Goose eventually ended up at the Evergreen Aviation & Space Museum, where it remains to this day. You can walk through the massive fuselage and, for an extra fee, sit in the cockpit wearing a Hughesian fedora.

Evergreen Aviation & Space Museum, 500 NE Captain Michael King Smith Way, McMinnville. Ⓝ 45.204228 Ⓦ 123.145140

ALSO IN OREGON

Enchanted Forest

Turner · This charmingly homemade fairy tale park is still going strong despite being dubbed "Idiot's Hill" by its detractors.

Tree Climbing Planet

Oregon City · This farm just south of Portland is dedicated to teaching people the professional art of tree climbing.

OLD MAN OF THE LAKE

CRATER LAKE

A floating log may not seem note-worthy, but this one is special. The Old Man is a 30-foot (9 m) hemlock tree that has been bob-bing along vertically in Crater Lake since at least 1896. Bleached by the sun and blown across the water by wind currents, the tree protrudes about 4 feet (1.2 m) from the lake's surface.

The Old Man likely arrived in Crater Lake—a 6-mile-wide, 1,946-foot-deep (9.6 km, 593 m) blue pool that fills the crater of a dormant volcano—via landslide. Rocks were probably lodged in his roots, pulling his base down and allowing him to float vertically. The clear, cold water has kept the Old Man looking relatively youthful.

In the summer of 1938, two naturalists plotted the Old Man's

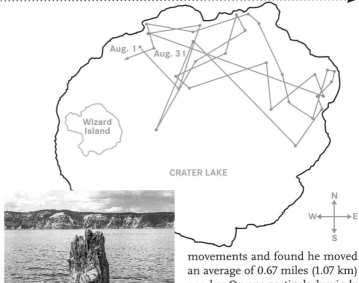

Tracking a month in the life of the floating log that has been drifting around Crater Lake since at least 1896.

movements and found he moved an average of 0.67 miles (1.07 km) per day. On one particularly windy August day he traveled 3.8 miles (6.11 km). When looking for him, keep his transient nature in mind. **Crater Lake National Park. The park is 80 miles (129 km) northwest of Medford. Ⓝ 42.868441 Ⓦ 122.168478**

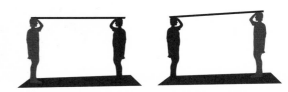

A simple optical illusion explains the slanting Mystery House's weird physics-flouting properties.

OREGON VORTEX

GOLD HILL

At the slanted House of Mystery on Gold Hill, balls roll upward and brooms stand unsupported. Their apparent defiance of the laws of physics is just one of many strange phenomena one can observe at the Oregon Vortex.

The House of Mystery was once the office of a local mining company. Built in 1890, it was aban-doned within 20 years, after which it slid off its foun-dations and landed at an odd angle. Officially, the house fell because of a mudslide. But the owners of the site have another explanation: magnetic vortex.

In 1930, the house opened to the public as part of the Fabulous Oregon Vortex, an attraction

designed to demonstrate the land's alleged physics-flouting properties. In addition to seeing the mys-terious balls and brooms, you can participate in height experiments: When walking from one side of a plank to the other, you appear to change size.

Guides work the vortex angle during each dem-onstration, throwing in the odd ghost story and paranormal explanation. It's a fun premise—and watching someone change size before your eyes is certainly mind-bending—but the real cause of all the weirdness is simple optical illusion.

4303 Sardine Creek Left Fork Road, Gold Hill. The House of Mystery is open between March and October. Ⓝ 42.493002 Ⓦ 123.084985

TILLAMOOK ROCK LIGHT

CANNON BEACH

Things did not begin well for the lighthouse that came to be known as Terrible Tilly. In 1878, the government decided that ships sailing around Tillamook Head needed more light and better guidance. There was no suitable place to build a lighthouse on the headland, so attention turned to Tillamook Rock, a big chunk of basalt 1.2 miles (1.9 km) offshore.

In 1879, mason John Trewavas set off for the island to do some surveying. When he attempted to land on the rock, his body was swept into the tumultuous sea. This would be but one of many deaths and calamities caused by Tillamook's wild waves and unforgiving landscape.

A team of quarrymen eventually managed to build the lighthouse in 575 days, but the process was arduous: They had to rig a line between their ship and the top of the rock and use it to transport both their tools and themselves. Less than three weeks before the lighthouse opened, a ship called the *Lupatia* sailed close to the shore in thick fog. The next morning, the bodies of all 16 crew members washed ashore.

Lighthouse operators at Terrible Tilly had their own trials to contend with. Hunkered down with six months of supplies, the men lived a mentally and physically challenging existence, beset by frequent storms and blaring foghorns.

The Tillamook Rock Light operated for 77 years before being decommissioned in 1957. Then came a fittingly macabre twist. In 1980, real estate developer Mimi Morissette bought Terrible Tilly and turned it into a columbarium: a storage house for urns full of cremated human remains. Dubbed "Eternity at Sea," the postmortem museum amassed a collection of over 30 urns before the Oregon Mortuary and Cemetery Board took away its license in 1999. (Less-than-stringent record-keeping and improper urn storage were among the issues.)

There are still urns full of remains inside the Tillamook building—"honorary lighthouse keepers," Eternity at Sea called them—but the rock is primarily a seabird nesting spot and can only be safely accessed by helicopter. **The lighthouse rock is part of the Oregon Coast National Wildlife Refuge Complex and therefore off-limits during seabird nesting season. Ⓝ 45.937225 Ⓦ 124.019055**

Built to keep sailors safe from harm, Tillamook's lighthouse has instead attracted death.

ARBORSMITH STUDIOS

WILLIAMS

Arborsculpture is the craft of shaping living trees into works of art and architecture. The term was coined by Richard Reames, one of its pioneering practitioners and the owner of Arborsmith Studios—a combination nursery, design studio, and al fresco art gallery.

By grafting, bending, and pruning, Reames has shaped trees into helixes, peace signs, lattices, chairs, and gazebos. It's a process that requires a lot of patience—larger projects can take a decade or more to complete.

Reames decided to beome an arborsculptor after seeing the Tree Circus, a roadside display of sculpted trees in Scott's Valley, California. Axel Erlandson, who created the Tree Circus in 1947, never called what he did "arborsculpture"—he died in 1963, decades before the term arose—but his creations, which included a ladder tree, a phone booth tree, and trunks that split and spiraled, certainly fit the category.

Reames says that he would love to see arborsculpture revolutionize construction methods. In his ideal world, people would grow houses out of living trees instead of building them from dead ones.

1607 Caves Camp Road, Williams. Arborsmith Studios isn't open to the public, but Reames occasionally allows visitors to tour his studios by prior arrangement. Ⓝ 42.184718 Ⓦ 123.330013

Richard Reames bends and prunes living trees to sculpt them into artwork.

NUTTY NARROWS BRIDGE

WASHINGTON

NUTTY NARROWS BRIDGE

An aerial crossing above Olympia Way provides safe passage for squirrels.

LONGVIEW

Look to the trees above Olympia Way at 18th Avenue and you'll see something unusual: a miniature suspension bridge made of aluminum piping and an old fire hose.

In 1963, Amos Peters would look out the window of his Olympia Way office and see squirrels struggling to cross the busy road. After witnessing a string of hit-and-runs, Peters came up with an idea: a bridge that would allow the skittish animals to make a midair run across the street.

The 60-foot (18 m) Nutty Narrows Bridge opened to squirrel traffic later that year. In the decades since, three similar bridges have popped up around Longview. The city has taken the squirrel theme and run with it—since 2011 it has hosted an annual Squirrel Fest with live music, model railroads, and squirrel-themed face painting.

Olympia Way (between 18th and 19th Avenues), Longview. Squirrel Fest happens annually in mid-August. Ⓝ 46.141424 Ⓦ 122.940344

DRY FALLS

COULEE CITY

Dry Falls is the location of the world's most spectacular waterfall. Unfortunately, you're 12,000 years too late to see it.

As the ice dams of Lake Missoula gave way at the end of the last Ice Age, trapped water began flooding Idaho, Washington, and Oregon. The thundering torrents of water and ice flowed at 65 miles (105 km) per hour, carving trenches into the basalt bedrock.

The floods cascaded over the 3.5-mile-wide, 400-foot-tall (5.6 km, 122 m) cliffs in central Washington now known as Dry Falls, creating the largest waterfall ever known. (By comparison, Horseshoe Falls—the largest of the Niagara waterfalls—is only 167 feet tall and half a mile wide [51 m, 0.8 km].)

With the ice sheets thawed and the waters neatly contained in rivers, Dry Falls is now desiccated. But walk out onto the lookout platform, gaze at the sculpted basalt cliffs, and imagine a waterfall so powerful that it flowed ten times faster than all the world's rivers combined. That's what you would have seen at the end of the last Ice Age.

Dry Falls is 7 miles (11 km) southwest of Coulee City.
Ⓝ **47.607205** Ⓦ **119.364223**

At the end of the last Ice Age, the world's most incredible waterfall cascaded down these cliffsides.

From tiny seeds grow mighty trees—even if there's a bike in the way.

THE BIKE TREE

VASHON

In a forest on Vashon Island, there is a bike in a tree. Not in the branches of a tree—embedded in its trunk, 12 feet (3.7 m) above the ground.

The mystery of how this Douglas fir ate a child-size bicycle has perplexed visitors for decades. A heart-tugging story tells of a teenage boy who chained his childhood bike to a tree and went off to fight in World War I, never to return. It's a compelling tale, but it's not true.

According to Don Puz, who grew up in the area during the 1950s, the bike once belonged to him. In 1954, when Puz was nine, his family's house burned to the ground, killing his father. The community responded with donations for the bereft boy, including a shiny new bicycle. But it was a little small, and Puz soon outgrew it. One day, after playing in the forest, he left the bike behind.

There may have been some human intervention in the ensuing decades—it would be an amazing tree indeed that could envelop a bike so neatly—but the image of a rusted child's bike embedded in a mighty Douglas fir has sparked many an imagination. In 1994, local author Berkeley Breathed published a children's book, *Red Ranger Came Calling*, which weaves the bike tree into a Christmas story.
Off Vashon Highway, Vashon Island, just north of SW 204th Street. Ⓝ 47.422995 Ⓦ 122.460085

A SOUND GARDEN

SEATTLE

In a park beside Lake Washington, on the campus of the National Oceanic and Atmospheric Administration, what looks like a collection of TV antennae or weather vanes rises. But there's something strange about these metal towers: They emit a low, haunting hum when the wind blows through them.

The structures—12 in all—form a public artwork called *A Sound Garden*. Installed by sculptor Douglas Hollis in 1983, the piece consists of wind-actuated organ pipes suspended vertically from steel towers. Seattle grunge band Soundgarden was so inspired by the sculpture that they adopted its name.
Warren G. Magnuson Park, 7600 Sand Point Way NE. Stop in at the guard station at NE 80th Street. Photo ID is required. Ⓝ 47.651034 Ⓦ 122.347323

THE FREMONT TROLL

SEATTLE

In a freeway underpass beneath Aurora Bridge is a giant cement troll clutching a Volkswagen Beetle. Long, bedraggled hair obscures his right eye. His left eye is a silver hubcap.

The 18-foot-tall (5.5 m) concrete monster has lived under the bridge since 1990, when the Fremont Arts Council held a public art competition to enhance the underpass. Local artists Steve Badanes, Will Martin, Donna Walter, and Ross Whitehead sculpted the troll, adding a real VW Beetle to make it look like he'd just swiped a car from the road above.

Within months of the troll's unveiling, vandals broke into the car and stole a plaster bust of Elvis, which had been placed inside as part of a time capsule. The car has now been filled with concrete.

Visitors are encouraged to climb on the sculpture, but modifying the troll's looks is frowned upon. (This doesn't stop people from painting his lips pink, giving him tattoos, and tagging him with their names.)

Under the Aurora Bridge at North 36th Street and Troll Avenue North. Ⓝ 47.680257 Ⓦ 122.253201

ALSO IN WASHINGTON

Mystery Soda Machine

Seattle · Feeling lucky? Hit one of the "mystery" buttons on this battered old soda machine and see what pops out.

Hall of Mosses

Forks · Hike along a gothic, untamed trail of trees that drip with strands of moss.

The beloved cement curmudgeon beneath Aurora Bridge gets redecorated regularly.

FOUR CORNERS AND THE SOUTHWEST
AND GREAT PLAINS

Four Corners and the Southwest

ARIZONA

TITAN MISSILE MUSEUM

GREEN VALLEY

When you go underground at the Titan Missile Museum just south of Tucson, you travel back in time to the Soviet-fearing days of "duck and cover," big red buttons, and mutually assured destruction. The museum, formerly known as Complex 571-7, is one of 54 subterranean missile sites across the country that were active during the Cold War. Decommissioned in 1982, the silo still contains

a 108-foot (33 m) missile—with the lethal bits removed.

The Titan II missile was capable of rapidly delivering a nine-megaton nuclear warhead to a target 6,300 miles (10,138 km) away. In other words, it could nuke Moscow within 30 minutes.

The launch process, simulated to exciting effect at the museum, began with a 35-character alphanumeric message from the president. The commander and officer at the launch center each copied down the message, then conferred with one another to make sure the codes they wrote were identical.

It was then time to open the Emergency War Order (EWO) safe, which contained authenticator

ARCOSANTI

MAYER

In 1970, Frank Lloyd Wright protégé Paolo Soleri took to the Arizona desert to break ground on an experimental community designed to forge a new way of urban living. Guided by principles of eco-friendliness, waste reduction, and what he called "elegant frugality," Soleri planned a hyperdense city: Arcosanti. The name incorporates "arcology," his concept for architecture-plus-ecology.

Intended as a test site for Soleri's urban development theories, the self-contained Arcosanti was designed to house 5,000 people. Though almost five decades have passed since the laying of the foundation stone, the city is still in its early construction stages. Lack of funding has kept Arcosanti small—it is now home to between 50 and 150 inhabitants, depending on the season.

Over the years, thousands of volunteers have helped construct apartments, storefronts, an outdoor amphitheater, and a visitor's center, all rendered in concrete with lots of arcs and semicircles. Landscaping workshops and internships are still available for

Arcosanti's experimental community is run by eco-loving volunteers.

people who want to be part of the urban experiment. Work at Arcosanti is funded by sales of bronze and ceramic wind chimes.

The buildings are a little rundown and shabby; Soleri passed away in 2013, so they'll likely stay that way. But Arcosanti is a fascinating look at one man's ambitious alternative to urban sprawl. **13555 South Cross L Road, Mayer. Arcosanti is located between Sedona and Scottsdale. Ⓝ 34.345418 Ⓦ 112.116278**

ALSO IN ARIZONA

Santa Claus

Mohave County · At this abandoned Christmas-themed town in the middle of the desert, the faded festive decor is covered with graffiti.

Antelope Canyon

Page · Due to flash-flood risks, the Southwest's most beautiful slot canyon can only be visited with a guided tour.

Pumpkin Spring Pool

Littlefield · Don't swim in this naturally formed limestone bathtub—its waters are laced with arsenic.

---

cards used to confirm that the message did indeed come from the president. Also in the safe: two launch keys, which the commander and officer would insert simultaneously at separate control stations.

Once the keys had been turned, there was no going back: 58 seconds later, the missile would be on its way to the preprogrammed target. Launch crews never knew the targets—it's easier to launch a nuclear warhead when you don't know who it will kill—and the three preprogrammed targets at this Arizona site are still classified.

In addition to the rare up-close-and-personal look at an intercontinental ballistic missile, the museum occasionally offers overnight stays. You and three friends can snuggle up in the old crew quarters, mere feet from one of the most murderous devices ever created.

1580 West Duval Mine Road, Green Valley. The museum is 25 miles (40 km) south of Tucson. Thirty days' notice is required to book a cozy overnight stay. Ⓝ 31.902710 Ⓦ 110.999352

Step inside a decommisioned missile silo for a taste of Cold War tension.

Built without blueprints, Bishop Castle has a fire-breathing dragon and a bridge to nowhere.

COLORADO

BISHOP CASTLE

PUEBLO

In 1969, at the age of 25, newly married Jim Bishop started constructing a stone cottage for his family. Over the decades, as he kept building, that stone cottage became a castle. Today it is a multilevel marvel with three towers, a grand ballroom, and a fire-breathing metal dragon guarding the main eave. And Bishop still isn't done.

The castle doesn't adhere to any building codes. There have never been any blueprints, and Bishop is quick to point out that while his father helped a tiny bit with initial construction efforts, he has done the vast majority of the work himself. (Just to make sure this point is crystal clear, the small section that his dad worked on is painted with the words "Jim started the castle, not his father, Willard.")

Bishop sees his castle as a symbol of American freedom. Signs surrounding the building tell of the local government's unsuccessful attempts to regulate his work. "They tried but failed to oppress and control my God-given talent to hand-build this great monument to hard-working poor people," one reads (in part—it's a long sign).

Bishop plans to keep building until he is no longer physically able. When visiting, you may see him carrying stones or making an impromptu speech from one of the towers—he is known to unleash his political views on visitors at high volume.

1529 Claremont Avenue, Pueblo. Admission to the castle is free—funding for ongoing construction comes from gift shop sales and donations. The castle is open during daylight hours. From I-25, take exit 74 at Colorado City toward the mountains. It's 24 miles (39 km) farther along Highway 165.
Ⓝ 38.240728 Ⓦ 104.629102 ➤

ALSO IN COLORADO

Dinosaur Best Western

Lakewood · This franchise hotel got a multimillion-dollar, dinosaur-themed makeover in 2013, complete with a T. rex skull, dino murals, and a pterosaur-shaped weather vane.

Paint Mines Interpretive Park

Calhan · The sandstone spires of this geological park run wild along the color spectrum, from creamy white to orange, purple, gray, rust, and chocolate brown.

Blue Mustang

Denver Airport · This reviled red-eyed, blue-skinned 32-foot (9.7 m) sculpture of a rearing bronco caused the death of its creator when a section of the 9,000-pound horse fell on him. Some Denver residents believe that the horse is cursed.

Solomon's Castle

Cano's Castle

➡ Other American Castles

MYSTERY CASTLE
PHOENIX, AZ

In the 1930s, a man named Boyce Luther Gulley left his wife and daughter in Seattle and fled to the desert to build this castle. But this was not a case of abandonment. Gulley had just discovered that he had tuberculosis, and intended to build the castle for his daughter, Mary Lou.

The 18-room, three-story building, created using rocks, salvaged scrap, adobe, mortar, cement, and goat milk, was passed on to Gulley's abandoned family members when he died in 1945. The mother and daughter moved into the property and Mary Lou led public tours until her death in 2010.

CANO'S CASTLE
ANTONITO, CO

Vietnam vet Donald "Cano" Espinoza built this set of four gleaming towers single-handedly, using beer cans, hubcaps, and other discarded metal. He has cited as his main influences Jesus and "Vitamin Mary Jane."

SOLOMON'S CASTLE
ONA, FL

In 1974, sculptor and pun lover Howard Solomon began constructing a castle out of aluminum printing plates salvaged from a local newspaper plant. The building is now a gleaming, medieval-style castle standing three stories high and incorporating a sculpture garden and a 60-foot (18.3 m) replica of a 16th-century Portuguese battleship.

RUBEL CASTLE
GLENDORA, CA

Embedded in the external walls of Rubel Castle are glass bottles, forks, motorcycle parts, and other trash yard treasures collected by builder Michael Clarke Rubel. Beginning in 1959 at the age of eighteen, Rubel spent decades constructing the castle using stones and scrap donated by friends. A drawbridge, dungeons, and cars from the 1920s are among the surprises hidden behind the 20-foot-high (6 m) wall surrounding the compound.

CORAL CASTLE
HOMESTEAD, FL

When 26-year-old Latvian Ed Leedskalnin was jilted by his sweetheart the day before their intended wedding, he dealt with his distress by moving to the US and making a castle out of coral. The 5-foot-tall (1.5 m), 100-pound (45 kg) Leedskalnin somehow hauled 1,100 tons (997 MT) of coral blocks around to create a monument to lost love. A 500-pound (227 kg) heart-shaped stone table is among the hints of heartbreak incorporated into the architecture. A plaque laid into one of the walls reads, cryptically, YOU WILL BE SEEING UNUSUAL ACCOMPLISHMENT.

Human error and geothermal pressure combined to create this rainbow wonder.

NEVADA
FLY GEYSER

GERLACH

These multicolored, knobby cones of calcium carbonate that spew water from multiple spouts are not natural—humans had a hand in their creation. In 1964, a geothermal energy company drilled a test well in the Black Rock Desert. The water they encountered was not hot enough to use, so they plugged up the well. But the seal didn't hold, and water began erupting into the air.

Over the years, the geyser grew as minerals from the water settled on its surface. Fly Geyser and the terraced mound on which it sits now measure 12 feet (3.65 m) tall. Thermophilic algae have turned the cones various shades of green, yellow, orange, and red, giving the geyser a Martian look.

State Route 34, Gerlach. The geyser is on private property but visible with binoculars from State Route 34 near Gerlach. Property owners offer tours a few times a year. Ⓝ 40.859318 Ⓦ 119.331908

ALSO IN NEVADA

Clown Motel

Tonopah · Oh, just a motel on the edge of the desert, decorated with thousands of clowns, conveniently located next to an abandoned graveyard.

Neon Boneyard

Las Vegas · A 3-acre plot filled with disused neon signs charts Vegas's luminous history.

THE BRISTLECONE PINES OF THE GREAT BASIN

BAKER

At high elevations in a region of the western United States known as the Great Basin, a species of pine lives a quiet, secluded, and exceedingly long life.

Few other plants can grow in the hard rock where bristlecones thrive. The pines can live at least 5,000 years—longer than any other non-clonal organism. The trees are exceptionally hardy, surviving freezing cold temperatures, deep snow, and bracing winds. In many cases, portions of the bristlecone pine can die off and allow the tree to conserve its limited resources. Because of this, bristlecones aren't very tall and often appear dead or extremely weathered.

In eastern Nevada, a tree known as Prometheus was studied by a group of conservationists. Scientists drill cores out of trees in order to count and measure the rings that chronicle their growth. Unfortunately, when Prometheus was drilled for a core, the tool used for this process broke off inside. The researcher needed his tool back, and to get it, he had to cut the tree down. Once he'd retrieved his tool and the core from Prometheus, he was able to determine that the tree had lived 4,862 years.

An older tree hasn't been found since, though one known as Methuselah in California is believed to be over 4,800 years old. Most of the living groves of these trees are under better protection now and can be visited in several locations throughout the Great Basin and California.

Great Basin National Park. There are several bristlecone pine groves in the park—the most accessible is on the northeast side of Wheeler Peak. In summer, rangers lead walks through the grove. Ⓝ 39.005833 Ⓦ 114.218100

Some of the trees in this national park are among the oldest in the world.

Peer through protective fencing to see inside the shallow cave and take photos.

Toquima Cave

AUSTIN

In the center of Nevada is a cave lined with colorful pictographs painted by hand, thousands of years ago, using pigments of bleached white, bright red, and a yellowy turmeric orange. Unlike petroglyphs, the images are not carved into the rock but were added to the surface, most likely with fingers, in circular, crosshatched, beautiful snaking patterns.

The images, over 300 in all, date from around 1300 to 600 BCE. They were created by members of the Western Shoshone, a group of local Native American tribes. The cave depth is fairly shallow, so if it was used as a dwelling, it was probably short-term. Given the sweeping views from the cliff, the site may have been used as a place marker or geographical guidepost for finding food sources, tracking hunting grounds, or managing other tribal movements.

Archaeologists and anthropologists have found it a challenge to decipher the images. They may be keys to the land or, as some have posited, keys to other, less worldly, places. Cave shelters are seen as portals by some Native peoples, providing access to commune with the earth, experience visions, and seek understanding of what lies beyond. Whether used for the temporal or spiritual world, Toquima Cave and the pictographs hold great power for the Western Shoshone.

The cave is at Pete's Summit in the Toquima Range, about a 30-mile (48.3 km) drive southeast of Austin, Nevada. From Toquima Cave Campground, you'll need to hike half a mile or so. Remember that this is a sacred site—take nothing and leave nothing. Ⓝ 39.187750 Ⓦ 116.790500 ➡

➡ Other Native American Sites

CAHOKIA MOUNDS
COLLINSVILLE, ILLINOIS
The largest pre-Columbian settlement in the Americas north of Mexico, Cahokia once had about 120 mounds, built beginning in the 9th century. One in particular, Mound 72, shows evidence of hundreds of sacrificial burials. Some of the skeletons were found with their fingers extended into the surrounding sand, suggesting to archaeologists that these people were alive when they were buried and tried to claw their way out.

At its height in roughly 1250, Cahokia was bigger than medieval London.

The settlement continued until perhaps the late 14th century, when it was abandoned for unknown reasons. Many theories have been proposed, including invasion and warfare, as well as lack of game animals and deforestation as a result of erosion.

Within the ceremonial complex there was a wooden monument built to mark the equinox and solstice.

Cahokia Mounds State Historic Site is just 8 miles from downtown St. Louis near Collinsville, Illinois.

Parowan Gap

LEO PETROGLYPH
RAY, OHIO

Nearly 40 images, collectively known as the Leo Petroglyph, are etched into a large slab of sandstone. It has familiar subjects like birds, fish, footprints, and stick-figure humans. Then there are the more abstract designs in the mix, including one that looks like a cartoon man with horns.

It's thought that people from the Native American Fort Ancient culture made the petroglyph around 1,000 years ago.

PAROWAN GAP
PETROGLYPHS
PAROWAN, UTAH

Just a few miles west of the small town of Parowan, and just at the edge of the dry "Little Salt Lake," lies a natural gap in the mountains, covered with hundreds of petroglyphs. They have been there for over 1,000 years.

Archaeologists have argued that the petroglyphs are a complex calendar system. Hopi and Paiute peoples have a variety of interpretations for the rock art as well, which includes representations of humans as well as depictions of animals and geometric shapes. Among the petroglyphs is a much more ancient reminder of the past: fossilized dinosaur footprints.

WASHINGTON HALEETS,
BAINBRIDGE ISLAND

Haleets, or Figurehead Rock, is about 100 feet (30.5 m) from the shore of Bainbridge Island, and it disappears underwater during high tide. The sandstone rock is inscribed with several petroglyphs believed to have been carved by the native Suquamish tribe between 1,000 BCE and 500. The date has been narrowed down thanks to a facial piercing depicted on one of the characters on the stone: That specific type of piercing—a labret—was no longer used by the local indigenous people after about the year 500.

The purpose or meaning of the stone itself is a lot harder to discern. As it's located near a body of water, the rock may have marked a boundary of some sort. Or, as one local amateur astronomer has claimed, the rock may mark the lunar or solar calendar.

MAP ROCK
MELBA, IDAHO

This large basalt rock is believed to have been carved by the Shoshone-Bannock people to map the area of the upper Snake River, possibly as long as 12,000 years ago. It depicts the Snake and Salmon Rivers, as well as the animals and tribes that inhabited the territories in between. No one can decisively say what the carving's purpose was, which only contributes to its mysterious, ancient allure.

Map Rock is beside the Snake River, about 6 miles northwest of Walters Ferry along Map Rock Road.

NEW MEXICO
Lightning Field

QUEMADO

A visit to the *Lightning Field* is not something you do on a whim. There are rules to be followed.

The field is a work of land art, installed in 1977 by sculptor Walter De Maria. It consists of 400 stainless steel, 2-inch-wide (5 cm) poles, each over 20 feet (6 m) high, arranged in a mile-wide grid.

To see it during the May to October visiting season, you must arrive in the small town of Quemado—a three-hour drive from Albuquerque—by 2:30 p.m. on your designated day. From there you'll be driven 45 minutes into the desert, where a rustic, six-person cabin sits beside the *Lightning Field*. The driver then departs, leaving you until 11 a.m. the next day to appreciate the art that surrounds you.

As the hours pass, the *Lightning Field* changes. The poles seem to shimmer. They turn gray and black, grow shadows, and reflect the brilliant orange of the setting sun. If you go in July or August, you

Walter De Maria's array of metal poles in the desert beckons fury from above.

may witness lightning—but despite the work's title, a storm isn't an essential part of the experience. **The Dia Art Foundation provides transportation to the *Lightning Field* from the tiny town of Quemado. Ⓝ 34.343409 Ⓦ 108.497650 ➤**

Ra Paulette's Caves

LA MADERA

"Manual labor is the foundation of my self-expression," says Ra Paulette, an artist who has spent decades carving intricately patterned caves into New Mexico's sandstone cliffs. Heading into the desert with a wheelbarrow strapped to his back, Paulette burrows into the rock to create chambers, arches, and columns, all of which he decorates with swirling flourishes.

Since 1990, Paulette has dug over a dozen caves, each one unique in design. Though some were created on commission, Paulette often veered away from the design requirements of his clients, instead carving out whatever designs his hands were instinctively drawn to. He is currently working on what he calls an "environmental and social art project": the Luminous Caves, a vast complex, lit via skylights, that will host gatherings and performances.

North of Santa Fe on US Route 285, at Ojo Caliente. The finished caves are not open to the public, but some of the interiors can be glimpsed through their skylights. Ⓝ 36.386554 Ⓦ 106.041037

The artist Ra Paulette has carved over a dozen whimsical caves into the New Mexico sandstone.

LORETTO CHAPEL STAIRS

SANTA FE

Visit Loretto Chapel and you're guaranteed to witness a miracle. In the back right corner of the small church is a spiral staircase supposedly built by a saint.

Around 1878, the chapel nuns were in need of a way to access the newly built choir loft. A conventional staircase would have taken up too much space, but a ladder would have been inappropriate for the nuns to use when wearing their habits.

Perplexed by the problem, the nuns apparently prayed to St. Joseph, patron saint of carpenters. Lo, on the ninth day, a mysterious man appeared at the chapel door with a donkey in tow and vowed to build the stairs. Within a few months he had crafted an inspiring feat of engineering, a spiral staircase with no visible supports and no central column.

When the nuns went looking for the benevolent stranger to pay him and thank him, he had disappeared. They concluded that the miracle staircase was built by St. Joseph himself. The fact that there are 33 steps—equaling the number of years Jesus lived—only adds to the legend.

The spiral stairs are structurally sound but a little bouncy owing to their springlike double helix shape. They have been closed to public foot traffic since the 1970s, but if you book your wedding in the chapel you can stand on them to get your photo taken.

207 Old Santa Fe Trail, Santa Fe.
Ⓝ 35.685387 Ⓦ 105.937637

ALSO IN NEW MEXICO

American International Rattlesnake Museum

Albuquerque · Glass cages of tail-shaking serpents line the walls of this museum, which is dedicated to showing the gentler, softer underbelly of the often-feared rattlesnake.

109 East Palace

Santa Fe · This innocuous-looking storefront was once the secret jump-off spot for Manhattan Project scientists working on the development of the atomic bomb.

The spiral staircase is said to have been constructed with the help of divine intervention.

LAND ART IN THE SOUTHWEST

SUN TUNNELS, LUCIN, UT

Four concrete cylinders big enough to walk in form *Sun Tunnels*, an outdoor artwork installed by Nancy Holt in 1976. Laid out in a cross configuration, the 18-foot-long, 9-foot-wide (5.5 m, 3 m) tunnels are positioned to frame the sunrise and sunset on the solstices. Each pipe has a cluster of holes drilled into its ceiling. They are not just random dots, but the patterns of constellations Draco, Perseus, Columba, and Capricorn. When the sun shines onto the tunnels, the constellations are projected on the interior walls.

Holt was part of the '70s land art movement along with her husband, Robert Smithson, whose 1970 work *Spiral Jetty* (shown below) is still in place at Great Salt Lake. The *Sun Tunnels* are a 4-hour drive west of Salt Lake City.

For over 30 years, *Spiral Jetty* (below) was hidden beneath the pink-tinged waters of Great Salt Lake. Artist Robert Smithson built the 1,500-foot-long (457 m) jetty in 1970, using mud, salt crystals, and basalt rocks. At the time, Salt Lake was experiencing a drought, and the water level was unusually low. When the rains returned, the lake swallowed the jetty.

It wasn't until 2002, during another drought, that the water lowered and *Spiral Jetty* reappeared. Unfortunately, Smithson wasn't around to see it return—he died in a plane crash three years after creating the work.

Encrusted with white salt and at the mercy of the elements—and the visitors who walk along it— *Spiral Jetty* looks a little worse for wear. It's a state of affairs Smithson would have appreciated. The artist, who coined the term "earthwork" for his brand of landscape-integrated art, had a fondness for entropy and the eroding power of nature. Rozel Point peninsula is on the northeastern shore of Great Salt Lake.

RODEN CRATER, FLAGSTAFF, AZ

In 1977, artist James Turrell purchased an extinct volcano northeast of Flagstaff in order to create a work of land art named *Roden Crater*. The centerpiece of this work, a naked-eye observatory housed in the 2-mile-wide (3.2 km) volcanic crater, is still under construction. Thus far, Turrell has installed a series of tunnels, viewing chambers, and a bronze staircase in the volcano. The projected completion date changes frequently, but something amazing is certainly in progress.

TEXAS

PRADA MARFA

VALENTINE

Along a silent stretch of Route 90, with only power lines and tumbleweeds for company, is a fully stocked Prada boutique.

Prada Marfa is a site-specific sculpture by Scandinavian artists Michael Elmgreen and Ingar Dragset. The shelves of the 15 × 25–foot (4.5 × 7.6 m) store are filled with Prada handbags and high heels, but none of the products are for sale, and the shop's door doesn't ever open.

The store was originally intended to be installed and left to the elements, so it would gradually deteriorate and be absorbed into the landscape. However, three days after the store's unveiling, vandals broke into Prada Marfa and stole every item on display. The products were replaced, the window glass thickened, and the concept tweaked: Representatives from the Art Production Fund and nonprofit cultural organization Ballroom Marfa now drop by periodically to pick up trash and paint over graffiti.

Route 90, 1.4 miles (2.25 km) northwest of Valentine. The Prada store is 37 miles (60 km) northwest of Marfa, a 2,000-resident town that has become a hub for contemporary art. Take a tour of the galleries and visit Donald Judd's installations at the Chinati Foundation.
Ⓝ 30.603461 Ⓦ 104.518484

An artist's look-but-don't-touch homage to luxury consumerism in the middle of the desert.

NATIONAL MUSEUM OF FUNERAL HISTORY

HOUSTON

"Any day above ground is a good one." So reads the slogan of the National Museum of Funeral History, which celebrates life by showing how we honor its loss.

Founded in 1992 by undertaker Robert L. Waltrip and attached to an embalming school, the museum displays the country's largest collection of funeral artifacts. Items range from 19th-century horse-drawn hearses to memorabilia from Michael Jackson's memorial service.

The exhibit on 19th-century mourning customs provides a fascinating look at the Victorian response to death. Among the items are a wooden clock that reminded family members to mourn on the hour, a quilt made from ribbons that bound the flowers at a funeral service, and jewelry made from the hair of the deceased.

Other exhibits cover the history of embalming, papal and presidential funerals, and "fantasy coffins." Look out for the Snow White–inspired glass casket and the roomy coffin built for three.

415 Barren Springs Drive, Houston. Ⓝ 29.989561 Ⓦ 95.430324

See some of history's great hearses up close.

Ozymandias on the Plains

AMARILLO

You wouldn't know it, but these legs are the shattered likeness of an Egyptian king. Ozymandias is the Greek name for Rameses II and was the inspiration and name of the 1818 poem by Percy Bysshe Shelley.

A plaque near the gigantic legs reads:

In 1819, while on their horseback trek over the Great Plains of New Spain, Percy Bysshe Shelley and his wife, Mary Shelley (author of Frankenstein), came across these ruins. Here Shelley penned his immortal lines, among them:

I met a traveller from an antique land

Who said: Two vast and trunkless legs of stone

Stand in the desert . . .

And on the pedestal these words appear:

"My name is Ozymandias, King of Kings,

Look on my works, ye mighty, and despair!"

The pedestal near the monument also asserts that the visage of the king was destroyed by Lubbock football players after losing a game to Amarillo.

The hoax sculpture has been vandalized numerous times, most notably with the addition of socks to the legs. Occasionally the vandalism is sandblasted off the sculpture, but the socks always reappear. The locals seem to prefer the king's legs be kept warm.

The socks are on the east side of I-27. Ⓝ 35.101703 Ⓦ 101.909135

A roadside pair of legs puts a modern spin on the Romantic poem "Ozymandias."

Hamilton Pool

AUSTIN

There are swimming holes and then there's Hamilton Pool, a glorious fairy grotto where you can paddle in a natural spring protected by a canopy of limestone.

The grotto formed when the dome of an underground river collapsed due to erosion. Maidenhair ferns and moss-covered stalactites crept along the overhang, creating a verdant oasis in the Texas desert. A 50-foot (15 m) waterfall pours from the overhang into the turquoise pool.

The pool is 23 miles (37 km) west of Austin, off Highway 71. There's a quarter-mile hike down to the water—wear closed-toe shoes. Due to the small size of the beach (really just a patch of sand on the edge of the spring), capacity is limited to 75 vehicles and a one-in, one-out policy applies. Ⓝ 30.345135 Ⓦ 98.135221

This emerald-green grotto not far from Austin may be America's most beautiful swimming hole.

MAGIC LANTERN CASTLE MUSEUM

SAN ANTONIO

Prior to the invention of cinema, audiovisual entertainment came in the form of the magic lantern, a slide projector invented in the middle of the 17th century.

Early magic lanterns used candlelight and hand-painted glass slides to project dim images onto walls and cloth. Better and brighter light sources became available during the 19th century: limelight, arc lamps, and, eventually, incandescent bulbs.

As illumination improved, so did the special effects. By layering slides and moving them independently, magic lantern operators could make a dog appear to jump through a hoop or a skeleton seem to talk. Live narration and musical accompaniment added to the experience. A genre of scary supernatural shows known as phantasmagoria developed, featuring cackling demons, eerie ghosts, and unsettling melodies.

The Magic Lantern Castle Museum was founded by Jack Judson, who discovered a passion for collecting lanterns and slides when he retired in 1986. **1419 Austin Highway, San Antonio. Visits are by appointment only. Ⓝ 29.492141 Ⓦ 98.438667**

UTAH

THE TREMBLING GIANT

RICHFIELD

Though this golden grove of quaking aspens may look like a forest, the whole thing is actually a single organism. Every tree—or stem, technically—is genetically identical, and the whole forest is linked by a single root system. The Pando aspens reproduce asexually by sprouting new stems from the root structure.

With over 40,000 stems and a total weight of 13 million pounds (approx. 6,000 MT), the Pando clonal colony is the heaviest known organism in the world. It's also among the oldest living things on the planet—the root system is an estimated 80,000 years old. **One mile southwest of Fish Lake on Utah State Route 25. Ⓝ 38.524530 Ⓦ 111.750346**

Though it may look like an ordinary forest, this grove of aspens is actually a single organism—the heaviest living thing in the world.

GRANITE MOUNTAIN RECORDS VAULT

SALT LAKE CITY

The history of your family is buried deep inside a mountain of solid rock just outside Salt Lake City. Granite Mountain Records Vault, or simply "The Vault," is a massive repository of genealogical records established in 1965.

The Church of Jesus Christ of Latter-day Saints built the vault to safeguard the genealogical information—such as records of births, deaths, and marriages—the church has been collecting on microfilm since 1938. The records collected are not just of church members, but everyone. This data is important to Mormons because they use it to trace their family trees in order to posthumously baptize their ancestors into the LDS faith.

Able to withstand natural disasters and nuclear blasts, the vault stores millions of microfilms in banks of 10-foot-high (3 m) steel cabinets, along with hard drives filled with digitized data. Heavy vault doors protect the six storage areas, which connect to one another via 25-foot-wide (7.6 m) tunnels. Natural conditions allow the temperature to stay at a steady 55°F (12.7°C) regardless of the season.

Due to the 200-year life span of microfilm, 60 full-time vault employees are currently converting the reels into digital images using microfilm scanners. This process was originally estimated to take 150 years, but technological advances and improved automation have reduced that to ten years. Much of the information stored in Granite Mountain is now remotely accessible via the Internet to anyone engaged in genealogical research.

The public may not visit the vault, purportedly due to concerns over contamination. The Church's history preservation department is particularly concerned about "blue jean dust"—the tiny fibers that fly into the air when pant legs brush against each other.

Little Cottonwood Canyon, Wasatch Range, Salt Lake City. The Vault is 20 miles (32 km) southeast of downtown Salt Lake City.
Ⓝ 40.570561 Ⓦ 111.762052

ALSO IN UTAH

USPS Remote Encoding Facility

Salt Lake City · If your penmanship is poor, your hand-addressed letters will end up here to be decoded and redirected to the rightful recipient.

Inside this mountain are millions of family history records.

SUMMUM PYRAMID

SALT LAKE CITY

Inside an orange pyramid, right beside the Lincoln Highway, is a religious group willing to mummify your corpse. The religion, Summum, was founded in 1975 by Claude Nowell (aka Corky Ra), who claimed to have been visited by advanced beings who revealed to him the nature of creation.

According to Summum philosophy, death does not snuff out a person's awareness or ability to feel. Though bereft of a body, our spirit, or essence, sticks around—and gets thoroughly confused by the change in circumstances.

The solution: mummification. By preserving your body, Summum provides a "home base" for your posthumous essence. Secure in this wrapped-up, chemically preserved corpse, your essence can safely communicate and make plans to move on to its next destination.

The entire mummification process takes four to eight months and ends with your gauze-wrapped body being hermetically sealed in a sarcophagus, or "mummiform." Summum offers lots of customization options for your mummiform. You can go with a traditional golden Egyptian look featuring ankhs and scarabs, or you can choose a simple, streamlined capsule for the final send-off.

Corky Ra himself became the first Summum mummy following his death in 2008. His mummiform, and that of his cat, Oscar, is on prominent display at the pyramid.

707 Genesee Avenue, Salt Lake City. The Summum Pyramid hosts publicly accessible readings and discussions on Wednesday nights.
Ⓝ 40.750707 Ⓦ 111.911651

Great Plains

IDAHO

Oasis Bordello Museum

WALLACE

For decades, Oasis Bordello was a busy brothel—one of five that catered to the small silver-mining town of Wallace. Then, in January 1988, the working women caught wind of an imminent FBI raid. Late one night, they fled, leaving their rooms in a state that still exists today—with a few creative modifications.

Michelle Mayfield, a Wallace native, purchased the bordello building and opened it as a museum in 1993. Go on a guided tour of the rooms and you'll see mannequins dressed in the lingerie left behind by the women who worked there. Strewn around the former bordello are dog-eared magazines, flimsy nightgowns, toiletries, and an Atari 5200, as

Vacated in a rush just before a 1988 police raid, this brothel has been left untouched as a museum.

well as a bag of groceries—Minute Rice sits on the kitchen table, untouched since 1988.

Three stops on the tour leave no doubt as to the nature of the business: A peek inside a cupboard full of red lightbulbs is followed by the handwritten menu of services on the wall, complete with time limits and prices. In a drawer are battered kitchen timers used to enforce the prescribed duration of appointments.

605 Cedar Street, Wallace. Off Interstate 90, between Spokane and Missoula. Make sure to get a souvenir garter belt at the gift shop. ℕ 47.472574 Ⓦ 115.923529

Museum of Clean

POCATELLO

Don Aslett is the King of Clean. As the ruler of an empire of cleaning products and janitorial services, Aslett is a big advocate of cleanliness, not just

in terms of keeping your shower mildew-free but also for purifying your mind and spirit.

The Museum of Clean, which Aslett opened in 2011, is an institution dedicated to this clutter-cleaning philosophy. Its mission statement is to "sell the idea and value of clean"—Aslett's dream is to create "clean homes, clean minds, clean language, clean community, and a clean world."

The museum aims to sweep visitors into an unsullied state of mind via exhibits of brooms, tubs, toilets, and vintage vacuum cleaners dating from 1869 to 1969. An art gallery displays cleaning-themed paintings and spotless sculptures.

711 South 2nd Avenue, Pocatello. ℕ 42.859605 Ⓦ 112.441706

The Museum of Clean targets dirt in all its literal and metaphorical forms.

KANSAS

STRATACA SALT MINE

HUTCHISON

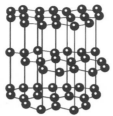

Six hundred and fifty feet (198 m) beneath the plains of Kansas lies a salt mine that's been operating since 1923. Some of the salt mined here ends up scattered on roads and highways in winter to prevent the buildup of ice. In an unused section of the mine you'll find a salt museum housing a crystal with a long and unlikely history.

In 1998, scientists discovered a strain of *Bacillus* bacterium lying dormant in a bubble within a 250-million-year-old salt crystal in New Mexico. At a lab in West Chester University, Pennsylvania, biologists were able to reanimate the bacterium by giving it fresh nutrients and putting it in a salt solution.

The discovery of this living bacterium from over 250 million years ago impacts our timeline of life on Earth—it is believed to be the world's oldest living organism.

3300 Carey Boulevard, Hutchinson (aka Salt City). Ⓝ 38.043184 Ⓦ 97.868482

Retired Civil War nurse Samuel Dinsmoor began sculpting his eccentric garden in 1905.

GARDEN OF EDEN

LUCAS

After serving in the Civil War as a Union nurse, Samuel P. Dinsmoor settled in Lucas, Kansas. It was there, in 1905, that he began building his own patriotic Garden of Eden.

The centerpiece of the garden was a "log" cabin that Dinsmoor sculpted out of limestone. Around the cabin he built tall, thin cement sculptures reflecting his Populist politics, Biblical interests, and distrust of authority. Adam and Eve, Cain and Abel, and the devil all make an appearance, as do snakes, angels, and a waving American flag. The most blatant political commentary is found in the "Crucifixion of Labor," a sculpture of a doctor, lawyer, banker, and preacher nailing a man labeled "Labor" to a cross.

In a corner of the garden is a pagoda-style mausoleum. Dinsmoor built it to house his own body and even posed for a double-exposed gag photograph depicting himself looking at his own corpse. When he died, Dinsmoor was embalmed and entombed behind glass, as per his plans. You can still see his smiling but slightly moldy face when you peer into the space.

305 East Second Street, Lucas. The tiny town is about 2 hours north of Wichita. Ⓝ 39.057802 Ⓦ 98.535061

MONTANA

TRANSCONTINENTAL AIR MAIL ROUTE BEACONS

WESTERN MONTANA

In the 1920s and '30s, prior to the development of radio navigation for powered flights, air mail pilots in the United States navigated their routes by following giant concrete arrows on the ground.

The 50- to 70-foot-long (15 to 21 m) arrows, painted yellow and installed alongside 50-foot-high (15 m) flashing beacons, lit up the national mail route, allowing pilots to take to the skies at night and greatly improving the delivery times for air mail. Hundreds of flashing beacons and concrete arrows were installed across the country. The system operated at full strength until radio-based navigation began to replace it in the mid-'30s.

During World War II, many of the navigational arrows and towers were destroyed out of fear that invading enemy bombers would use them to find their way to heavily populated areas. There are, however, some leftovers strewn across the States—most

Before radio navigation, huge concrete arrows guided pilots home.

of them sporting cracks in the concrete and faded paint jobs. Montana is the only state that continues to make use of its lighted towers. In the state's western mountain region, a series of 17 well-maintained beacons helps guide pilots home.
The beacons are spread along the mountainous ranges west of Helena. Ⓝ 46.229605 Ⓦ 112.781044

BERKELEY PIT

BUTTE

At the Berkeley Pit you can pay to stand on a viewing platform and gaze at a mile-long lake of toxic waste. The former open-pit copper mine operated from 1955 to 1982, after which groundwater began to fill the 1,780-foot-deep (542.5 m) hole. The lake is now poisoned with a toxic cocktail of heavy metals and chemicals, including copper, iron, arsenic, cadmium, zinc, and sulfuric acid. The reddish, iron-rich water at the surface shifts to a vibrant lime green below, where the copper concentrations are higher.

The water level in the pit is rising at the rate of roughly 0.7 feet (21.3 cm) per month. If it reached 5,410 feet (1.6 km) above sea level, pit water could contaminate the nearby groundwater of the Butte valley, home to more than 30,000 people. A water treatment plant was built in 2003 to prevent the Berkeley Pit water from ever reaching that critical level.

Though measures have been put in place to prevent the toxic pit from harming humans, other creatures have perished in its murky waters. In 1995 a flock of over 300 migrating snow geese landed at the pit and promptly dropped dead.
Exit 126 off I-90, Butte. The pit can be seen from a viewing platform on the Southwest rim, open from March to November. Ⓝ 46.017266 Ⓦ 112.512039

ALSO IN MONTANA

American Computer Museum

Bozeman · Founded in 1990 by husband-and-wife tech enthusiasts George and Barbara Kremedjiev, the American Computer Museum includes an original NASA moon mission navigation computer, ancient adding machines, and the room-size computers of yesteryear.

One of Butte's most striking sights is a pit full of toxic waste.

RINGING ROCKS

PIPESTONE

The most fun you can have 18 miles (29 km) east of Butte involves a pile of rocks and a BYO hammer. The Ringing Rocks, a half-mile-wide assemblage of angular stones, make pleasingly sonorous sounds when you strike them. The rocks sound different notes according to their size and shape. If removed from the pile, they no longer resonate with their bell-like ring.

The exact cause of the rock music is unknown, but it has something to do with the layout and density of the stone mass, which formed when magma cooled just below the Earth's surface around 78 million years ago. Uplift and erosion over the eons exposed the rocks and gave them their squared-off edges, which make it easy for visitors to tap out some tunes.

The rocks are off Interstate 90 in Deerlodge National Forest, on the southwest side of Dry Mountain. The roads aren't great, and you'll need to clamber up a few hills on foot. Watch out for bears along the way.
Ⓝ 45.943491 Ⓦ 112.237504

NEBRASKA

CARHENGE

ALLIANCE

England has Stonehenge. Nebraska has Carhenge. It's pretty much the same thing, give or take a few millennia of technological advancement.

After studying Stonehenge in England, Carhenge creator Jim Reinders returned to the family farm in Alliance with a plan: to create a replica of the prehistoric monument using junkyard automobiles. In the summer of 1987, with the help of family members, Reinders assembled 39 cars, all spray-painted gray, into a 96-foot-wide (29.2 m) circle—roughly the same dimensions as Stonehenge. The creation was inaugurated on the summer solstice with champagne, song, poetry, and a play written by members of the Reinders family.

In addition to providing an intriguing roadside stop for travelers, Carhenge is a memorial to Reinders's father, who once lived on the farm where the monument stands.

Three miles (4.8 km) north of Alliance, on Highway 87.
Ⓝ 42.142229 Ⓦ 102.857901

Built to the same dimensions as its Neolithic predecessor in England, only with cars.

SITE OF A JAPANESE BALLOON BOMB EXPLOSION

OMAHA

In the evening hours of April 18, 1945, a Japanese balloon bomb exploded in the evening sky above the Dundee district of Omaha.

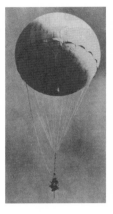

The bomb was one of more than 9,000 balloons the Japanese launched during a 6-month period at the end of World War II, and one of the nearly 300 that were found or observed in the United States. Also called "fire balloons," they were filled with hydrogen and carried bombs varying from 11 to 33 pounds.

Few people in Omaha knew a bomb had gone off at the corner of North 50th and Underwood. Some saw a flash of light, and others heard noises they thought were fireworks. One witness described "a ring of fire" in the sky. The plaque commemorating the incident notes: "The incendiary device flared brightly in the night, but caused no damage."

Japanese military leaders deployed these balloon bombs with the hopes of creating panic and widespread media attention, which would allow them to chart courses for future attacks. There were a few reports of explosions published in various outlets; however, the US Office of Censorship—a wartime agency set up during World War II to censor communications coming in or out of the US—sent messages to all media outlets asking them not to publish news of the balloon bombs, and the Dundee explosion, as well as several others, was not reported until after the war had ended.

The plaque commemorating the explosion is on the building at the southwest corner of 50th and Underwood Avenue. Look for the big green clock on the sidewalk. Ⓝ 41.266825 Ⓦ 95.987766

Not an Egyptian mausoleum, but a Cold War–era missile defense system.

NORTH DAKOTA

THE NEKOMA PYRAMID

NEKOMA

In a field behind a fence lies a 79-foot-tall (24 m), sinister-looking pyramidal frustum—a pyramid with the top cut off. This gray structure, and the clusters of exhaust towers beside it, look like an occult monument, or perhaps a bit of Egyptian architecture misplaced in the Great Plains. In fact, they are the remnants of an antimissile complex constructed during the Cold War.

The Stanley R. Mickelsen Safeguard Complex, as it was known, was built to house anti-ballistic missiles capable of intercepting incoming Soviet rockets. Radar inside the topless pyramid scanned the skies while 100 missiles underground sat ready to launch against a Russian attack.

Built at great expense, the Safeguard Complex was nonetheless short-lived. It became operational on October 1, 1975—one day before Congress voted to end the Safeguard program and decommission the Nekoma site.

North of Nekoma, where Highway 1 meets 81st Street. Ⓝ 48.589529 Ⓦ 98.356503

OKLAHOMA

MUSEUM OF OSTEOLOGY

OKLAHOMA CITY

In 1986, skull collector Jay Villemarette turned his hobby into a profession by establishing Skulls Unlimited, a business dedicated to cleaning, mounting, and selling animal skulls. (The cleaning process is fascinating: Flesh-eating dermestid beetles are let loose on the animal heads to chew away skin, muscle, and fat, leaving a dry and gleaming white skull.)

In order to showcase some of the more striking specimens from Skulls Unlimited, Villemarette opened the Museum of Osteology next door. Here you'll find over 300 carefully articulated specimens, including the skeletons of a Komodo dragon given by Indonesia to President George W. Bush, a rare Javan rhinoceros found in a shop in Paris, and a two-faced calf.

10301 South Sunnylane Road, Oklahoma City. Ⓝ 35.364772 Ⓦ 97.441840

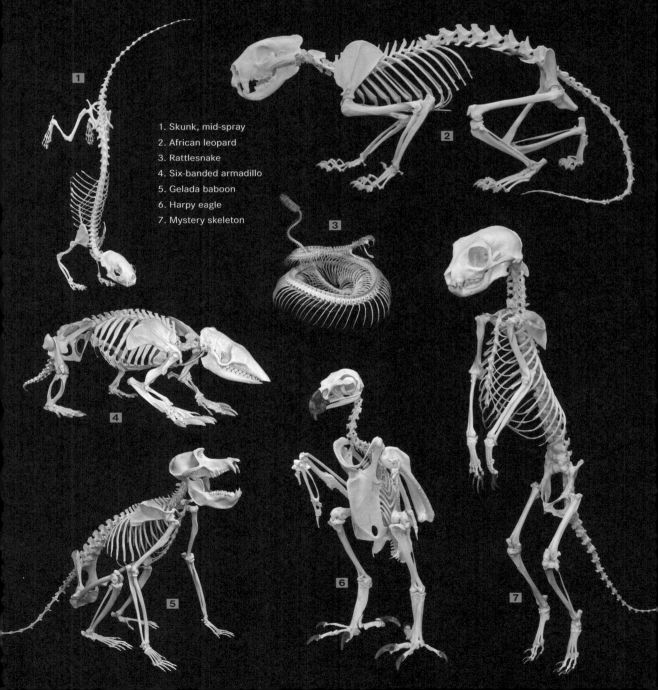

1. Skunk, mid-spray
2. African leopard
3. Rattlesnake
4. Six-banded armadillo
5. Gelada baboon
6. Harpy eagle
7. Mystery skeleton

WICHITA MOUNTAINS BUFFALO HERD

LAWTON

In 1907, 15 bison rode a train from New York to Oklahoma. Their 650 descendants now roam the plains of the Wichita Mountains Wildlife Refuge, a 92-square-mile (238 km²) area dedicated to protecting endangered animals.

The story of why six male and nine female buffaloes took a train trip across the country begins in 1905 with the formation of the American Bison Society. At the time, hunting and settlement had reduced the country's bison population to approximately 1,000—down from 60 million in 1830.

Spearheaded by William T. Hornaday and under the auspices of President Theodore Roosevelt, the society aimed to save the buffalo from extinction, increase the population, and provide a safe place for them to roam. Thankfully, the Wichita Mountains Wildlife Refuge had been established four years earlier—and the newly opened Bronx Zoological Park in New York was willing to spare a herd of bison.

The journey from New York to Oklahoma took six days by train and wagon. When they arrived at the preserve, the bison, who traveled in individual crates, received a rapturous welcome from curious crowds and members of the Native American Comanche tribe.

The animals have roamed free in the prairies ever since. The North American bison population is now around half a million, including the 650 of the Wichita Mountains Wildlife Refuge.

Visitors Center, Cache Meers Road, Lawton. The refuge is 59,000 acres. Your best bets for spotting buffalo are to bring binoculars and have a little patience. Organized nature tours, held throughout the year, get you closer to the action. Ⓝ 35.750961 Ⓦ 98.682064

ALSO IN OKLAHOMA

Center of the Universe

Tulsa · Speak a few words inside this concrete circle, and your voice will become amplified for no discernible reason.

45th Infantry Museum

Oklahoma City · See the mirror from Hitler's Berlin bunker and a Mickey Mouse gas mask made for kids.

THE GRAVE OF ELMER McCURDY

GUTHRIE

It was 1976. Crew members from the TV show *The Six Million Dollar Man* were preparing to shoot on location at the Pike Amusement Park in Long Beach, California. The plan was to capture Steve Austin, the titular pricey fellow, riding in one of the cars along the track of a spooky ride called the "Laff in the Dark." The ride featured a tunnel in which ghouls, demons, and skeletons would pop up and scare you as your car jolted from side to side in the dark.

While sprucing up the set, a stagehand spotted a mannequin hanging from a noose in the corner. He reached for the mannequin's arm and was surprised when it broke off in his hand. Looking at the dismembered limb, the worker was astonished to see what looked like bone beneath desiccated layers of skin. This was no mannequin. This was a man.

The hanging corpse in question belonged to Elmer McCurdy, an outlaw who died 65 years earlier. In 1911, the mischief-making vagabond robbed a train near Okesa, Oklahoma, then took his spoils—$46 and two jugs of whiskey—north, where he holed up in a barnyard on the Kansas border. Police pursued him and ended up killing him in a shootout.

McCurdy's body was taken to a funeral home in Pawhuska, but no one claimed it. Seeing a money-making opportunity, the undertaker embalmed him and allowed visitors to view the preserved corpse if they placed a nickel in its mouth.

Five years into this lucrative scheme, a carnival runner turned up at the funeral home claiming to be a long-lost relative of McCurdy and requested to take the body so it could be laid to rest properly. He was, of course, lying through his teeth. Within weeks, the McCurdy corpse was the star attraction of a traveling carnival.

For 60 years, McCurdy's mummy made the rounds of carnivals, wax museums, and haunted houses, until it turned up, inexplicably, at the Pike in Long Beach. By this time, the legend of Outlaw McCurdy was long forgotten, and the body assumed to be fake. After the *Six Million Dollar* discovery, police identified McCurdy and sent the body to Summit View Cemetery in Guthrie, Oklahoma, for long-delayed interment.

McCurdy's grave is marked by a stone that lists his death date as 1911 and burial date as 1977, with no elaboration on the matter.

Summit View Cemetery, North Pine Street, Guthrie. McCurdy's grave is in the Boot Hill section near Wild Bunch bandit Bill Doolin. Ⓝ 35.878937 Ⓦ 97.425318

ELMER McCURDY
SHOT BY SHERIFF'S POSSE
IN OSAGE HILLS
ON OCT. 7, 1911
RETURNED TO GUTHRIE OKLA.
FROM LOS ANGELES COUNTY
CALIF.
FOR BURIAL APR. 22, 1977

SOUTH DAKOTA

THUNDERHEAD FALLS

RAPID CITY

To see this 30-foot (9 m) waterfall, you will need to journey 600 feet (183 m) into a mountain. Thunderhead Falls is located in the tunnels of a disused gold mine established in the 1870s. To the miners tunneling for treasure, the gushing torrent that spewed forth was an inconvenient surprise amid more general disappointment— the mine never yielded any gold, and was abandoned by the time the 20th century rolled around.

With no gold in them hills, Thunderhead Falls was forgotten until 1949, when Vera Eklund boarded a sightseeing train from Rapid City to Mystic and noticed a stream of water flowing down the side of a mountain. Eklund and her husband, Albert, returned to the site to track down the source of the stream and found the underground waterfall. The Eklunds acquired the land and opened Thunderhead Falls to the public as a tourist attraction the following year. With a tromp through the musty tunnels, you too can experience the highs and lows of the gold-rush days.

10 miles (16 km) from Rapid City along Highway 44 West. Ⓝ 44.066968 Ⓦ 103.409214

Water rushes through an old gold mine deep inside a mountain.

ALSO IN SOUTH DAKOTA

Petrified Wood Park

Lemmon · A park the size of a city block is filled with 100 conical sculptures created out of petrified wood in the early 1930s.

Crazy Horse Memorial

Custer · Begun in 1948, this sculpture of the Oglala Lakota Chief, carved out of a mountain, is still in progress. Completion is a long way off, but the ultimate vision is to create the world's largest sculpture, at 641 feet (195 m) wide and 563 feet (172 m) high. For now, you can see the fully carved head, which is 87 feet (26 m) tall. (For comparison, the presidential heads at Mount Rushmore are 60 feet [18 m] tall.)

WYOMING

PHINDELI TOWN

BUFORD

Until 2013, PhinDeli was known as Buford. The town sign provided a unique photo opportunity: Planted beside the dusty main road, it read BUFORD; POP: 1; ELEV: 8000. That one crucial person tallied was Don Sammons, a Vietnam vet who moved to Buford in 1980.

Founded in 1866 during the construction of the First Transcontinental Railroad, Buford reached a peak population of around 2,000 people. As the rail line moved west, however, so did the workers. When Sammons, his wife, Terry, and son arrived in Buford in 1980 hoping for a quiet life, they got it: The trio comprised the entire population of Buford. In 1992, the family bought the town—consisting of a gas station, convenience store, modular home, garage, and surrounding land—for $155,000.

After Sammons's wife died and his son moved to Colorado, it was time for a change. In 2012, Buford went up for auction and was snatched up for $900,000 by mystery investors from Vietnam. The next year, the plan for the town was revealed: The PhinDeli Corporation, makers of Vietnamese coffee, intended to capture a share of the US market by establishing a branded town in the American heartland.

The town of PhinDeli now sells Vietnamese coffee in its convenience store. Though Sammons has moved to Colorado to be closer to his son, the population is still one: a caretaker who lives in the town's only house.

Interstate 80 between Laramie and Cheyenne. No visit to PhinDeli is complete without a trip to its only attraction: the Buford Trading Post. It has a restroom, gas, and, of course, coffee. Ⓝ 41.123688 Ⓦ 105.302292

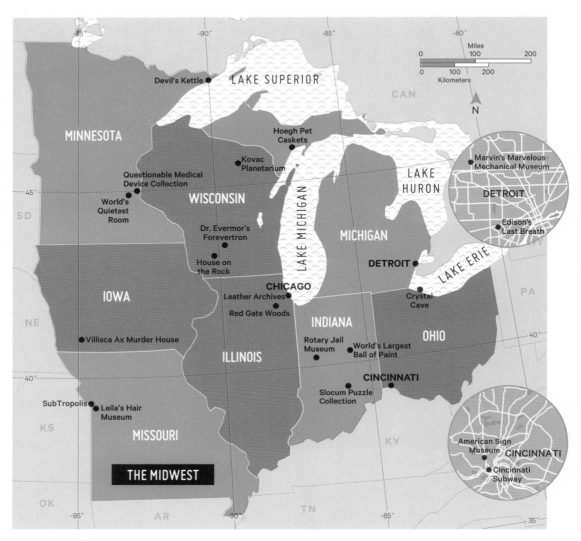

ILLINOIS

LEATHER ARCHIVES & MUSEUM

CHICAGO

The sign above the door of this former church says LA&M in large white letters. A picture of a black boot to the left gives a tiny hint of what's inside, but no passerby would guess that this is a museum devoted to alternative sexual practices.

The Leather Archives & Museum documents the leather lifestyle, a subculture mostly associated with gay men, fetish, and BDSM—though the museum encompasses all sexualities and genders. The collection includes clothing, books, photos, and provocative paintings like *The Last Supper in a Gay Leather Bar with Judas Giving Christ the Finger*. The Dungeon displays fetish and BDSM items such as a stainless-steel male chastity device and a red leather spanking bench.

6418 North Greenview Avenue, Chicago. The museum is open from Thursday to Sunday. You must be 18 or older to visit the galleries. Ⓝ 41.998637 Ⓦ 87.668273

Galloping Ghost Arcade

Brookfield • Hundreds of arcade games await your coin-inserting, button-mashing presence in this suburban den.

Shit Fountain

East Village • This oversize bronze coil of feces is both a tribute to doggie defecation and a reminder to pick it up.

Lizzadro Museum of Lapidary Art

Elmhurst • Cut and polished stones, some carefully sculpted into mini boats and temples, are on display at this jade-lover's paradise.

U-505

Hyde Park • After sustaining much damage during World War II, the most unlucky U-boat in the German fleet is on display at the Museum of Science and Industry.

Oz Park

Lincoln Park • This *Wizard of Oz*–themed urban oasis features statues of the Tin Man, the Scarecrow, the Cowardly Lion, Dorothy, and Toto.

Busy Beaver Button Co.

Logan Square • Ever wanted to visit a museum dedicated solely to pinback buttons? Busy Beaver's got what you need.

Chicago Cultural Center's Tiffany Dome

The Loop • Completed in 1897, this 38-foot-wide, zodiac-themed Tiffany dome is made of about 30,000 pieces of glass.

Chicago Temple

The Loop • Those affected by vertigo may have trouble worshipping at the First United Methodist Church of Chicago, which is located on top of a 23-story skyscraper.

Money Museum

The Loop • See a money pit and stand in the shadow of a million bucks at this museum dedicated to the almighty dollar.

Pritzker Military Museum & Library

The Loop • The citizen soldier is the focus of this library, which opened in 2003. Along with more than 40,000 volumes on the subject, it offers military posters, recruitment art, soldiers' journals, and Civil War memorabilia.

SS *Eastland* Memorial

The Loop • Learn how a ship that sank in just 20 feet (6 m) of water, a mere 20 feet from shore, resulted in more than 800 deaths.

Bohemian National Cemetery

North Park • Established by members of the city's Czech community in 1877, this cemetery features a rare glass-fronted columbarium—a structure that stores the ashes of the deceased.

Pullman Historic District

Pullman • Take a stroll through the first planned industrial community in the United States, established in 1880.

Eternal Silence

Uptown • A shrouded statue in Graceland Cemetery imbued with creepy legends.

Garfield–Clarendon Model Railroad Club

Uptown • A small, long-running miniature train club has built one of the largest model rail lines in the United States.

Inez Clarke Monument

Uptown • Legend has it that this 19th-century statue of a wide-eyed young girl regularly comes to life and explores the surrounding Graceland Cemetery.

Fountain of Time

Washington Park • This concrete depiction of 100 humans at various stages of life is a solid reminder of the inescapable nature of time.

RED GATE WOODS

LEMONT

Standing on a grassy clearing in Red Gate Woods just outside Chicago is a grave-stone with a most unusual inscription: CAUTION—DO NOT DIG.

The grave marker pays tribute not to a person, but a project: nuclear research. Buried beneath the stone is radioactive waste from Chicago Pile-1, the world's first artificial nuclear reactor.

Nicknamed CP-1, the reactor was built in 1942 as part of the Manhattan Project, the United States initiative to develop an atomic bomb during World War II. The reactor was a literal pile: In a squash court beneath the stands of Stagg Field, the University of Chicago's football field, Italian physicist Enrico Fermi and his team of scientists built a stack of uranium pellets and bricks of graphite, interspersed with cadmium control rods. On December 2, the control rods were removed and the reactor went critical.

Following the initial testing, CP-1 was disassembled in 1943, moved to Red Gate Woods, and rebuilt, with a radiation shield for safety, as a new reactor named CP-2. Another experimental reactor, CP-3, followed in 1944. When the Manhattan Project scientists were done with these reactors, they dismantled them and buried the remains in the woods, at spots marked Plot M and Site A. Both are now marked by granite monuments—Plot M has the headstone marked DO NOT DIG.

Though the buried waste is radioactive, the site poses no threat to public safety—Geiger counter readings in the area are consistent with standard background radiation levels.

Archer Avenue, Lemont. Ⓝ 41.699599 Ⓦ 87.921223

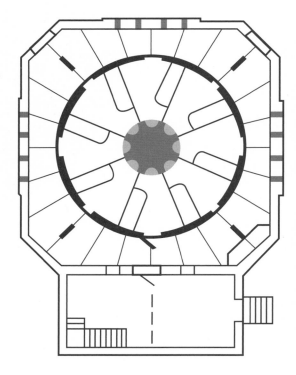

Spinning jails, built across the Midwest in the 1880s, rotated like carousels. Unfortunately, they also sometimes crushed prisoners' hands between the bars.

INDIANA

ROTARY JAIL MUSEUM

CRAWFORDSVILLE

In the spring of 1881, architect William H. Brown and iron foundry owner Benjamin F. Haugh, both of Indianapolis, filed a patent for a most ingenious innovation: a jail with revolving cells.

Their design consisted of a two-tier cylindrical cell block with a central column that served as both support and plumbing for the individual toilets in the cells. Each tier had eight wedge-shaped cells, but the surrounding structure had only one door. When a guard rotated a hand crank, the cell block spun, sending the prisoners on a disorienting carousel ride past the lone access point.

Brown and Haugh's invention quickly became a reality. In 1882, the first spinning jail, a two-tiered, 16-cell institution known as Montgomery County Rotary Jail, opened in Crawfordsville, Indiana. Other states in the Midwest soon got in on the idea—the three-tiered, rotating Pottawattamie County Jail, nicknamed the "Squirrel Cage Jail," opened in Iowa in 1885, followed by a single-story spinning jail in Gallatin, Missouri, in 1889. As many as 18 rotary jails were built in the United States, mostly in the Midwest.

Unfortunately, Brown and Haugh's novel, almost whimsical design had its flaws. Chief among them was the fact that a prisoner standing at the front of a cell with his hands resting on the bars had a decent chance of getting an arm crushed when the rotary mechanism was engaged. Natural light was scant, ventilation was poor, and mechanical problems could interfere with the operation of a jail. In the case of a fire, all the prisoners whose cells weren't aligned with the access door would likely be doomed.

In light of these problems, many rotary jails had their turntables immobilized during the 1930s. After operating in a modified state for decades, Montgomery County Jail closed for good in 1973. Pottawattamie County Jail sent its prisoners away in 1969, while the Gallatin jail shut up shop in 1975. All three now operate as museums. Montgomery County is the only one that still spins.

225 North Washington Street, Crawfordsville.
🄽 40.043839 🅆 86.901742

WORLD'S LARGEST BALL OF PAINT

ALEXANDRIA

Suspended from the rafters in a small shed behind the Carmichael family home is a 4,000-pound (1.8 MT) ball. Buried at its core is a baseball—the rest is paint.

When Michael Carmichael was a teenager in the 1960s, he and a friend were tossing a baseball in a paint shop and knocked over a gallon of paint. The ball was covered, and Michael's head was overtaken by an ambitious idea. Every day during his last two years of high school he applied a new coat of paint to the ball. He made it to 1,000 coats before losing interest. But a decade later, the idea resurfaced. In January 1, 1977, Michael rustled up a fresh baseball and presented it to his three-year-old son, who brushed on a coat of blue paint. Thus began the odyssey of the second, and current, Ball of Paint.

That ball has now been painted about 25,000 times. Michael, his wife Glenda, their children, their grandchildren, and a stream of visitors have all applied layers. Anyone is welcome to pick up a brush, and there is only one rule: Each new coat must be a different color than the last.

10696 North 200 West, Alexandria. The ball is located at the Carmichael family residence—call ahead to schedule a viewing or painting appointment.
Ⓝ 40.258752 Ⓦ 85.709122

14-foot (4.3 m) circumference

What began as a painted baseball has turned into a spherical tribute to the power of persistence.

SLOCUM MECHANICAL PUZZLE COLLECTION

BLOOMINGTON

"Mechanical puzzles" are those brain-teasers that must be physically manipulated to achieve a specific outcome. Jerry Slocum began collecting mechanical puzzles as a child, eventually becoming the unofficial authority on the subject with the publication of his 1986 book, *Puzzles Old and New*. By 2006 he had accumulated over 40,000 mechanical puzzles, thanks in part to the International Puzzle Party, an annual private get-together for mechanical puzzle enthusiasts and traders, which Slocum inaugurated in 1978.

In 2006, he donated over 30,000 of the puzzles to the Lilly Library at Indiana University to create

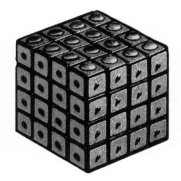

Around 30,000 manually operated mind-benders make up the Slocum puzzle collection.

the Slocum Mechanical Puzzle Collection. In addition to the staggering number of puzzles, Slocum also donated thousands of books about puzzles. Among the pieces on display (only a few hundred out of the thousands in the collection) are an archaic Rubik's Cube with differing sizes of nails on each side, called a "texture cube"; a trick cup that seems normal until its drinker fills it too full and it drains away into the base; and more whimsical amusements like a Coke bottle with a wooden arrow through it. There are also countless intricate wooden geometrical curiosities that must be twisted and shifted together and apart.

Today, visitors to the library can actually try out a number of the puzzles and see countless others sitting in displays, just waiting to be solved.
1200 East Seventh Street, Bloomington. Ⓝ 39.167906
Ⓦ 86.518973

IOWA

VILLISCA AX MURDER HOUSE

VILLISCA

For a unique overnight stay, book a night in the home where eight people were once slaughtered in their beds as they slept.

In 1912, this house belonged to the Moore family: married couple Josiah and Sarah and their young children, Herman, Katherine, Boyd, and Paul. On the night of June 9, the family returned from Children's Day services at the local church, accompanied by two of the girls' friends: eight-year-old Ina Stillinger and her twelve-year-old sister, Lena.

Sometime between midnight and 5 a.m., an unknown person entered the house and murdered every person inside by striking their heads with an ax. From observations at the crime scene it appears that all were asleep at the time they were killed, apart from Lena, who exhibited a defensive wound on her arm and was positioned across her bed.

Over a century later, the case remains unsolved. The main suspect, traveling minister Reverend George Kelly, had taught at the church on June 9 and left town at approximately 5 a.m. the next day. He was tried twice but never convicted.

The Villisca Ax Murder House, as it is now bluntly known, was purchased in 1994 by Darwin and Martha Linn and restored to its 1912 state. You can tour the home by day, then spend the night in a room that was once a blood-soaked crime scene.

508 East 2nd Street, Villisca. All eight victims of the ax murders are at Villisca Cemetery, a 15-minute walk north of the house. Ⓝ 40.930704 Ⓦ 94.973316 ➤➤

➤➤ Other Murder Houses

LIZZIE BORDEN BED & BREAKFAST
FALL RIVER, MA

The tagline of this eight-room hotel is "Where everyone is treated like family!" Ordinarily that might sound like a good thing, but getting treated like Lizzie Borden's family means copping a hatchet to the head.

On the morning of August 4, 1892, the bodies of Lizzie's father, Andrew, and stepmother, Abby, were found in separate rooms of the family home. Both had been bludgeoned in the head with an ax. Lizzie, who had been in or near the house during the time of the murders, was arrested a week later and stood trial in June the following year.

Despite contradicting herself during her testimony and incurring

the ire of lead prosecutor Hosea Knowlton, Lizzie received an acquittal from the jury.

The house in which Abby and Andrew met their bloody ends is now a Borden-themed bed and breakfast. You can stay in the very rooms where Lizzie's parents took their last gasping breaths. Good night, sleep tight, don't let the ax murderer strike.

AMITYVILLE HORROR HOUSE
AMITYVILLE, NY

On the evening of November 13, 1974, 23-year-old Ronald "Butch" DeFeo Jr. murdered his parents and four younger siblings in their home at 112 Ocean Avenue. DeFeo then went to a bar down the road, where he ran in screaming for help, claiming his parents had been shot by a mob hit man.

When the deaths were confirmed and DeFeo was brought into the local police station for questioning, inconsistencies began to appear in his story. The following day, DeFeo confessed to the murders, telling police that once he started, he couldn't stop.

After DeFeo was given six consecutive sentences

Gruesome murders turned this family home into a morbid attraction.

of 25 years to life, the Lutz family, consisting of George and Kathy Lutz and their three children, moved into the house on Ocean Avenue. Less than a month later, they vacated the place permanently, claiming paranormal activity had made it impossible to live there.

The Lutzes' supernatural claims, which include descriptions of mysterious voices, slime oozing from the walls, and red glowing eyes appearing in the dark, were dramatized in the 1977 book *The Amityville Horror*, which has since been adapted into multiple films. The "based on a true story" tagline has never been corroborated with evidence.

The house at 112 Ocean Avenue (now 108 Ocean Avenue), an iconic part of American pop culture, is still private property—view it from afar if you like, but there's no need to warn the current inhabitants of the supposed supernatural peril. They've heard it all before.

MICHIGAN
HOEGH PET CASKETS

GLADSTONE

A tour of Hoegh Pet Caskets, established 1966, begins in the showroom. There, lined up on plinths against the walls, are blue, pink, white, and camouflage caskets ranging from 10 to 52 inches (25–132 cm) in length.

After seeing the finished products, it's time to see the factory, where 18 caskets are made every hour. The tour concludes at Hoegh's mock pet cemetery, where the recently bereaved can get ideas for conducting a final farewell to Spanky or Mittens.

The caskets, which are shipped all over the world, are not just for animals—they are also purchased by amputees who want to lay their severed limbs to rest.

311 Delta Avenue, Gladstone. Tours available daily.
Ⓝ 45.849229 Ⓦ 87.011343

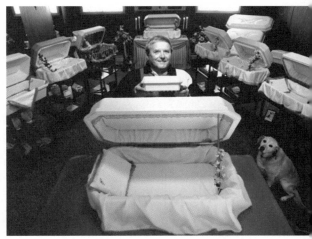

Departed critters and amputated limbs are lovingly laid to rest in Hoegh's plush coffins.

MARVIN'S MARVELOUS MECHANICAL MUSEUM

FARMINGTON HILLS

At this overstimulating penny arcade you can play tic-tac-toe with a chicken, witness a mechanical depiction of the Spanish Inquisition, and marry a friend via the AutoWed machine.

The crowded lineup of vintage and new games, automata, and oddities belongs to Marvin Yagoda, who has been collecting coin-operated machines since 1960. The standard pinball, Skee-ball, and fortune-teller machines are interspersed with more atypical fare like Dr. Ralph Bingenpurge, a mechanical food inspector who vomits with violent and disturbing realism.

Marvin's is a good place to visit if you're unsure of what to do with your life—get a reading from a mechanical fortune-teller, then consult the career machine, which will tell you whether your ideal job is chorus girl, bootlegger, or soda jerk.

31005 Orchard Lake Road, Farmington Hills. Look for the big clock. Ⓝ 42.525442 Ⓦ 83.361727

Coin-operated oddities dance, dole out prizes, and, in one case, vomit continuously.

EDISON'S LAST BREATH

DEARBORN

In a display case at the Henry Ford Museum is a sealed test tube labeled "Edison's Last Breath?" The question mark is key: The great inventor didn't wheeze his final exhalation straight into the tube before falling back into the pillows and expiring. But the real story is just as compelling. And it begins with a long friendship between two prolific inventors.

In 1891, Henry Ford got a job as an engineer at the Edison Illuminating Company. The future car manufacturer regarded Edison as a personal hero, but the two men didn't meet until 1896. That year, Ford built his first vehicle: the four-wheeled, gas-powered Ford Quadricycle. Edison's positive response to the vehicle encouraged Ford to keep working on gas-powered cars. He left the Edison Illuminating Company and, in 1908, debuted the revolutionary Model T.

Edison and Ford's friendship remained strong well into the 20th century. In 1916, Ford purchased the property beside Edison's vacation home in Fort Myers, Florida. When Edison needed a wheelchair due to his failing health, Ford bought a matching one so they could race around the grounds.

In 1931, Edison died at his New Jersey home in the presence of his son, Charles. Beside the inventor's death bed was a rack of test tubes. Charles took one, had it sealed with paraffin, and sent it to Ford as a final memento of his dear friend.

20900 Oakwood Boulevard, Dearborn. The test tube is near the front door. Ⓝ 42.303109 Ⓦ 88.229686

MINNESOTA

WORLD'S QUIETEST ROOM

MINNEAPOLIS

You may think silence is peaceful until you visit Orfield Laboratories. The lab is home to an anechoic chamber: a room that has no echo. 99.99 percent of all sound made inside the chamber is absorbed by bouncy 3-foot-thick foam wedges that cover every surface.

Within seconds of entering, you will notice sounds you don't usually hear: The beat of your heart, the flow of your breath, and the gurgling of your digestive system start to become unnerving. Visitors often become disoriented, especially if lab founder Steven Orfield turns out the lights. Most don't last beyond a few minutes. Half an hour is unimaginable.

Manufacturers use the quiet room during product testing to gauge the volume of switches, displays, and other components.

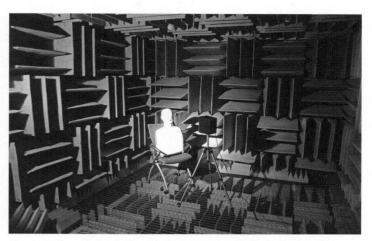

After just a few minutes in the noiseless room, aural hallucinations set in.

The chamber has been Guinness-certified as the quietest place in the world, with an ambient noise level of –9 decibels. (A quiet bedroom at night is around 30 decibels.) **2709 East 25th Street, Minneapolis. Ⓝ 44.957042 Ⓦ 93.232773**

ALSO IN MINNESOTA

House of Balls

Minneapolis · Local sculptor Allen Christian established this funhouse of found art as a physical incarnation of his idea that "we all possess the creative impulse and we owe ourselves the balls to express it."

QUESTIONABLE MEDICAL DEVICE COLLECTION

ST. PAUL

During the 1930s and '40s, American parents needing to take their kids shoe shopping could provide them with a fun incentive. Many stores at the time had shoe-fitting fluoroscopes, which used X-rays to show how a child's feet fit into their new shoes. The kid would insert his or her shoe-clad foot into the four-foot-high wooden box, and a salesperson would peer into a peephole to look at the position of the bones.

It wasn't until 1957—after the long-term effects of radiation exposure became better understood—that shoe-fitting fluoroscopes started getting banned in the US. (They made shopping for boots a real kick, but they also leaked radiation.)

The foot fluoroscope is one of hundreds of fraudulent and dangerous medical devices on show at the Science Museum of Minnesota. You'll also find a vibratory chair from 1900 (it shakes violently to stimulate digestion), a foot-operated breast enlarger pump from the mid-'70s, and the Relax-A-Cizor, a 1960s weight-loss device that delivered electric shocks to muscles.

Roger's Vitalator Violet Ray medical appliance was used in electrotherapy to cure a wide array of ailments.

Science Museum of Minnesota, 120 West Kellogg Boulevard, St. Paul. Ⓝ 44.953703 Ⓦ 93.089958

The right half of the waterfall goes into the river, while the left half just disappears.

THE DEVIL'S KETTLE

GRAND MARAIS

A large rock bisects the Brule River in Judge C.R. Magney State Park. The water that flows to the east tumbles 50 feet (15 m) down a cliff and continues toward Lake Superior. The water that flows to the west enters a hole and disappears.

Known as the Devil's Kettle, this water portal to nowhere has long puzzled Minnesotans. In attempts to trace the underground flow, researchers have dropped objects such as ping-pong balls and dye into the hole. So far, all efforts to map the water's path have proven fruitless.

Judge C. R. Magney State Park (Highway 61). Ⓝ 47.828945 Ⓦ 90.049609

MISSOURI

SubTropolis

KANSAS CITY

Beneath the limestone bluffs on the north side of the Missouri River lies a vast underground city. Known as SubTropolis, the 2-square-mile man-made cave is used for corporate workspaces and storage. According to its creators, it is the world's largest underground business complex.

Limestone mining began in the bluffs during the 1940s. Two decades later, the Hunt Midwest company began renting out office and storage space in the carved-out cliffs. Today, SubTropolis houses everything from the original *Gone With the Wind* and *Wizard of Oz* film reels to USPS commemorative stamps.

The unchanging ambient conditions—temperature of 65° to 70°F (18–21°C); humidity of 40 to 50 percent—assist with preservation and save energy by making it unnecessary to heat or cool the space.
8300 Northeast Underground Drive. See inside SubTropolis by participating in the annual Groundhog Run, a 5K foot race held in the complex each January.
Ⓝ 39.157638 Ⓦ 94.478478

An eco-friendly mega warehouse hidden underground.

Leila's Hair Museum

KANSAS CITY

The wreaths, bouquets, and pieces of jewelry in Leila's Hair Museum are made from woven strands of human hair, shorn from Victorian heads.

Peaking in popularity in the mid-19th century, hair art and jewelry functioned as tokens of mourning, family heirlooms, and gifts between friends or lovers.

Locks of hair became personal mementos and served as "portraits" of an individual or family in an era when photography was not yet widely accessible. Some of the wreaths adorning the museum walls incorporate hair from multiple generations, twisted into flowers and vines that reflect the family tree. Friends and relatives of the recently departed wore bracelets made from the deceased's hair, or lockets with hair tucked inside.

Museum founder and cosmetology school owner Leila Cohoon began collecting hair art in 1949. The thousands of pieces on exhibit comprise the world's only museum dedicated to this near-extinct art form.
1333 South Noland Road, Independence. The museum is 1.5 miles (2.4 km) southeast of the Independence Amtrak station.
Ⓝ 39.076007 Ⓦ 94.413452

Also in Missouri

Glore Psychiatric Museum

St. Joseph · The Lunatic Box and the Tranquilizer Chair are among the exhibits that illustrate the starkly different way mental health was treated in the 18th and 19th centuries.

Leila Cohoon celebrates an extensive collection of Victorian jewelry made from human hair.

OHIO
CINCINNATI'S LOST SUBWAY

CINCINNATI

In the early years of the 20th century, Cincinnati's streets were clogged with slow-moving street-cars, crowds of pedestrians, horse carriages, and a few of those new-fangled contraptions known as automobiles.

At the time, the Boston, New York, and Philadelphia sub-ways had just started operating. Cincinnati needed a rapid transit system, and an underground rail-road seemed the obvious solution. The city raised money via bonds, and Cincinnati residents voted in favor of the subway in April 1917.

Unfortunately, this was precisely the month that the US became involved in World War I. Suddenly, bonds stopped being issued and construction had to be halted. When the war ended 19 months later, building costs skyrocketed. The city resumed the subway project in 1920, but had to stop work in 1927 when the money ran out. More than 2 miles (3.2 km) of tunnels had been built, as well as seven stations, but tracks were never laid.

Cincinnati still doesn't have a subway system. The stock market crash of 1929, the increasing popularity of the automobile, and the United States's involvement in another global war all contributed to the abandonment of the project.

The three above-ground stations have been demolished, but the tunnels and four underground stations remain. Appearance-wise, little has changed since the '20s—parts of the Liberty Street station were converted into a nuclear fall-out shelter during the 1960s, and a water main was laid in the tunnel in 1957, but besides that it's just dusty platforms, musty smells, and stairs leading to nowhere.
The tunnel runs under Central Parkway for 2 miles (3.2 km), beginning at Walnut Street and ending just north of the Western Hills Viaduct. Tours of the tunnel are conducted once a year. Accessing the subway system at other times constitutes trespassing. Ⓝ 39.107302 Ⓦ 84.512853

Despite decades of planning and construction, Cincinnati's underground train tunnels have never been used.

CRYSTAL CAVE

PUT-IN-BAY

Ohio's Crystal Cave is not a cave at all, but rather a single rock. It is the world's largest known geode, and its walls are lined with huge white-blue celestite crystals, some as much as 3 feet long.

This hidden gem was discovered about 40 feet (60.4 m) below the ground in 1897, by workers digging a well for the Heineman Winery. Upon exploring the cavern, winery owner Gustav Heineman found that his business was sitting on top of a vug, or large cavity within a rock. Emerging from the limestone walls of the cavity are countless crystals, composed of strontium sulfate, better known as the mineral celestite.

Heineman opened up the giant geode to visitors, and this unique attraction sustained the winery through the years of Prohibition. Today, Crystal Cave is even larger but perhaps less stunning than it once was, as many of the crystals were mined over the years to manufacture fireworks. Yet it remains a natural wonder unlike anything else.

The cave is best explored with a glass of wine or grape juice, as part of the overall Heineman Winery tour. ℕ 41.646648 Ⓦ 82.826842

Come for the wine; stay for the world's largest geode.

AMERICAN SIGN MUSEUM

CINCINNATI

Some people rage against the indignities of middle age by buying a sports car. Tod Swormstedt founded a sign museum.

Swormstedt's self-proclaimed "midlife crisis project" began in 1999 as a real-world version of *Signs of the Times*, a trade magazine about sign making and outdoor advertising that he edited and published.

In 2005, his newly named American Sign Museum opened to the public. It is now crammed with fiberglass mascots, neon marquees, and hand-painted 19th-century signs advertising cobblers, druggists, and haberdashers. Tours

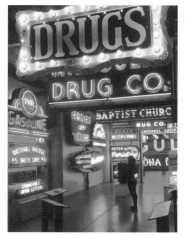

Get buzzed on neon at this ode to old-school Americana.

are accompanied by an ever-present buzz from all the neon. Visit on a weekday to get a glimpse of the neon shop, where employees from Neonworks of Cincinnati demonstrate how they make signs.
1330 Monmouth Avenue, Cincinnati. ℕ 39.145530 Ⓦ 84.539904

ALSO IN OHIO

Haserot Angel

Cleveland · Due to the effect of weathering and erosion, this bronze angel at Lakeview Cemetery appears to be weeping black tears.

WISCONSIN
HOUSE ON THE ROCK

SPRING GREEN

Opened to the public in 1960, this attraction is a window into the unusual mind of Alex Jordan, who built the home as a weekend retreat.

Jordan was a collector with eclectic tastes. Over decades, he filled the house with an astounding array of objects: pipe organs, dollhouses, antique weapons, coin-operated music machines, chandeliers, and miniature circuses, to name a few. Jordan died in 1989, but his house of treasures lives on.

The collection takes hours to walk through and can induce sensory overload. Two parts of the house really stand out: the Carousel Room and the Infinity Room. The Carousel Room contains the world's largest merry-go-round, featuring 269 animals—none of which are horses. The Infinity Room, built in 1985, is a cantilevered, glass-walled hallway that extends 218 feet (66.5 m) out over the valley. A walk through the house raises a lot of questions: Is this all real? Why is that massive sea monster battling a giant squid, and how does it relate to the robot orchestra? Don't look for answers. Just marvel.

5754 State Road 23, Spring Green. The town is about 2 hours from Milwaukee. Ⓝ 43.090644 Ⓦ 90.131808

By turns creepy, campy, and beautiful, House on the Rock features a carousel tricked out with angels, 20,000 string lights, and 182 chandeliers.

KOVAC PLANETARIUM

RHINELANDER

On an October afternoon in 1996, Frank Kovac and his fellow Boy Scouts ventured to Mud Creek Observatory and waited for the sun to set so they could gaze at the universe. But when it was finally dark, cloud cover made stargazing impossible.

Annoyed at the clouds for, in his words, "obscuring the universe," Kovac decided to create his own celestial sphere. He spent a decade building a 2-ton, 22-foot-wide (1.8 MT, 6.7 m) rotating globe and painting the interior with every northern hemisphere star visible to the naked eye.

The Kovac Planetarium is only the fourth mechanical globe ever created. Kovac treats visitors to a sky show that lives up to its motto: "Kovac Planetarium, Where the Universe Revolves Around You."

2392 Mud Creek Road, Rhinelander. Ⓝ 45.573826 Ⓦ 89.065458

DR. EVERMOR'S FOREVERTRON

NORTH FREEDOM

Nestled in the trees on a remote stretch of Highway 12 is the Forevertron, a 19th-century space-craft built by an eccentric man named Dr. Evermor.

Dr. Evermor didn't exist until 1983. That's when former industrial wrecking and salvage expert Tom Every retired and assumed the alter ego of a Victorian professor and inventor. Having amassed a personal collection of beautiful old machinery components during his wreck-and-salvage days, Every set about sculpting a scrap-metal spacecraft with an 1890s aesthetic.

The Forevertron comes with a story: Its purpose is to launch Dr. Evermor into the heavens on a magnetic lightning beam. The big

glass egg at the top of the sculpture, latticed with copper, is the doctor's personal space capsule. An elevated gazebo beside the main sculpture allows royalty to watch the launch from a safe vantage point. The Celestial Listening Ears are designed to allow visitors to hear voices from space.

There is no set launch date for the Forevertron. Even if it never blasts into the heavens, it has already earned an impressive distinction: At 50 feet (15 m) high and 120 feet (36.5 m) wide, it's the largest scrap metal sculpture in the world.

US 12, North Freedom. The Forevertron is 5 miles (8 km) south of Baraboo, on a gravel road behind Delaney's Surplus Sales.
Ⓝ 43.375669 Ⓦ 89.768549

ALSO IN WISCONSIN

FAST Fiberglass Mold Graveyard

Sparta · The molds once used to make roadside sculptures, from menacing sharks to goofy giant mice to a towering Santa Claus, have been strewn across this grassy lot, making it look like the ruins of a particularly quirky civilization.

If all goes according to plan, this scrap metal sculpture will blast into space carrying its creator.

DELIGHTFULLY GOOFY DINOSAUR PARKS

Whether created for the purposes of entertainment, education, or some combination thereof, dino parks are a proud part of the great American road trip. Just don't rely on them for anatomical accuracy.

DINOSAUR PARK
RAPID CITY, SD
The half-dozen concrete dinosaurs perched on a hill overlooking Rapid City look pretty absurd. The round-jawed, smiling tyrannosaur appears to have double-jointed forelimbs. The duck-billed trachodon—now known as an anatotitan—stands upright and awkward, forced into a bipedal stance and looking resentful about it.

A stegosaurus, a triceratops, and a brontosaurus (since reclassified as an apatosaurus) stand nearby, looking as though they arrived from the same cartoonlike alternate dimension.

The dinosaurs were built in 1936 at the height of the Great Depression. Aside from an occasional fresh coat of bright green paint, they haven't changed since. Though their design diverges wildly from the fossil record, the Rapid City dinos are imposing enough—until a car bound for Mount Rushmore stops by and kids clamber all over the stegosaurus spines.

CABAZON DINOSAURS
CABAZON, CA
Since the 1980s, travelers barreling down Southern California's Interstate 10 have been greeted by a 150-foot-long (45.7 m) apatosaurus and a 65-foot-tall (20 m) T. rex.

Theme park artist and sand sculptor Claude K. Bell began building the steel-and-concrete dinosaurs, known as Dinny and Mr. Rex, in 1964. The main aim was to create hollow, climbable structures that would attract more customers to his Wheel Inn diner next door. But Bell also had a more

Argentinosaurus

personal motive: After spending years building sand sculptures and watching them disappear in the wind, he wanted to create something permanent.

Bell finished the apatosaurus in 1973 and began construction on the T. rex in 1981. He died in 1988, but not before seeing his dinosaurs achieve fame—they were featured in commercials, music videos, and the 1985 film *Pee-wee's Big Adventure*.

Since being sold in the mid-1990s, the Cabazon dinosaurs have changed dramatically, but not in a way that's visible from the outside. In short, they've found God. Dinny's belly now contains a gift shop and museum operated by creationists. Signs and displays espouse the view that dinosaurs were created along with humans 6,000 years ago.

Beside Dinny and Mr. Rex is an open-air robotic dinosaur museum. It features a medieval knight jousting with a velociraptor.

DINOSAUR LAND WHITE POST, VA

Somehow, sculptures of King Kong, a giant cobra, and a monstrous praying mantis ended up here among the T. rexes and pterodactyls. And then there are the walk-through shark and octopus.

None of it makes much sense, but Dinosaur Land, built in the 1960s, still has enough dinosaurs to justify its name. Many have been sculpted in attack poses, making them particularly suitable for creative photo composition. Make sure you stop by the giganotosaurus that is casually chomping a pterodactyl out of the sky.

Pterodactyl

Tyrannosaurus rex

Triceratops

THE SOUTHEAST

ALABAMA

SPEAR HUNTING MUSEUM

SUMMERDALE

Eugene Morris was the greatest spear hunter in the world, according to Eugene Morris. From 2006 until his death at 78 in 2011, Morris maintained the Spear Hunting Museum, dedicated to celebrating his own accomplishments. Though the painted exterior—emblazoned with "Eugene Morris: The Greatest Living Spear Hunter in the World" next to a rendering of the man in action—has had to change, the museum lives on under the administration of his wife, Heather.

Inside the building are many of the 500 animals Morris has killed with spears, including buffalo, lions, zebras, bears, alligators, turkeys, and deer.

Photos of Morris on the prowl line the walls, as do the very spears he once launched into the abdomens of those he hunted down.

Formerly a gun hunter, Morris switched methods in 1968 because shooting animals had become too easy. He first tried bow hunting, then two-handed spears, and finally took to the wilderness with a spear in each hand, poised to strike whatever living creature crossed his path. Sometimes he threw both spears at once, killing two animals simultaneously.

Morris died sitting up, spear in hand. His will specified that the slogan on the museum wall be altered to read "Gene Morris: The Greatest Spear Hunter in Recorded History."

20216 Highway 59, Summerdale. The museum is between Robertsdale and Summerdale.
Ⓝ 30.519810 Ⓦ 87.707679

Passengers' lost property becomes shoppers' treasured bargains.

Unclaimed Baggage Center

SCOTTSBORO

Every item being sold at this store was once lost in transit. The Unclaimed Baggage Center buys lost luggage from US-based airlines by the truckload and sells the contents to the public in a building the size of a city block.

The clothing, accessories, electronics, and luggage itself are the main offerings, but the staff has also unpacked 3,500-year-old Egyptian burial masks, a stuffed Canada goose, and Hoggle, the curmudgeonly dwarf puppet from the 1986 film *Labyrinth*. He and the salvaged ancient artifacts were never put on sale—they have a permanent home at the center's museum.

509 West Willow Street, Scottsboro, next to Cedar Hill Cemetery. Ⓝ 34.673176 Ⓦ 86.044589

ALSO IN ALABAMA

Monument to Hodges Meteorite

Sylacauga · This marble sculpture pays tribute to a meteorite that, in 1954, crashed into a local home.

ARKANSAS

Ozark Medieval Fortress

LEAD HILL

The plan was ambitious: A team of medieval enthusiasts would spend 20 years building a 13th-century castle using only tools and techniques from the Dark Ages.

The Ozark Medieval Fortress was intended to attract visitors who would pay for the privilege of observing a historic construction site. Opened to the public in 2010, the fortress lasted almost two years before lack of attendance forced its closure. Even the allure of falconry demonstrations and stone carving contests did not make up for the fact that it's pretty dull to stand around watching people build a fortress by hand.

The castle foundations have sat idle since 2012, awaiting a medieval-obsessed investor to get the project going again. There is a precedent for success: France's Guédelon Castle, which is based on the same concept as the Ozark fort, was founded by the same group in 1997. It is scheduled for completion in the 2020s and is flourishing as a tourist destination.

1671 Highway 14 West, Lead Hill. The fortress foundations and construction equipment are still in place, but walking among them requires trespassing beyond a fence. Ⓝ 36.438990 Ⓦ 93.036066

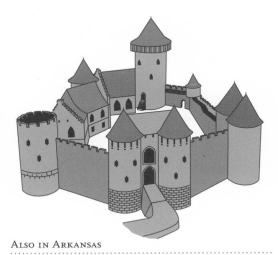

ALSO IN ARKANSAS

Gurdon Light

Gurdon · There's got to be a scientific explanation for this mysterious light that hovers above the railroad tracks, but inventing tall tales is more fun.

Billy Bass Adoption Center

Little Rock · Billy Bass was an animatronic singing fish that was a hugely popular gag gift in the early 2000s. The seafood diner Flying Fish has rows of Billies on its walls, each one given up for adoption by a customer in exchange for free food.

FLORIDA
UNIVERSITY OF FLORIDA BAT HOUSES

GAINESVILLE

When a fire destroyed the University of Florida's Johnson Hall in 1987, the bats living in the attic became homeless. In need of new eaves to inhabit, they nestled into the crannies of the running track bleachers and the tennis stadium. This was less than ideal for athletes and spectators.

In 1991, the university captured thousands of campus bats and relocated them to a giant, newly built, elevated bat house beside a lake. The bats escaped their new home within days. Four years later, they finally came around and moved in for good. Now around 300,000 bats live in the house and its neighboring barn, which was added in 2010.

With a combined capacity of 750,000, the house and barn are the world's largest occupied bat houses. At sunset, the bats emerge and stream across the sky in search of tasty bugs.

Museum Road between Village Drive and Radio Road, across from Lake Alice, University of Florida, Gainesville. The best time to see the bats is on dry, warm evenings in spring through early summer. Beware of falling guano. Ⓝ 29.644211 Ⓦ 82.362859

300,000 bats, cooped up in the world's largest bat houses, wait for sunset.

In shallow waters a mile off Miami stand the former party shacks of Stiltsville.

STILTSVILLE

MIAMI

In 1933, a Miami man known for his chowder—"Crawfish" Eddie—constructed a wooden shack on stilts a mile offshore. Built during the last gasp of Prohibition, the shack soon attracted drinkers and gamblers, who chased their beers and poker hands with a bowl of Eddie's famous chowder.

Crawfish Eddie's was the first shack in Stiltsville, a hodgepodge collection of offshore houses that became a playground for Miami's coolest sinners.

The Quarterdeck Club, an invite-only gentleman's establishment, appeared in 1941. Fishing buddies would anchor between buildings to drink, dine, and cast their lines into the shallow waters from the Quarterdeck's front dock.

The early 1960s was boom time for Stiltsville. Twenty-seven shacks crowded the sand flats, and the new Bikini Club offered free drinks for women in two-pieces and a secluded sundeck for naked lounging. The boom, however, was swiftly followed by busts both governmental and meteorological. The vice squad raided the Bikini Club and shuttered it for operating without a liquor license and possessing 40 unauthorized, undersize, out-of-season crawfish. Then came Hurricane Betsy in 1965, which swept away all but half a dozen shacks.

Many were rebuilt, but by this time the state's patience for Stiltsville shenanigans was wearing thin. Florida issued formal leases to the shack owners with an expiration date of 1999, declaring that any remaining houses would need to be removed. When the crucial date finally arrived—after Hurricane Andrew had destroyed all but seven of the stilt houses—community support helped Stiltsville owners negotiate a deal with the government. The shacks were saved and are now available to rent for parties.

The houses now lie within the boundaries of Biscayne National Park, and are co-managed by the National Parks Service and the Stiltsville Trust. Ⓝ 25.651305 Ⓦ 80.174227

The charmingly kitschy underwater mermaid ballet has been performed since the 1940s.

WEEKI WACHEE MERMAIDS

WEEKI WACHEE

Mermaids have lived in the waters of Weeki Wachee Springs since 1947. That's when swim coach Newt Perry developed an underwater mermaid show, performed at a depth of 20 feet (6 m) by glamorous women in bikini tops and fish tails.

To watch the mermaid show, spectators file into a 400-seat underwater theater built into one side of the spring. The lights dim, gentle music starts playing, and a faded blue curtain rises to reveal the water behind a 100-foot-wide (30.5 m) glass wall. As a voiceover describes the mermaids' world—"water as blue as the loveliest cornflower and as pure as the clearest glass"—one of the mythical creatures swims the length of the window, smiling and waving as the tendrils of her tail waft behind her.

The ensuing show is a lip-synced, music-enhanced retelling of *The Little Mermaid*, complete with handsome prince, evil sea witch, and a terrestrial wedding once the mermaid gets her legs. To breathe underwater, the scuba-certified performers inhale from free-flowing air hoses that snake around the spring. Fish and turtles are unwitting co-stars.

Since the arrival of Disney World in 1971, Weeki Wachee has struggled to attract visitors. During the 1960s there were nine shows a day. Now there are three.

Weeki Wachee Springs State Park, US 19 at State Road 50. Ⓝ 28.491034 Ⓦ 82.632138

ALSO IN FLORIDA

Pinecraft

Sarasota · Between December and April, busloads of Amish and Mennonite vacationers swap their frosty climes for the resort neighborhood of Pinecraft, where popular activities include shuffleboard and lounging—fully dressed—on the beach.

Robert the Doll

Key West · Held in a glass box at the Fort East Martello museum, this innocent-looking doll in a sailor uniform has a reputation for masterminding evil deeds.

Bok Tower Gardens

Lake Wales · A neo-Gothic 205-foot (62.5 m) tower on Iron Mountain houses 60 singing bells.

Cassadaga

Lake Helen · A community of spiritualists offers healing services, psychic readings, and nighttime spirit photography.

Skunk Ape Research Headquarters

OCHOPEE

Most Bigfoot sightings occur in the Pacific Northwest, but the Florida Everglades claim a local variant: the Skunk Ape. Bipedal, 6 to 7 feet (2 m) tall, and exuding an odor of rotten eggs mixed with moldy cheese, the Skunk Ape apparently stalks the grounds of the Big Cypress National Preserve.

No one claims to know more about Skunk Apes than Trail Lakes Campground owner and Skunk Ape Research Headquarters proprietor Dave Shealy. His obsessive quest to prove the ape's existence began at the age of ten when he spotted the Skunk Ape while on a hunting trip. Ever since, Shealy has scoured the park swamps looking for tracks, droppings, and the elusive creature itself. He claims a total of three Skunk Ape sightings.

Some of Shealy's findings are in his *Everglades Skunk Ape Research Field Guide*, which includes tips on planning an expedition and setting bait. ("If you

Dave Shealy's center is devoted to the search for the Everglades' most elusive primate.

plan to use deer liver, remember that it should be kept on ice until your site is chosen.") You can pick up a copy at the gift shop.

40904 Tamiami Trail East, Ochopee. Skunk Ape headquarters is on the Trail Lakes Campgrounds.
Ⓝ 25.892642 Ⓦ 81.279830

Neptune Memorial Reef

KEY BISCAYNE

A few miles east of Miami lies an underwater city. A pair of lions guard its entrance, which lead the way to stone roads, soaring gates, and crumbling ruins. Did an ancient civilization once live here? No. The "city" is a cemetery, and it was built in 2007.

Conceived as a living reef and modeled after the lost city of Atlantis, the site was created by cremation-services provider the

Neptune Society. Anyone wishing to bury their loved one at the city can hand over the cremated remains—in person or by mail—to be mixed with cement and sand, poured into a shell- or starfish-shaped mold, and added to the reef. Family members are welcome to participate in the process, either by scuba diving or watching from a boat above. Postburial, they may visit the reef at any time for free.

If you want your remains to become part of the underwater city, you'll need to decide which

part of the reef to join. There are 15 burial locations, ranging from standard placements (road railings; the "Fish Habitat Bench") to premium (gatekeeper columns) to exclusive (the Welcome Feature Centerpiece). Shipwreck diver Bert Kilbride, once listed as the oldest living scuba diver in the Guinness World Records, is interred in a place of honor at the top of one of the entry columns at the reef gate.

The Neptune Society has big plans for its undersea cemetery. Designed to attract fish and promote the growth of coral and marine organisms, the memorial reef will gradually take on a more authentic ancient-city look. The society's ultimate goal is a 16-acre metropolis containing the remains of 125,000 people. Currently, the reef is a quarter of an acre and is the final resting place of a few hundred.
The reef is just over 3 miles (5 km) east of Key Biscayne, 40 feet (12 m) under the sea. You are welcome to visit by boat and even dive at the site, but fishing is not permitted.
Ⓝ 25.692940 Ⓦ 80.102861

At this underwater artificial reef, human remains are mixed into the architecture.

GEORGIA
CRYPT OF CIVILIZATION

ATLANTA

A pair of stockings, a bottle of beer, and a voice recording of Hitler: three of the artifacts by which future generations will judge 20th-century civilization. These three items sit alongside thousands of others in what is considered the world's first time capsule, sealed inside a former indoor swimming pool on May 28, 1940.

This "crypt of civilization" was the idea of Dr. Thornwell Jacobs, then president of Oglethorpe University. Frustrated by a dearth of primary sources during his research on ancient Egypt, Jacobs saw it as his "archaeological duty" to create an enduring, detailed record of life in the 20th century. When Jacobs made his plans known in 1936, the earliest date recorded in human history was equivalent to 4241 BCE. In accordance with this date from 6,177 years before, Jacobs decided his crypt should be opened in 8113—6,177 years into the future.

The question of what items would represent the sum of all human knowledge and experience was a tricky one to answer. After being inundated with unsolicited suggestions over three years, Jacobs and his crypt archivist, Thomas Peters, narrowed the options down to a few thousand essentials. A small sample of the inventory:

- *Gone With the Wind* script
- Microfilm reader and microfilms containing over 800 works, including fiction, the Bible, textbooks, news photographs, and drawings of human inventions
- Voice recordings of Hitler, Stalin, Mussolini, and Franklin D. Roosevelt
- Set of false eyelashes
- Potato masher
- Fishing rod
- Kodak camera
- Male and female mannequins
- Dental floss
- Toaster
- Donald Duck toy
- Steel plates used to print the *Atlanta Journal,* featuring articles on World War II

The time capsule, sealed behind a steel door in a 20-foot-long, 10-foot-wide (6 × 3 m) reinforced room, is scheduled to be opened on May 28, 8113. Jacobs and his staff took steps to ensure that 82nd-century individuals will be able to decipher the contents, regardless of societal changes—at the entrance of the crypt is a machine that teaches the English language.

Phoebe Hearst Hall, Oglethorpe University, 4484 Peachtree Road Northeast, Atlanta. You can't see inside the crypt until 8113, but you may visit its tightly sealed stainless-steel door. Ⓝ 33.874984 Ⓦ 84.333672

ALSO IN GEORGIA

Dotson Runway Graves

Savannah · When a new runway was laid at Savannah Airport in the 1980s, the preexisting graves of Richard and Catherine Dotson became part of the tarmac. Their two headstones, which lie side by side, flush with the runway, are clearly visible from departing and landing planes.

National Tick Collection

Statesboro · Scrutinize the differences among 850 species of tick at this overwhelming assemblage of specimens.

Sealed in 1940, Oglethorpe University's room-size time capsule is set to be cracked open in 8113.

GEORGIA GUIDESTONES

ELBERTON

When in need of life guidance, some turn to a higher power. Others look to four granite slabs in rural northeast Georgia.

In 1979 a man using the name R.C. Christian approached Georgia's Elberton Granite Finishing Company with plans to build a monument of four granite slabs, each almost 20 feet (6 m) tall, arranged in a cluster and topped with a smaller, horizontal slab. Each of the four vertical stones was to be inscribed with the same ten precepts for humanity, carved in eight languages.

These guidelines ranged from common-sense advice ("Balance personal rights with social duties") to New Age-y maxims ("Prize truth—beauty—love—seeking harmony with the infinite") to downright impractical instructions, served up with a hint of genocide ("Maintain humanity under 500,000,000"). The placement of the stones was carefully configured to align with solstices and equinoxes.

Armed with the detailed

No one knows who commissioned these mysterious megaliths, engraved in eight languages.

blueprints, the Elberton Granite Finishing Company duly created this mysterious monument and installed it in a field off Highway 77. A granite tablet was placed a few feet from the monument to provide some context for the Georgia Guidestones, as they have come to be known. It reads, in part, "Let these be guidestones to an Age of Reason."

The guidestones opened to the public in March 1980 and immediately became a magnet for conspiracy theorists. Over the years, visitors to the guidestones have left their marks by scrawling symbols and commentary on the monument, creating yet more intrigue.

There is still no definitive explanation for what the Georgia Guidestones mean, who actually commissioned them, or when the Age of Reason is due to be ushered in.

Guidestone Road Northwest. The stones are 7 miles (11 km) north of Elberton, just off Highway 77. Ⓝ 34.232056 Ⓦ 82.894389

KENTUCKY

VENT HAVEN MUSEUM

FORT MITCHELL

Vent Haven Museum is the only place in the world where you can walk into a room and see rows of ventriloquist figures sitting in chairs, all staring goggle-eyed in the same direction.

The museum's eerie displays grew out of the personal collection of William Shakespeare Berger (1878–1972), former president of the International Brotherhood of Ventriloquists. ("Vent" is slang for the profession.)

The most popular dummy at the museum is a replica of Charlie McCarthy, a figure in a top hat, tux, and monocle. Charlie and his human partner, ventriloquism pioneer Edgar Bergen, appeared on the *Chase and Sanborn Hour* radio show from 1937 to 1956. (The fact that a ventriloquist act achieved prolonged success in a nonvisual medium is one of the great mysteries of showbiz.)

Charlie's banter with Mae West on the December 12, 1937, show led to the actress being banned from NBC radio for 12 years. Her description of him being "all wood and a yard long" and teasing remarks about him giving her splinters the night before were deemed vulgar and obscene by the Federal Communications Commission.

33 West Maple Avenue, Fort Mitchell. The museum is open by appointment from May through September. Ⓝ 39.053008 Ⓦ 84.551937

ALSO IN KENTUCKY

Cumberland Falls

Corbin · On a clear night, with a full moon above, a moonbow, or lunar rainbow, appears in the mist above these waterfalls.

HEIGOLD HOUSE

LOUISVILLE

Built in the 1850s, Heigold House still stands today as a testament to its owner's enduring belief in the American experiment. Well, at least its facade does.

Christian H. Heigold was a successful stonecutter who immigrated to Louisville from Germany in 1850, settling in a thriving neighborhood called The Point. There he quickly went about building a mansion, using his ample stonecutting skills to adorn the house with images of his family in scenes of Americana.

Louisville in the mid-19th century was a crucible of anti-immigrant sentiment, particularly targeting those of German Catholic and Irish descent. These tensions came to a head on August 8, 1855, when a Protestant mob attacked over 100 predominantly Irish and German businesses and homes, killing 22 people. No one was ever prosecuted for the violent riot, which came to be known as Bloody Monday.

The patriotic scenes Heigold cut into his home were a way of publicly proclaiming his American identity amid the rampant anti-immigrant sentiments. Even without the three-quarters of the original mansion, the facade abounds with details depicting Heigold's American idealism. A bust of James Buchanan is flanked on either side by the words HAIL TO BUCHANAN, NOW AND FOREVER and THE UNION FOREVER. HAIL TO THE UNION FOREVER; NEVER DISSOLVE IT. On the decorative lintel over the door, Heigold cut a scene depicting George Washington flanked by two figures, the Lady of Justice and the Lady of Culture, along with the words HAIL TO THE CITY OF LOUISVILLE.

After a laborious construction, Christian Heigold was not able to enjoy his home for long. He died in 1865, leaving the house to his son Charles. The passing of the elder Heigold marked the beginning of nearly a century and a half of unrest for the house and surrounding neighborhood.

By 1953, the city of Louisville began to buy and demolish properties in The Point to expand a dump site that had been encroaching on the neighborhood. The Point had virtually disappeared by then, having been flooded often by a diverted river. Buildings were demolished or covered in city waste, and the mayor had the facade of the Heigold House moved across River Road. After a developer bought the adjacent land to build luxury condominiums in 2007, the 70,000-pound facade was carefully moved again to its current location.

449–495 Frankfort Ave, Louisville. At night, spotlights illuminate the facade. It's a nice bit of drama.

Ⓝ 38.264259 Ⓦ 85.725211

A testament to one man's American identity in the face of anti-immigrant sentiment and unrest.

The inmates of Louisiana State Penitentiary put on a public rodeo and craft fair.

LOUISIANA

ANGOLA PRISON RODEO

ANGOLA

Every Sunday in October, inmates from the maximum-security Louisiana State Penitentiary host a publicly viewable rodeo. The prison—a former plantation nicknamed "Angola" after the original home of its slave workers—opens its doors to thousands of visitors who come to watch the bucking bulls, hear prison bands, and purchase crafts made by inmates.

The Angola rodeo tradition began in 1965 when a small group of prisoners and staff built an arena, intending to hold a rodeo purely for their own entertainment. What started as a way to pass the time turned into a huge public event. By 1969, visitors were cramming into a brand-new 4,500-seat stadium to watch inmates attempt their six-second bull rides. The current arena holds 10,000 people.

Inmates don't just participate in rodeo events—they are the ones selling hot dogs and candy apples, playing music for the crowds, and running market stalls featuring their own art, jewelry, leather goods, and woodwork. Easily distinguished by their black-and-white stripes, the prisoners chat freely with visitors but don't handle any money. The rodeo offers a rare chance to be among the public—three-quarters of Angola's 5,000-strong inmate population are serving life sentences.

End of Highway 66, approximately 22 miles (35 km) northwest of St. Francisville (Highway 61). Cameras are not permitted. Ⓝ 30.955436 Ⓦ 91.593903

HISTORIC VOODOO MUSEUM

NEW ORLEANS

Though all its offerings are crammed into just two dusty rooms and a hallway, the New Orleans Historic Voodoo Museum leaves a lasting impression. Founded in 1972 by local artist Charles Gandolfo, the museum focuses on Louisiana voodoo, which evolved from traditional West African vodun.

West Africans brought voodoo to Louisiana during the trans-Atlantic slave trade of the early 18th century. By the mid-19th century, the culture of New Orleans had begun to transform the spiritual practice. Voodoo spirits merged with Catholic saints, rituals gave way to processions, and Creole voodoo queens like Marie Laveau rose to prominence.

In 1932, a poorly acted, hastily shot horror movie—*White Zombie*—featured Bela Lugosi as an evil Haitian voodoo master with a crew of murderous zombies. The perverted pop-cultural portrayal of voodoo thrust it into the public eye. New Orleans stores capitalized on the trend by selling potions and voodoo dolls, while those who practiced genuine voodoo went underground.

Through its artifacts, the museum aims to convey the spiritual and historical context of Louisiana voodoo. Visitors can see traditional dolls and gris-gris pouches (amulets believed to provide luck or protection), and get a psychic reading from Dr. John, the resident voodoo priest. The gift shop sells love potions, snake skins, and chicken feet, which are used as protective charms.

724 Dumaine Street, New Orleans. Ⓝ 29.959903 Ⓦ 90.063851

St. Roch Chapel

NEW ORLEANS

There's a cemetery in the neighborhood of St. Roch (pronounced "rock"). At the center of that cemetery is a chapel. Inside that chapel, in a small room behind an iron gate, rows of prosthetic legs hang from the peeling walls. On shelves beneath sit plaster feet and false teeth, and a few pairs of artificial eyeballs.

Dedicated in 1867, the chapel honors St. Roch, who is associated with good health and healing. Born in the 14th century in Montpellier, Majorca—now part of France—St. Roch is said to have cared for and cured plague victims in Italy.

When a yellow fever epidemic hit 19th-century New Orleans, Reverend Peter Thevis, the pastor of Holy Trinity Catholic Church, prayed to St. Roch for relief and promised to build a shrine to him if the members of his parish were protected from the disease.

Though 40,000 New Orleanians succumbed to yellow fever, Father Thevis's community recorded no losses. The reverend held up his end of the bargain and built the St. Roch chapel and the surrounding cemetery. The gates opened to the public in 1876.

The room in the chapel has since become filled with offerings left by those in need of healing—as well as people who have prayed to St. Roch and recovered. Bricks on the ground are inscribed with the word "thanks" and littered with coins. Children's

The faithful leave tributes to St. Roch, including coins, flowers, and prosthetic limbs.

polio braces, crutches, and false limbs line the walls, interspersed with praying hands, rosaries, and figurines.

1725 St. Roch Avenue, New Orleans. Ⓝ 29.975445 Ⓦ 90.052018

Lake Peigneur

NEW IBERIA

Originally just 10 feet (3 m) deep, Lake Peigneur was once a popular but unremarkable fishing and recreation spot. However, on the morning of November 21, 1980, that all changed, when one of the largest man-made whirlpools in history flushed the lake and 65 acres of surrounding land—along with barges, big-rig trucks, and a Texaco drilling platform—down an enormous vortex.

Early that morning, the Wilson Brothers drilling crew knew something was amiss when their 14-inch drill bit became stuck and the entire platform shook. What they didn't know was that they had mistakenly drilled through the ceiling of the huge salt mine below. Wisely, they

abandoned the structure and, once safely on shore, watched in horror as their entire 150-foot rig sank like a magic trick into the shallow lake. Meanwhile, hundreds of feet below, 50 miners were scrambling to escape as water poured into the mine. Miraculously, there were no fatalities or serious injuries.

The damage from the all-consuming whirlpool was catastrophic and continues today. The freshwater lake became permanently salinated, forever altering the local ecosystem. Brackish water from Delcambre Canal and Vermilion Bay poured in through a newly formed 50-foot waterfall, while compressed air from the mine shafts created 400-foot geysers. The mine, which had been in operation for over 100 years, closed in 1986.

You can see views of Lake Peigneur and watch a short film about the drilling disaster at the Rip Van Winkle Gardens on Jefferson Island, located on the eastern shore of the lake. Ⓝ 29.978065 Ⓦ 91.984900

The remains of a flooded house.

The Army Corps of Engineers used Italian and German prisoners of war to build a scale model of 15,000 miles of waterway.

MISSISSIPPI

MISSISSIPPI RIVER BASIN MODEL

JACKSON

In the precomputer 1940s, when engineers needed to model a complex system, they would build an amazingly elaborate scale model.

The Army Corps of Engineers, the federal agency in charge of developing and maintaining the nation's water resources, built many such models, but none were on the scale of the Mississippi River Basin Model. Made in response to a series of catastrophic river floods, it simulated the effects of weather and flooding on the more than 15,000 miles (24,000 km) of waterways that make up the Mississippi River Basin. It was created at a scale of 1:100 vertical and 1:2,000 horizontal and covered over 200 acres of Buddy Butts Park.

Work on the model began in 1943 by Italian and German prisoners of war, who had been shipped over from North Africa. Though projected to be completed by 1948, the model took much longer than

expected to build. It wasn't truly finished until 1966, a full 23 years after it was started. Six years later, it was flooded for the last time.

By the early 1970s, the push toward computer modeling had begun, and by the 1980s, the model had become a burden for the Army Corps. In 1990, the site was transferred to the city of Jackson but was too expensive for them to maintain, so the city simply abandoned it.

The river basin model now sits surrounded and hidden by overgrown woods in Buddy Butts Park. It is open to the public to visit, but the tiny concrete banks of the rivers are now overgrown with comparatively giant foliage.

Buddy Butts Park, 6180 McRaven Road, Jackson. Ⓝ 32.305984 Ⓦ 90.315903

ALSO IN MISSISSIPPI
....

USS *Cairo*

Vicksburg · This iron-and-wood Civil War river gunboat was the first vessel to be sunk by a torpedo.

NORTH CAROLINA

STANLEY REHDER CARNIVOROUS PLANT GARDEN

WILMINGTON

Named in memory of Wilmington horticulturalist and carnivorous-plant lover Stanley Rehder, who died in 2012 at age 90, this garden is chock-full of meat-hungry flora. You'll see pitcher plants—which are known to swallow the odd frog or shrew—as well as Rehder's favorites: Venus flytraps.

In 2013, the garden took a major hit when thieves made away with approximately 1,000 flytraps—90 percent of the population. Fortunately, enhanced security and patient cultivation of replacement plants have helped the flesh-eating garden to flourish once more.

2025 Independence Boulevard, Wilmington. The garden is in the Piney Ridge Nature Preserve just

behind Alderman Elementary School. Ⓝ 34.205827 Ⓦ 77.907280

ALSO IN NORTH CAROLINA
....

The Can Opener

Durham · The Gregson Street Railroad Trestle, or the Can Opener, is a truck driver's nightmare—the low-clearance rail bridge, built before the implementation of minimum height standards, regularly shaves the tops off trucks.

DEVIL'S TRAMPING GROUND

BENNETT

According to local legend, Satan likes to venture forth from the underworld and spend some quality time plotting the downfall of humanity at a camping spot in Chatham County.

A dusty circle of land in a forest northwest of Harpers Crossroads has come to be known as the Devil's Tramping Ground, based on stories of the Dark One that date back to the 1880s. Apparently, the devil regularly turns up at his favorite spot in the forest to pace in a circle, which accounts for the 20-foot-wide (6 m) patch of barren ground.

Beyond preventing the growth of vegetation, Beelzebub's cameos are said to have cursed the ground so that it causes objects placed there to move or disappear. There have also been reports of dogs whimpering and running when they encounter the cursed spot.

Signs bearing the name of the Devil's Tramping Ground Road are regularly stolen—presumably to serve as kitschy mementos. Unless, of course, it's a darker force that's "making" them disappear. **The ground is about 10 miles (16 km) south of Siler City on State Road 1100. Ⓝ 35.584783 Ⓦ 79.487017**

Where the devil comes to dance.

SOUTH CAROLINA

PEARL FRYAR'S TOPIARY GARDEN

BISHOPVILLE

When Pearl Fryar moved into a new house in 1981 he set himself a goal: to win the "Yard of the Month" award from the local gardener's club. Though he had no training in horticulture and no experience tending to plants, Fryar was determined to prove that he could grow a glorious garden. He visited local nurseries, took their discarded plants, and surrounded his house with them.

As the plants grew, Fryar began trimming them into diamonds, spirals, spheres, and cones. He won that coveted Yard of the Month award in 1985. By that time, he was well on his way to becoming the world's most celebrated topiary artist.

Fryar's garden now has 400 plants and trees, all pruned into fantastical shapes. **165 Broad Acres Road, Bishopville. Ⓝ 34.206793 Ⓦ 80.271868**

Originally motivated by a "Yard of the Month" award, Pearl Fryar has created a green wonderland with hundreds of ornamental shrubs.

OYOTUNJI AFRICAN VILLAGE

SHELDON

On the road into Oyotunji Village, a sign states, "You are leaving the United States. You are entering Yoruba Kingdom." Established in 1970, Oyotunji is an intentional community based on the culture of the Yoruba ethnic group of West Africa, particularly Nigeria. Its inhabitants live according to traditional Yoruba values and honor the Supreme Being Olodumare and the ancestral spirits through ritual dance, music, and ceremonies.

During its early days, Oyotunji was home to about 200 people. Now there are fewer than ten families living there, but the tight-knit community continues to host festivals—14 of which are open to the public—and a trading bazaar.

Oyotunji also offers Yoruba spiritual services, such as African naming ceremonies, divination readings, and communication with family members back home.

56 Bryant Lane, Sheldon. Oyotunji, just off Highway 17, is open for tours daily. Ⓝ 32.608852 Ⓦ 80.803306

ALSO IN SOUTH CAROLINA

Neverbust Chain

Columbia · This massive steel chain linking two office buildings was installed without permission but instantly beloved.

UFO Welcome Center

Bowman · Jody Pendarvis's homemade, rickety UFOs are intended to provide a rest stop for bewildered extraterrestrial visitors.

Oyotunji residents live according to the traditions of West African Yoruba and Fon cultures.

MARS BLUFF CRATER

MARS BLUFF

On March 11, 1958, an atomic bomb more powerful than the one dropped on Nagasaki fell out of a B-47 jet and onto Walter Gregg's back lawn.

The US Air Force plane had taken off from Savannah, Georgia, and was bound for England. There it would participate in Operation Snow Flurry, a series of mock bomb attacks. A real nuclear weapon was on board, just in case the war with the Soviet Union suddenly turned hot from cold.

Shortly after takeoff, a red warning light in the cockpit flashed to indicate that the bomb was not secured properly. Flight navigator Bruce Kulka went to inspect the situation and accidentally pressed the emergency release button, sending the 6-kiloton bomb plummeting toward Earth.

Mercifully, the bomb was not yet loaded with its uranium and plutonium core. But it did contain 7,600 pounds (3.4 MT) of payload-triggering explosives, resulting in significant damage to the rural ground on which it landed. When the dust cleared, a 50 × 70-foot wide (15 × 21 m), 30-foot-deep (9 m) crater scarred Gregg's yard. His house had been lifted off its foundation and was riddled with pockmarks and gaping holes. There were no human fatalities, but a few chickens perished in the accident.

The crater is still visible, but less dramatic—decades of forest growth have turned it into a peaceful little pond surrounded by trees.

Crater Road, Mars Bluff. The crater is on private property, but a historical marker sits by the access road. Ⓝ 34.200940 Ⓦ 79.657117

TENNESSEE

THE BODY FARM

KNOXVILLE

There is a 2.5-acre forest full of rotting human corpses at the University of Tennessee, but it's nothing to worry about—the bodies are meant to be there. The Knoxville institution is home to the country's first "body farm," a facility where forensic anthropologists and FBI agents get a close-up view of what happens when you die.

Anthropologist Dr. William M. Bass established the body farm in 1971 to advance the study of decomposition. Bodies are buried or left exposed within the wooded plot so they can be observed at various stages of the postmortem process.

By analyzing the effects of weather and insect activity on decomposition, students and researchers are better able to estimate the time of death. Law enforcement agencies also use the facility to sharpen criminal investigation skills—FBI agents, for example, practice exhuming and identifying human remains.

There are around 40 bodies in the farm at any one time, and donations are welcome (provided they're properly sourced). If you're interested in someday becoming part of the farm, make your intentions known to the university.

Don't mind the smell.

1924 Alcoa Highway, Knoxville. The body farm is not open to tours, but its staff gives talks, and the facilities are open to researchers. Ⓝ 35.940308 Ⓦ 83.942340

SYNCHRONIZED FIREFLIES OF THE GREAT SMOKY MOUNTAINS

ELKMONT

Like their bioluminescent counterparts in Malaysia, the little lightning bugs of the Great Smoky Mountains flash in sync to find a mate. After maturing from their larval stage, male fireflies stop eating and live for just three weeks. Given a mere 21 days to attract a partner, the fireflies flash en masse in order to help the females of the species find them.

Peak mating season lasts for about two weeks, between mid-May and mid-June. Visit the national park on a clear summer night and you'll find yourself in the middle of a romantic natural light show.

Elkmont, Great Smoky Mountains National Park. Firefly shuttles run from the Sugarlands Visitor Center to the Elkmont viewing area on evenings during mating season. Ⓝ 35.685606 Ⓦ 83.536598

ALSO IN TENNESSEE

Concrete Parthenon

Nashville · The trouble with the Parthenon in Athens is that it's falling apart. You could use your imagination to fill in the gaps, or you could head to Nashville's Centennial Park to see an intact, full-scale version.

Inside Luray Caverns, the cave itself becomes a musical instrument.

VIRGINIA

THE GREAT STALACPIPE ORGAN

LURAY, SHENANDOAH VALLEY

Deep underground in Luray Caverns is an unusual musical instrument designed by Leland W. Sprinkle. With its four-keyboard console, it looks like a standard variety church organ, with one crucial difference: There are no pipes. The "pipes" are stalactites and the instrument is a lithophone—a device that produces music by striking rocks of differing tones.

Sprinkle, a mathematician and electronics scientist at the Pentagon, came up with the idea for the "stalacpipe organ" after touring the caverns in 1954. He spent three years searching for stalactites that corresponded to the required musical notes, sanding them down to be pitch-perfect, and running 5 miles (8 km) of wires between the console and the rubber mallets that would strike each stalactite.

When the organ was first installed, Sprinkle himself took to the keys to entertain visitors. He even released a vinyl record of songs—it was promoted as "musical gems from solid rock" and sold in the Luray Caverns gift shop. These days the organ entertains cave visitors with automated renditions of such classics as "America the Beautiful," Moonlight Sonata, and, during the festive season, "Silent Night." **101 Cave Hill Road, Luray. Ⓝ 38.664094 Ⓦ 78.483618**

ALSO IN VIRGINIA

The Raven Room

Charlottesville · Edgar Allan Poe's old dorm room is now a shrine to the author's legacy.

CIA MUSEUM

LANGLEY

At the CIA's Virginia headquarters—a building formally known as the George Bush Center for Intelligence—there's a museum full of secret spy stuff. Its five galleries are hidden from public view, accessible only to CIA employees and special guests with security clearance.

In these rooms lie artifacts from decades of espionage dating back to World War II, when the CIA's predecessor, the Office of Special Services (OSS), was established. A German Enigma enciphering machine sits beside a letter written by an OSS officer on Hitler's personal stationery nine days after the dictator's suicide.

The Al Qaeda–focused gallery contains equipment and models used in SEAL training exercises in the lead-up to the 2011 raid on Osama bin Laden's Abbottabad compound. There's a scale model of bin Laden's hideout, as well as a wall from the full-size mock compound that was constructed for practice attacks. Bin Laden's AK-47, found beside his body, is on display, as is a brick from the real compound. And an Al Qaeda rocket-launching manual, scarred with burn marks.

Among the most fascinating exhibits are the unmanned vehicles and spy cams. Pigeon-mounted cameras, dragonfly drones, and robotic fish are a few of the devices that were trialed for stealthy surveillance. (The dragonfly drone, or insectothopter, was less than effective—developed during the 1970s, the remotely controlled insect had to be scrapped when it proved too susceptible to crosswinds.) **1000 Colonial Farm Road, McLean. The museum is not open to the public. Ⓝ 38.951791 Ⓦ 77.146607**

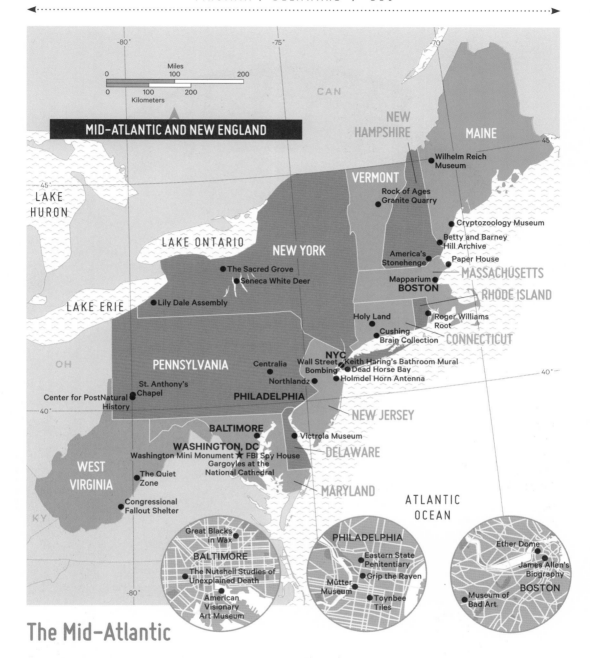

MID-ATLANTIC AND NEW ENGLAND

The Mid-Atlantic

DELAWARE

Johnson Victrola Museum

DOVER

The phrase "Put a sock in it!" originates from the days when people would stuff a sock into their Victrola horns to lower the volume. That's but one of the many fascinating facts you can learn when browsing this delightful collection of phonographs, also known as gramophones.

The museum is named after Eldridge R. Johnson, a Dover man who founded the Victor Talking Machine Company in 1901. Johnson's line of Victrola phonographs, which housed the device's horn inside a sleek wooden cabinet, became hugely popular when they were introduced in 1906.

In addition to the many phonographs on display, the museum has a "Nipper Corner"—Nipper being the terrier who served as the canine model for the legendary His Master's Voice logo.

375 S. New Street, Dover. Ⓝ 39.156455 Ⓦ 75.527210

The misfits and outsiders of the art world are celebrated at Baltimore's American Visionary Art Museum.

MARYLAND

AMERICAN VISIONARY ART MUSEUM

BALTIMORE

This museum honors the work of self-taught artists with an obsessive bent—those driven to paint and sculpt and create despite a lack of formal training.

Museum founder Rebecca Alban Hoffberger got the idea for the museum while working in a Baltimore hospital program aimed at reintroducing psychiatric inpatients to the community. Inspired by the intensity and imagination in the patients' artwork, Hoffberger created a home for visionary, outsider art.

Opened in 1995, the museum displays work that is playful, dark, funny, disturbing, and sometimes awe-inspiring, including a crocheted "horse dress" with giant equine eyes at the nipples and a 55-foot-tall

(16.7 m) whirligig. The permanent collection is supplemented with temporary exhibits by guest curators. In these, too, the mood is varied: "The Art of Storytelling" featured Holocaust survivor Esther Krintz's tapestries of Nazi labor camps, while "What Makes Us Smile" featured a "Toot Suite" that honored the role of flatulence in the arts.

800 Key Highway, Baltimore. Each May the museum hosts the Kinetic Sculpture Race, in which human-powered vehicles traverse a 15-mile (24 km) course over land and water. Ⓝ 39.280035 Ⓦ 76.606742

ALSO IN MARYLAND

Urology Museum

Linthicum · See a collection of wince-inducing but beautiful bladder stones.

Inspiring figures immortalized in wax include Joe Lewis, Jackie Robinson, and Jesse Owens.

GREAT BLACKS IN WAX

BALTIMORE

The name of this museum may be light and quirky, but a visit to Great Blacks in Wax is a powerful, confrontational, and ultimately uplifting experience—even if the figures on display aren't quite up to the standard of Madame Tussaud's when it comes to verisimilitude.

Elmer and Joanne Martin founded the National Great Blacks in Wax Museum in 1983 to stimulate interest in black history and provide strong role models for African American youth. A tour of the

exhibits, arranged in chronological order, begins with a walk through a replica slave ship where tightly packed rows of black slaves are chained by their necks. The Lynching Room, signposted with a graphic-content warning, contains life-size figures being hanged, beaten, and eviscerated.

Amid these vivid exhibits are over 100 wax figures of black civil rights leaders, entertainers, athletes, writers, and other luminaries—from Billie Holiday and Barack Obama to lesser-known figures such as Granville T. Woods, known as the "Black Edison."
1601 East North Avenue #3, Baltimore. Ⓝ 39.311749 Ⓦ 76.596842

THE NUTSHELL STUDIES OF UNEXPLAINED DEATH

BALTIMORE

In 1943, Frances Glessner Lee began working on a series of detailed, dollhouse-style dioramas. Working at a scale of 1 inch to 1 foot, the 65-year-old forensics expert filled her minirooms with hand-sewn textiles, color-coordinated furniture, and bottles with hand-painted labels. Posed in each room was a minicorpse exhibiting just the right degree of decay.

Lee's dioramas were known as the Nutshell Studies. Created to assist forensic science students at Harvard, they depicted murder scenes, suicides, and lethal accidents. The clues in Lee's nutshells—such as blood spatter patterns, the position of the body, and the items found around it—allowed students to analyze and determine the nature of each grisly death.

To make each scene authentic, Lee read crime reports and police

Tiny forensic dollhouses train detectives to solve crimes.

interviews to compile details. She was unflinching in her interior decoration: In one scene, "The Three-Room Dwelling," a husband and wife lie dead in their bedroom. Crimson footprints separate them from the corpse of their blood-spattered baby.

All 18 nutshell dioramas are displayed at the Maryland Medical Examiner's Office. The gory tableaus are still studied by detectives in training.
900 West Baltimore Street, Baltimore. ⓝ 39.289109 ⓦ 76.632637

NEW JERSEY

HOLMDEL HORN ANTENNA

HOLMDEL

In 1965, Arno Penzias and Robert Wilson accidentally discovered one of the greatest secrets of the universe.

The radio astronomers were using the Bell Labs horn antenna to scan for radio waves being bounced off NASA communications satellites. To Penzias's and Wilson's annoyance, an ever-present low hum interfered with their data collection. They checked their equipment, shooed away some pigeons that had been nesting in the antenna, and listened again. Still the hum persisted.

The noise was not coming from the antenna, or anywhere in New Jersey, or anywhere on Earth. It came from the universe itself. Penzias and Wilson had just stumbled upon cosmic microwave background.

Penzias's and Wilson's discovery provided the first observational evidence that the universe began with a Big Bang. The discovery earned them a

Once the pigeons were shooed away, scientists were able to detect the faint echoes of the Big Bang.

Nobel Prize in Physics. The decommissioned horn antenna they used for their explosive discovery is now a National Historic Landmark.
Holmdel Road and Longview Drive, Holmdel. ⓝ 40.390760 ⓦ 74.184652

NORTHLANDZ

FLEMINGTON

Over 8 miles (13 km) of miniature train tracks run through this building, making it the largest model train layout in the world. Creator Bruce Zaccagnino began building the miniature railroads in his basement in 1972. After twice-annual open houses drew big crowds, Zaccagnino decided to open Northlandz to the public as a permanent attraction. It debuted in 1996 in a 52,000-square-foot (4,831 m²) building along Route 202.

The 135 trains of Northlandz cross over 400 bridges and pass by miniature cities, mountains, half a million trees, and crowds of teeny people. Then there are the odder sights, such as an outhouse factory and a crashed plane with survivors standing on its wings. There's a layer of dust on everything, and the lighting's a little dim, but the vastness and detail of the scenery transcend the moth-eaten vibe.

The world's largest model railroad began in a basement. It now incorporates a 30-foot mountain.

495 US Route 202, Flemington. About an hour north of Trenton. Ⓝ 40.517085 Ⓦ 74.819335

NEW YORK

DEAD HORSE BAY

BROOKLYN

Stroll along the beach at Dead Horse Bay and you'll be stepping on the faces of broken porcelain dolls, glass soda bottles from the 1950s, and fragments of 19th-century horse bones.

The debris strewn on the sand tells the story of the bay's past. Formerly part of Barren Island, Dead Horse Bay acquired its unsavory name in the 1850s, when it became the location for multiple horse-rendering plants. New York's former carriage-pulling equines arrived here to be transformed into glue and fertilizer. Having no use for their chopped, boiled bones, plant workers dumped them in the water. A horrid smell hung in the air.

The introduction of the automobile brought an end to carriage horses, and thus an end to horse-rendering plants. In the late 1920s the factories shut down and the city poured sand, silt, and garbage into the channel separating Barren Island from the mainland. The area functioned as a garbage dump until 1953, when the landfill was capped. Decades of erosion have uncovered the artifacts that dot the shore today.

Glass bottles make up most of the debris—hence the bay's nickname, Bottle Beach—but you

First a horse-rendering plant, then a landfill, Brooklyn's bottle-covered beach is now a scavenger heaven.

may also find saddle fragments, equine teeth, art-deco cosmetics cases, and broken toys. All are from 1953 or earlier, making the bay a garbage-strewn portal to the past.

Flatbush Avenue at Aviation Road, Brooklyn. Get a Brooklyn-bound 2 train to the last stop, Flatbush Avenue–Brooklyn College. From there, take the Q35 bus to Floyd Bennett Field. The bay is on the opposite side of Flatbush Ave. Ⓝ 40.581689 Ⓦ 73.898504

A little-remembered 1920 terrorist attack left holes along Wall Street.

SCARS OF THE WALL STREET BOMBING

MANHATTAN

The corner building at 23 Wall Street bears scars from a terrorist attack—one that occurred back in 1920. Just after noon on Thursday, September 16, a bomb placed on a parked horse-drawn wagon exploded, blasting 500 pounds (226 kg) of small iron weights into the air. Thirty-eight people died in the attack. One hundred and forty-three were seriously injured.

At the time, 23 Wall Street was the headquarters of J.P. Morgan & Co., the nation's most powerful bank. The perpetrators of the attack have never been officially identified, but anti-capitalist Italian anarchists were likely responsible.

The pockmarked limestone on the facade of 23 Wall Street was never repaired. These little marks are the only on-site hint of the attack—there are no signs or plaques to commemorate the bombing. **23 Wall Street, Manhattan. Ⓝ 40.706795 Ⓦ 74.010480**

KEITH HARING'S *ONCE UPON A TIME* BATHROOM MURAL

NEW YORK

In May 1989, artist Keith Haring created the bathroom mural *Once Upon a Time*. He was 31, and it was his last major mural before his death in February 1990 of AIDS-related complications. The piece was created for The Center Show, a celebration of the 20th anniversary of the Stonewall riots, which are regarded as the start of the gay liberation and LGBT rights movement. The Center Show called upon LGBT artists to create site-specific works of art in the building of LGBT community hub The Center. Haring chose to create his provocative work in the second-floor men's bathroom.

The mural covers four interior walls with Haring's signature black-on-white line drawings. These are not the Haring images we are used to seeing on buttons, magnets, puzzles, and clothes; this is the private, sexual Haring come to life in a grandiose and unapologetic celebration of gay sexuality.

Unfortunately, time and the elements gradually took their toll on the mural. The bathroom was used as The Center's meeting room until 2012, when funds were raised for a major restoration of the work. The mural was restored and opened to the public in 2015.

Since Haring's works have sold at auction for up to $5.6 million, it is quite possible that this is the most valuable bathroom in America. Luckily, you can see it for free.

208 West 13th Street, Manhattan. The Keith Haring Bathroom and other Center Show works are open to the public during regular business hours at the Center. Ⓝ 40.738152 Ⓦ 74.001057

A one-of-a-kind mural that is as graphic as it is masterful.

Bronx Zoo Bug Carousel

The Bronx • Right next to the butterfly exhibit at the Bronx Zoo is the first—and likely only—carousel of hand-carved and elaborately painted insects.

High Bridge

Harlem River Drive • Connecting the boroughs of Manhattan and the Bronx, this 2,000-foot arched aqueduct soars above the Harlem River. The oldest bridge in the city, it reopened as a pedestrian walkway after four decades of neglect and abandonment.

Panorama of the City of New York

Queens • It took three years and more than a hundred model makers to build this panorama in time for the 1964 World's Fair. It knocked out the crowds with 10,000 square feet of every building in all five boroughs, each re-created in miniature.

Houdini's Grave

Queens • There's a Houdini museum in Manhattan, but in Queens you can visit the final resting place of the greatest illusionist of all time. Fans still gather to wait for his escape from the shackles of death, and they leave behind stacks of stones and packs of cards.

Castle William

Governors Island • In 1996 the Coast Guard—the last of the military tenants of Governors Island—finally folded up their charts, and the centuries-old military outpost opened to the public, including a historic fort that housed a hundred cannons.

Corpse Flower at the Brooklyn Botanic Garden

Brooklyn • Time it right, and lucky visitors may catch a whiff of one of the worst smells on the planet: the scent of the giant bloom of the Corpse Flower that the Brooklyn Botanic Garden calls Baby. Even when not in full bloom, the *Amorphophallus titanum*, or "titan arum," lives up to its name ("titan" that is, but Baby is awful cute).

Weeksville Heritage Center

Brooklyn • Brooklyn's largest African American cultural institution, this sparkling facility celebrates the little-known 19th-century community of Weeksville, intentionally created for free blacks and former slaves, with exhibitions, performances, and guided tours of the settlement's remains.

Fort Wadsworth

Staten Island • The traditional starting line of the New York Marathon, America's longest-manned military fort is now an abandoned hulk of ruins along the Verrazano Narrows, part of the Gateway National Recreation Area and perfect for melancholy picnics and beach bird-watching.

Floyd Bennett Field

Brooklyn • Part of the Gateway National Recreation Area of the National Park Service and named after the first person to fly over the North Pole, New York City's first local airport is now home to the Historic Aircraft Restoration Project.

Historic Richmond Town

Staten Island • A sizable piece of New York's late 17th- and early 18th-century past is preserved in a settlement of 30 structures in the city's farthest borough, including its oldest continuously operating farm, plus a 350-year-old home on its original site, one of the oldest in the country.

Roosevelt Island Smallpox Hospital Ruins

Roosevelt Island • These haunting ruins are a stone's throw from some of the most exclusive and expensive real estate on the planet. With a design by the architect of St. Patrick's Cathedral, this crumbling 19th-century relic maintains a sense of grandeur despite the suffering it once housed.

World's Largest Chess Board

Manhattan • Stuck to the side of an apartment building is the world's largest chess match. With pieces that measure more than two feet each, and with just one move a week (it takes a cherry picker to do it), it's also probably the world's slowest.

Holographic Studios

Manhattan • Bringing new life to a former blacksmith's forge, the world's oldest hologram gallery and laser laboratory has been creating, selling, displaying, and teaching the art and science of lasers since the 1960s.

Obscura Antiques and Oddities

Manhattan • In the back room of this charming store, with its astonishing variety of medical antiques, turn-of-the-20th-century taxidermy, and Victorian mourning jewelry, there is an exquisite 19th-century anatomical model (not for sale).

Angel Orensanz Foundation

Manhattan • The oldest surviving synagogue in the city is now a soaring art and concert space, where the music of Philip Glass has echoed through the pillars, and the designs of Alexander McQueen have shone under the filtered light of stained glass.

Dream House

Manhattan • Billed as a sound and light environment, the Dream House is a collaboration of a modern composer and a visual artist, harkening back to a time when Tribeca and Soho were neighborhoods with cheap rents and vibrant creative communities.

Mmuseumm

Manhattan • This tiny museum in a freight elevator specializes in the "overlooked, dismissed, or ignored," with unpredictable rotating exhibits that range from the shoe thrown at George W. Bush to the taxonomy of Corn Flakes.

City Hall Station

Manhattan • A lavish and abandoned subway station from 1904, complete with chandeliers and intricate skylights, greets the patient passengers who stay onboard the 6 train as it makes its downtown loop.

The High Bridge and its water tower date back to the 19th century.

THE SACRED GROVE

PALMYRA

In the Spring of 1820, a 14-year-old boy went to a quiet grove near his log cabin home to ask a big question. Confused over which religion he ought to join, the young man prayed to God for guidance. According to the boy, both God and Jesus then appeared, surrounded by white light and bearing a message: Do not join an existing religion, for their doctrines are incorrect.

Ten years later, that boy, Joseph Smith, published the Book of Mormon and formally established the Church of Jesus Christ of Latter-Day Saints. Smith's experience in the grove, referred to in LDS lore as the First Vision, was the moment he became a prophet.

The Sacred Grove, as it is known to Mormons, is now open to visitors year-round. The exact location of Smith's vision is unknown, but you can reenact his adolescent experience by walking from the replica log cabin to the forest and praying to the deity of your choice. A tour guide will gladly accompany you while telling tales from Smith's action-packed life.

843 Stafford Road, Palmyra. Start at the Hill Cumorah Visitors Center, located at 653 State Route 21. Ⓝ 43.040884 Ⓦ 77.239877

LILY DALE ASSEMBLY

LILY DALE

There's a saying in Lily Dale: "If you believe, you will receive." Small, peaceful, and remote, Lily Dale is a community of spiritual mediums who offer their psychic services. In order to move in, mediums must demonstrate their powers to the Lily Dale board of directors. If their messages from beyond the grave are of satisfactory accuracy, they are invited to join.

Visitors come to Lily Dale for readings, during which they hope to receive or convey messages to departed loved ones. The 40 or so resident mediums operate independently and set their own prices— visitors walk around the community, shopping for the spiritualist who suits them best.

Whether or not you believe in the residents' psychic abilities, Lily Dale is a tranquil place to spend a day. Besides the individual readings there are group

gatherings at the Inspiration Stump, during which a medium relays short messages to select people in the crowd. Stop by the pet cemetery, wishing tree, and healing temple to make the most of the entrance fee.

5 Melrose Park. Lily Dale is best visited in the summer months, when there are daily and weekly events for guests. Ⓝ 42.350730 Ⓦ 79.325898

SENECA WHITE DEER

SENECA COUNTY

For years, rumors circulated about the strange herd of white deer living in the former Seneca Army Depot. Some speculated that the "albino" breed was the result of an army experiment gone wrong, while others attributed the animals' appearance to an underground supply of radioactive military weapons.

The white deer were first spotted around 1941, when the US Army fenced off 24 square miles (62 km²) of land for a munitions storage site. Protected by the fencing, the deer population thrived, as did the recessive gene for all-white coloration.

As the white deer population proliferated through the 1950s, the US Army decided to protect the unique herd. Aiding the process of artificial selection, a depot commander hunted the brown deer population while forbidding GIs from shooting any white deer. Since then, the white deer population has grown to approximately 300, making it the largest herd of its kind in the world.

The Seneca Army Depot shut down in 2000 and has been closed to the public ever since. A nonprofit group has been fighting to turn the area into a conservation park and Cold War museum. Until then, dozens of deer remain visible from the highway, frolicking among the hundreds of abandoned bunkers.

A good vantage point for spotting deer is Route 96A, 6 miles south of Geneva. Ⓝ 42.747692 Ⓦ 76.858960

The largest Catholic relic collection outside the Vatican is located in Pittsburgh.

PENNSYLVANIA

ST. ANTHONY'S CHAPEL

PITTSBURGH

Pittsburgh may seem an unlikely pilgrimage destination, but the Catholic faithful flock to this chapel to see its collection of relics. St. Anthony's has thousands of saintly bone fragments and clothing remnants, held in jeweled reliquaries. Together they comprise the largest relic collection outside the hallowed halls of the Vatican.

The collection was established by Father Suitbert Mollinger, the Belgian-born priest who designed and built the chapel in 1880. Among the more notable relics are a sliver of wood from the Last Supper table, five fragments of the cross on which Jesus was crucified, and a particle from the Virgin Mary's veil. Mollinger, who died in 1892, used the relics during healing masses, which were aimed at restoring health and vitality to the infirm.

1704 Harpster Street, Pittsburgh. You can take a tour of the relic collection, but remember that the chapel is a place of worship. Photos are not permitted and silence is preferred. Ⓝ 40.464911 Ⓦ 79.983664

CENTER FOR POSTNATURAL HISTORY

PITTSBURGH

A natural history museum tweaked for the 21st century, the Center for PostNatural History exhibits organisms that have been altered by humans via processes like selective breeding and genetic engineering.

The museum aims to examine the ways in which nature, culture, and biotechnology intersect. Its collection of living and preserved organisms includes genetically engineered glowing fish—which contain genes from bioluminescent jellyfish and coral—and a preserved "BioSteel" goat, which was genetically modified to produce spider silk proteins in its milk.

4913 Penn Avenue, Pittsburgh. The center is open on Sundays and the first Friday evening of every month. Ⓝ 40.465432 Ⓦ 79.944659

Mutant vegetables, transgenic mosquitoes, and a spider goat are exhibited at this unconventional natural history museum.

CENTRALIA

CENTRALIA

In October 2013, the eight remaining residents of the once 2,700-strong town of Centralia won a long court battle over the right to stay in their homes. The inhabitants of the former mining settlement are now free to keep living in an overgrown field crisscrossed by cracked roads that belch carbon monoxide from their crevices.

Located on a rich seam of anthracite coal, Centralia was settled as a mining town in the mid-1800s. In 1962, something happened that would transform Centralia from a quaint and lively small town into a bleak and hazardous wasteland: A fire in one of the underground mines began to burn out of control.

The exact cause of the fire is still disputed—some argue that it resulted from the volunteer fire department's annual burning of the landfill, while others claim that a coal fire from 1932 was never fully extinguished, and had been slowly spreading toward an abandoned strip-mine pit.

The scale of the problem did not become widely apparent until 1979. That year, mayor and gas-station owner John Coddington was checking the fuel levels of his underground tanks when he discovered that the gasoline had been heated to 172°F (77.7°C).

A real shocker came in 1981, when the ground tried to swallow Todd Domboski. The 12-year-old was walking in his backyard when the earth gave way and he fell 8 feet (2.4 m) into a smoking sinkhole. Domboski was able to hold on to tree roots at the sides of the hole until he was pulled to safety. The sinkhole was later found to be about 150 feet (46 m) deep and filled with lethal levels of carbon monoxide.

Following these troubling incidents, the government began claiming Centralia properties under eminent domain, condemning them, and relocating residents. The population fell from 1,017 in 1980 to 21 in 2000. Centralia lost its zip code in 2002.

Today, the remaining residents of Centralia have been granted the right to stay there for the rest of their lives, after which the government will take possession of their properties. Meanwhile, the fire rages on.

Centralia is 2.5 hours northwest of Philadelphia.
ⓝ 40.804254 ⓦ 76.340503

WARNING - DANGER

UNDERGROUND MINE FIRE

WALKING OR DRIVING IN THIS AREA COULD
RESULT IN SERIOUS INJURY OR DEATH

DANGEROUS GASES ARE PRESENT

GROUND IS PRONE TO SUDDEN COLLAPSE

Commonwealth of Pennsylvania
Department of Environmental Protection

GRIP THE RAVEN

PHILADELPHIA

The stuffed raven perched on a log in the display case of the Central Library's Rare Book Department is named Grip the First. Until his death and preservation in 1841—after which he was replaced by Grip the Second and Grip the Third—the raven was the beloved pet and sometime muse of Charles Dickens. So important was Grip, who had a habit of biting children's ankles, that the author paid tribute to him by including him as a character in his 1841 mystery novel, *Barnaby Rudge*.

Enter Edgar Allan Poe. In his review of *Barnaby Rudge*, Poe wrote that Dickens ought to have made Grip more ominous and

Barnaby Rudge with his raven.

symbolic. Four years later, clearly influenced by the novel, Poe published "The Raven," in which a "stately Raven of the saintly days of yore" induces grief and madness with its cry of "Nevermore." The raven's tapping in the poem echoes a line in *Barnaby Rudge*—when Grip first makes a sound, a character asks, "What was that—tapping at the door?"

Following Dickens's death, Grip ended up in the hands of Poe memorabilia collector Richard Gimbel. In 1971, Gimbel donated the bird, along with Poe's handwritten copy of "The Raven," to the library.

1901 Vine Street, third floor, Philadelphia. Get the subway (Broad Street line) to Race-Vine and walk 5 blocks west. ⓝ 39.959605 ⓦ 75.171023

Eastern State Penitentiary

PHILADELPHIA

Prior to 1829, prisons were chaotic, unruly institutions where criminals of all ages and sexes lived in the same cells. Then came Eastern State Penitentiary. Influenced by Enlightenment thinking, the prison was the first to implement the "Pennsylvania System," a philosophy that kept prisoners isolated from each other and the outside world in the hope that their solitude would induce profound regret.

Eastern State's design was based on Jeremy Bentham's panopticon, with cell-block "spokes" radiating from a central observation post. Each cell was equipped with a bed, flushing toilet, skylight, and Bible. All other reading material, including letters from family members, was forbidden. A door on the back wall of each cell led to a small exercise yard, where inmates could spend up to one hour per day. During any activity requiring a prison guard escort, prisoners had to wear a hood to prevent eye contact with another human being.

After visiting the site in 1842, Charles Dickens wrote in *American Notes for General Circulation* that the solitary confinement system was "cruel and wrong." Decades of criticism—and a burgeoning prison population—led to the gradual relaxation of Eastern State's strict solitude policies.

Suffering from deteriorating mechanical and electrical systems, Eastern State Penitentiary closed in 1971. For the next 15 years it sat abandoned, an urban forest springing up among its stone walls. In 1994, the prison reopened for public tours. The cell blocks remain in a state of neglect—mint-green paint flakes from the walls and a shaft of light from each cell's skylight illuminates rubble, rusted bed frames, and stray old boots.

2027 Fairmount Avenue, Philadelphia. Look for Al Capone's cell, which he occupied for 8 months after being arrested for carrying a concealed weapon in 1929. Equipped with rugs, lamps, a writing desk, and radio, it re-creates the luxury to which he was accustomed. Ⓝ 39.968327 Ⓦ 75.172720

Also in Philadelphia

Camac Street

Take a stroll down the last of the city's woodblock-paved streets.

Mummers Museum

This museum displays extravagant costumes from the annual Mummers folk parade.

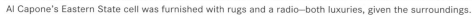

Al Capone's Eastern State cell was furnished with rugs and a radio—both luxuries, given the surroundings.

TOYNBEE TILES

MULTIPLE LOCATIONS (UNITED STATES, SOUTH AMERICA)

Mosaic rectangles with enigmatic messages, known as the Toynbee tiles, have been appearing on city streets for decades, and their origin and purpose are still a mystery. The tiles usually feature a cryptic message along the lines of: TOYNBEE IDEA—IN KUBRICK'S 2001—RESURRECT DEAD—ON PLANET JUPITER. Several hundred of these license plate–size plaques have been discovered since the 1980s, scattered randomly around major cities. They first appeared around Philadelphia and New York, and soon spread across the United States and into South America.

No one knows for sure why or how these tiles were laid.

The meaning of the "tiles," better described as adhesive patches that stick to concrete sidewalks and asphalt streets, has been interpreted as everything from a futurist utopian vision to a secret message from playwright David Mamet. (He denies this.) According to one popular theory, they were created by a Philadelphia carpenter who started a group that hoped to colonize Jupiter by resurrecting the dead there. Not surprisingly, his idea was not taken seriously by the media, and a few longer tiles have been found with side messages such as MURDER EVERY JOURNALIST I BEG OF YOU. Connecting the dots, it would appear that the tiler felt targeted for his beliefs of resurrection and chose to spread his message anonymously on the streets.

There are many theories on who the Toynbee tiler truly is and what their intent may be. But no concrete evidence has been found, and no one has ever laid eyes on the tiler. Some speculate that the typical tiles are laid simply by being tossed out of a hole in the floorboard of a car, which could explain the puzzling placement of many of the messages. **Many original tiles have been almost obliterated by constant foot and vehicle traffic, and some have been paved over in recent years. Others are well-preserved and easily read. There is an interactive online map of known tiles at toynbeeidea.com/portfolio/where. Ⓝ 39.951062 Ⓦ 75.165631**

MÜTTER MUSEUM

PHILADELPHIA

Named for Dr. Thomas Dent Mütter, who donated his collection of medical specimens in 1858, this museum is an enthralling mix of human skulls, diseased body parts, and anatomical models designed to educate aspiring doctors.

A highlight on the top floor is the Hyrtl Skull Collection: 139 19th-century crania arranged in eight rows. Each skull bears a description of the deceased's name, age, occupation, location, and cause of death. Though brief, the information is highly evocative—one 28-year-old Hungarian man died of "suicide by gunshot wound of the heart, because of weariness of life," while a Bosnian fellow was "killed in battle with Austrian sharpshooters."

Downstairs you'll find examples of exceedingly rare afflictions.

A skeleton in one of the display cases looks to be melting—the bones are the remains of Harry Eastlack, a man who suffered from a connective tissue disease known as fibrodysplasia ossificans

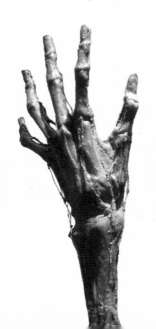

progressiva. The condition gradually turns muscle and tendons into bone, immobilizing joints and imprisoning the sufferer in his own body.

A few steps away, in a glass display case, is what looks like a giant papier mâché snake. Brown, bulbous, and 9 feet (2.7 m) long, it represents the "mega colon" of a man whose large intestine accumulated 40 pounds (18 kg) of feces due to a nerve disorder. Turn around and you'll see a tiny skeleton curled into a sphere. This is a lithopedion, or "stone child," an extra-uterine fetus that died and calcified inside the mother, remaining undiscovered for 24 years.

19 South 22nd Street, Philadelphia. The museum is a short walk from 30th Street Station, Philadelphia's main train hub. Photography is not permitted. Ⓝ 39.953201 Ⓦ 75.176637

WASHINGTON, DC

GARGOYLES OF THE WASHINGTON NATIONAL CATHEDRAL

The sixth-largest cathedral in the world, this building is an elegant blend of flying buttresses, neo-Gothic spires, and a carefully sculpted Darth Vader.

The masked head of Anakin Skywalker sits between two arches high on the northwest tower. Unlike the dinosaur of Ta Prohm temple, there is no ambiguity about this bust—during the 1980s, while the church was still in its 83-year construction process, kids entered a competition to design a decorative sculpture for the cathedral. Twelve-year-old Christopher Rader submitted a crowd-pleasing drawing of Darth Vader.

The sci-fi villain is not the cathedral's only unconventional architecture. Other decorative sculptures

Grab binoculars to spot a Sith Lord on the northwest tower.

include a raccoon, a girl with braces, and a grinning man holding an umbrella. One of the stained glass windows honors the Apollo 11 moon landing and contains a fragment of lunar rock.

3101 Wisconsin Avenue, NW, Washington. You will need binoculars to see Darth Vader. From inside the cathedral, go through the wooden doors near the Abraham Lincoln statue. Stand on the grass to your right and turn around to look at the closest tower. You'll find him beneath the central point at the first peaked roof. Ⓝ 38.930655 Ⓦ 77.070747

FBI SPY HOUSE

WASHINGTON, DC

The three-story house at 2619 Wisconsin Avenue had a big front porch and was located in a residential neighborhood. Perfect for a family. Or the FBI.

In 1977, the Soviet Embassy moved into a new building complex across the road from the house. The FBI and the NSA soon snapped up 2619 Wisconsin Avenue to set up a secret spying station to keep tabs on Russian business.

This inocuous house hides a secret.

The house's cover was negligible. The curtains were always drawn and no mail was ever delivered, yet people were frequently seen coming and going. Cameras could be clearly spotted in the windows, filming all those who entered the Russian embassy.

The FBI's uses for the house may have gone beyond playing paparazzi. Operation Monopoly was a secret plan to dig a tunnel beneath the embassy to record conversations taking place within the

building in the hopes of gleaning secret information. The problem was that the FBI had little knowledge of the embassy's layout. The agency hoped the tunnel would run underneath a room where people congregated to spill secrets, but it was just as likely to end up beneath a storage closet.

Though the FBI acknowledges the existence of the tunnel, they have never revealed which house in the neighborhood they began digging it from. Speculators believe it was either this observation house on Wisconsin Avenue or an abandoned house around the side of the embassy on Fulton Street. The truth may never be known for sure, as the tunnel has been sealed.

2619 Wisconsin Avenue, Washington, DC. In February 2018, the street separating the house and the embassy was named Boris Nemtsov Plaza, after the Russian physicist and Putin critic who was assassinated in 2015. Ⓝ 38.924258 Ⓦ 77.072811

WASHINGTON MINI MONUMENT

WASHINGTON, DC

Near the Washington Monument is a manhole. And in that manhole is another Washington Monument.

The mini monument is a 12-foot-tall (3.7 m) replica of the 555-foot (169 m) original. It serves surveyors as a geodetic control point—a marker that provides a starting point for maps and measurements. It's part of the network of a million control points across the country that helps the National Geodetic Survey synchronize all the government's maps.

Geodetic control points are often metal caps or rods that are driven down into the ground, but this quirky control point mirrors the form of its next-door neighbor. It's been used for surveys since the early 1900s. Outside of surveying circles, it's largely forgotten.

The survey marker is underneath a manhole just south of the Washington Monument. Speak to a park ranger before trying to see it. Ⓝ 38.889150 Ⓦ 77.035211

Officially known as "Bench Mark A," this miniature is ¹⁄₄₆ᵗʰ the size of the original.

WEST VIRGINIA

CONGRESSIONAL FALLOUT SHELTER

WHITE SULPHUR SPRINGS

Guests staying at the luxurious Greenbrier resort during the 1960s often saw TV repairmen walking around the west wing of the hotel. What they didn't know was that these apparent maintenance workers were actually government employees tasked with tending to the secret on-site nuclear bunker.

As the Cold War heated up in the 1950s, it became clear to the Eisenhower administration that Congress needed a place to escape to in the event of nuclear war with the Soviet Union. The Greenbrier, long a favorite getaway of presidents and located just a couple of hours from Washington, was chosen to house the 1,100-person congressional fallout shelter. In 1958, under the cover of building a new wing of the resort, work began on a 112,000-square-foot (10,405 m²)

In case of nuclear attack, Congress would convene in a subterranean bunker beneath a swanky hotel.

bunker buried 720 feet (219 m) into the side of a hill.

Completed just in time for the 1962 Cuban missile crisis, the facility contained dormitories—with name-tagged bunk beds for Congress members—a clinic, a decontamination chamber, and a television broadcast center with a large, soothing backdrop of Capitol Hill. Some of the shelter's 53 rooms were hidden in plain sight. The Greenbrier's seemingly unremarkable Exhibition Hall was actually part of the bunker.

In the event of nuclear attack, concealed blast doors would seal off the rooms from the outside world, allowing the government to continue to function.

Like other Cold War bunkers, the Greenbrier's congressional fallout shelter was never used for its intended purpose. In 1992, journalist Ted Gup revealed the secret facility in an article for the *Washington Post*, resulting in its decommissioning.

Much of the bunker is now a private data storage facility, but a section is open to visitors by guided tour. The drab, utilitarian furnishings are quite the contrast to the five-star rooms located directly overhead.

The Greenbrier, 300 West Main Street, White Sulphur Springs. Amtrak's Cardinal train, running between Chicago and New York three days a week, stops at White Sulphur Springs. The resort is a 5-minute walk from the station. Ⓝ 37.785946 Ⓦ 80.308166

Green Bank's ban on radio waves has turned it into a haven for those who say they have electromagnetic hypersensitivity.

THE QUIET ZONE

GREEN BANK

Cell phones and wi-fi are not allowed in Green Bank. Since 1958, the small town has been part of the National Radio Quiet Zone, a 13,000-square-mile (33,670 km²) patch of land in which all electromagnetic radiation on the radio part of the spectrum is banned. This drastic measure is all in the name of science: Green Bank is home to an observatory with the world's largest fully steerable radio telescope. Radio waves in the vicinity would interfere with its operations.

The remote town with a population of around 150 offers a refuge for people who believe they are hypersensitive to electromagnetic transmissions. These "wi-fi refugees" report that their symptoms of aches, pains, and fatigue disappear in Green Bank. Despite these claims, the scientific community does not recognize electromagnetic hypersensitivity as a medical condition.

Whether its soothing effect on the electromagnetically overstimulated is scientifically verifiable or not, Green Bank offers something special for every visitor: the bizarre sight of cows, farmhouses, and vast green pastures dwarfed by a 485-foot (148 m) telescope.

National Radio Astronomy Observatory, West Virginia
28. The Green Bank observatory is open year-round for tours. ℕ 38.432896 Ⓦ 79.839717

New England

CONNECTICUT

HOLY LAND USA

WATERBURY

In the early 1950s devout Roman Catholic John Baptist Greco had a vision of a roadside theme park devoted to God. By the end of the decade, his vision had been realized. He called it Holy Land USA.

The theme park included a miniature Bethlehem, a re-creation of the Garden of Eden, biblical-themed dioramas, and various tributes to the life and work of Jesus Christ. But it was best known for its Hollywood-style sign reading HOLY LAND USA and its 56-foot (17 m) steel cross that could be seen for miles, especially when lit up at night. It is a town joke that Waterbury kids grow up thinking Jesus was electrocuted on the cross.

By the 1960s, Holy Land was attracting 50,000 visitors a year. But its popularity waned, and in 1984, the park—rundown, dated, and in need of a spruce-up—was closed for renovation. Greco had hopes of expanding the site to attract more tourists, but it never happened. He died in 1986.

Responsibility for the park passed to a group of nuns. For a while, they tried to keep the park clean and neat, but despite their efforts, Holy Land attracted vandals and graffiti artists. Statues were beheaded, dioramas destroyed, and tunnels blocked. Overgrown, dilapidated, and strewn with garbage, Holy Land has acquired an unholy reputation—a status sealed by the murder of a young woman there in 2010.

Slocum Street, Waterbury. Holy Land is down a dead-end road. It is up to you whether to believe the signs regarding surveillance cameras and prosecution of trespassers. Once inside, stick to the paths.
ℕ 41.548636 Ⓦ 73.030328

CUSHING BRAIN COLLECTION

NEW HAVEN

Arrayed on shelves in a lush wood-paneled room at Yale's medical library are 550 jars filled with human brains. The collection once belonged to pioneering neurosurgeon Harvey Cushing, who preserved the brains from 1903 to 1932 as part of his tumor registry. When Cushing died in 1939, his undergraduate alma mater inherited the brains.

Cushing was among a handful of doctors operating on the brain during the early 20th century. At the time, about a third of patients who underwent brain tumor surgery did not survive the operation. Cushing introduced practices that dramatically lowered the mortality rate, such as monitoring blood pressure during surgery and operating with local anesthesia instead of ether. He was also the first to use X-rays to diagnose brain tumors.

Before they were restored and put on display in the medical library in 2010, the leaky jars holding Cushing's brain collection were locked in a basement under Yale's med student dorms. During the 1990s, students seeking a thrill would sneak into the dark, dusty storage room to view the fabled brains. Though these expeditions were unauthorized, students treated the specimens with care and apparently never swiped any. When moving the brains in 2010, Yale employees found a poster with names scrawled on it bearing the words "Leave Only Your Name. Take Only Your Memories."

Whitney Medical Library, Yale University, 333 Cedar Street, New Haven. Ⓝ 41.303218 Ⓦ 72.934003

ALSO IN CONNECTICUT

Crypt at Center Church on the Green

New Haven · When Center Church was built on a portion of New Haven's burial ground, 137 graves ended up in the basement.

A meeting of the minds—hundreds of them.

MAINE

WILHELM REICH MUSEUM

RANGELEY

A stone building with blue trim, once used as a laboratory, now holds much of the legacy of Wilhelm Reich, a psychoanalyst who believed that orgasmic energy could control the weather.

Reich began his career in 1919, working alongside Freud in Vienna. Influenced by Freud's theories on libido, Reich became fixated on what he called "orgastic potency": the complete release of energy and tension during orgasm. According to Reich, all neuroses—and even diseases like cancer—result from the inhibition of sexual energy.

Reich's work in the 1920s and '30s was unconventional— his "vegetotherapy" approach

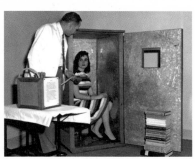

A woman sits in an orgone accumulator, designed by Wilhelm Reich to channel the omnipresent libidinal life force.

involved asking patients to strip to their underwear, after which he massaged them to loosen up their "body armor" until they screamed or vomited. During a series of bioelectricity experiments, Reich wired patients up to an oscillograph and observed changes in their bioelectricity as they engaged in sexual contact with each other.

Two weeks before the outbreak of the World War II, Reich moved to New York. It was here that he declared his discovery of "orgone," an omnipresent libidinal life force responsible for gravity, weather patterns, emotions, and health. Reich began building orgone accumulators: wooden booths lined with metal in which a subject could sit naked to absorb orgone energy.

Ten years later, in accordance with his new belief that the atmospheric accumulation of orgone radiation caused drought, Reich designed a cloudbuster. The machine consisted of a row of tubes aimed at the sky, attached to hoses immersed in water. Reich believed that orgone energy would get sucked from the sky down through the tubes into the water, causing rain. In the mid-1950s, Reich switched his attention to UFOs, which he believed were spraying orgone radiation in an attempt to destroy Earth. He and his son traveled to Arizona, where they used cloudbusters as "space-guns," aiming them at UFOs in an attempt to drain their energy.

By this time, Reich had attracted the attention of the FDA, which obtained an injunction to prevent him from shipping orgone accumulators out of Maine, where he then lived. When one of Reich's associates violated the injunction, the FDA ordered the destruction of Reich's accumulators, pamphlets, and books. Reich himself received a two-year prison sentence—the admitting psychiatrist at Danbury Federal Prison observed him to be experiencing delusions of grandeur. Eight months later Reich died in his cell bed after suffering a heart attack.

The museum at Orgonon, the idyllic site of Reich's Maine lab, contains equipment used in his eccentric experiments, as well as orgone accumulators, personal memorabilia, and original editions of publications burned by the FDA. Outside, a short walk into the woods, is a cloudbuster aimed directly at Reich's tomb.

19 Orgonon Circle, Rangeley. To really soak up the Orgonon energy, stay at Tamarack, one of the site's rental cottages, once the living quarters of the Reich family. Ⓝ 44.965682 Ⓦ 70.642710

INTERNATIONAL CRYPTOZOOLOGY MUSEUM

PORTLAND

An 8-foot-tall (2.4 m) Sasquatch guards the door of this museum, whose 10,000-item collection includes hair samples of the Abominable Snowman, fecal matter from a yeti, and a life-size mold of a coelacanth, a fish once thought to be extinct but rediscovered in 1938. Owner Loren Coleman, a lifelong cryptic enthusiast, is happy to talk to you about mothmen, chupacabras, tatzelwurms, and his own travels on the bigfoot-hunting trail. Pick up a yeti finger puppet or bigfoot-shaped air freshener in the gift shop.

11 Avon Street, Portland. Ⓝ 43.654222 Ⓦ 70.265869 ➤

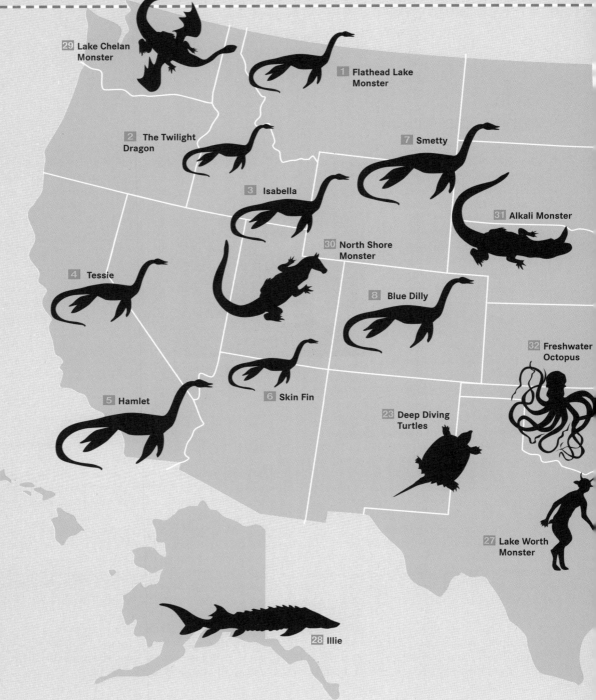

LAKE MONSTERS OF THE USA

29 Lake Chelan Monster

1 Flathead Lake Monster

2 The Twilight Dragon

7 Smetty

3 Isabella

31 Alkali Monster

30 North Shore Monster

4 Tessie

8 Blue Dilly

32 Freshwater Octopus

5 Hamlet

6 Skin Fin

23 Deep Diving Turtles

27 Lake Worth Monster

28 Illie

9 Pepie

35 Mishebeshu

25 The White Monkey

15 Poco

14 Champ

11 Rocky

16 Gloucester Sea Serpent

12 Lake Michigan Monster

13 Bessie

10 Obojoki

17 Kipsy

24 Beast of Busco

18 Chessie

34 Herry

33 Whitey

19 Normie

20 Altamaha-ha

26 Tainted Keitre

21 Tarpie

22 Muck Monster

Whether scaly, slimy, or furry, America's mythical lake monsters have inspired a whole lot of small-town tales. Here are some of the fabled creatures that have made their mark on imaginations around the country.

GIANT TURTLES

23. Deep Diving Turtles
Bottomless Lakes, NM

24. Beast of Busco
Fulks Lake, IN

WEBBED HOMINIDS

25. The White Monkey
Saco River, ME

26. Tainted Keitre
Honey Island Swamp, LA

GOAT MAN

27. Lake Worth Monster
Lake Worth, TX

MONSTER FISH

28. Illie
Iliamna Lake, AK

WINGED ALLIGATOR-SNAKE

29. Lake Chelan Monster
Lake Chelan, WA

HORSE-HEADED ALLIGATOR

30. North Shore Monster
Great Salt Lake, UT

HORNED ALLIGATOR

31. Alkali Monster
Alkali Lake, NE

GIANT KILLER OCTOPUS

32. Freshwater Octopus
Lake Thunderbird, OK

HORNED BEAST

33. Whitey
White River, AR

GIANT EEL PIG

34. Herry
Herrington Lake, KY

AQUATIC LYNX MONSTER

35. Mishebeshu
Lake Huron, MI

MASSACHUSETTS

MAPPARIUM

BOSTON

In 1930, Boston architect Chester Lindsay Churchill was commissioned to design the new Christian Science Publishing Society headquarters. His orders: Build something that could compete with the other grand newspaper headquarters of the day. The *New York Daily News* building had its famous gigantic spinning globe. The *Christian Science Monitor* had to do one better.

Enter the Mapparium, a three-story-tall, inside-out stained-glass globe that is bisected by a glass walkway. Once illuminated with hundreds of lamps, it now glows with the light of LEDs.

The Mapparium is the only place in the world in which the surface of the earth can be seen without distortion. Even when looking at an accurate globe, the relative sizes of the continents are distorted by perspective, as the spherical shape causes different regions to appear at different distances from the eye. But with a view from the very center of a globe, the eye is the same distance from every point on the map.

It is fascinating to view Earth this way. Africa is huge. North America, Europe, and Asia are all jammed up against the North Pole. You have to look nearly straight up to see them. Sizes and locations of continents and countries you've always taken for granted are suddenly unfamiliar.

While the relative size and position of the continents are correct, the map's political boundaries are long out of date. The Mapparium hasn't changed since 1935. It's still possible to find Siam and French Indochina, but not Israel or Indonesia.

The Mary Baker Eddy Library, 200 Massachusetts Ave, Boston. While inside the Mapparium, stand in the middle and listen. You'll be able to hear the voices of anyone else inside with perfect clarity, regardless of where they're standing or how loudly they're talking. Ⓝ 42.345130 Ⓦ 71.086294

Test out the enhanced acoustics while standing at the center of the globe.

JAMES ALLEN'S BIOGRAPHY

BOSTON

Among the rare books, maps, and manuscripts of the Boston Athenæum, a private library established in 1807, is a small publication bound in a leathery light gray material. Titled *Narrative of the Life of James Allen*, it is a well-preserved example of autoanthropodermic bibliopegy: a book bound in the skin of its author.

James Allen was a New England bank robber and highwayman. His brazen ambushes during the 1830s landed him in Massachusetts State Prison, where he died of tuberculosis in 1837 at the age of 28. As ill health tightened its grip on the illiterate criminal, he began to dictate his memoirs to a warden, instructing him to deliver a skin-bound copy of the finished book to a man named John Fenno.

Though Allen had only met Fenno once, the circumstances of the encounter ensured he would never forget him. In 1834, Allen attacked Fenno on the Salem Turnpike. To the surprise of the highwayman, Fenno fought back, only fleeing after Allen scrambled for his gun and fired a shot, hitting the victim's torso. Miraculously, a buckle on Fenno's clothing deflected the bullet.

Arrested in 1835 based on a tip by Fenno, Allen nonetheless maintained an admiration for the target of his attempted robbery. As a token of respect, Allen bequeathed a unique gift: his life story, bound in his own tanned, dyed skin.

10½ Beacon Street, Boston. Get the T (Red or Green line) to Park Street station. Guided tours for nonmembers are available on Tuesday and Thursday afternoons. N 42.357945 W 71.062029

It is rare to see a book bound in the skin of its author.

MUSEUM OF BAD ART

BOSTON

Woman Riding Crustacean, one of the works in the Museum of Bad Art's 600-strong collection, is a portrait of a faceless, handless, footless naked woman astride a giant lobster. For reasons unknown, an amorphous black blob encases the woman and her animal. Like the other works in this museum's collection, there is a glaring gap between the artist's sincerity and skill level.

The museum began with a single painting found between two trash cans on a Boston street in 1993. Antiques dealer Scott Wilson spotted a portrait of an elderly woman dancing in a field of flowers under a yellow sky. In one hand she holds a freshly picked bouquet. In the other she holds a red armchair. Using this heartfelt but poorly rendered painting as a foundation piece, Wilson and his friend Jerry Reilly established a bad-art collection.

Since its inception, the museum has had rigorous standards. Nine out of every ten submissions are rejected on the grounds that they display too much artistic competence. Those chosen—acquired via donations, thrift stores, yard sales, and trash heaps—exhibit wonky perspective, confusing symbolism, and lurid color combinations. Artwork depicting humans often omits hands and feet due to their being difficult to draw.

Around 30 items from the museum's collection are on show at each of its two Boston locations. Official museum commentary accompanies each artwork—*Juggling Dog in Hula Skirt* is described as "a fine example of labor-intensive pointlessism"—but guests are welcome to contribute their own interpretations in the visitor's book.

46 Tappan Street, top floor, Brookline. Get the T (Green line) to Brookline Hills station. There is another gallery at the Somerville Theatre, located at 55 Davis Square. Admission is free with every movie ticket. N 42.331603 W 71.127618

Ninety percent of submissions to the museum are rejected for being too good.

ETHER DOME

BOSTON

In the 1830s, surgery was a fast, brutal affair. Patients—plied with opium or whiskey, or punched unconscious if they were lucky—would be held down while doctors sawed off a leg or excised a tumor as quickly as possible. Top Scottish surgeon Robert Liston was famed for his ability to lop off a limb in under three minutes. (Once he cut off a leg in under three minutes, but was so caught up in the excitement of the feat that he accidentally removed the patient's testicles.) Diethyl ether, a volatile liquid narcotic, was used recreationally at the time—college students and bored socialites would take a big sniff and go on giggly "ether frolics"—but its pain-erasing properties had not yet been discovered.

The first public demonstration of ether as a surgical anesthetic occurred in 1846 at a Massachusetts General Hospital operating theater. A crowd of onlookers peered from the tiered seats as dentist William Morton administered ether vapors to Edward Gilbert Abbott. A few minutes later, with Abbott listless and unresponsive, surgeon John Warren sliced into a tumor on Abbott's neck. The absence of screams was encouraging.

When Abbott awoke and confirmed he had felt nothing, Warren turned triumphantly to the audience. "Gentlemen," he said, "this is no humbug!"

The Ether Dome, as the operating theater came to be known, was in use from 1821 to 1867.

Today the restored dome is open to the public when not being used for lectures and meetings. In addition to its ether-infused history, the room contains a collection of 19th-century surgical instruments, a skeleton, and Padihershef, an Egyptian mummy donated to the hospital in 1823. **55 Fruit Street, Bulfinch Building, 4th floor, Boston. While you're there, also take a look at the dome's architecture and the historic oil painting depicting the use of ether in an operating theater. Ⓝ 42.363154 Ⓦ 71.068833**

CITY GUIDE: More to Explore in Boston

Dutch House

Brookline • Built for the 1893 Chicago World's Fair, this four-story Dutch Renaissance–style abode was dismantled and reassembled in Brookline.

Infinite Corridor (MIThenge)

Cambridge • The 825-foot (251.4 m) corridor that threads straight through multiple MIT buildings has garnered the nickname MIThenge—thanks to its east-west positioning, it aligns with the setting or rising sun twice a year.

Mark I at Harvard's Science Center

Cambridge • This 51-foot-long (15.5 m) World War II calculator harks back to the days when "computer" was a job title.

Metropolitan Waterworks Museum

Chestnut Hill • See the steam-powered pumping engines that supplied Boston's water back in the 1880s.

Madonna Queen National Shrine

East Boston • A 35-foot statue (11 m) of "Madonna, Queen of the Universe"—that's Mary, mother of Jesus, not the Material Girl—was built in 1954.

Granary Burying Ground

Historic Downtown • Wander among rows of 18th-century gravestones, many decorated with skulls and bones, at this burial ground founded in 1660.

Jamaica Pond Bench

Jamaica Plain • Installed in 2006 as guerrilla art, a U-shaped, seatless park bench has earned a permanent place in Parkman Memorial Park.

All Saints Way

North End • A wall in a narrow alleyway has become one man's shrine to Catholic saints.

Molasses Flood Plaque

North End • Pay your respects at the site of the Great Molasses Flood of 1919, during which the slow, sweet sludge killed 21 people and injured 150.

Venetian Palace Diorama

North End • In the main room of the Boston Public Library's North End branch is a miniature replica of the Doge's Palace, a grand bit of Venetian architecture first built in the 14th century.

Franklin Park Zoo Bear Dens

Roxbury • The bear dens of the first Franklin Park Zoo, built in 1912, have been left in their original wooded location for explorers to discover.

Museum of Modern Renaissance

Somerville • A former Masonic hall has been transformed into a mystical temple of art, complete with murals of flowers, mermaids, and druids.

Steinert Hall

Theater District • Since it closed in 1942, this ornate 19th-century concert hall, located four floors beneath a piano store, has sat waiting for the music to return.

Even the furniture is made of rolled-up newspapers.

PAPER HOUSE

ROCKPORT

The phrase "paper house" may invite images of something fragile, but the Paper House of Rockport has been standing strong since the 1920s. In 1922, mechanical engineer Elis F. Stenman started building a house with newspapers. Initially, it was a hobby. Stenman pressed the papers together to form the walls, and applied glue and varnish to make them sturdy.

When it came time to furnish the house, Stenman continued the newspaper theme, rolling papers into tiny logs, stacking them to form chairs, bookcases, and desks, and sloshing varnish on top for a lacquered wood effect.

Stenman spent summers in his newspaper house until 1930, after which the place opened to the public as a museum. Look through the layers of shellac and you'll see headlines and tales from the 1920s, including, on one desk, accounts of Charles Lindbergh's pioneering transatlantic flight. **52 Pigeon Hill Street, Rockport. The house is around 40 miles (64 km) northeast of Boston, in the coastal town of Rockport. It's open from spring through fall. Ⓝ 42.672947 Ⓦ 70.634617**

NEW HAMPSHIRE

AMERICA'S STONEHENGE

SALEM

The name of this place sets you up for disappointment—it's nothing like the original. America's Stonehenge, formerly known as Mystery Hill, is a collection of small stone walls, modest rock arrangements, and underground chambers, all of which were constructed for undetermined reasons by persons unknown.

Radiocarbon dating of charcoal pits at America's Stonehenge reveals that people occupied the area during the second millennium BCE. Prevailing wisdom points toward a Native American presence at that time, but some people, including the site's current owners, suggest that the stone structures were the work of pre-Columbian Europeans. How such people might have ended up in New Hampshire millennia before the arrival of Columbus is anyone's guess.

A tour of America's Stonehenge incorporates "The Sacrificial Table," a hefty slab of granite with a groove around its perimeter—handy for draining the blood spilled during a ritual killing. This table looks very similar to the lye-leaching stones used during America's colonial era, which only confuses things further. The family of llamas that saunters around the site make for an adorable diversion when the muddled history becomes too frustrating. **105 Haverhill Road, Salem. Ⓝ 42.842852 Ⓦ 71.207217**

BETTY AND BARNEY HILL ARCHIVE

DURHAM

Betty and Barney Hill were the first people to claim to have been abducted by aliens, back in 1961.

Late at night on September 19, 1961, married couple Betty and Barney Hill were driving home to Portsmouth, New Hampshire, from a vacation in Montreal when they spotted a moving light in the sky. Intrigued by the unusual sight, they pulled over. And that, according to the couple's later statements, is when things took a turn for the freaky-deaky.

The Hills said that as the light approached, they realized it was a spaceship. They could tell because they looked in the window of the craft and saw a dozen gray-skinned Reticulans staring back. During hypnosis sessions conducted after the incident, Betty and Barney spoke of being taken aboard the ship, subjected to invasive medical experiments, and returned to their car several hours later with scant recollection of the experience.

Abduction by gray-skinned aliens has become a sci-fi trope, but the now-clichéd imagery originated with the Hills' tale. The Betty and Barney Hill Archive at the University of New Hampshire contains correspondence, personal journals and essays, newspaper clippings, photos, slides, films, and audiotapes relating to the couple's alleged alien abduction. Standout items include transcripts of hypnosis sessions, the purple dress Betty was wearing on the night in question, and notebooks in which Betty documented all her UFO sightings from 1977 to 1991.

Dimond Library, University of New Hampshire, 18 Library Way, Durham. Ⓝ 43.135515 Ⓦ 70.933210

RHODE ISLAND
ROGER WILLIAMS ROOT

PROVIDENCE

Embedded behind glass in a wall at the John Brown House is the tree root that ate Roger Williams.

Williams, the founder of Rhode Island, died in 1683 and was buried in an unmarked grave on the family farm. There he remained for 177 years, until Providence community leader Zachariah Allen led an effort to locate and disinter Williams's remains in order to create a more fitting memorial to the esteemed man.

When the grave was dug up and the coffin opened, however, not a trace of Williams remained. In his place was the vaguely anthropomorphic root of an apple tree. Naturally, those who came upon the root assumed that it had grown into the coffin and eaten Williams, after which (having swallowed his essence) it assumed a stick-figure form.

This tree root, which may or may not have fed on the body of the state's founding father, is now on display at the John Brown House, Providence's oldest mansion.

52 Power Street, Providence. While in town, visit Roger Williams Park. Ⓝ 41.822778 Ⓦ 71.404444

ALSO IN RHODE ISLAND

John Hay Library

Providence · Books bound in human skin and H. P. Lovecraft's letters are among a few of the treasures this library has to offer.

Gun Totem

Providence · This 12-foot-high (3.6 m) pillar was made with 1,000 guns recovered during a 2001 firearms buyback program in Pittsburgh.

VERMONT

ROCK OF AGES GRANITE QUARRY

BARRE

The Rock of Ages is the world's largest deep-hole granite quarry. A minibus transports visitors up to the top of the quarry, where you can view this mammoth mine safely from behind a gate. Although much of its depths are under a well of milky-green water, the crater is astoundingly huge, plunging nearly 600 feet (183 m) deep.

Next to the quarry is an enormous cutting facility that has been in operation since 1885. In this 160,000-square-foot (14.865 m²) space, huge blocks of granite are moved around, cut, polished, and engraved for tombstones and monuments.

Before leaving, be sure to roll a few frames at the outdoor granite bowling alley. The Rock of Ages company experimented with granite bowling lanes in the 1950s, but the concept never caught on. A prototype from that trial period is on display at the quarry site and has been restored for family fun.

The bus ride to the quarry passes piles and piles of granite blocks, where quarry workers have simply dumped pieces of rock with fractures or cracks over the years. Called "grout piles" from the Scottish word for scrap (many Scots worked in the quarry in its early days), these piles can be seen all over the town.
Ⓝ 44.156731 Ⓦ 72.491400

Also in Vermont

Dog Chapel

St. Johnsbury · This small village church celebrates the spiritual bond between canines and humans.

Most of America's granite headstones come from this massive crater.

Alaska and Hawaii

ALASKA
SPIRIT HOUSES

EKLUTNA

The cemetery behind St. Nicholas Orthodox Church in Eklutna has more than 100 tiny, colorful houses that look like chicken coops. Built to cover graves, the miniature buildings combine Russian Orthodox tradition and Native American practices.

Eklutna, located about 25 miles (40 km) outside of Anchorage, was the site of many Dena'ina Athabascan Indian villages about 800 years ago. Russian Orthodox missionaries arrived in the area around 1830 and the two communities slowly integrated.

Before the missionaries arrived, Athabascans cremated their dead. As the population became assimilated into Russian Orthodoxy, which forbids cremation, they began burying their deceased in the cemetery of St. Nicholas Church.

Spirit houses, an Athabascan tradition, provide a place for the deceased soul to dwell during the 40 days it is believed to linger in this world. When a body is buried, stones are piled on the grave and

covered with a blanket to provide symbolic warmth and comfort to the person. Then the spirit house is placed over the blanket, and relatives paint it in colors that represent the family.

The final touch is an Orthodox symbol: a wooden three-bar cross. The bars represent, from the top, the sign placed on the cross during Christ's crucifixion, the bar to which his arms were nailed, and the footrest that supported his body.

Eklutna Historical Park, Eklutna Village Road, Anchorage. The cemetery is a 30-minute drive from downtown Anchorage and is open May through September.
Ⓝ 61.460946 Ⓦ 149.360985

A mix of Russian Orthodox and Native American traditions, small graveside houses keep the souls of the dead safe.

Skee-ball, igloos, a giant furry boot, Santa's "rocket ship"—all can be found at Mukluk Land.

MUKLUK LAND

TOK

Try as you might, it's difficult to determine the unifying theme of Mukluk Land. "Stuff from Alaska" is about as close as it gets.

Retired schoolteachers George and Beth Jacobs established Mukluk Land in 1985 as a way to share their Alaskan memorabilia with the public. The park, which touts itself as "Alaska's most unique destination," contains a junkyard, a room full of arcade games, an impressive collection of beer cans, a minigolf course, a giant cabbage, and a vintage red-and-white bus known as "Santa's Rocket Ship."

And then there are the dolls. A log cabin houses hundreds of them—in rows on the floor, seated on shelves, stuffed side-by-side into open suitcases, and crammed into a red plastic convertible. All face a window through which you can peer into the room. Entering the cabin is forbidden, a fact asserted by the open bear trap on the floor.

Once you've perused Engine Alley, Heater Heaven, and the rest of the rusting machinery on the grass, pose for a photo in front of the park's main attraction: the giant mukluk. The big red boot, festooned with white pompoms, is suspended at head level from the front gate.

Milepost 1317 Alaska Highway, Tok. The park is 3 miles (5 km) west of Tok (which rhymes with "smoke"). It's open June through August.
Ⓝ 63.343807 Ⓦ 143.098213

ALSO IN ALASKA

Mendenhall Ice Caves

Juneau · Take a walk inside a 12-mile-long (19 km), partially hollow glacier blessed with brilliant blue walls.

Lady of the Lake

North Pole · An abandoned WB-29 weather reconnaissance aircraft sits submerged in a lake in the Alaskan wilderness.

Musk Ox Farm

Palmer · This livestock farm harvests the strong, soft wool of the Alaskan musk ox, a horned, 600-pound (272 kg) species whose males emit a pungent musky odor during mating season.

Aurora Ice Museum

Fairbanks · Mind-blowing, neon-lit ice carvings and frozen fantasy scenes grace the largest year-round ice environment in the world.

City of Whittier

Whittier · Just one building houses the vast majority of this town's 217 residents.

ADAK NATIONAL FOREST

ADAK

There's no chance you'll get lost in the forest on Adak Island, a 275-square-mile (712 km²) splotch of tundra located toward the farthest end of the Aleutian Island chain that stretches west from the Alaskan peninsula. Adak National Forest—the smallest in the United States—consists of just 33 pines, all huddled together in the middle of a sprawling, treeless plain. A sign at the edge of the grove reads: "You are now entering and leaving the Adak National Forest."

The modest forest came about when members of the US military, stationed at Adak's air base during World War II, participated in a Christmas-tree-planting program to improve morale. In homage to the original purpose of the pines, the 300 residents of Adak decorate the forest with Christmas lights every December.

Off Hillside Boulevard, near Bayshore Highway, Adak. Flights to Adak from Anchorage take about 3 hours.
Ⓝ **51.906106** Ⓦ **176.658055**

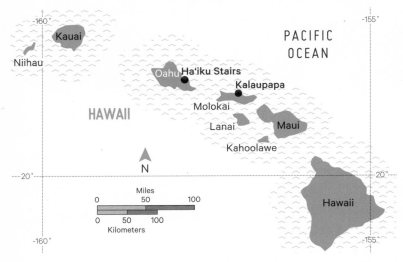

HAWAII

HAʻIKU STAIRS

KANEOHE, OAHU

To score a spectacular view of the sunrise—and avoid being chased away by a security guard—you'll need to start your hike up Haʻiku Stairs in the dead of night. The 3,922 steps, installed in 1942 in order to string antenna cables across the valley, became off-limits in 1987, when vandals destroyed three sections of the stairway. Undeterred, thrill-seeking hikers soon took to sneaking up the stairs, incurring the ire of local residents.

Though the city repaired the broken and rusted segments in 2002 with the intent of opening the Haʻiku Stairs to the public, community complaints and liability concerns have kept them closed. At just 18 inches (46 cm) wide, and more like a ladder in the steepest sections, the steps require climbers to travel single-file. A handrail on each side offers protection from tumbles down the mountain, but those rails—and the stairs—are often wet and slippery due to high humidity.

Each day at about 5 a.m., a security guard arrives at the foot of the stairs to enforce the no-trespassing rule. If you can find your way through the pitch-black jungle before the guard arrives, you'll be able to hike to the summit and back without getting into trouble. (As long as you are quiet and polite, the guard may even congratulate you or take a group photo when you arrive back at the bottom.) The reward for completing the adrenaline-fueled, mist-shrouded ascent is a glorious 360-degree view of Oahu.

Kuneki and Makena Streets, Kaneohe. Just a reminder: This climb is illegal. But if you're going to do it, bring a headlamp, water, a waterproof jacket, and gloves with grip. You'll see a gate—go past it on the right side and down the concrete road. Make a left at the first fork and a right at the second. You'll see an opening leading to a path on your left. Walk along it until you arrive at another paved road. Go left on the road and walk until you see the security guard's blue tent. Make a right into the jungle and go past the next gate on its left side. You've arrived.
Ⓝ **21.410265** Ⓦ **157.818364**

Nicknamed the Stairway to Heaven, the Ha'iku Stairs ascend into the clouds above Kaneohe.

Isolated from the world by 1,600-foot cliffs, Kalaupapa was home to more than 8,000 exiled Hawaiian lepers.

KALAUPAPA

MOLOKAI

For over a century the island of Molokai served as a beautiful prison for people forcibly removed from society. In 1865, Hawaii's King Kamehameha V authorized the Act to Prevent the Spread of Leprosy. At the time, the condition, also known as Hansen's disease, was regarded as highly contagious and incurable. Those afflicted with the disfiguring disease faced exclusion from their communities, due to the belief that leprosy was a punishment.

The 1865 act allowed the Hawaiian government to arrest and exile anyone thought to be affected by leprosy. From 1866 until 1969, over 8,000 people were transported to the isolated colony of Kalaupapa on the north coast of Molokai. Most had leprosy. Some were mistakenly diagnosed. All spent their days surrounded by ocean on one side and 1,600-foot (487 m) sea cliffs on the other.

A lack of food and shelter made the colony's early years particularly miserable. In 1873 a Belgian Catholic priest named Father Damien de Veuster arrived at Kalaupapa. For the next 16 years, Father Damien improved conditions at the colony by arranging medical services, building homes, and attending to the sick. His close contact with patients proved fatal: He died from leprosy in 1889, at age forty-nine.

The introduction of sulfone drugs in the 1940s rendered leprosy noncontagious. Though compulsory isolation ended in 1969, many patients chose to stay at Kalaupapa. A few members of the colony still live among the historic community structures, which include a movie hall, a group home, and 14 cemeteries. These places are preserved so visitors can better understand the experiences of the people once confined to this prison in paradise.

Kalaupapa National Park, Molokai. Molokai is a 30-minute flight from Honolulu. Mule is the preferred mode of transport along the trail to Kalaupapa. All visitors must be over sixteen and join a tour offered by residents of Kalaupapa. Ⓝ 21.166395 Ⓦ 157.105464

ALSO IN HAWAII

Pineapple Maze at the Dole Plantation

Wahiawā · Race through the largest plant maze in the world.

Kamilo Beach

Naalehu · Formerly a stretch of pristine white sand, this beach has become a trash trap of the Great Pacific Garbage Patch.

Latin America and the Caribbean

South America
ARGENTINA · BOLIVIA · BRAZIL · CHILE · COLOMBIA PERU · URUGUAY · VENEZUELA

Mexico

Central America
BELIZE · COSTA RICA · EL SALVADOR GUATEMALA · HONDURAS · NICARAGUA · PANAMA

Caribbean Islands
BAHAMAS · BARBADOS · BERMUDA · CAYMAN ISLANDS CUBA · CURAÇAO · DOMINICA · DOMINICAN REPUBLIC GUADELOUPE · HAITI · MARTINIQUE · MONTSERRAT PUERTO RICO · ST. KITTS AND NEVIS · TRINIDAD AND TOBAGO ST. VINCENT AND THE GRENADINES

15°

HND

CARIBBEAN
SEA

-80°   -75°   -70°   -65°   -60°   -55°   -50°   -45°   -40°

15°

LV

NIC

CRI

PAN

10°

Everlasting Lightning ●   ★ CARACAS

Pablo Escobar's
Hippos

VENEZUELA

● Drowned Church

ATLANTIC
OCEAN

10°

5°

Armero

BOGOTA ★

Guayabetal Zip Lines

Caño Cristales ●

COLOMBIA

● Sarisariñama

SUR   GUF

GUY

● Amazon Stonehenge

5°

EQUATOR

ECU

Amazon River

Nazi
Graveyard

● Amazon Bore Surfing

EQUATOR

0°

0°

-5°

Sarcofagi
of Carajía   ● Gocta Falls

Chan Chan ●

PERU

● Boiling River
of the Amazon

Fordlândia

● Lençóis
Maranhenses

-5°

Santo Daime ●

BRAZIL

-10°

LIMA ★

● Ica Stones

● Last Incan Grass Bridge

-10°

-15°

Uros ●

North Yungas Death Road

LA PAZ ★ Witches' Market

BOLIVIA

★ BRASÍLIA

-15°

Nitrate
Towns

● Salar de Uyuni

Glowing Termite
Mounds of Emas ●

-20°

-20°

Mano del
Desierto ●

PRY

SOUTH PACIFIC OCEAN inset:

EQUATOR

SOUTH
AMERICA

● The Unfinished
Giants of Easter Island

SOUTH
PACIFIC
OCEAN

Miles
0   1,000
0   1,000
Kilometers

-25°

● Snake Island

-25°

-30°

Robinson
Crusoe
Island ●

● Ischigualasto
Provincial Park

URUGUAY

-30°

SANTIAGO ★

● Laguna del
Diamante

BUENOS
AIRES ★

★
MONTEVIDEO

● Laguna Garzón Bridge

ATLANTIC
OCEAN

-35°

-35°

PACIFIC
OCEAN

CHILE

ARGENTINA

The Girl Who
Died Twice ●

-40°

El Ateneo Grand ●
Splendid

BUENOS AIRES

-40°

-45°

● Marble Cathedral

-45°

SOUTH AMERICA

Miles
0   500   1,000

-50°

0   500   1,000
Kilometers

N

-50°

Falkland Islands

-55°

-85°   -80°   -75°   -70°   -65°   -60°   -55°   -50°   -45°   -40°

South Georgia

This historic, palatial theater is now one of the world's most beautiful bookstores.

ARGENTINA
EL ATENEO GRAND SPLENDID

BUENOS AIRES

With frescoed ceilings, ornate theater boxes, elegant rounded balconies, detailed trimmings, and plush red stage curtains, the El Ateneo Grand Splendid is hardly your average bookstore. Built in 1919, the majestic building began as a theater featuring tango legends, then became a cinema—the first in Buenos Aires to show films with sound. While the titles on offer tend to be more expensive than those of other city bookstores, the staggeringly opulent interior is reason enough to pay a visit.

Avenida Santa Fe 1860, Buenos Aires. Ride the subway to the Callao stop and walk three blocks north to Avenida Santa Fe. Ⓢ 34.595907 Ⓦ 58.394185

ALSO IN BUENOS AIRES

Castillo Naveira

Buenos Aires · This enormous neo-Gothic castle, hidden from the public, has a hall of armor and a stable of vintage cars.

Floralis Genérica

Buenos Aires · Every day this 105-foot-wide (32 m) giant metallic flower blooms anew.

Laguna Epecuén, Carhué

Buenos Aires Province · This Argentinean lake swallowed an entire village in 1985.

Museo del Mar

Buenos Aires Province · This museum holds a vast collection of seashells, fossils, and marine invertebrates.

Pedro Martín Ureta's Forest Guitar

Buenos Aires Province · A man's lost love inspired this giant guitar made entirely of living trees.

Xul Solar Museum

Buenos Aires · See the collected work of an artist of alternate worlds, inventor of languages, and dreamer of utopias.

TOMB OF THE GIRL WHO DIED TWICE

BUENOS AIRES

Most visitors to the beautiful baroque Recoleta Cemetery come to see the tomb of Eva Perón. But just a short walk south lies another curious sight: the crypt of Rufina Cambaceres, the girl who died twice.

In 1902, on her 19th birthday, Rufina was getting ready for a night out when she lost consciousness and collapsed. Three doctors declared her dead. The young socialite was placed in a coffin, given a funeral, and sealed in a Recoleta mausoleum.

A few days later, a cemetery worker, noticing signs of a break-in, entered the mausoleum. Suspecting a grave robber, he opened the casket to find Rufina's remains and scratch marks on the inside of the coffin. Rufina had been buried alive. Awakening in her coffin, she struggled to escape, clawing at the lid before she died of cardiac arrest.

Like many "buried alive" stories, it is difficult to separate truth from fiction, and Rufina's tale is told with varying layers of embellishment. In one version, her initial "death" is caused by the scandalous revelation that her boyfriend had been sleeping with his own mother.

Whether any of the stories are true or if they are the result of overactive imaginations, Rufina's tomb is worth visiting for its beauty alone. It features a full-size statue of the girl gazing out at the cemetery while holding shut the door to the very mausoleum that entrapped her.

Junín 1790, Buenos Aires. Pick up a free map at the entrance to Recoleta Cemetery. Rufina's grave is number 95, about three blocks south of Eva Perón's in the southwestern section.
Ⓝ 34.588328 Ⓦ 58.392408 ➥

➥ Historical Methods of Preventing Premature Burial

During the 18th and 19th centuries, doctors understood that many people were accidentally being buried alive, but didn't yet know how to tell the difference between someone who was dead and someone who merely looked it. To solve this quandary, a variety of unusual methods were used to test for signs of life.

TOBACCO SMOKE ENEMAS were commonly used in 18th-century Europe to try to resuscitate the apparently dead. Smoke—blown through a pipe into the rectum using a bellows or from the mouth of an unflinching rescuer—was thought to bring people back from the brink of death. Drowning victims were often subjected to smoke enemas after being hauled from the water, with occasional (and probably only coincidental) success.

FOOT TORTURE was occasionally employed postmortem. Well-meaning physicians cut corpses' soles with razors, shoved needles under their toenails, and applied red-hot irons to their soles, all to ensure the person was really, truly, undeniably dead.

WAITING MORTUARIES, in which bodies were kept until they showed signs of putrefaction, were popular in Germany in the late 1800s. They were essentially hospitals for the dead, their wards of corpses watched over by nurses. In order to mask the smell of rotting human flesh and organs, flower arrangements were placed beside each bed.

SAFETY COFFINS addressed the fear of premature burial by incorporating features such as air tubes, strings linking the body's hands and feet to an aboveground bell, flag, or lights, and for coffins installed in vaults, spring-loaded lids. Despite the rash of safety coffin patents, there are no reported cases of the mistakenly dead being saved by such contraptions.

Safety coffins allowed the mistakenly buried to make their living status known.

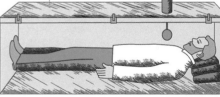

ESMA Museum

Núñez • A haunting and horrific museum of Argentina's "Dirty War," this former naval officers' school, one of 400 detention centers for political prisoners in the late 1970s and early '80s, opened its doors in 2008 to keep the memory of the military junta's "disappeared" from fading.

Remembrance Park

Belgrano • Along the Rio de la Plata, the Parque de la Memoria pays tribute to the tens of thousands lost to state-sponsored terrorism and violence in the late 1970s and early '80s—with emotionally expressive sculpture, sweeping lawns, wide ramps, and a wall of remembrance etched with the names of the victims.

Tierra Santa

Belgrano • Join the crowds flocking to catch the pop-up 40-foot Jesus, eat a "Last Supper" with some mannequin apostles, see a recreation of the Book of Genesis (did they have animatronic hippos back then?), and grab a selfie with Judas Iscariot.

Carlos Gardel Tomb

Chacarita • Like a tale from one of his own sad songs, tango composer and singer Gardel died in a plane crash at the height of his powers. His tomb at La Chacarita Cemetery features a life-size tuxedoed statue of the great entertainer, surrounded by dozens of bronze plaques from around the globe expressing undying love and gratitude.

The Weather Indicator at the Botanical Garden

Palermo • A marble obelisk topped with a bronze globe etched with zodiac symbols sits near a pond in the Carlos Thays Botanical Garden. A plaque tells you what it's called (a "Weather Indicator"), when it was made (1910), and who made it (Jose Markovich)— what it's supposed to do, or

Dance down the alley of El Caminito.

how it's connected to the weather, is a quiet mystery.

Castillo de Naveira

Lujan • Long kept hidden from public view, this sprawling neo-Gothic castle seems to rise out of the Argentine countryside and into a Grimm's fairy tale—it's a good hour outside the city, but worth the journey back to another time and place.

Recoleta Cemetery

Recoleta • A few of the mausoleums in this cemetery for Buenos Aires's upper crust have fallen on hard times, visited more often by the resident cats, but it's still one of the world's most beautiful resting places. The rows and rows of nearly 5,000 ornate vaults (including the tomb of Eva Perón) make for an extraordinary city of the dead.

The Golden Hall

San Nicolás • Teatro Colón is universally considered one of the three or four greatest opera houses in the world. There is more than a little touch of the Versailles in its Salón Dorado, a gallery positively dripping in gilt, and lined with mirrors of infinite reflection.

Barolo Palace

Monserrat • Until 1935, this 22-story office tower was the tallest building in South America. Its allegorical design—from Hell in the lobby, up through Purgatory, and finally to Heaven at the top—is modeled after Dante's *Divine Comedy*.

Block of Lights

Monserrat • Known as the Illuminated Block, or Manzana de las Luces, this 17th- and 18th-century complex of buildings, churches, and secret tunnels has played several crucial roles over the course of Buenos Aires history— educational, missionary, legislative, military—but the extent and original purpose of those tunnels remain a mystery.

Puente de la Mujer

Puerto Madero • An improbably cantilevered pedestrian crossing, the "Bridge of the Women" is, according to its architect, intended to evoke a pair of tango dancers, but its futuristic needle-nose design would be more at home on the set of *Star Trek* than in the dance hall—stick around to catch its 90-degree twirl as it opens for passing ships.

Urban Birding at Costanera Sur Ecological Reserve

Puerto Madero • Nature triumphs along the Rio la Plata at the huge Reserva Ecológica Costanera Sur, where a wide array of species have repopulated the reclaimed nature preserve, including roadside hawks, bar-winged cinclodes, vermilion flycatchers, scissor-tailed nightjars, and white-faced whistling ducks.

Russian Orthodox Cathedral

San Telmo • With its cobalt-blue onion domes, the Russian Orthodox Cathedral is very likely the only building in Buenos Aires that was designed in Saint Petersburg in the spirit of 17th-century Russia.

El Caminito

La Boca • This alley-long outdoor museum and permanent street fair is a riot of tango color and culture, lightheartedly carrying a solemn promise to keep the national dance a way of life.

Creole Museum of the Corrals

Mataderos • This small museum covering the history of the old meatpacking district is a little moth-eaten, but its charming, dusty quality is well suited to its focus on gaucho life—with cowboy artifacts, hundreds of antique lassos and knives, and some startling vintage slaughterhouse photography.

Hundreds of stone spheres dot the ball court.

ISCHIGUALASTO PROVINCIAL PARK

SAN JUAN, SAN JUAN

Nicknamed the Valley of the Moon on account of its odd geological formations, this park is home to giant petrified tree trunks and some of the world's oldest dinosaur remains. Among the odd stone formations is one called the "ball court," a field of hundreds of stone spheres formed over millions of years by wind and erosion.

The closest cities are San Juan (5 hours south) and La Rioja (3 hours northeast). Charter vans and buses depart from both locations.
Ⓢ 30.513765 Ⓦ 67.582397

ALSO IN ARGENTINA

Museo Rocsen	House of Plastic Bottles	Perito Moreno Glacier	Petrified Forest	The Campo del Cielo Meteorite
Nono · Photography, insects, toys, religious relics, fossils, and a Peruvian mummy are among the 25,000 objects in this eclectic 20-room museum.	*Puerto Iguazú* · One man built this one-bedroom cottage out of plastic drink bottles, CD cases, and milk cartons.	*Santa Cruz* · A growing glacier causes havoc when it dams Lago Argentino and then bursts—with spectacular results.	*Santa Cruz* · The two largest petrified trees on the planet are in this Patagonian park.	*Santiago del Estero* · The heaviest meteorite ever recovered on Earth landed at this site.

LAGUNA DEL DIAMANTE

SAN RAFAEL, MENDOZA

There are plenty of reasons why life should not exist at Laguna del Diamante. Located within one of the world's largest calderas—the bowl-shaped formations resulting from land collapsing after a volcanic eruption—the lagoon is surrounded by sulfur-spewing vents. The hyper-alkaline lagoon is five times saltier than seawater and has levels of arsenic that are 20,000 times higher than the amount deemed safe for drinking by the EPA.

Despite these inhospitable conditions, which mimic those of the early Earth, millions of bacteria known as "extremophiles" have managed to flourish. Scientists hope that the mysterious microorganisms can be used to discover new antioxidants or enzymes and may someday help to explain how life on Earth began.

The lake is 4 hours south of Mendoza, near the Chilean border. The road there is only passable from December to March. Four-wheel-drive tours depart from nearby San Rafael and Mendoza. Expect a bumpy ride. Ⓢ 34.149999 Ⓦ 69.683333

A toxic lake, on an active volcano, frequented by flamingos.

Stock up on llama fetuses at Bolivia's premier market for occult goods.

BOLIVIA

WITCHES' MARKET

LA PAZ

Dozens of vendors line this street, selling folk remedies, dried reptiles, and llama fetuses that are said to bring prosperity and good luck. "Witches" wander through the market offering fortune-telling services, spiritual advice, and traditional medicine cures. If you'd like a glimpse into your future, or help with a physical or spiritual ailment, look for the people wearing dark hats.

Calle Linares, La Paz. Taxis, micros (minivans), and *trufis* (cars with specific routes) are all cheap ways of getting to this cobblestone street between Sagarnaga and Santa Cruz. Ⓢ 16.496624 Ⓦ 68.138655

During the wet season, a thin layer of water transforms the salt plain's surface into a seemingly endless mirror.

SALAR DE UYUNI

UYUNI, POTOSÍ

Hotel Luna Salada is built almost entirely out of the most abundant local resource: salt.

The walls are salt bricks held together with salt mortar. You can eat at a salt table, sleep on a salt bed, and watch the sun set over the stark white of the Salar de Uyuni, the largest salt flat in the world.

Salar de Uyuni was created after a lake dried up thousands of years ago. It is a place where nothing seems to make sense. In the wet season, a thin layer of water turns the Salar de Uyuni into a mind-bending mirror, its 3,800 square miles (9,842 km²) of white desert reflecting the sky and creating the illusion of infinity.

Fierce sun during the day gives way to freezing temperatures at night. (This can be a nuisance at Hotel Luna Salada, due to limited hot water, so bring warm clothing and ask at the front desk about showering times.) Giant cacti are the only form of vegetation, but pink flamingos gather by the thousands at nearby Laguna Colorada.

NASA has used the Salar de Uyuni to calibrate its ICESat, a satellite that measures the elevation of ice sheets in Antarctica and Greenland. Because the surface of the salt flat is stable year-round, NASA scientists can calibrate their instruments by measuring the time it takes for the satellite's laser pulse to reflect off the ground.

The surreal landscape of Salar de Uyuni is complemented by a collection of rusted train cars in the desert. Uyuni was a transport hub for locomotives carrying minerals to the Pacific coast in the late 19th century. The decline of the mining industry in the 1940s resulted in many trains being abandoned in the desert. There has been talk of establishing a museum for the cars, but for now they sit in a line under an unrelenting sun.

Uyuni is a 10-hour overnight bus ride from La Paz. Buses also leave from Sucre in the morning. They stop in Potosí and then go on to Uyuni, where they arrive in the late afternoon. Ⓢ 20.280265 Ⓦ 66.982512

A hotel built entirely of salt.

NORTH YUNGAS DEATH ROAD

LA PAZ

To one side is solid rock; to the other, a 2,000-foot (610 m) abyss. Between is a two-way, 12-foot-wide (3.65 m) highway known as Death Road.

One of the world's most dangerous routes, North Yungas Road was cut into the side of the Cordillera Oriental Mountain chain in the 1930s. Its steep descent, lack of guardrails, and tendency to be shrouded in fog and covered with falling rocks and mudslides explain why an estimated 200 to 300 people are killed there every year.

Though vehicles keep to the right side of the road elsewhere in Bolivia, North Yungas drivers keep their cars on the left so they can see the cliff's edge more clearly. The drop isn't the only thing to be concerned about—travelers must also watch out for the intrepid cyclists who arrive determined to barrel down North Yungas for thrills.

North Yungas Road is 43 miles (69.2 km) long, leading from La Paz, the capital, to the town of Coroico. Several companies in La Paz organize Death Road bike rides. Be prepared to sign multiple forms releasing them from responsibility for your demise.
§ 16.221092 Ⓦ 67.754724

Bolivia's infamous Death Road has killed so many people, it's become a tourist attraction.

ALSO IN BOLIVIA

Horca del Inca

La Paz · Above Lake Titicaca is a pre-Incan, 14th-century astronomical observatory.

Museo de la Coca

La Paz · A museum dedicated to the sacred leaf of the Andes.

Valle de las Ánimas

La Paz · Needlelike rock formations are densely packed into this "Valley of the Souls."

Great Train Graveyard

Potosí · On the outskirts of a desert trading village, high on the Andean plain, steel giants have been destroyed by salt winds.

Potosí Silver Mines

Potosí · A mountain of unimaginable riches bankrolled the Spanish Empire.

Fort Samaipata

Samaipata · Remnants of Inca and Mojocoyas culture can be seen in the rock carvings, plaza, and agricultural terraces of this hill.

Cal Orck'o

Sucre · Visit a limestone wall covered with 5,000 preserved dinosaur tracks from the Cretaceous period that was discovered in 1994 near a cement factory. It is the largest site of fossilized dinosaur tracks in the world.

Laguna Colorada

Sur Lípez · Explore a red lake 14,000 feet (4,267 m) above sea level that's home to extremophile bacteria and rare flamingos.

BRAZIL

AMAZON STONEHENGE

CALÇOENE, AMAPÁ

In 2006, archaeologists digging on the banks of the Rego Grande river in northern Brazil discovered a strange grouping of 127 giant stones. The megaliths, each standing over 10 feet (3 m) high, were arranged in circles in an open field. By analyzing ceramic shards found nearby, the archaeologists estimated the arrangement of stones is between 500 and 2,000 years old. The placement of the stones appears to be astronomically based—the shadow of one of the blocks disappears during the winter

The vestiges of an ancient observatory, according to archaeologists.

solstice—suggesting it might have been built as an observatory.

Anthropologists have long argued that large, complex civilizations could not have existed in the Amazon, as the poor-quality soil could not have supported the agriculture necessary to establish large communities. But the Amazon Stonehenge and other recent findings have cast doubt on this assertion and opened up the possibility that thriving metropolises existed in the jungle thousands of years ago—it's just a matter of finding them.

The stones are in Calçoene, 240 miles (386 km) north of Macapá.
Ⓝ 2.497778 Ⓦ 50.948889 ➤➤

➤➤ Don't Follow That Man

The origin of the Amazon Stonehenge remains a mystery, but its discovery lends credence to the hypothesis that the South American rain forest may be teeming with the remains of lost cities. Such notions consumed the minds of explorers like Colonel Percy Fawcett, who ventured into the wilderness of Brazil's Mato Grosso region in 1925 in search of the city he dubbed "Z." Fawcett, his son, and his son's best friend all vanished without a trace.

In the decades following Fawcett's disappearance, over a dozen expeditions were launched in the hope of discovering his fate. None found conclusive evidence, and it is believed over 100 explorers have perished in the jungle looking for him. Here are a few who tried and failed:

Swiss trapper Stefan Rattin arrived at the British Embassy in São Paulo in 1932, claiming that he had encountered a long-haired, animal-skin–clad Fawcett five months earlier while hunting near the Tapajós River. According to Rattin, "Fawcett" told him he was being held captive and pleaded for help from the embassy.

With the blessing of Fawcett's wife, Rattin set off on a rescue mission with two men, walking

through the jungle for weeks and building canoes out of bark. A later dispatch reported that the trio was about to enter hostile Indian territory. No one heard from them again.

Shortly afterward, English actor Albert de Winton, bored with Hollywood life, decided to become a genuine jungle explorer. Vowing to find Fawcett, he ventured into the wilderness, his publicist issuing a press release about his heroism. Nine months later, he emerged, thin and dressed in rags, and posed for photographs to send to the *Los Angeles Times*.

After replenishing himself in Cuiabá, de Winton headed back into the jungle. The only subsequent sign of him came a few months

later, when an Indian messenger brought a crumpled note out of the forest declaring he had been captured. The unfortunate story of his demise followed years later: Members of the Kamayurá tribe had found him in a canoe, naked and deranged, and clubbed him to death.

In 1947, missionary Jonathan Wells warned New Zealand schoolteacher Hugh McCarthy against venturing into the Mato Grosso, but McCarthy was determined. The ever-cautious Wells gave him seven carrier pigeons for the journey. Over the next few months, three of the birds brought messages. The first reported a leg injury but

remained optimistic. The second said McCarthy had abandoned his rifle and canoe and was living on berries and fruits, having exhausted his food supply. The third and final missive was simple and resolute: "My work is over and I die happily."

Percy Fawcett, doomed jungle explorer.

Nazi Graveyard

AMAPÁ

On a small island on a tributary of the River Jary stands a 9-foot-high (2.7 m) wooden cross with an unusual decoration. Marked on the cross are the words "Joseph Greiner died here on 2.1.1936" and, above that, a swastika.

Greiner, a Nazi soldier, arrived in Brazil in 1935. He was accompanied by fellow scientist and SS officer Otto Schulz-Kampfhenkel. Their mission, known as the "Guayana Project," was to evaluate the area's suitability for colonization by the Third Reich.

A Third Reich grave marker in the Amazon.

Despite Schulz-Kampfhenkel's encouraging reports that the area offered "outstanding possibilities for exploitation" for "the more advanced white race," the Nazi colonization of Brazil obviously never took place. (Schulz-Kampfhenkel, however, was able to put aside his racial views long enough to father a child with an indigenous woman.)

Today, all that remains of this monomaniacal plan are the rotting grave of the Nazi soldier who perished in pursuit of it and a short film made while on the expedition, called *Rätsel der Urwaldhölle*, or "Riddles of the Jungle Hell."

Head southwest from Macapá on highway 156 until you hit the Amapá–Para border. The grave site is a little farther north. S 0.623461 W 52.577819

Amazon Bore Surfing

SÃO DOMINGOS DO CAPIM, PARÁ

Every year in February and March, during the spring tides of a new or full moon, the world's longest wave comes tumbling down the Amazon River at speeds of up to 20 miles (32.1 km) per hour. Referred to colloquially as *pororoca*, or "great destructive noise" in the Tupi language, its roar can be heard half an hour before it arrives.

Though the *pororoca* has the power to destroy trees, houses, and livestock, surfers from all over the world converge to compete in the annual Brazilian National Pororoca Surfing Championship. Braving wave heights of up to 12 feet (3.65 m), winners have experienced the ride of a lifetime, surfing the bore for more than 30 continuous minutes. However, Amazon bore surfing has its risks, including murky water, floating trees, poisonous snakes, and hungry alligators.

Surfers meet at São Domingos do Capim, a 2-hour drive east of Belém. The contest is usually held between February and April, but includes a lot of waiting, as the exact moment of the *pororoca* is always unknown. S 1.675948 W 47.765834

Santo Daime Ayahuasca Ceremonies

BOCA DO ACRE, AMAZONAS

For followers of the Santo Daime religion, the violent expulsion of bodily fluids is both a regular and a desirable experience. Founded in the 1930s, Santo Daime mixes Christianity, shamanism, African animism, and the ceremonial ingestion of a psychoactive vine known as ayahuasca. Boca do Acre, Brazil, is the religion's psychedelic mecca, with people traveling from around the world to attend the ceremonies held here. Because ayahuasca is used for the purpose of healing, self-enlightenment, and spiritual communion, the Brazilian government has deemed it legal.

During ceremonies, which begin in the evening and last all night, participants are divided by sex, age, and—occasionally—virgin or non-virgin status, and given cups of brewed ayahuasca. As leaders sing and pray, the drug takes effect, causing out-of-body experiences, visions, loss of motor skills, and, most importantly, the "purge." Violent vomiting, diarrhea, and wailing are common and looked upon favorably; within Santo Daime, these expulsions signify evil spirits leaving the body.

Ayahuasca tourism brings visitors to Brazil, Peru, and Ecuador, where they are guided through a ceremony in groups under the instruction of local shamans. Experiences differ markedly, but many people report revisiting childhood traumas, letting go of their egos, and waking up the next morning with a sense of peace and clarity. Others just report vomiting.

Boca do Acre is a 5–10 hour drive from Rio Branco, depending on the state of the poorly maintained dirt roads. There is a local airport, but no commercial flights—you must charter a plane for the 25-minute trip from Rio Branco. S 8.740689 W 67.384081

ENTHEOGENS

Entheogens, or psychotropic drugs used to enhance religious experience, have long been part of the spiritual practices of indigenous people, particularly in South America. Used in rituals to attain self-enlightenment, commune with nature, and enhance the senses, here are a few of the most fascinating:

1 VIROLA TREES contain a hallucinogenic resin in their bark. Colombian, Venezuelan, and Brazilian shamans prepare a snuff by collecting shavings from the inner layer of the bark and reducing them to a powder or paste. During Virola ceremonies, men and older boys use long tubes to blow the hallucinatory drug up each other's nostrils, then hop and crawl along the floor before losing consciousness.

2 THE SAPO is a large green tree frog whose skin secretions induce a racing heart, incontinence, and vomiting within minutes, before leveling off into a state of listlessness and, finally, euphoria. The Matsés Indian hunters traditionally apply the substance by burning their arms and rubbing the wounds with a stick dipped in the secretions. After the initial effects wear off, hunters are left with improved stamina and strength, a decreased appetite, and keener senses, all of which allow for stealthier stalking of animals.

3 HUACHUMA, or San Pedro cactus, is used in Peru for guidance, decision-making, and healing. It is ingested in the form of a bitter, dark-green liquid and begins to take effect within an hour or two. Drowsiness, a sense of detachment, and the feeling of being connected to all things throughout time may be felt for up to 15 hours. Sight and hearing may be enhanced for days afterward.

4 AFRICAN DREAM ROOT
is used by the Xhosa people of
South Africa to induce vivid and
supposedly prophetic lucid dreams.
The powdered root is mixed with
water and drunk in the morning,
allowing its effects to take hold at
bedtime. The Xhosa believe that
their deceased ancestors speak to
them during dreams, and regard
the dream root as a divination tool.

5 SALVIA DIVINORUM, a plant
native to Oaxaca, Mexico, has been
used in spiritual healing sessions
by Mazatec shamans for centuries.
According to Mazatec beliefs, salvia
is an incarnation of the Virgin Mary
who speaks to those who drink
the juice of her leaves. Due to the
mildness of salvia, which usually
produces only feelings of floating
and dizziness, light and noise are
said to chase the Virgin away.

6 IBOGA is the centerpiece of the
Bwiti religion, practiced in Gabon
and Cameroon. The root bark is
often ingested in large quantities
during initiation rituals in order to
bring about visions of the world
that lie beyond death. Iboga has
been found to reduce withdrawal
symptoms from other drugs, and
is used outside of Africa to treat
substance abuse.

SNAKE ISLAND

ILHA DE QUEIMADA GRANDE, SÃO PAULO

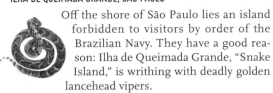

Off the shore of São Paulo lies an island forbidden to visitors by order of the Brazilian Navy. They have a good reason: Ilha de Queimada Grande, "Snake Island," is writhing with deadly golden lancehead vipers.

At its peak, the density of snakes on the island was believed to be about one per 11 square feet (1 m²), many of the reptiles being found overhead, hanging from the trees. In other words: death, every 3 feet, in all directions. Over the last decade, the golden lancehead population has declined—there are no mammals on Ilha de Queimada Grande, so the snakes must feed on visiting birds or resort to cannibalism. The serpents, found only on the 0.16-square-mile (.41 km²) island, are now classified as critically endangered. Estimates of the current count vary from 2,000 to 4,000—still one of the highest population densities of any snake.

Locals in the coastal towns near the Queimada Grande happily recount grisly tales about the island.

In one, a fisherman unwittingly wanders onto the island to pick bananas and is bitten only moments after stepping ashore. The fisherman manages to stumble back to his boat before dying in a pool of his own blood. Unique to Snake Island, the golden lancehead is believed to have the fastest-acting venom of any lancehead viper. The effects of its bite include bleeding orifices, brain hemorrhaging, and kidney failure.

To visit Ilha de Queimada Grande you must get official permission from the Brazilian Navy. Ⓢ 24.487922 Ⓦ 46.674155

ALSO IN SÃO PAULO

Carandiru Penitentiary Museum

Now closed, this prison was famous for its terrible conditions and a prison break in which over 100 inmates escaped through a tunnel.

Instituto Butantan

A biomedical research center world-renowned for its collection of poisonous snakes, including 407 varieties of cobra.

LENÇÓIS MARANHENSES NATIONAL PARK

BARREIRINHAS, MARANHÃO

Known as "the bedsheets of Maranhão," Lençóis Maranhenses is an area packed with sand dunes, 15 miles (24.1 km) inland from the Atlantic in northeast Brazil. During the rainy season, the valleys between the dunes fill with water, resulting in an odd sight: a desert full of blue and green lagoons. Look closely and you'll see fish between the dunes, hatched from eggs transported from the sea by birds.

Entry to the park is via the town of Barreirinhas, about 4 hours east of São Luís. Tour buses depart daily from the São Luís Bus Terminal. Ⓢ 2.485938 Ⓦ 43.128407

In a park without greenery, lagoons and sand dunes sit side by side.

Henry Ford's $20-million jungle utopia didn't quite go according to plan.

FORDLÂNDIA

SANTARÉM, PARÁ

Traveling through thick Brazilian jungle up the Tapajós River, one arrives at a shockingly out-of-place tableau. Amid the monkeys and macaws stand the overgrown ruins of an abandoned American suburb, complete with houses surrounded by white picket fences, fire hydrants, and a golf course. It's Pleasantville, dropped in the middle of the rain forest.

Industrialist Henry Ford created his slice of Americana in the Amazon in the late 1920s. Troubled by the high price of rubber, Ford decided to build his own rubber plantation.

He bought over six million acres of Brazilian land and shipped in employees from Michigan to manage the model town. He named his settlement Fordlândia, and the workers—both American and Brazilian—were forced to live according to Ford's strict, teetotaling rules. This meant no smoking, no drinking, and attending wholesome poetry readings and sing-alongs.

Workers quickly became disgruntled. Local Brazilians didn't appreciate having to wear nametags, eat hamburgers, and learn square dancing, while the Midwestern managers of the plantation had trouble adjusting to the jungle climate and ever-present malaria. Strikes, knife fights, and mayhem became the rule. In 1930, the Brazilian workforce had had enough and rioted, chasing the American managers out of Fordlândia with machetes.

Worst of all, the rubber saplings planted by Ford—without the help of a trained botanist—were barely growing. Those that had taken root were soon hit by a catastrophic leaf blight. Fordlândia was officially a failure.

Henry Ford retired from the rubber industry in 1945, having spent $20 million—equivalent to over $200 million in today's dollars—without producing a single piece of rubber worthy of his cars. **Fordlândia receives few visitors but can be reached by a 10-hour river voyage via charter boat from Santarem. ⑤ 3.830107 Ⓦ 55.497180**

ALSO IN BRAZIL

Gruta do Lago Azul

Bahia · A brilliant blue lagoon hides in a limestone cave rich in fossils.

Curitiba Botanical Gardens

Curitiba, Paraná · Designed to resemble French royal gardens, Curitiba's lawn has one of the world's most incredible greenhouses.

Teatro Amazonas

Manaus · Brazil's 19th-century rubber barons built a grand opera house in the middle of the rain forest. Then the money ran out. The Teatro lay dormant for 90 years until an injection of government funds in 2001 brought the music back.

Victoria Amazonica

Manaus · The leaves of this gigantic plant, found in the shallow waters of the Amazon, grow up to eight feet (2.5 m) in diameter. Many are strong enough to support the weight of a child.

World's Longest Street

Pará · Lined with houses for its entire length, this street stretches for 311 miles (500 km).

New Jerusalem Theater

Pernambuco · The world's largest open-air theater spans 24 acres and hosts a massive re-creation of the Passion of Christ.

Escadaria Selarón

Rio de Janeiro · Jorge Selarón created these vibrant ceramic steps.

Largo do Boticário

Rio de Janeiro · Remnants of Rio's colonial past lie in this square tucked behind Corcovado ("Hunchback") Mountain, in Cosme Velho.

GLOWING TERMITE MOUNDS OF EMAS NATIONAL PARK

ALTO PARAÍSO DE GOIÁS

Termites thrive all over the world, building sprawling nests and wreaking occasional havoc on local lumber supplies. In Brazil, they build tall towers of cement-like Earth.

These termite mounds can grow quite large, with diameters nearing 100 feet (30.5 m) and towering heights of 16 feet (5 m) or more. Not only do they provide a home for up to several million termites; they're also nesting sites for many birds and home to hundreds of glowing *Pyrophorus* beetle larvae. At night, the termite mounds look like they're wrapped in Christmas lights.

Beetle larvae give the mounds a bioluminescent quality.

While the adult beetles eat plants, the young are carnivorous, and their lights are a lure. Unsuspecting insects will make their way toward the pretty lights, only to be seized for a meal by the hungry larvae. And, yes, termites are a favorite food. In fact, the larval growth cycle is timed to take advantage of termite migrations, and many a termite will meet its end in the jaws of an inconsiderate houseguest.

The glowing mounds are best seen in the savannas during the summer, though they can be found in the jungles as well. The terrain can be quite rugged, but there are several private tour outfits who can guide you. ⓢ 14.005634 ⓦ 47.684606

CHILE

ROBINSON CRUSOE ISLAND

JUAN FERNÁNDEZ ISLANDS, VALPARAÍSO

In 1704, Scottish sailor Alexander Selkirk made a rash decision. Feuding with his cocaptain over their ship's seaworthiness as they sailed along the western coast of South America, Selkirk declared he would rather the vessel continue its journey without him. Taking him at his word, the crew dumped him on Más a Tierra, an island 419 miles (674 km) off the coast of Chile. (Selkirk's suspicions proved correct—the ship, *Cinque Ports*, sank shortly afterward, sending many of its sailors to the bottom of the sea.)

For the next four years and four months, Selkirk roamed the island alone, eating shellfish, chasing after goats with a knife, and digging up parsnips. To stave off boredom, he read his Bible and taught cats to dance. After hiding twice from Spanish privateers, he was finally rescued by English sea captain Woodes Rogers in 1709.

The most renowned castaway of all time, Selkirk is likely the inspiration for the title character in Daniel Defoe's novel *Robinson Crusoe*. In 1966, Más a Tierra was renamed Robinson Crusoe Island. Now home to a few hundred people, it is a place of extreme beauty, with coral reefs, white-sand beaches, blue lagoons, and abundant tropical fruits—not a bad place to be stranded.

Two-hour flights from Santiago depart several times per week. Hike to Selkirk's Lookout—a 3-hour trek that the marooned sailor made daily to watch for rescue ships. ⓢ 33.636666 ⓦ 78.849588

Only a third of Easter Island's *moai* were completed—hundreds of unfinished statues lie in the Rano Raraku quarry.

THE UNFINISHED GIANTS OF EASTER ISLAND

ISLA DE PASCUA, VALPARAÍSO

Between 1400 and 1600 CE, the Polynesian inhabitants of Easter Island carved 288 stone statues, or *moai*, and hauled them across the island, where they were installed on ceremonial pedestals. These representations were erected between the village and "chaos"—the ocean—as a wall of protection. Remarkably, the 288 figures represent less than a third of the statues that were created. The others lie either "in transit" at various places on the island or in the Rano Raraku quarry where they were carved. Most notable among the figures is "El Gigante,"

an unfinished 72-foot-tall (21.9 m) moai that surpasses the weight of two full 737 airplanes. It is questionable whether it could have even been moved using the wooden sleds, log rollers, and ropes that were presumably used to transport the island's other moai.

The mass of unfinished stone figures leaves many questions. What were the plans for El Gigante and the rest of the moai? Why carve so many and just leave them in a pile? It may have been a case of ambition eclipsing resources. Anthropologists have argued that the Easter Islanders used up all of their island's resources in the process of building their society. The two major tribes of Easter Island lived in a tropical rain forest, a

paradise of food and fishing, with plenty of time to put into the "great work" of statues.

According to Easter Island's resident archaeologist, Edmundo Edwards, the Polynesians used to sail back and forth across great distances among the Pacific islands, but eventually they used up all the large trees, thereby losing the ability to build large canoes. At this point, they became effectively trapped. The old middens (dumps for domestic waste) show that fish bones got progressively smaller, as the Polynesians could no longer sail out to deep fisheries.

Flights to Easter Island leave from Santiago. The trip takes about 5 hours. Ⓢ 27.121191 Ⓦ 109.366423

ALSO IN CHILE

El Tatio Geysers

Antofagasta · With over 80 active geysers, some of which you can bathe in, El Tatio is the third-largest geyser field in the world.

Sewell

Cachapoal · Founded in 1904 around a copper mine and once home to 15,000, this town was abandoned during the 1970s.

Villa Baviera

Linares · Formerly a cult known as Colonia Dignidad, Villa Baviera was a secretive German-Chilean community surrounded by barbed wire and led by ex-Nazi Paul Schäfer before his 2005 arrest.

Magic Mountain Hotel

Panguipulli · Located in a forest, this moss-covered nine-room hotel is shaped like a volcano and spews water from its roof. Entrance is via a cable bridge.

Villarrica Caves

Pucón · Travel hundreds of feet into this active volcano and see the hardened remains of a lava flow that once oozed out of the mountainside.

World's Largest Coca-Cola Logo

Arica · 60,000 empty bottles of Coke comprise a 30-year-old ad seen only from the air.

When "white gold" was no longer needed, Chile's mining towns fell apart.

NITRATE TOWNS

IQUIQUE, TARAPACÁ

Until 1909, Chile had something very rare and valuable: large deposits of sodium nitrate. Also known as "white gold" or "Chile saltpeter," sodium nitrate is used in the production of fertilizer and explosives. So valuable was this "white gold" that Chile went to war with Peru and Bolivia in 1879 over areas containing the chemical compound.

At the turn of the century, Chile's northern Tarapacá region was full of sodium nitrate mining towns. Workers from South America, Europe, and Asia formed communities around the mines. In the words of UNESCO, each town became a "distinct urban community with its own language, organization, customs, and creative expressions."

But something on the horizon would change all this. In 1909, German scientists Fritz Haber and Carl Bosch figured out how to chemically fix nitrogen—that is, how to make white gold on an industrial level. The discovery was disastrous for the Chilean saltpeter towns. By 1960, all lay abandoned. Today, their remains stand as rusting ruins in the inhospitable Atacama desert.

The Humberstone and Santa Laura Saltpeter Works are an hour east of Iquique on Route 16.
ⓈÂ 20.205805 Ⓦ 69.794050

MANO DEL DESIERTO

ANTOFAGASTA, ANTOFAGASTA

The barren monotony of the Atacama desert is shattered by what looks like a buried giant reaching out for help. Mano del Desierto, a 36-foot-tall (10.9 m) hand protruding out of the sand, is the work of the Chilean sculptor Mario Irarrázabal, and was built in the early 1980s. Irarrázabal's work is known for its portrayal of human vulnerability and helplessness—two concepts that certainly come across when viewing the unnerving, half-submerged palm.

The hand is about an hour drive south of the town of Antofagasta, along the Pan-American Highway.
Ⓢ 24.158514 Ⓦ 70.156414

The Hand of the Desert has acquired graffiti on its palm.

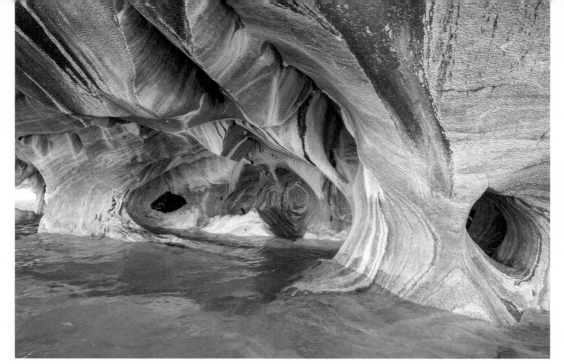

Swirling arches of rock carved out over the millennia.

THE MARBLE CATHEDRAL AT LAKE GENERAL CARRERA

PUERTO RÍO TRANQUILO, GENERAL CARRERA

Within the banks of the deepest lake in South America is a marble "cathedral" formed by natural erosions. When sun shines onto the caves, the pale blue water reflects against the gray-and-white-striped marble, turning the whole scene aquamarine. Eaves and arches in the rock, created by water lapping at the marble, complete the cathedral effect.

Boat tours leave from the small town of Puerto Río Tranquilo, on the western shore of the lake. The nearest city is Coihaique, 5 hours north. ⑤ 46.475690 Ⓦ 71.291650

....

COLOMBIA

BURIED CITY OF ARMERO

ARMERO, TOLIMA

Visit the eerily quiet former town of Armero and you'll see only a few buildings surrounded by weeds. Then you will realize that you are looking at the upper levels of those buildings. The ground floors are buried by mud that smothered the town more than 30 years ago.

Armero was home to almost 30,000 people on November 13, 1985. That was the day the nearby Nevado del Ruiz volcano erupted, sending torrents of mud and debris down its slopes at 40 miles per hour (64.3 km). Soon, a 15-foot (4.5 m) layer of the sludge had covered the town, trapping and killing 23,000 of its inhabitants.

Those who escaped instant death faced an agonizing 12-hour wait for relief workers to arrive. The mud, which pulled at their feet like quicksand, made rescue efforts frustratingly slow. People who had been buried up to their necks watched, helpless, as workers tried and failed to dig them out.

One such victim was 13-year-old Omayra Sánchez. Imprisoned in the mixture of water, mud, and the concrete remains of her own home, she smiled, sang, talked to rescue workers, and was interviewed. Volunteers made multiple attempts to dig Omayra out, but her legs were pinned by concrete. Sixty hours after becoming trapped, Omayra died. A photograph of her staring, helpless, into the camera while immersed in neck-high sludge was published around the world and became the symbol of the disaster.

Covered in mud and with more than two-thirds of its inhabitants wiped out, Armero became a ghost town. Survivors created a kind of cemetery, constructing tombs in place of the old houses and honoring their former residents with epitaphs. The Armando Armero foundation has established a Memory Interpretation Center on the site, where visitors can learn about the buried city and the volcano that destroyed it.

The ruins of Armero are a 5-hour drive west of Bogotá, Colombia's capital. Ⓝ 4.966666 Ⓦ 74.827318

CAÑO CRISTALES

LA MACARENA, META

From September to November, during the period between the wet and dry seasons, the Caño Cristales river, in Colombia's remote Meta province, becomes a liquid rainbow. The riverbed is carpeted with *Macarenia clavigera*, a species of river weed found nowhere else on earth, and the source of the Caño Cristales's notoriety. Pale green under the shade of riverbank foliage, the *Macarenia* turns a stunningly intense magenta under the full sun. Jet-black rocks, white water coursing over cascades, and the occasional crater of yellow sand complete the colorful tableau.

Until 2009, the river was off-limits to visitors due to the presence of FARC guerrillas in the region. Now the tourists are arriving again, brought to Caño Cristales by local guides who stick to authorized paths.

Vibrant river weeds give Colombia's "River of Five Colors" its rainbow effect.

Charter a flight from Villavicencio to the village of La Macarena, once a guerrilla stronghold. From town, it's a 15-minute ride in a motorized canoe up the Guayabero River, followed by a long hike down a FARC-built dirt road to the river. **Ⓝ 2.182991 Ⓦ 73.785850**

PABLO ESCOBAR'S HIPPOS

PUERTO TRIUNFO, ANTIOQUIA

Notorious Colombian cocaine baron Pablo Escobar once lived on this sprawling estate, spending his days riding his hovercraft over its many lakes, wandering among his collection of vintage cars, and strolling through his zoo filled with hippopotamuses and exotic birds.

Escobar was killed in a hail of gunfire by the Colombian police in 1993, and the hacienda became dilapidated. It has since been revived—this time as a theme park. There have been some problems, however. Locals are known to sneak in—on one occasion with a backhoe—and dig up the lawn and floors in search of treasure they believe Escobar buried.

Meanwhile, the four hippos Escobar imported from Africa have thrived and, in one case, escaped the confines of the estate. In 2009, a hippo named Pepe was found 62 miles (100 km) away and killed by the Colombian military. The group of wild hippos is now estimated at around 40. They are ruled by an alpha male named Pablo.

The hacienda is about 4 hours east of Medellin on Route 60. Visitors can feed baby hippos, tour the drug

The drug baron is long gone, but his hippos live on.

lord's personal effects, and even see the Cessna that carried Escobar's first load of cocaine to the US. **Ⓝ 5.886187 Ⓦ 74.642486**

GUAYABETAL ZIP LINES

GUAYABETAL, CUNDINAMARCA

If you want to cross the Rio Negro valley near Guayabetal, you have two options: Hike a steep forest path for four hours, or hitch yourself to a steel cable suspended 1,300 feet (396 m) above ground and hold on tight. Most locals opt for the latter.

The tools for crossing using the zip lines are simple: a length of rope used to form a seat, a steel roller with a hook to attach the rope, and a wooden yoke that straddles the cable and serves as a brake. The half-mile (0.8 km) trip takes less than 30 seconds.

The journey is not without its dangers. According to some reports, 22 people have plunged to their deaths. The most recent fatality occurred in 2004, when a 34-year-old man was decapitated while trying to cross with two other passengers lashed to him (they both survived). He'd already used the cable earlier that day to pass two beds, a doghouse, chickens, a television, a stereo, and three chairs across the valley without a hitch.

In the late 1990s, a Bogotá television news program aired an exposé on the cables, which included a clip of a six-year-old boy who lived on the far side of the valley zipping across the line to get to school. A public outcry ensued, provoking calls from Bogotá for the cables to be torn down. The lines were technically illegal, but because the Rio Negro Valley

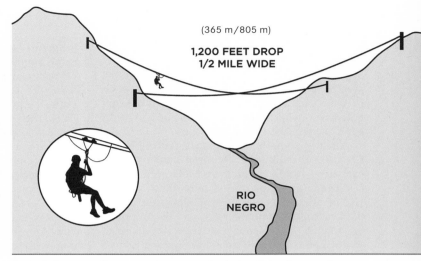

(365 m/805 m)

**1,200 FEET DROP
1/2 MILE WIDE**

**RIO
NEGRO**

Proto zip lines once flew commuters across the Andes. Only a few remain.

spans two Colombian states—Meta and Cundinamarca—it had never been clear who held jurisdiction over them. In 2001, the government finally ordered 18 of the cables removed.

Locals saw the move not only as a threat to their livelihoods but as disrespectful of an old tradition that had served the community well for generations. The mayor of Guayabetal reportedly received death threats saying that if he complied with orders to remove the cables, he'd be "carried out in wooden pajamas"—a euphemism for a coffin. In an op-ed published in *El Tiempo*, a man who'd used one of the cables for 20 years without incident wrote, "If the authorities are so concerned with safety, they should take away airplanes, since not long ago one

crashed in the United States, killing all 260 passengers." To protest the impending removal, a local woman slid out to the middle of one of the lines and dangled there for hours.

In the end, all but four of the cables were removed. A school was built on the far side of the valley so that children no longer have to zip back and forth each day. Instead, a teacher from Guayabetal now crosses over at the beginning of the week and comes back to town for the weekend.

The zip lines are 2 hours south of Bogotá. If you can find the relatively well-hidden cables, you may be able to convince one of the locals to take you across, but you will be risking both their life and yours. Ⓝ 4.220892 Ⓦ 73.816551

ALSO IN COLOMBIA

Museo el Fósil	Malpelo Island	Peñol Stone	Las Lajas Sanctuary	Salt Cathedral
Boyacá · A bus-size crocodile-like fossil is the centerpiece of this roadside museum.	*Choco* · Dive for hammerheads and silky sharks in this biodiversity hotspot.	*Guatapé* · From afar, this 10-million-ton stone looks like two halves of a giant rock messily stitched together. Get closer and you will realize that the "stitching" is actually a spiral staircase.	*Nariño* · Built within a gorge, this Gothic Revival church is located at the site where the Virgin Mary supposedly appeared during the 18th century.	*Zipaquirá* · Explore a subterranean church in an old mine featuring floors that crunch underfoot, a large cross carved into the back wall, and blue and purple lighting.

PERU

SARCOFAGI OF CARAJÍA

CHACHAPOYAS, AMAZONAS

Staring down from a cliffside above a river gorge, the vertical Sarcofagi of Carajía (or Karijia) kept watch over the Utcubamba Valley for hundreds of years before researchers were able to climb up and investigate the mysterious mummies.

Created sometime in the 15th century by the Chachapoya civilization, the seven standing burial capsules—formerly eight; one collapsed during a 1928 earthquake—are located almost 700 feet (213 m) above the valley floor. While a great deal of the Chachapoya culture was lost after the Inca came to conquer, the sarcophagi survived largely intact due to their seemingly impossible location. Each of the figures stands a remarkable 8 feet tall (2.4 m) and change, constructed out of grass and clay. Some of the graves even retain the human skulls that were installed atop the sarcophagi.

It was not until the mid-19th century that researchers were able to scale the cliff face and examine the mummies, dating them and speculating as to their construction. It is believed that the original architects of these graves worked from natural outcroppings that were later destroyed, either deliberately or naturally. While the sarcophagi are largely protected from the elements by the rock walls around them, birds and other small animals have done some damage. The researchers removed the contents of the sarcophagi to preserve the ancient innards from any further predation.

Hike or rent a horse from Cruz Pata, where you can also rent boots for navigating muddy trails.
Ⓢ 6.163243 Ⓦ 78.019354

Standing burial capsules topped with human skulls overlook the Utcubamba Valley.

THE BOILING RIVER OF THE AMAZON

PUCALLPA, UCAYALI

The 200°F river is believed to be sacred and have healing powers.

Hidden in the dense jungle of the Peruvian Amazon is a percolating, roiling river. The steaming turquoise waters, which can reach up to 200°F (93°C), are guided by ivory-colored stones and guarded by 60-foot (18 m) walls of lush forest and vegetation.

The headwaters of the Boiling River are marked with a boulder in the shape of a snake's head. According to legend, a giant serpent spirit called Yacumama, or "Mother of the Waters," gives birth to hot and cold waters and heats the river. The geothermal feature is unusual since it isn't located near a high-energy heat source such as an active volcano—usually a prerequisite for a boiling river.

Researchers led by geothermal scientist Andrés Ruzo believe that a fault-led hydrothermal feature causes the river to reach such temperatures. The water seeps deep into the earth, heats up underground, and resurfaces through faults and cracks.

It may look like a natural hot tub, but don't jump in—you'll get third-degree burns within seconds. **Ⓢ 8.820001 Ⓦ 74.730011**

CHAN CHAN

TRUJILLO, LA LIBERTAD

Chan Chan is a giant, elaborate sand castle. Built by the Chimú people around 850 CE, the adobe city—the world's largest—is now an archaeological site. The 28-square-mile (72.5 km²) city, comprised of temples, houses, kitchens, gardens, and cemeteries, had a population of around 30,000 before falling to the Incan Empire in 1470.

The Chimú addressed the desert city's lack of water by building a complex irrigation system that included a 50-mile (80.4 km) canal to the Moche and Chicama rivers. Ironically, Chan Chan now faces the opposite problem: too much water. The city is slowly dissolving, as El Niño storms become fiercer and more frequent. Wind and torrential rains have substantially weakened the city, washing away the intricate animal friezes that decorate its walls. Peru's National Institute of Culture is supporting efforts to protect and preserve Chan Chan, but the city's substantial size makes it difficult to shelter from the elements. Time is running out, and the rains keep coming.

Off Avenida Mansiche, Trujillo. Only about 10 percent of the city is open to explore—a tour guide will be able to explain what lies beyond the publicly accessible portion. **Ⓢ 8.105999 Ⓦ 79.074537**

Conservationists struggle to maintain the art and architecture at Chan Chan, once the largest adobe city on Earth.

ICA STONES

ICA, ICA

For his 42nd birthday in 1966, Peruvian physician Javier Cabrera received an unusual present: a stone carved with what looked like an extinct fish. Over the next few years, Cabrera sought out more of the stones from a local farmer, who claimed to have found them in a cave. Eventually, the physician amassed a collection of more than 10,000. In addition to featuring animals, the carvings depicted strange scenes of ancient people battling Tyrannosaurs, conducting kidney transplants, and gazing through telescopes. Were these artifacts proof of a sophisticated ancient culture, or even evidence that humans and dinosaurs coexisted? Not quite.

Basilio Uschuya, the farmer who sold Cabrera the stones, confessed in 1973 that he had forged them. Then things got complicated: Uschuya retracted his admission soon afterward, claiming he lied to avoid being arrested for selling archaeological artifacts.

What ancient civilization—or modern charlatan—carved this stone?

Over 20 years later, when Cabrera opened a museum showcasing the stones, Uschuya changed his story yet again. This time he said he had carved most, but not all, of them. Cabrera was unfazed by the revelation, believing in the authenticity of the stones until his death in 2001.

Though the collection of carvings can't be dated due to its lack of organic material, archaeologists regard the stones as a hoax. This has not dissuaded many creationists and extraterrestrial enthusiasts from hailing the stones as the creations of "Gliptolithic Man"—an ancient, highly intelligent people who shared Earth with the dinosaurs before departing from the planet in spacecraft they built themselves.

Calle Bolivar 174, Ica. Ica is a 4-hour drive south of Lima, the capital. 11,000 of the stones are currently on display at Ica's Cabrera Museum, which is open to visitors by appointment. ⓢ 13.450437 ⓦ 76.150840

FLOATING ISLANDS OF THE UROS

PUNO, PUNO

Though the origins of the Uros are shrouded in anthropological mystery, their basic story goes something like this: At some point in the distant, pre-Columbian past, a tribe of comparatively dark-skinned people migrated out of the Amazon and found themselves on the shores of Lake Titicaca. Oppressed by the local population and unable to find land of their own to tend, they ended up moving out into the middle of the lake on small floating islands they constructed from layers of cut totora, a thick reed that grows like kudzu in the lake.

In the middle of frigid Titicaca, the Uros found relative peace and scraped by for centuries as bird hunters and fishermen while living one of the most unique lifestyles on the planet. Today, some 1,200 Uros still live on an archipelago of around 60 artificial islands, strung out like a necklace near the city of Puno.

Stepping foot onto a floating island is a strange feeling, like walking on a giant sponge that squishes underfoot. Though the mats of reeds are up to 12 feet (3.6 m) thick, there is a persistent feeling that one could step right through to the cold lake below.

The Uros Islands are a half-hour boat trip from Puno. Disarmingly cute local children may sing for money during the ride. ⓢ 15.818667 ⓦ 69.968991

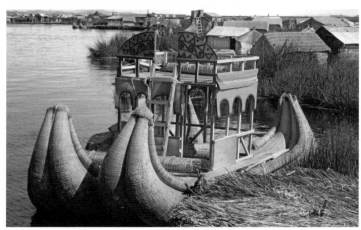

In the middle of Lake Titicaca, a civilization that predates the Incas survives on floating islands of reeds.

Until 2005, the world's third-tallest waterfall was known only to those who lived beneath it.

GOCTA FALLS

COCACHIMBA, AMAZONAS

In May 2005, a German economist named Stefan Ziemendorff went for a hike in the remote Utcabamba valley of Peru. In the distance, he spied what looked to be an impossibly tall two-tiered waterfall that hadn't appeared before on any map. The following March, after returning with proper surveying equipment and measuring the falls at 2,531 feet (771 km), Ziemendorff held a press conference to announce that he had discovered the third tallest waterfall in the world. The rank of third-tallest has been hotly contested since then, but that's not the only debate.

Like many of geography's most heralded "discoveries," Ziemendorff's wasn't news to everyone. While the waterfall may have been a total secret to the outside world, there were 200 residents in an isolated village called Cocachimba who not only knew all about it, but lived almost directly underneath it. For 53 years, since the hamlet's founding, Cocachimbans had awakened each morning to one of the world's most picturesque views—and apparently never mentioned it to anyone. As it turned out, the locals had good reason not to bring it up—they were afraid of it.

A legend about the waterfall had been passed down for generations. According to locals, once upon a time, a man named Gregorio told his wife that he was taking off for a short trip. Not realizing that his suspicious wife had decided to trail him through the forest, he made his way toward the base of Gocta. There, the wife caught Gregorio cavorting with a beautiful blond mermaid at the foot of the falls and flew into a jealous rage. The frightened siren grabbed Gregorio and pulled him into the waterfall with her. He never reemerged, and locals came to believe that anyone brave enough to hike to the falls was chancing a run-in with dangerous, supernatural forces.

According to a town official, it took the safe return of dozens of tourists before the residents of Cocachimba shook their phobia of Gocta. The town has since come around to the benefits and beauty of the natural wonder that looms overhead.

Again, for waterfall enthusiasts, the question of whether Gocta even deserves its stature as the bronze medalist of waterfalls has been a subject of rancorous debate. It all comes down to your definition of what a waterfall is. If there's a break in the drop, as there is in Gocta, does that count as more than one waterfall? What if the water cascades over the side of an inclined cliff, rather than spilling off vertically, as is the case with several towering falls in Norway? And what if the water slows to only a trickle during the dry season? No matter which criteria you use, everyone agrees that Angel Falls in Venezuela tops the charts, particularly in terms of awe-inspiring spectacle.

Cocachimba is 5 hours east of Trujillo. Resident guides are happy to take you on a hike to the falls.
Ⓢ 6.028728 Ⓦ 77.888125

ALSO IN PERU

Toro Muerto

Arequipa · Multiple ancient cultures created this petroglyph field.

Sacred City of Caral-Supe

Barranca · "The oldest town in the new world" dates back to 3,000 BCE, the same era as the first dynasty of ancient Egypt.

Kuelap Fortress

Chachapoyas · The "Machu Picchu of the North," constructed around 500 CE by the Chachapoyas people, now consists of 400 ruined buildings enclosed by a huge, high-walled fortress.

Manú National Park

Cusco · One of the most biodiverse places in the world, this park is home to over 1,000 species of birds.

Qenqo Temple

Cusco · Death rituals and sacrifices were once commonplace inside this Peruvian megalith.

Band of Holes

Ica · This barren rock near Pisco Valley is dotted with neat rows of mysterious holes that stretch over mountain terrain.

Moche Pyramids

Lambayeque · At these pyramids, warrior-priests conducted rituals involving costumed battles and human sacrifice.

Huayllay National Sanctuary

Pasco · With its unusual formations in open fields, this rock forest is a dream destination for climbers.

OLLANTAYTAMBO RUINS

OLLANTAYTAMBO

Dating back to the reign of 15th-century Emperor Pachacuti, who conquered the region, the town of Ollantaytambo contains some of Peru's best-preserved Inca ruins.

Ollantaytambo was home to the Inca elite. The town's primary attraction is the fortress on its outskirts, in a section known as the Temple Hill. Though originally built for worship, the fortress served as the last Inca stronghold against the Spanish conquistadors and is a site of one of the few battles in which the Inca successfully repelled Spanish forces. Other nearby attractions include the Temple of the Sun and the Princess Baths, both of which feature examples of Incan carvings.

Buses and combis run from Urubamba, about 30 minutes away. Ⓢ 13.258048 Ⓦ 72.263311 ➙

The Inca town's distinctive layout includes quarries, terraces, and storehouses.

➥ Other Inca Ruins

Ingapirca

INGAPIRCA, GUAPÁN, ECUADOR

Ingapirca is Ecuador's largest site of Inca ruins. The city was home to the Inca and the indigenous Cañari, and after initial conflict, they merged to form a hybrid community.

The largest structure still standing on the site is the Temple of the Sun. It appears to have been built at such an angle that the sun would have shone directly into the doorway during the solstices. Advanced design is also on display in the startling underground aqueduct system that supplied the community with water.

A bus from Cuenca takes 2 hours to reach the ruins. While there, say hi to the llamas, who roam about freely.

CRADLE OF GOLD, CHOQUEQUIRAO, PERU

Known as the other Machu Picchu, Choquequirao ("Cradle of Gold") is filled with ruined buildings and terraces that sit below a flattened hilltop ringed with stones.

Only about one third of Choquequirao, a once vital link between the Amazon and Cusco, has been excavated. What has been uncovered to date follows traditional Inca construction: A temple and some administrative buildings are positioned directly around a central square, with living quarters farther out.

One of the most impressive features found in and around Choquequirao are two terraces that incorporate figures of llamas or alpacas. The shapes of the animals have been set into the large terraces using carefully carved white rock.

Choquequirao is a 2-day hike from Cusco. Experienced climbers only—this one's rough and tough.

HUÁNUCO PAMPA, PROVINCIA DE DOS DE MAYO, PERU

Though it was abandoned around 1539, Huánuco Pampa is remarkably well preserved. Apart from the strategic mountainous location, this site was probably quite unremarkable in its time. Its lack of grandeur may have been its saving grace. The conquistadores by and large neglected Huánuco Pampa, which helped it to survive and become the prime archaeological site it is today.

The site was an administrative center, built on a plateau conveniently defended by the steep slopes marking its perimeter. Huánuco Pampa was also a foodstuff storage hub. Its nearly 500 storehouses and numerous food processing centers show the massive planning and organizational skills the Inca needed to sustain their sprawling empire.

Huánuco Pampa

On a stroll through Huánuco Pampa today, you'll come across the remains of the baths, the main palace, and an enormous plaza with a pyramid.

Huánuco Pampa is a 5-hour drive from the city of Huánuco.

Choquequirao

The Keshwa Chaca has been rewoven from fresh grass every year since Incan times.

THE LAST INCAN GRASS BRIDGE

HUINCHIRI, CUSCO

The Incas never invented the wheel, never figured out the arch, and never discovered iron, but they were masters of fiber. They built ships out of fiber (you can still find reed boats sailing on Lake Titicaca). They made armor out of fiber (pound for pound, it was stronger than the armor worn by the conquistadors). And their greatest weapon, the sling, was woven from fibers and powerful enough to split a steel sword. They even communicated in fiber, developing a language of knotted strings known as *quipu*, which has yet to be decoded. So when it came to solving a problem like how to get people, animals, and goods across the steep gorges of the Andes, it was only natural that they would turn to fiber.

Five centuries ago, the Andes were strung with as many as 200 suspension bridges braided from nothing more than twisted mountain grass and other vegetation, with cables sometimes as thick as a human torso. Three hundred years before Europe saw its first suspension bridge, the Incas were spanning longer distances and deeper gorges than anything that the best European engineers, working with stone, were capable of.

Over the centuries, the empire's grass bridges gradually gave way and were replaced with more conventional works of modern engineering. The most famous Incan bridge—the 148-foot (45 m) bridge immortalized by Thornton Wilder in *The Bridge of San Luis Rey*—lasted until the 19th century, but it too eventually collapsed. Today, there is just one Incan grass bridge left, the Keshwa Chaca, a sagging 90-foot (27.4 m) span that stretches between two sides of a steep gorge, near Huinchiri. According to locals, it has been there for at least 500 years.

Despite its seemingly fragile materials, modern load testing has found that in peak condition, the Keshwa Chaca can support the weight of 56 people spread out evenly across its length.

In 1968, the government built a steel truss bridge just a few hundred yards upstream. Though most locals now use it instead of the grass bridge to cross the valley, the yearly tradition of rebuilding the Keshwa Chaca has not abated. Each June, it is renewed in an elaborate three-day ceremony. Each household from the four surrounding towns is responsible for bringing 90 feet (27.4 m) of braided grass cord. Construction takes place under the supervision of the all-important bridge keeper, or *chacacamayoc*. The old bridge is then cut down and thrown into the river. Because it has to be willfully, ritually regenerated each year, the Keshwa Chaca's ownership passes from generation to generation as a bridge across not only space, but also time.

The bridge crosses Apurimac Canyon, 5 hours south of Cusco. Ⓢ 14.383056 Ⓦ 71.493333

URUGUAY
LAGUNA GARZÓN BRIDGE

MALDONADO, MALDONADO

When Uruguay's Rocha and Maldonado counties sought to replace the tried-and-true way of crossing Laguna Garzón—via single-vehicle float rafts—they ended up with one of the most immediately recognizable bridges in the world today.

After six years of public hearings and negotiations with the local communities, Rafael Viñoly Architects found a way to meet the strict needs of the natural and cultural communities connected by the new bridge. The project's success hinged on its ability to incorporate the environmental laws of the area, which stipulate that just 35 percent of the stretch between the lagoons may be developed, while 50 percent must be devoted to green areas.

These prolonged negotiations resulted in the circular Laguna Garzón Bridge, which opened in 2015. Its design was partly motivated by safety concerns—the road's one-way half circles force motorists to slow down, thereby breaking up a mile-long stretch that otherwise might have been a tantalizing speedway.

The bridge is just under an hour's drive east of Maldonado. There is an observation deck and a fishing pier, so bring your rod if you fancy some roadside angling.
Ⓢ 34.802470 Ⓦ 54.572100

ALSO IN URUGUAY

Vizcaíno Creek Fossil Bed

Canelones · Thousands of fossils have been discovered at this site, including remains of the glyptodont, an extinct armadillo the size of a Volkswagen Beetle.

Valle del Hilo de la Vida

Lavalleja · Ninety cone-shaped mounds built from rock and believed to be over 1,000 years old dot this hill.

The Hand

Punta del Este · A giant hand emerging from the sands of Brava Beach is a monument to the drowned.

More than 1,000 vehicles cross this circular bridge daily.

Dense jungle gives way to cavernous sinkholes.

VENEZUELA

SARISARIÑAMA

SUCRE

One of the most remote locations in all of Venezuela, Sarisariñama is also one of the most bizarre landscapes in the world. High in the clouds on a *tepui*, or "tabletop mountain," are four giant sinkholes, each a near-perfect circle.

The largest of the sinkholes is 1,150 feet wide (350 m) and 1,000 feet deep (305 m). Adding to the mystery of the place is the local legend that gave the tepui its name. According to the Ye'kuana people, a flesh-eating evil spirit lived on the mountain, and it made the sound "sari sari" when consuming a meal of human meat.

Although you may not encounter actual person-eating spirits on the mountain, the desolation still adds a level of eeriness to the surroundings. Visitors to the area are often shocked at how the dense jungle pushes all the way to the sinkhole edges, making their presence more dramatic and even more unnerving. **The sinkholes are remote. Really remote. No roads go near Sarisariñama, and exploration of the tepui is best left to scientists. N 5.008121 W 64.147789**

....

DROWNED CHURCH OF POTOSÍ

POTOSÍ, TÁCHIRA

In 1985, the residents of sleepy, bucolic Potosí received some unwelcome news: The government intended to flood their entire town in order to create the Uribante Caparo hydroelectric dam. The evacuation order came swiftly, and residents were relocated to nearby towns. The dam was completed and the town was submerged, but not all of Potosí disappeared quite so willingly. At low tide, the cross at the top of the 82-foot-tall (25 m) church spire would emerge from the water, a reminder of Potosí's former existence.

In early 2010, the El Niño weather phenomenon caused droughts across Venezuela and the water behind the Potosí dam gradually dried up, revealing the church in its entirety for the first time since flooding.

The site began to attract visitors and, with the area's water stores reaching critically low levels, locals gathered at the church to hold a mass and pray for rain. Their wish was granted—the skies opened and the Potosí church disappeared beneath the water once again. **Potosí is a 7-hour drive east of Caracas, the capital. The church steeple is now visible to varying degrees, according to rainfall and dam levels—check ahead. N 7.948304 W 71.653638**

THE EVERLASTING LIGHTNING STORM

CONGO MIRADOR, ZULIA

There's something strange in the air where the Catatumbo River flows into Lake Maracaibo. For 260 nights out of the year, often for up to ten hours at a time, the sky above the river is pierced by almost constant lightning, producing as many as 280 strikes per hour. Known as the *relampago del Catatumbo* ("the Catatumbo lightning"), this everlasting lightning storm has been raging for as long as people can remember.

In 1595, Sir Francis Drake's attempt to take the city of Maracaibo by night was foiled when the lightning storm's flashes gave away his position to the city's defenders. This happened again during the Venezuelan War of Independence in 1823, when Spanish ships were revealed by the lightning and fell to Simón Bolívar's upstart navy.

In fact, the lightning, visible from 25 miles (40.2 km) away, is so regular that it's been used as a navigation aid by ships and is known among sailors as the Maracaibo Beacon. Interestingly, little to no sound accompanies this fantastic light show, as the lightning moves from cloud to cloud, far above the ground.

It's still unknown exactly why this area—and this area alone—should produce such regular lightning. One theory holds that ionized methane gas rising from the Catatumbo bogs meets cold air pouring down from the Andes, helping to create the perfect conditions for a lightning storm. **The best place to see the storm is from Congo Mirador, a village built on stilts on Lake Maracaibo. Head to Encontrados to make arrangements. Ⓝ 9.563214 Ⓦ 71.382437**

There is a dazzling lightning show over Lake Maracaibo almost every night of the year.

ALSO IN VENEZUELA

Colonia Tovar

Aragua · For over a century, residents of this German village in the cloud forest outside of Caracas spoke German and married only within their village.

Cerro Sarisarinama Tepui

Bolívar · Unique forests are found at the bottom of massive sinkholes on this tabletop mountain in one of the most remote places in the country.

Pedernales

Delta Amacuro · Explore mud volcanoes that are constantly bubbling up wet earth.

Médanos de Coro National Park

Falcón · A desert of massive sand dunes constantly shifts in the winds.

MEXICO

CENOTE ANGELITA

TULUM, QUINTANA ROO

It seems like a riddle: How can there be an underwater river?

In the thick jungle just outside the ruins of the Mayan city of Tulum, a series of sinkholes and caves lead to an amazing submerged world. One such opening is Cenote Angelita. The 200-foot-deep (60 m) pool cavern was created by the crumbling of porous limestone as water crept in and hollowed out the space. The gaping cavern was once even worshipped as a holy site by ancient Mayan cultures, hence its name; *cenote* is a derivation of a Mayan word meaning "sacred well," and *angelita* means "little angel."

The flooded Angelita cave has a unique quality: It seems that a separate river runs near the bottom of the water-filled pit. This illusion is a product of the water's chemistry. Different portions of the water in the caverns have different levels of salinity, causing the denser water to sink to the bottom, where it looks like a misty underwater river all its own.

The cenote is just southwest of the town of Tulum. Guides and transport may be arranged through tour agencies in Tulum. Ⓝ 20.137519 Ⓦ 87.577777

A flooded cave appears to hide an underwater river.

POMUCH CEMETERY

POMUCH, CAMPECHE

When a Pomuch resident dies, he or she is temporarily buried at the town graveyard. After three years, family members come to disinter the bones, clean them, and place them in a wooden box for permanent display. Each year after that, the families return on the Day of the Dead to participate in the ritual bone-cleaning.

In addition to reuniting families and allowing them to confront the pain of death, the tradition is tied to a belief that deceased relatives will become angry and wander the streets if their bones are not cared for properly.

Pomuch is a small town east of Campeche City. ℕ 20.137530 𝕎 90.174339

LAS POZAS

XILITLA, SAN LUIS POTOSÍ

Las Pozas is the creation of Edward James, an eccentric English poet, artist, and patron of the Surrealist movement. James sponsored Salvador Dalí, allowed René Magritte to use his London home as a studio, and was acquainted with such luminaries as Dylan Thomas, Sigmund Freud, D. H. Lawrence, and Aldous Huxley. Huxley introduced James to Hollywood types, who in turn introduced him to spiritualist Western visionaries, who then introduced him to the wilds of Mexico.

Won over by the country's lush vegetation and leisurely pace, James purchased a coffee plantation in 1947 and spent the next ten years cultivating orchids and tending to exotic animals. After an unprecedented frost in 1962 destroyed many of his plants, James started building the extraordinary sculpture garden that remains on the site today. The design of Las Pozas was inspired by both James's orchids and the vegetation of the jungle of La Huasteca, combined with architectural elements taken from the Surrealist movement.

Construction on Las Pozas began in 1962, and carried on for over 20 years. The gardens feature concrete structures with Surrealist names, like the "House on Three Floors Which Will in Fact Have Five or Four or Six," the "Temple of the Ducks," and the "House With a Roof Like a Whale." Stairs spiral up into the air, mismatched columns support uneven floors, and decorative arches range from ornately finished to seemingly incomplete.

In the 1960s and 1970s, James dedicated more and more of his resources to his "Surrealist Xanadu," as he referred to it, spending millions of dollars and employing hundreds of masons, artisans, and local

A Surrealist Mexican sculpture garden built by an eccentric English poet.

craftsmen. By the times James died in 1984, he had built 36 sculptures spread out over more than 20 acres of tropical jungle. Over the years, trunks and vines have snaked their way among the structures, adding to the surrealism of the scene.

20 de Noviembre, Xilitla. The closest international airport is in Tampico, a 3-hour drive from Xilitla. ℕ 21.396710 𝕎 98.996714

Minichelista

Hogar y Seguridad • In this café created by artists, you might find yourself sipping spicy coffee in a closet, eating off a typewriter, sharing fries on a glass-topped motorcycle, or hitting up a hookah alongside a robot.

Soumaya Museum

Granada • With a whopping 66,000 pieces of pre-Columbian, Mexican, and European art, this anvil-shaped museum was paid for, built by, and filled with the collection of Mexican magnate Carlos Slim Helú, one of the world's richest men.

Alameda Arts Lab

Centro • Art, science, and technology collide inside a bright-yellow 16th-century chapel, with futuristic electronic exhibits and artistic experiments.

Man, Controller of the Universe

Centro • Diego Rivera's original version of this controversial mural and its social-realist clash of capitalism and communism was done thousands of miles away in New York City's Rockefeller Center, but it so angered the Rockefeller family that it was completely painted over—only to be re-created here.

Palacio Postal

Centro • The city's main post office is everything a mail palace should be, a century-old gilded heaven for philatelists and architecture freaks alike.

Meteorites at the Palace of Mining

Centro • There are four meteorites on display at this early 19th-century school for mining engineers (there were five, but one went to live at the Institute of Astronomy), including the giant *El Morito*—at 14 tons, it's one of the largest chunks of space debris ever discovered on Earth, and the first one recorded in the Western Hemisphere.

Museum of Mexican Medicine

Centro • Not for the squeamish, this scientific repository houses wax figures sporting goiters, boils, and other ailments that are lit as if sculpted by Rodin, befitting the graceful 18th-century century palace they call home.

Caricature Museum

Centro • Housed in a Baroque building, this museum is dedicated to the promotion and preservation of Mexican cartooning, stretching back to the politics of the late 19th century and still happily thumbing noses and tweaking egos right up to today.

Room 2 at Templo Mayor

Centro • Most of the great temple of the Aztec city of Tenochtitlan was destroyed by the Spanish to make way for a cathedral, but along with the ruins many sacrificial artifacts have been found and are held in the museum's Room 2, including decorated skulls, pots of cremated bones, and altar vessels made to hold human hearts.

National Sanctuary of the Angel of the Holy Death

Centro • Three days a week, despite condemnation from the Vatican, believers gather in a modest storefront church to attend an untraditional mass, worshiping Nuestra Senora de la Santa Muerte—Our Lady of the Holy Death—Mexico's most cherished folk saint next to Santa Maria de Guadalupe.

Deportes Martinez

Doctores • In Mexico, freestyle wrestling, aka *lucha libre*, goes back more than 150 years, and the *luchadores'* go-to shop for spandex and iconic *máscaras* is Don Martinez's sports shop, just around the corner from Arena Mexico, the Friday-night fight venue.

MODO

Roma Nte • The space in this Art Nouveau mansion is tight so the 30,000 items in its collection are rotated in and out for an ever-changing display—some rare, some everyday, some edible, some wearable, all a feast for the eyes.

Old Toy Museum of Mexico

Doctores • This cacophony of a collection ranges from the 19th century through the 1980s, a vast jumble of miniatures and masks, vintage tin toys and action figures, plastic novelties and one-of-a-kinds: It's like wandering through the daydreams of a distracted schoolkid.

Museo Cabeza de Juárez

Cabeza de Juarez III • Built in 1972, the 100th anniversary of the death of president Benito Juárez, this giant head sculpture/museum is made from sheet metal and steel rods, and stares out gloomily from the middle of a traffic rotary.

La Casa Azul

Del Carmen • The deep azure-blue childhood home of Frida Kahlo, one she later shared with her husband, Diego Rivera, and, for a time, with Leon Trotsky, is now a museum of Kahlo's life and work.

Leon Trotsky Museum

Del Carmen • The Russian revolutionary lived in Mexico City for the last two years of his life. Visit the home he shared with his wife with its backyard tombs, bullet holes in the hallway, ominous brick guard towers, and Trotsky's nearly untouched study, the site of his 1940 assassination.

Admission is always free at the Soumaya Museum.

Dirt-smeared dolls hanging from trees form a creepy memorial for a drowned child.

THE ISLAND OF THE DOLLS

MEXICO CITY

As a gondolier steers you along Teshuilo Lake to La Isla de las Muñecas, you'll see two giant teddy bears sitting sentry on the shore. Beyond them are the main attraction—hundreds of dirt-encrusted dolls nailed to trees, strung along wires, and pinned to a dilapidated wooden shack.

Some dolls are missing limbs. Others have spider webs forming in their eye sockets. Their faces have been bleached and discolored by the sun, and their hair is stringy and matted. Their clothes are gradually rotting away. Most are attached by their necks, their heads sagging forward, giving them the appearance of having been hanged.

Dolls began appearing on the island in the 1950s, when a man named Don Julian Santana Barrera, ostracized from his hometown for his religious preaching, left his wife and children and moved there to live in isolation. A local legend told of a girl who had drowned in the surrounding lake. Santana Barrera became fixated on her and was convinced that her spirit lingered on the island. In order to appease her, and to protect himself from any evil spirits lurking in the lake, he began collecting dolls from the trash and arranging them into makeshift memorials.

Over the next five decades, Santana Barrera collected hundreds of plastic children. An avid gardener, he traded produce for dolls and suspended them carefully from trees, wires, and the walls of his wooden hut. He continued to be haunted by the spirit of the drowned girl—though there is no evidence that the girl ever existed.

In 2001, Santana Barrera's nephew, Anastasio Velazquez, came to the island to help his uncle plant pumpkins. As they fished in the canal, Santana Barrera, then 80, sang passionately, claiming that mermaids in the canal were beckoning to him. Velazquez left briefly to work on the garden. When he returned, he found Santana Barrera lifeless, lying facedown in the canal in the spot where the girl is said to have drowned.

Though the troubled man behind the dolls is gone, his unsettling creations live on. Velazquez keeps the private island open to visitors, many of whom bring dolls of their own as tributes for the "girl of the lake."

Teshuilo Lake, Xochimilco, Mexico City. Get the metro line to Tasqueña station, then the light rail to Xochimilco, a district of canals and artificial islands. From there, walk to Cuemanco landing, where you can hire a gondola. Make sure the gondolier is willing to ferry you to the island, as it's not part of the standard route. The trip takes about 2 hours. Ⓝ 19.272847 Ⓦ 99.096510

GREAT PYRAMID OF CHOLULA

SAN PEDRO CHOLULA, PUEBLA

When the Spanish arrived at the city of Cholula in 1519, they were pleased to find a large hill just itching to have a Catholic church built on top. What they didn't realize was that it was no mere hill—beneath the overgrown grasses was a pyramid with a volume larger than that of the Great Pyramid of Giza.

Construction on Cholula's great pyramid began around the first century BCE. During the many pre-Columbian power shifts in Mexico, each conquering culture—the Olmecs, the Toltecs, and the Aztecs—built its own additions to the structure, creating a stack of pyramids in different architectural styles.

At some point before Hernán Cortés and his army arrived in Cholula, the pyramid fell out of favor as a place for religious ritual. It became overgrown, slowly transforming into what looked like a big hill. Veiled by nature, the pyramid avoided the fate of surrounding temples and sacred sites, which Cortés destroyed and replaced with churches per colonial policy. The Church of Our Lady of the Remedies, built on top of the "hill" in 1594, is still there.

The pyramid was not revealed again until 1910, when diggers preparing the construction of an asylum at its base uncovered the foundations. Archaeologists have since excavated the pyramid's stairways, platforms, altars, and over 5 miles (8 km) of tunnels snaking through the structure's innards.
Av 8 Norte #2, Centro, Puebla. Hike to the hilltop church for a great view of Puebla. Ⓝ 19.105270 Ⓦ 98.225566

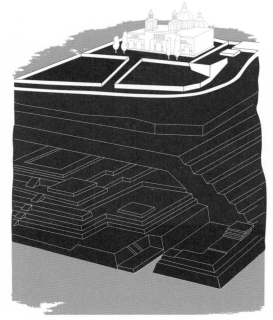

Cholula's ancient pyramid hides under a grassy hill.

A 16th-century Spanish church sits atop a structure larger than the Great Pyramid of Giza.

MAPIMÍ SILENT ZONE

MAPIMÍ, DURANGO

According to legend, there is a patch of desert 2.5 miles (4 km) from San Ignacio where cell phones don't work, the animals are mutants, and aliens do flybys in UFOs. Mapimí Silent Zone, so called due to a belief that radio waves can't be transmitted within the area, is Mexico's answer to the Bermuda Triangle. The fact that both zones—as well as the Egyptian pyramids—are located between the 26th and 28th parallels only fuels the fire of conspiracy theorists' imaginations.

Mapimí's supposed magnetism can be traced back to July 11, 1970, when a US Air Force Athena test rocket lost control, accidentally invaded Mexican airspace, and landed in the desert region of Durango, hundreds of miles from its planned New Mexico destination. The rocket was carrying two small containers of cobalt-57, a radioactive element.

A team of covert specialists brought in to find the fallen rocket conducted aerial searches lasting three weeks. When they finally recovered the rocket, a road was built to transport the wreckage, as well as a small amount of contaminated topsoil. All these operations were carried out under tight security, spurring rumors among the locals.

Rumors became legends: People claimed that radios didn't work, reported sightings of very tall people in "tight silver suits" (which may have held some truth, as during the missile cleanup people might have seen men in silver bio-suits), and spoke of concentrated "Earth energy," "light spheres," and UFOs. A seemingly malformed species of desert tortoise with triangles on its shell sparked cries of mutation. (In fact, the patterns are a normal variation among the Bolson tortoise population of Mexico.)

People now come from all over to explore the area, look for the elusive "silent zones," and sometimes attempt communication with otherworldly beings. Though visitors may be surprised to find their compasses and radios working just fine, an experienced guide will remind them that since the zones move, it can be hard to locate them.

Unfortunately, these new age and paranormal enthusiasts, known locally as *zoneros* or *silenciosos*, are now having an adverse effect on the desert area that contains the Silent Zone. By collecting and keeping both natural and historical artifacts they find in the desert, they are depleting the area of its finite resources.

The zone is east of Federal Highway 49, in the Chihuahuan Desert. Part of the Silent Zone is located in the Mapimí Biosphere Reserve; please respect the reserve and keep any supernatural activities subtle.
Ⓝ 26.738181 Ⓦ 103.722721

LA MONA

TIJUANA, BAJA CALIFORNIA

In 1987, a scruffy part-time art student named Armando Garcia approached Tijuana city officials with a simple plan: to build a humongous naked woman in the middle of the city to mark the 1989 Tijuana centennial. Unsurprisingly, they declined the offer.

Garcia was undeterred, simply relocating the work to his neighborhood, an overlooked ghetto of Tijuana. Two years later, despite the doubts of his professors and classmates, he had built it: a 5-story, 18-ton nude woman rising from hovels and a trash dump. Her right arm, pinky finger extended, points toward the sky. It is a sly gesture that mimics the location of Tijuana on a map of Mexico.

The huge woman, officially named *Tijuana III Millennium* by Garcia but known locally as *La Mona*, or "the doll," was modeled after one of Garcia's ex-girlfriends. For several years, Garcia lived inside the hollow woman with his wife. Their bedroom was located in the woman's breasts, the study in her head, the kitchen in her stomach, and the bathrooms, appropriately, in her behind.

Garcia has since moved to another house in Puerto Nuevo called *La Sirena*. It, too, is in the shape of a giant nude woman.
Ensenada Street, Aeropuerto, Tijuana. The Aeropuerto neighborhood is just southwest of Tijuana International Airport. Local taxi drivers will know how to find La Mona. Ⓝ 32.539038 Ⓦ 116.993191

Armando Garcia's woman-shaped home is hard to miss.

HIERVE EL AGUA

SAN PABLO VILLA DE MITLA

Sometimes nature creates counterfeits: a fish that looks like a plant, a fruit that looks like a vegetable, or, in the case of Mexico's Hierve El Agua, a rock formation that looks like a waterfall.

At a distance, this enormous structure looks exactly like a frozen waterfall, a seeming impossibility in the hot temperatures of San Pablo Villa de Mitla, the town closest to the rock. The rocks are, in fact, mineral deposits on top of a limestone mountain. On the mountain's ledges sit two freshwater pools noted for their medicinal properties and having springs that are saturated with calcium carbonate and magnesium. Water from the pools drips through the cliffs, depositing the minerals onto the side of the mountain. Over time, these deposits have accumulated in staggered columns.

The name of the place, which means "the water boils" in Spanish, comes from the way the water bubbles as it travels through the spring. The Zapotec people, who lived in the area more than 2,000 years ago, revered these pools and directed the spring waters to irrigate their plants. Over thousands of years, their canals have petrified into this unusual rock sight. **Take a bus from Oaxaca or Mitla. The park's hours aren't consistent, so make sure to call ahead before setting out. Ⓝ 16.865684 Ⓦ 96.276006**

A freshwater pool tops off what appears to be a frozen waterfall.

ALSO IN MEXICO

San Juan Parangaricutiro

Angahuan · The towers and top floor of this church are all that protrude from a layer of lava that smothered the town in 1944.

The Mummies Museum

Guanajuato · With their mouths gaping open, their arms clutching at themselves, and their tissue-thin skin ripped to reveal bone, these 118 mummies do not appear to be resting in peace.

Statue of José Maria Morelos

Janitzio · A staircase spirals up inside this 131-foot (40 m) statue of a Mexican independence hero. Above the stairs are murals depicting the life of Morelos and his role in the country's history.

Little Boy Zero

La Gloria · Edgar Hernandez, the five-year-old patient zero of 2009's global swine flu pandemic, is immortalized in this bronze statue.

Museum of Perversity

Manzanillo · The graphic dioramas here that depict torture, violence, and cruelty are intended to turn visitors into human-rights activists.

Mercado de Sonora

Mexico City · This market is a one-stop shop for herbal medicine and occult supplies. Stock up on dried skunk, amulets, and ingredients for love spells.

City of Books at Biblioteca Vasconcelos

Mexico City · A jaw-dropping megalibrary contains the complete, perusable personal book collections of five of Mexico's greatest thinkers.

Quetzalcoatl's Nest

Naucalpan · Visit condos designed in the shape of a feathered Aztec snake god, complete with gaping maw.

Cosmovitral Botanic Garden

Toluca · Stained-glass windows bathe this artful garden in cosmic light.

Tree of Tule

Tule · The Montezuma cypress's spectacular girth earns it a place in the record books, while its gnarled bark inspires the imagination. Visitors have found likenesses of human faces, lions, jaguars, and elephants.

Festival of the Exploding Hammers

San Juan de la Vega · Each February, brave and/or reckless locals strap explosives to sledgehammers and throw them down with all their might.

CAVE OF THE CRYSTALS

NAICA, CHIHUAHUA

In 2000, workers at Naica, Mexico's largest mine, were excavating a new tunnel 1,000 feet (305 m) underground when they broke into an extraordinary cave. This chamber, now known as the Cave of the Crystals, contains some of the largest crystals ever found. Its crisscrossing shards of selenite measure up to 39 feet (12 m) long.

The unusually immense crystals formed over half a million years in water that was a steamy and stable 136°F (57.7°C). These conditions allowed a particular mineral in the cave, anhydrite, to absorb water and transform into gypsum, its lower-temperature, stable form. Gypsum deposits gradually built up, forming the giant selenite crystals. When mining operations began in the area, workers pumped water from the surrounding caves, inadvertently draining the crystal cave and exposing its treasures.

Conditions in the Cave of the Crystals are very hazardous to humans. Ambient temperature is around 125°F (51.6°C), and the 90 to 99 percent humidity creates a stifling environment in which higher brain functions quickly deteriorate and breathing becomes difficult. The terrain is uneven and the smaller crystals have sharp edges, making walking tough.

Scientists and researchers have been exploring the cave since 2006. They do so wearing cold-water respirators and suits lined with ice. Even with this special equipment, the explorers can only stay in the cave for 45 minutes at a time.

Naica's gigantic crystals were discovered in 2000.

Unfortunately, the water-draining process that revealed the Cave of the Crystals is also destroying it. Selenite deteriorates when exposed to air. Currently, the mine's water-pumping operations keep the cave dry. To preserve the crystals and allow them to grow larger, the cave must be flooded again, which would cut off human access. The dilemma is whether "saving the cave" means maintaining scientific access to it or allowing its crystals to grow unobserved.

The mine and cave are located in the southwest part of Naica, a town with around 5,000 residents. Due to its fragility and oppressive conditions, the cave is only accessible to researchers.

Terreros 7, Naica. Naica is a 2-hour drive south of Chihuahua City. Ⓝ 27.850833 Ⓦ 105.496389

THE CAVE OF SWALLOWS

AQUISMÓN, SAN LUIS POTOSÍ

If you fell into the Cave of Swallows, it would be at least 10 seconds before you hit the floor. The limestone sinkhole is 1,093 feet (333 m) deep. That's deep enough to fit the Chrysler Building with the Statue of Liberty balancing on top.

BASE jumpers used to fling themselves into the abyss regularly, using a mechanical winch to make the long journey back up. Now, however, BASE jumping and motorized winches are no longer allowed. The noise and constant high-speed descents were disrupting the resident birds as well as the nearby humans.

To see down to the the bottom of the cave—which, incidentally, is covered with guano and crawling

Rapelling into the unknown.

with insects, snakes, and scorpions—you'll need to throw a rope into the abyss and rappel down. The hard part is climbing back up, which takes 40 minutes if you're super humanly fit, and 2 hours if you've been known to skip a few sessions at the gym. If you'd rather not descend into the fathomless chasm, you can tie a safety rope around your waist and peer over the edge.

The Cave of the Swallows is named after the thousands of birds who nest in the cave walls, spiral out of the sinkhole every morning, and return every night. They are not actually swallows—they're a combination of white-collared swifts and green parakeets.

The cave is a 30-minute drive west of the small town of Aquismón. Ⓝ 21.599836 Ⓦ 99.098964

Central America

BELIZE

THE CRYSTAL MAIDEN

SAN IGNACIO, CAYO

To reach the caverns of Actun Tunichil Muknal, you'll need to endure a bumpy 45-minute drive, a 45-minute hike through the jungle, three river crossings on foot, and a claustrophobia-inducing swim into the narrow mouth of the waterlogged cave. But at least you'll make it out alive.

During the late Classic period of their civilization (700 to 900 CE), the Maya came to the cave to perform sacrificial rituals, believing it to be a gateway to Xibalba, the underworld. Carrying flaming torches, burning incense, and holding ceramic pots containing edible offerings, they led people inside to be killed and offered to the gods.

Archaeologist Thomas Miller discovered Actun Tunichil Muknal, also known as ATM, in 1986. Later excavations uncovered the bones of 14 people, including six children under the age of three, in a large chamber called the Cathedral. Cranial trauma and the positioning of the remains indicated they were fatally struck on the head and thrown to the

ground. Left undisturbed in the dripping cave for approximately 1,200 years, the bones are coated in calcite crystals, giving them a sparkling, puffy appearance. Hundreds of pieces of broken pottery surround the human remains—following sacrificial ceremonies, the Maya would smash the ceramic pots to release the energy contained within.

A full skeleton, lying on its back with its head raised as if to look at you, stands out among the scattered bones. Calcification has softened its harsh contours, and the crystals dusting the bones glitter in the light of your headlamp. This is the "Crystal Maiden." She died probably at 18—in a particularly violent way, judging by her two crushed vertebrae.

Due to site preservation requirements you must go with a guide—you can find several in San Ignacio. Cameras are no longer allowed, after a tourist dropped one on a skull estimated to be a thousand years old in 2012. Bring dry socks to wear in the Cathedral, where shoes are not permitted. Wear shoes you can swim and hike in. You'll be given a helmet with a headlamp.

The caverns are a 2-hour drive southwest of Belize City.
Ⓝ 17.117496 Ⓦ 88.890467

COSTA RICA
TERRITORIO DE ZAGUATES (LAND OF STRAYS)

CARRIZAL ALAJUELA

Over one million stray dogs roam Costa Rica, and the numbers tick higher every day. Many street dogs are taken into shelters, but purebreds are more likely to be adopted than mutts. One rescue organization is different: Up in the mountains, *Territorio de Zaguates*, or the Land of Strays, celebrates the unique mix of each dog it cares for.

Veterinarians at Territorio de Zaguates do their best to analyze the physical traits of each of the hundreds of rescue dogs that live at the free-range shelter and make guesses at their breeds' lineage. They then give every mutt its own unique pedigree, with names like the Long-Legged Irish Schnaufox or the Fire-Tailed Border Cocker.

When the shelter's head vet appeared on television to talk about the special breeds bestowed upon the canines, he emphasized their uniqueness, boasting that "these dogs exist only in our country." Dog lovers went wild, calling in to reserve the Bunny-Tailed Scottish Shepterrier or the Furry Pinscher Spaniel for themselves. The shelter grew in popularity, leading to more adoptions of these one-of-a-kind pups.

Territorio de Zaguates formerly sponsored public hiking events where visitors could frolic in the mountains with hundreds of pups. The park is currently closed to the public but there are plans to reopen the sanctuary for tours and visits with the pack. Ⓝ 10.096143 Ⓦ 84.156100

ALSO IN CENTRAL AMERICA

COSTA RICA

727 Fuselage Home

Quepos · Stay in a beachfront hotel repurposed from an old Boeing 727 fuselage.

GUATEMALA

Mapa en Relieve

Guatemala City · For a smaller, steeper, more-turquoise version of Guatemala, head to the open-air relief map at Minerva Park.

NICARAGUA

Footprints of Acahualinca

Managua · Over two thousand years ago, a dozen or so people went for a stroll through Managua. Their footprints, embedded in volcanic ash which then solidified, are still visible today.

PANAMA

Ancón Hill

Panama City · Surrounded on all sides by Panama City, a tiny patch of jungle wilderness remains protected amid a rapidly industrializing and burgeoning urban center.

A young woman sacrificed around 1,200 years ago has transformed into a glittering skeleton.

STONE SPHERES

SAN JOSÉ

Clearing land along the Diquis Delta for future banana plantations in 1939, the United Fruit Company uncovered something unexpected: hundreds of spherical stones, measuring up to eight feet (2.4 m) across.

To this day, it is unclear who shaped the stones, when they did it, and for what purpose, but their perfect roundness and smooth surfaces indicate they are man-made, rather than naturally formed. They were likely shaped into balls from larger boulders, using a combination of fracturing, chiseling, and grinding.

Almost all of the stone spheres have been moved from their original locations, making it even more difficult to determine their

Formed by an ancient civilization, Costa Rica's mysterious stone balls have become prized lawn ornaments.

archaeological context. Treasure seekers have blown up some of the balls, acting on the belief that there is gold hidden inside. There is not.

Six of the spheres are in the courtyard of Costa Rica's national museum in San José, Calle 17, Cuesta de Moras. Ⓝ 9.932609 Ⓦ 85.071967

EL SALVADOR

FIREBALL FESTIVAL

NEJAPA, SAN SALVADOR

Every August 31, once the sun goes down, young men with painted faces crowd the streets of Nejapa to hurl flaming fuel-soaked rags at one another. Hundreds of onlookers cheer as the *bolas de fuego*, or "fireballs," fly through the air.

For over 30 years, the annual fireball festival has commemorated the 1658 eruption of the El Playon volcano, which buried the town of Nexapa and forced its residents to build a new settlement: Nejapa. The

fireballs, which are made from rolled-up rags soaked in barrels full of kerosene, represent the flaming rocks spewed from the volcano. The festival also honors San Jerónimo, the town's patron saint, who is said to have fought the devil in a fiery struggle.

To protect themselves from burns, fireball festival combatants wear gloves and waterlogged clothes. Even with these precautions in place, dozens of people each year have to be treated for burns.

Nejapa is a small town 30 minutes north of San Salvador, the capital. Ⓝ 13.819263 Ⓦ 89.233773

Once a year, flaming rags are flung in the air to commemorate a volcanic eruption.

Hidden in the dense Guatemalan jungle is one of the world's largest pyramids.

GUATEMALA

LA DANTA

CARMELITA, PETÉN

An enormous stone structure pokes above the trees in the northern Guatemalan jungle. It looks like a strange, lonely volcano randomly plopped in the vegetation. But the structure is actually one of the largest pyramids on the planet, which stands within the ruins of El Mirador, the "lost city of the Maya."

The ancient city of El Mirador thrived between the 6th and 3rd centuries BCE, before being abandoned and swallowed up by the jungle. The site laid dormant for hundreds of years, until archaeologists began excavating. Thousands of structures were unearthed within the pre-Columbian city, but perhaps none are as impressive as the La Danta temple.

La Danta stands a staggering 230 feet tall (70 m). It has a total volume of nearly 99 million cubic feet, making it one of the world's largest pyramids and among the most enormous ancient structures on Earth. It's been calculated that 15 million days of labor were needed to construct the gigantic monument. A staircase leads up the temple's eastern face, rewarding those who climb it with views of a vibrant canopy of trees stretching as far as the eye can see.

You can reach El Mirador from Flores via helicopter. It's also possible to hire a guide and hike from the village of Carmelita. Hiking to the lost city takes several days, and you'll walk through jungle inhabited by the venomous fer-de-lance snake. Ⓝ 17.756152 Ⓦ 89.919017

The arduous journey to Semuc Champey ends in a natural paradise of turquoise pools and limestone caves.

SEMUC CHAMPEY

LANQUIN, ALTA VERAPAZ

Tucked away in the densely forested mountains of Alta Verapaz lies an idyllic limestone paradise. Virtually inaccessible to all but the heartiest four-wheel drive vehicles, the Semuc Champey natural monument boasts six stunning tiered turquoise pools and an extensive cave network, complete with underground waterfalls. The shockingly blue pools rest on top of a natural limestone bridge covering a portion of the Cahabón River.

The site is located far from any large settlement down rough, rock-strewn roads. However, the difficulty of getting to this unique natural landscape is rewarded many times over by the sights and experiences awaiting guests upon arrival. If you choose to take a guided tour, you'll start with an optional rope swing leap into the river before grabbing a candle and wading through a series of watery caves. The above-ground portion of the tour meanders through the forest before dropping guests at their final activity: a relaxing swim in the limestone pools. **While it is possible to rent a car and drive on your own, the easiest and safest way to get to Semuc Champey is by booking a minibus through a travel agency in one of Guatemala's larger cities. Minibuses will drop you off at Lanquín, the town closest to Semuc Champey, and from there you can either walk for about 2.5 hours or take a 4x4 pickup truck taxi to the entrance of the falls. Ⓝ 15.533352 Ⓦ 89.961379**

HONDURAS

RAIN OF FISH

YORO, YORO

The *Lluvia de Peces,* or "Rain of Fish," is said to occur at least once and sometimes twice a year in the small town of Yoro: During massive rainstorms, hundreds of small silver fish supposedly cascade from the heavens.

The fantastical story says that from the 19th century onward, every May or June, a large storm has rolled through the town bringing heavy rain. When the maelstrom passes, the streets are alive with gasping fish.

In the 1970s, a team from the National Geographic Society actually witnessed the flapping fish—one of the few credible sightings of the phenomenon. They did not, however, find proof that the fish came from the sky.

The "animal rain" weather phenomenon has been reported around the world for centuries, though the scientific understanding of it remains sketchy. The simplest explanation for these falling fauna is that large rainstorms flood rivers and force certain animals out of their homes, causing them to fill the streets. Another simple explanation is that a flash flood can deposit fish far from their waters before quickly receding, leading bystanders to believe that the animals must have come down with the rains.

While it's extremely rare, animals do occasionally fall from the sky when they get picked up by waterspouts. Waterspouts are small tornadoes that form over a body of water. Though they do not suck up water into the air (the "spout" is actually condensation), the whirlwind has the power to lift small animals from the water and release them over land.

Among the most mysterious elements of Yoro's "Rain of Fish" is that the fish themselves are not local to the area and may come from as far away as the Atlantic Ocean, some 140 miles (200 km) to the north. A less exciting theory postulates that the fish live in underground rivers and are forced up onto the streets rather than falling down onto them. This hypothesis is supported by the 1970s National Geographic team's discovery that the silver fish are completely blind.

In Yoro, the explanation is often religious rather than scientific. In the 1860s, Father José Manuel Subirana, a Catholic priest living in the area, prayed for sustenance for the hungry. At the end of his marathon prayer session, it is said to have rained small fish. Since 1998, an annual festival has been held in honor of the Lluvia de Peces miracle. It includes a parade, during which revelers carry effigies of Father José Manuel Subirana through the streets. **Buses run between the Honduran capital of Tegucigalpa and Yoro—a 125-mile (201 km) journey. The festival is usually held in June, coinciding with the first major rainfall of the season. Ⓝ 15.133333 Ⓦ 87.142289**

HIEROGLYPHIC STAIRWAY OF COPÁN

COPÁN

The city of Copán in what is now western Honduras served as a political, civil, and religious center of the Mayan civilization for over 400 years. The site is host to a number of marvelous ruins, the most striking of which must be the epic stairway in the temple-pyramid of Structure 26.

This construction, which forms the longest discovered Mayan text, was originally commissioned by the 14th governor of Copán, K'ak Joplaj Chan K'awiil, and completed about six years after his death in the year 755. The pyramid is nearly 100 feet (30.5 m) high and etched with around 2,000 glyphs. This collection of symbols offers a rare window into the rich history of the Copán Valley and the culture that ruled it for so many years.

Researchers, first stumped by the hieroglyphs, came to realize that the staircase is a record of the royal history of Copán, listing the names of kings, their births, their deaths, and the defining events of their rule. The happy realization that the stones were arranged chronologically was somewhat tempered by the fact that archaeologists in the 1930s—not 100 percent clear on Mayan syntax—had liberally rearranged the stone blocks in an attempt at reconstruction. Only the bottom 15 stairs remain in their original positions.

Despite the jumble, modern archaeologists have figured out that the stairs document the rule of 16 kings, beginning with Yax K'uk Moh at the bottom step and ending with the death of a ruler known as "18-Rabbit" at the top. It is also believed that there is special emphasis on the story of the 12th king, K'ak Uti Ha K'awiil, whose burial plot was discovered inside the pyramid that supports the staircase.

The ruins are a 3-hour bus ride from San Pedro Sula. Ⓝ 14.837331 Ⓦ 89.141511

A statue of a Mayan ruler dressed as a Teotihuacan warrior sits at the base of the stairs.

NICARAGUA
CERRO NEGRO
LEÓN, LEÓN

Cerro Negro, South America's youngest volcano, is the world's first location to offer ash boarding. Also known as volcano surfing or volcano boarding, the sport involves strapping a wooden plank to your feet and coasting down the 1,600-foot (488 m) ash-and-pebble slope of Cerro Negro. Boarders wearing gloves, goggles, and jumpsuits reach speeds of up to 50 miles (80.5 km) an hour, kicking up clouds of dust on the way down. Those of a more timid disposition can opt to sit on the board or simply run down the steep slope.

The ascent is less thrilling—it's an hour-long hike—but the summit brings its own rewards. A stunning

Surf down the slopes of an active volcano.

360-degree panoramic view reveals the chain of active and dormant volcanoes, lined up one after the other, surrounded by blue skies and lush green foliage. Cerro Negro itself, which first appeared in 1850, is an active volcano whose crater often emits smoke. It has erupted 23 times, most recently in 1999. **Boarding tours leave from León, an hour southwest of the volcano. Ⓝ 12.506864 Ⓦ 86.703906**

PANAMA
DARIÉN GAP
DARIÉN PROVINCE

The Pan-American Highway stretches all the way from the northern shore of Alaska to the southern tip of South America with just one break: a 54-mile (87 km) missing piece on the Panama–Colombia

The only gap in 30,000 miles of Pan-American Highway is a very dangerous place.

border known as the Darién Gap. Visit and you will encounter a wild jungle oasis that is home to many rare plants and wildlife. You also may not make it out alive.

The Darién region is home to members of the Revolutionary Armed Forces of Colombia (FARC), a Marxist–Leninist guerrilla group engaged in a decades-long armed conflict with the Colombian government. The overgrown jungle provides plenty of hiding places for the storage and trafficking of drugs—activities that earn FARC hundreds of millions of dollars each year. Many travelers attempting to cross the Darién Gap have gone missing, been kidnapped, and been held hostage by rebels on both sides of the border.

The current Colombian and Panamanian governments have no desire to build a road that would complete the Pan-American Highway. It's too expensive and too dangerous, and development would damage the jungle's fragile ecosystem. In the words of explorer Robert Young Pelton, who was kidnapped and held hostage in the Darién Gap in 2003, "It's probably the most dangerous place in the Western Hemisphere . . . Everything that's bad for you is in there." **The thick jungle and militia presence make the Darién Gap a destination for only the most foolhardy. To travel between Central and South America more safely, buy a plane ticket or hop aboard one of the many yachts that make the trip from Panama to Colombia. Ⓝ 7.868171 Ⓦ 77.836728**

Caribbean Islands

BAHAMAS

SWIMMING PIGS

BIG MAJOR CAY

"Uninhabited island full of feral pigs" may not sound like a delightful vacation destination. But consider this: These pigs are friendly, and they want nothing more than to frolic with you in the clear, refreshing waters of Big Major Cay, also known as Pig Beach.

The swimming pigs hang out during the day on the beach and in the waters of the appropriately named Pig Beach. When a boat approaches from the adjacent resort island of Fowl Cay, the eager swine throw themselves into the water and paddle up to greet their guests. A steady stream of visitors ensures they are well fed and well loved.

It is unclear how pigs got to Big Major Cay in the first place, but they were likely left there by sailors

Friendly wild pigs float in the warm Bahamian waters.

stopping off on the way to bigger bits of land.

To get to Big Major Cay from Nassau, take a plane to Staniel Cay, then a boat north. The pigs like it when you bring potatoes. Ⓝ 24.183874 Ⓦ 76.456411

OCEAN ATLAS

NASSAU, NEW PROVIDENCE

The world's largest underwater sculpture is submerged beneath the sea in the crystal Bahamian waters off the coast of Nassau. Entitled *Ocean Atlas*, the giant sculpture is a contemporary take on the ancient Greek myth of Atlas. But rather than a Titan condemned to carry the heavens on his back for eternity, it depicts a young Bahamian girl supporting the ceiling of the ocean on her shoulders.

Created by artist, naturalist, and diver Jason deCaires Taylor in 2014, *Ocean Atlas* is over 16 feet tall and weighs 60 tons. Located just off the coast of New Providence, the most populous island in the Bahamas, the watery wonder is meant to deter tourists from endangered reefs and encourage coral colonization. The artwork is forged from sustainable, pH-neutral materials designed to kick-start the growth of local coral. It's an environmental gesture intended to portray the positive potential of human interaction with the natural world, even as the Earth's oceans face numerous threats from climate change and human activity.

The artist has designed three other submerged museums and installations around the world: Vicissitudes, the world's first underwater sculpture park, located in Grenada's Molinere Bay; MUSA (Museo Subacuático de Arte), an underwater sculpture park in Cancún; and Museo Atlántico, the first submerged contemporary art museum in the Atlantic Ocean, off the coast of Lanzarote, Spain.

***Ocean Atlas* is submerged just off the coast of Nassau within Clifton Heritage National Park. You can book a snorkeling tour of the sculpture online. Ⓝ 25.013896 Ⓦ 77.551996**

A modern Atlas sits beneath the sea.

BARBADOS
PROJECT HARP SPACE GUN

SEAWELL, CHRIST CHURCH

Put simply, Project HARP was established to create a cartoonishly large gun to shoot satellites into space. Short for High Altitude Research Project, the 1960s experiment was a joint initiative between the United States and Canada to study the use of ballistics to deliver objects into the upper atmosphere and beyond. Numerous space guns were built, but the most impressive surviving relic of the program is the gigantic, abandoned gun barrel in Barbados.

The gun was designed by the brilliant and controversial ballistics engineer Gerald Bull, who spent his life in passionate pursuit of his dream to build a long-range super-gun (an obsession that would later, after Project HARP was shut down, lead him to design weapons for the Iraqi government). Bull was assassinated in 1990 inside his apartment. His killing remains unsolved.

Engineer Gerald Bull

The Barbados gun was built from a 65-foot-long (20 m) naval cannon, the kind that might be seen on a battleship. That cannon was later joined to another barrel, extending the length to 118 feet (36 m) and making the gun too big for effective military application but seemingly perfect for satellite delivery. At its peak in 1963, this giant piece of artillery was able to fire an object a staggering 111 miles (178 km) into the sky, setting the world record.

The Barbados gun was abandoned by 1967 and left to rust on its original launch site on a small cliff overlooking the Atlantic Ocean. After years of neglect, it looks more like a painted sewer pipe than a mammoth space cannon.

The abandoned space gun is located on an active military base and is only accessible with permission.
Ⓝ 13.077242 Ⓦ 59.475511

BERMUDA
THE UNFINISHED CATHEDRAL

ST. GEORGE'S, ST. GEORGE'S

St. Peter's will never be completed, but the open-air ruin now has its own appeal.

Grass grows where the pews ought to be, half the support pillars have crumbled, and the roof is long gone. The unfinished church at Somers Garden is the result of conflicts in the congregation, money troubles, and one almighty hurricane.

Construction on the Protestant cathedral began in 1874. The building, designed to seat 650, was intended to replace St. Peter's Church, an Anglican place of worship established shortly after the 1612 English settlement of St. George's.

St. Peter's Church still stands, but the new cathedral remains unfinished. The first hurdle came when the congregation split and a group of former parishioners left to build their own Reformed Episcopal Church. In 1884, a cathedral in nearby Hamilton burned down, requiring funds to be diverted from the construction project. By 1894, with the unfinished cathedral having suffered from financial setbacks, storm damage, and squabbles within the Anglican community over its legitimacy, the congregation decided they would rather renovate St. Peter's than complete the new cathedral.

Thirty years later, a hurricane caused substantial damage to the western end of the unfinished cathedral, sealing its fate as a modern ruin. Though the building has no ceiling, no floor, and no windows, it is a popular site for wedding ceremonies.

Blockade Alley, St. George's. Public buses run to nearby Somers Garden. Ⓝ 32.382350 Ⓦ 64.676251

ALSO IN BERMUDA

Somerset Bridge

Somerset Island · The world's smallest drawbridge is just wide enough for a sailboat mast.

CAYMAN ISLANDS

UGLAND HOUSE

GEORGE TOWN, GRAND CAYMAN

There are nearly 100,000 corporate entities worldwide that use the Cayman Islands' zero percent tax rate to dodge corporate taxes, a number higher than the territory's population itself. These corporations could together form a miniature city or a major financial district, but thousands of their offices have the exact same address.

The five-story Ugland House in George Town covers a mere 10,000 square feet of land but is the official home of a whopping 18,857 corporate entities. That equals one corporation registered for every three square feet of space in the building.

The Cayman Islands' image as a tax-free paradise has its own mythological origins. In 1794, a convoy of 10 British ships en route from Jamaica met disaster on the treacherous reefs of Grand Cayman. As the passengers and crew struggled to survive amid the breaking waves, island residents from the East End and Bodden Town, having heard the ships' distress signals, paddled out to the reefs in canoes to attempt a rescue. In the darkness and pounding surf, the Caymanians saved 450 of the stranded souls. Amazingly, only 6 people lost their lives in the disaster.

The heroism of the Caymanians in rescuing the English sailors and passengers fueled a legend that lingers to this day. The story goes that one of the passengers rescued from the wrecked ships was a son of King George III. When the king learned of the islanders' bravery, he rewarded them by decreeing that the Cayman Islands would forever be free of taxation and war conscription. However, there is no record that any member of the royal family was on one of the ships or that the king ever issued such a decree.

121 S Church Street, George Town. When you're done scoping out Ugland House (you can't go inside), head to the East End, where you'll find a simple memorial commemorating the maritime disaster. Ⓝ 19.292199 Ⓦ 81.385544

CUBA

PRESIDIO MODELO

NUEVA GERONA, ISLA DE LA JUVENTUD

During their four decades of operation, the circular cell blocks of Presidio Modelo housed political dissidents, counterrevolutionaries, and even Fidel Castro. Cuban president-turned-dictator Gerardo Machado oversaw the prison's construction in 1926. Modeled after Jeremy Bentham's Panopticon design, with tiered cells surrounding a central observation post, the prison provided constant surveillance of its inmates.

Fidel Castro spent two years at Presidio Modelo after leading a 1953 attack on Moncada Barracks that killed dozens and ignited the Cuban Revolution. The future Communist leader and icon spent his sentence penning "History Will Absolve Me," a revolutionary manifesto that formed the basis for his junta.

When Castro seized power in 1959, the prison soon became overcrowded with enemies of the socialist state. Designed to hold 2,500 inmates, Presidio Modelo housed over 6,000 inmates by 1961. Riots and hunger strikes broke out regularly, leading to the prison being closed for good in 1966.

Today, the buildings are open as a museum and national monument. **Isla de la Juventud is south of the main island of Cuba. A boat from the port in Batabanó takes around 2 hours, or you can get a half-hour flight from Havana. The prison is just east of Nueva Gerona. Ⓝ 21.877609 Ⓦ 82.766451**

Presidio Modelo's circular cell blocks made prisoner surveillance easy.

JURAGUA NUCLEAR POWER PLANT

JURAGUA, CIENFUEGOS

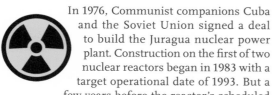

In 1976, Communist companions Cuba and the Soviet Union signed a deal to build the Juragua nuclear power plant. Construction on the first of two nuclear reactors began in 1983 with a target operational date of 1993. But a few years before the reactor's scheduled completion, the USSR collapsed. The flow of crucial Soviet funds ceased, 300 Russian technicians went home, and Cuba was forced to suspend construction on its badly needed power plant.

Lacking nuclear fuel and without the primary components installed, the plant sat in limbo until December 2000, when Russian president Vladimir Putin paid a visit to Cuba. Putin offered Fidel Castro a belated $800 million to finish the first reactor. Despite Cuba's reliance on imported oil for power, Castro declined. Project status: officially abandoned.

The unfinished plant, a huge, domed concrete structure, sits on the Caribbean coast, across the bay from the city of Cienfuegos. You can get a closer look by going to Castillo de Jagua, an 18th-century Spanish fortress two miles from the plant. Access to the plant itself is prohibited.

The plant is a few minutes away from the town of Jagua.
Ⓝ 22.066660 Ⓦ 80.513275 �

➤ Abandoned Nuclear Power Plants

HANFORD SITE
WASHINGTON
Located on Washington State's Columbia River and surrounded by a wildlife refuge, the 560-square-mile (1,450 km²) Hanford site is the largest radioactive waste dump in the US.

Selected as the location for a plutonium production complex in 1943, the first nuclear reactor at the site served up its initial batch of plutonium in November 1944. By February 1945, Hanford had a trio of identical reactors producing plutonium.

Plutonium produced at Hanford ended up in Fat Man, the bomb that killed approximately 80,000 people in Nagasaki. Almost all of the 50,000 construction workers who built the plant had no idea they had contributed to the creation of a nuclear weapon until news of the Hiroshima bomb reached them.

Decommissioned after the Cold War, the Hanford complex left 53 million gallons of radioactive waste stored in leaking underground tanks. A multidecade cleanup is currently underway, with a projected completion date of 2040.

EXPERIMENTAL BREEDER REACTOR I (EBR-I)
IDAHO
Built as a research facility in 1951, this plant in Idaho's Arco desert became the first nuclear reactor to generate electricity. Deactivated in 1964, the site is now a museum. Visit to see two nuclear reactors, press buttons and flip switches in the control room, and use mechanical manipulator claws that once handled radioactive waste.

In the parking lot are two prototype reactors designed to power nuclear bomber aircraft. Both the United States and the Soviet Union researched nuclear-powered planes during the Cold War, but neither country produced a working model.

WUNDERLAND KALKAR
GERMANY
The swing ride at the Wunderland Kalkar amusement park north of Düsseldorf is in the cooling tower of a former nuclear power plant.

Wunderland Kalkar offers amusement park thrills in the shadow of a nuclear cooling tower.

Construction on the plant and reactor SNR-300 began in 1973 and took 12 years to complete. During its construction, the community was vocal about its concerns over nuclear power. In 1985, the reactor began partial operation. Then came the Chernobyl disaster of April 26, 1986. Concerned about safety and wanting to avoid high operating costs, the state government halted the opening of the plant.

Five years later, with SNR-300 officially canceled and its valuable parts sold and shipped away, Dutchman Hennie van der Most purchased the land. He then took the obvious next step: turning a nuclear power plant into a family amusement park.

Kernwasser Wunderland ("Corewater Wonderland") opened in 2001 with over 40 rides, a 437-room hotel, bars, restaurants, and a bowling alley. The star attraction is the cooling tower, now painted to resemble a snowy mountain landscape. A rock-climbing trail snakes up its outer wall, while the inside is home to a swing ride and "Echoland"— just shout and you'll understand why.

CURAÇAO
KLEIN CURAÇAO

WILLEMSTAD

Measuring just 1.2 square miles in size, the flat, abandoned coral atoll of Klein Curaçao is home to a deserted lighthouse, the shipwrecks of vessels it failed to save, several collapsed ancient stone buildings, and a burial site.

The northern side is a tropical paradise of palm trees, white beaches, and crystal waters. The southern, windward side, more exposed to the elements, is an unforgiving shoreline of pounding surf and ragged coral reefs. Here lie the shipwrecks of Klein Curaçao. There's about half left of the oil tanker *Maria Bianca Guidesman*, which ran aground in the 1960s. Next to the *Maria Bianca* are the remnants of a once glorious 30-foot (9 m) luxury boat that smashed against the sharp coral reefs.

Crumbling stone structures, perhaps former homes to passing fishermen, dot the island. On

A long-abandoned lighthouse sits on the southside of paradise.

the northeastern end there is also a simple burial site, marked by a bleached, plain wooden cross without a name.

Standing alone at the center of the island is the lighthouse. Once painted a vibrant coral pink, it was built in 1850. Its wooden stairs are still intact, as are the two stories of rooms where the lightkeepers lived.

In 1888, the German navy attempted to build a base at Klein Curaçao as part of a hoped-for colonization of the Caribbean, but they were driven off by the windswept conditions of the remote outpost when a tropical storm swept away the first foundations of a wharf.

Klein Curaçao is a 2-hour sail southeast from neighboring Curaçao. Several companies offer day trips by catamaran. Ⓝ 11.984563 Ⓦ 68.644221

DOMINICA
BOILING LAKE

ROSEAU

The lake on Watt Mountain is not a good place for a refreshing dip: Drinking or bathing in the water will result in death, or at least severe burns.

Morne Watt (Watt Mountain) is a stratovolcano, its boiling lake a flooded fumarole—a direct line to the molten subsurface of Earth, with vents pumping scalding steam and gases into the water. A thick cloud of vapor rises from the bubbling blue-gray lake. Water temperature is around 194°F (90°C).

You need to be fit to visit the lake—it's a 3-hour hike from the nearest road, over difficult terrain. En route, you'll pass through the Valley of Desolation, a sulfurous expanse of volcanic vents, hot springs, and bubbling mud. Ⓝ 15.333608 Ⓦ 61.324139

This mountaintop lake churns, bubbles, and steams in a perpetual boil.

The cruciform building that may (or may not) hold Christopher Columbus's remains.

DOMINICAN REPUBLIC
THE COLUMBUS LIGHTHOUSE

SANTO DOMINGO ESTE, SANTO DOMINGO

In 1506, Christopher Columbus died in his mid-fifties in Spain. But death was not the end of his adventures—his body continued to travel in a centuries-long shell game.

Immediately after his death, Columbus was buried in Valladolid, the town in which he spent his last days. Then, on the wishes of his son Diego, the body was moved to Seville. But that wasn't Columbus's final destination—Diego wanted a grander tribute for his father. He headed back to the Dominican Republic, where in 1514 he laid the cornerstone for the Cathedral of Santa María la Menor that would hold Columbus's remains. Unfortunately, frequent traveler Diego died in 1526 in Montalbán, Spain, before the cathedral was finished. His body was transported to Seville and interred next to his father.

The bodies of father and son stayed in Seville for the next 16 years, but when the Cathedral of Santa María la Menor was completed in the Dominican Republic, Diego's widow put the wheels in motion to have both bodies moved there. In 1542, the remains sailed the ocean blue again, joining the body of Christopher's brother, Bartholomew, who had died in Santo Domingo the year before.

There they remained, for more than 200 years, until the Spanish were ousted from the Dominican Republic in 1795. On their way out, Spaniards took the explorer's body with them to the other Spanish stronghold in the Caribbean: Havana, Cuba.

Back in the Dominican Republic, nearly a century later, a construction worker working on the cathedral renovation uncovered a lead box—unimpressive, save for the inscription on the inside of the lid: "The illustrious and excellent man, Don Colón, Admiral of the Ocean Sea."

At first pass, it seemed obvious that the Spanish must have, in their haste, taken the wrong box. But there's a catch—both father, Christopher, and son, Diego, were known as Don Colón in their lifetimes, and both held the title Admiral of the Ocean Sea.

In 1898, the Spanish left newly independent Cuba. They took the (assumed) remains of Columbus back to Seville and placed them in an elaborate cathedral tomb. Back in the Dominican Republic, a 1931 design competition resulted in the 688-foot-long (210 m) cruciform memorial complex that allegedly now holds the boxed remains of the explorer.

Thus far, science has not solved the mystery of the mixed-up remains. DNA analysis of the Seville remains in 2003 was inconclusive, and Santo Domingo authorities will not permit any exhumations.

Though it's referred to as a lighthouse, the building that (maybe) houses Columbus's remains is actually a blocky, gray, seven-story museum. It is also home to Pope John Paul II's robes and a popemobile.

Avenida Faro a Colón, L213, Santo Domingo Este. The lighthouse is in the Sans Souci part of Santo Domingo, near the cruise terminal. Buses marked "Corredor Independencia" and "Ave Las Americas" will get you within a 20-minute walk.
Ⓝ 18.478714 Ⓦ 69.866531

A stone circle on the slope of a volcano commemorates the leader of an anti-slavery uprising.

GUADELOUPE
FORT DELGRÈS MEMORIAL

BASSE-TERRE

In 1802, Louis Delgrès, a free man of color born in Guadeloupe, led a doomed rebellion against Napoleon's General Antoine Richepanse to prevent the return of slavery in the French Caribbean. Now the fort where he made his last stand bears his name.

Delgrès was an idealist and a distinguished soldier in many battles for the French Republic. He was even captured and sent to England as a prisoner once or twice. When, in 1802, Napoleon Bonaparte sent General Richepanse to Guadeloupe to restore it to its "pre-1789" state

(i.e., to reinstate slavery), Delgrès led an armed rebellion of civilians and soldiers of color.

Unfortunately, the rebels were no match for the French army. They retreated into this fort, where Delgrès issued a proclamation "to the entire Universe" explaining what he was fighting for. Then, when it became clear there was no hope of victory, Delgrès and 400 of his followers holed up in a plantation on the volcano's slope and blew themselves up, along with as many French soldiers as they could.

Slavery was reinstated—though some say the rebellion's failure motivated the successful

liberation struggle in Haiti. As for General Richepanse, he got yellow fever and died a few weeks later and is now buried in the military cemetery at the very top of the fort complex.

The fort's memorial to Louis Delgrès is a sort of cross between a meditation maze and Stonehenge, with Delgrès's head in the center. If you venture into the stone spiral, you'll be able to pick out some carvings on one of the rocks that read LIBERTÉ and JUSTICE.

Basse-Terre's southwest coast is where you will find the fort. Informational signs are in French.
Ⓝ 15.988921 Ⓦ 61.722995

ALSO IN THE CARIBBEAN

ANTIGUA AND BARBUDA

Kingdom of Redonda

Redonda · Multiple kings claim to rule this hotly contested island micronation that was founded on an uninhabited island in 1865.

BAHAMAS

Bimini Road

Bimini Islands · These evenly spaced underwater stones raised hopes that Atlantis had finally been found. Further examination showed that the "road" was a naturally occurring rock formation.

CAYMAN ISLANDS

Hell

Grand Cayman · A patch of jagged limestone formations has come to be known as Hell. Hell postcards with Hell postmarks are available from the local post office.

GRENADA

Mopion

Resembling the prototypical cartoon desert island, this tiny sandbar, supposedly the Caribbean's smallest island, is home only to a single umbrella.

HAITI
SAUT-D'EAU WATERFALLS

HAUT SAUT D'EAU, CENTRE

According to Haitian Catholics, the Virgin Mary once appeared in a palm tree next to the Saut-d'Eau waterfalls. According to Vodou practitioners, the apparition was the *Iwa*, or spirit, Erzulie Dantor. Though the tree was chopped down, members of both faiths have trekked to the waterfalls every July for over a century in search of spiritual and physical healing.

An annual pilgrimage takes place during the festival of Our Lady of Mount Carmel, from July 14 to the 16th. (July 16, 1843, is when the Virgin is said to have appeared in the tree.) The sick and needy come to the 100-foot-tall (30 m) waterfall to pray, bathe, and cleanse themselves with medicinal herbs, hoping to feel the presence of the Virgin Mary—or be possessed by the spirit of Erzulie Dantor. Hundreds of pilgrims crowd the base of the falls, arms raised in reverence toward the torrent of water. Some fall into a trance and must be supported by others so they don't drown.

Pilgrimages to Saut-d'Eau increased following the catastrophic Haitian earthquake of 2010. **The waterfalls are near Mirebalais, an hour drive north of Port-au-Prince. Ⓝ 18.816902 Ⓦ 72.201512**

Thousands of Haitians travel to the falls each year to seek the healing powers of the Virgin Mary of Mount Carmel and her Vodou counterpart, Erzulie.

MARTINIQUE
PRISON CELL OF LUDGER SYLBARIS

ST. PIERRE, ST. PIERRE

A glance at the windowless stone cell that Ludger Sylbaris once occupied may elicit pity for the man— until you learn that this building saved his life.

On May 7, 1902, the town troublemaker ended up in solitary confinement after being arrested for drunk and disorderly conduct. The next day, Mount Pelée, a volcano just north of St. Pierre, erupted, sending a cloud of superheated gas and dust racing toward the city. Within a minute, St. Pierre was leveled. Thirty thousand people burned to death instantly. There were just three survivors: a shoemaker who lived at the edge of town, a girl who escaped on a boat, and Ludger Sylbaris.

Trapped in his cell, Sylbaris couldn't fully escape the intense heat as the ash came flying in through the tiny slot in the door. Suffering from burns and desperate to cool down, Sylbaris urinated on his clothes and stuffed them into the slot. Four days later, rescuers freed him from his prison.

Having survived the worst volcanic disaster of the 20th century, Sylbaris became a celebrity, even touring the world with Barnum & Bailey's circus. Posters billed him as "the only living object that survived in the 'Silent City of Death.'" St. Pierre, once the cultural capital of Martinique, is now a modest town, home to less than 5,000 people.

The stone prison cell that saved a man from the deadliest volcanic eruption of the 20th century.

Rue Bouille, St. Pierre. The closest international airport to St. Pierre is at Le Lamentin, a 45-minute drive away. Ⓝ 14.742798 Ⓦ 61.174719 ➥

➥ Other Sole Survivors

JULIANE KOEPCKE
It was noon on Christmas Eve of 1971, and Juliane Koepcke had just attended her high school graduation ceremony in Lima, Peru. Accompanied by her mother, a German ornithologist, Koepcke boarded LANSA Flight 508, bound for Pucallpa, a city in eastern Peru.

The standard flight time was 50 to 60 minutes. Forty minutes into the journey, a bolt of lightning struck the plane. The fuel tank on the right wing ignited, causing the plane to burst apart.

Juliane Koepcke woke up under her seat on the floor of the Amazon rain forest approximately 20 hours later. Concussed, suffering from a broken right collarbone, and squinting through blood-red eyes due to ruptured capillaries, Koepcke spent half a day fading in and out of consciousness until she managed to stagger to her feet. Her first concern was finding her mother. Despite a full day of searching, she never found her. But Koepcke did encounter some of the 92 passengers who died in the crash.

Though Koepcke could hear rescue planes in the sky, she had no way of signaling to them through the dense foliage. Realizing she would have to rely only on herself to survive, she found a stream and followed it. After nine days of floating and wading, she finally found a motor boat moored on the water and a path leading to an empty shack.

Koepcke was lying in the shack, too exhausted to move, when she heard voices. A trio of locals appeared, utterly shocked to see her after hearing of the crash on the radio. They gave Koepcke food, tended to her wounds, and took her to the hospital at the nearest small town.

It was several more days before rescuers located the wreckage and confirmed Koepcke was the only survivor.

RANDAL MCCLOY

Early in the morning on January 2, 2006, the first day back at work after the New Year's break, 26-year-old Randal McCloy began his shift at the Sago coal mine in West Virginia.

At 6:30 a.m., 2 miles (3 km) inside the mine, a huge explosion rumbled the earth. A sealed passage, blocked due to its high methane levels, had blown open, sending clouds of toxic methane and carbon monoxide rushing through the mine tunnels.

McCloy and his 12 coworkers attempted to drive to the surface, but fallen debris from the explosion made the journey impossible. Each miner carried an emergency "self-contained, self-rescuer" pack, which provided an hour of breathable air, but 4 out of the 13 packs weren't working. Out of options, the men hunkered down in a tunnel section, creating a makeshift tent by nailing plastic sheeting to the ceiling and weighing it down with coal.

The miners took turns ramming an 8-pound (3.6 kg) sledgehammer into a wall bolt to signal their location to rescuers. They listened for blasts above ground—signals that would indicate they had been found. None came. It would be 12 hours before surface crews could even begin the rescue mission, due to the high levels of carbon monoxide and methane.

Four and a half hours after the explosion, the miners were weak and disoriented from carbon monoxide poisoning. They recited prayers together and borrowed one another's pens to scrawl notes to their families. Junior Toler, a 51-year-old section foreman who had spent 32 years in coal mines, wrote, "Tell all I see them on the other side. I love you. It wasn't bad. I just went to sleep."

One by one, the men lost consciousness and collapsed. Forty-one hours after the explosion, a rescue crew finally reached the tunnel, where they found McCloy barely alive beneath the body of a coworker. One of the rescuers relayed the news of the other miners' deaths to the surface via walkie-talkies, but the distance garbled the signal, resulting in an awful miscommunication: The media and miners' families, gathered at a nearby church, were told that all of the trapped miners had survived. Three hours of singing and rejoicing followed before they finally learned the horrific truth.

McCloy spent weeks in a medically induced coma while receiving hyperbaric oxygen therapy. With damage to his brain, heart, kidneys, and liver, as well as a collapsed lung, the prognosis was poor. But following months of inpatient rehabilitation, McCloy was walking, speaking, and able to return home.

VESNA VULOVIĆ

Vesna Vulović has no memory of what occurred on January 26, 1972. That afternoon, the 22-year-old Serbian flight attendant boarded JAT Yugoslav Airways Flight 367 in Stockholm, bound for Belgrade with a stopover in Copenhagen. She wasn't scheduled to work—she had been mixed up with a different Vesna—but took the opportunity to visit Denmark and add some miles to her tally.

The Stockholm-to-Copenhagen leg of the journey was unremarkable. Then came the two-hour flight to Belgrade. About 40 minutes after takeoff, the plane exploded. Twenty-seven people died. One survived: Vesna Vulović.

Vulović fell 33,330 feet (10,159 m), earning her a place in the Guinness Book of Records for surviving the highest fall without a parachute. The wrecked fuselage landed in Srbská Kamenice, in what is now the Czech Republic. A former World War II medic found Vulović pinned under a food cart in the middle section of the plane. She had sustained a fractured skull, three broken vertebrae, and two broken legs, and was temporarily paralyzed from the waist down.

Three days later, Vulović awoke from a coma. With no recollection of the flight or the crash, she was astonished to read the details in the newspaper. It would be 10 months before she could walk again.

The official cause of the explosion on Flight 367 was a suitcase bomb in the front luggage compartment, planted by the Croatian terrorist group Ustaša. In 2009, however, a Czech journalist and two German journalists challenged this explanation based on newly acquired documents. Their assertion—that Czechoslovakia's air force shot down the plane at low altitude after mistaking it for an enemy aircraft—has been refuted by Vulović herself.

The city of Plymouth was buried in ash when Montserrat's volcano erupted.

MONTSERRAT
ABANDONED PLYMOUTH

PLYMOUTH

On February 11, 2010, vacationers flying from Toronto to St. Lucia on a Boeing 737 heard the pilot make an unexpected announcement: "Ladies and gentlemen, if you look to the left of the plane, you'll see a volcano erupting."

The volcano that was indeed hurling a plume of ash into the sky was the Soufrière Hills volcano on the Caribbean island of Montserrat. It was a fantastic sight, but not an uncommon one. Soufrière Hills began to erupt in 1995—for the first time since the 17th century—sending lava flows and ash falls over the 10-mile-long (16 km) island. The affected areas were evacuated and no one was killed. However, just two years later, the volcano went off again, this time killing 19 people. Continuing eruptions destroyed the capital city of Plymouth and covered the entire southern half of the island in a thick layer of ash.

The south side of the island was declared uninhabitable and remains an exclusion zone. Over half of Montserrat's 12,000 residents never returned after being evacuated. Those who stayed live on the north side of the island and have grown accustomed to the ongoing eruptions at Soufrière Hills. The island's economy is even benefiting from volcano-focused tourism, such as helicopter flights over devastated Plymouth and boat rides offering views of the smoking mound.

Montserrat is a 15-minute plane ride or 2-hour ferry ride from Antigua. The southern part of the island is off-limits, but you can see the volcano by boat. Ⓝ 16.707232 Ⓦ 62.215755

PUERTO RICO
MONKEY ISLAND

CAYO SANTIAGO

Half a mile off the eastern coast of Puerto Rico is Cayo Santiago, an island teeming with free-ranging Rhesus monkeys. Researchers from Harvard, Yale, and the University of Puerto Rico's Caribbean Primate Research Center visit the island to study the monkeys' behavior, development, communication, and physiology.

The simian population numbers around 800. All monkeys on the island are descendants of the 409 monkeys imported from India in 1938 to establish the facility. Cayo Santiago has no human inhabitants. Visitors are not permitted, and with good reason: Rhesus monkeys may carry Herpes B, a virus that can be fatal to humans.

Kayak trips to Monkey Island leave from Punta Santiago. You'll be required to stay 30 feet (9 m) from the island, which is still close enough to spot some monkeys running wild. Ⓝ 18.156404 Ⓦ 65.733832

THE TANKS OF FLAMENCO BEACH

CULEBRA

With its soft white sand, turquoise-tinted water, and proximate empanadas, Flamenco Beach is considered one of the best beaches in the world. But there is one incongruous sight in this paradise: a pair of rusting battle tanks, left on the sand as a souvenir from the US Navy.

In 1901, following Spain's ceding of Puerto Rico to the US, President Theodore Roosevelt allocated all of Culebra's public land to the Navy. Soon, troops were conducting test landings and ground maneuvers on the island. In 1939, the Navy began using Culebra for bombing practice. Bombardment reached its peak during 1969, when pilots trained for the war in Vietnam. Missiles hit the island on 228 days of that year.

By 1970, the 700 residents of Culebra were well and truly fed up with the Navy using their home as a bombing ground. Unexploded ordnance littered the island, and the ground bore craters and pockmarks from the shelling. A naval attempt to evict the entire population of Culebra was the last straw. In the summer of 1970, residents began a series of nonviolent protests, aiming to rid the island of naval occupation.

Locals continue to add new art to these abandoned tanks.

After seven months of marches, sit-ins, and human blockades of naval sites, the Culebra activists succeeded. In January 1971, the Navy agreed to stop using the island as a test location by 1975.

Though it's been decades since the Navy left, the tanks on the beach, painted over and over by locals, remain.

Tanks are back in the brush and on the shoreline on the far west side of the beach. Ⓝ 18.316951 Ⓦ 65.290032

ARECIBO OBSERVATORY

ARECIBO

Few telescopes are as awe-inspiring as the one at Arecibo Observatory. At 1,000 feet (304 m) across and 167 feet (51 m) deep, the Arecibo dish is the largest and most sensitive radio telescope in the world. Built into a natural limestone sinkhole, it is made of nearly 40,000 perforated aluminum panels.

Cornell University professor William E. Gordon opened the observatory in 1963, aiming to study the scattering of radio waves off molecules in the Earth's upper atmosphere. Since then, significant astronomical discoveries have been made at Arecibo, including the detection of the first planets outside our solar system. The telescope has also been at the center of several projects in the search for extraterrestrial intelligence.

In 1974, astronomers Frank Drake and Carl Sagan wrote the Arecibo message, a binary string that was beamed from the telescope toward the star cluster M13 some 25,000 light-years away. If the message is eventually decoded by an intelligent race, the extraterrestrial recipients will be greeted with a 23-by-73-pixel bitmap image depicting a human being, chemical formulas, the solar system, and even the telescope itself.

PR-625, Arecibo. While you can't walk on the telescope, you can look at it from a viewing platform and see space-related exhibits at the visitor center. Ⓝ 18.346318 Ⓦ 66.752819

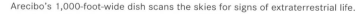

Arecibo's 1,000-foot-wide dish scans the skies for signs of extraterrestrial life.

The museum is named for (and shaped like) its subject.

CEMI MUSEUM

JAYUYA

To the indigenous Taíno people of the Caribbean, the *cemi* is an ancestral spirit as well as a small, usually three-pointed object that holds that spirit. Fashioned from stone and other materials, cemis have a central point representing a mountain peak on which sits Yaya, the Creator. The mouth-like point represents Coabey, the land of the dead. Finally, the third point represents the land of the living.

The Cemi Museum building takes the shape of one of these sacred symbols, and when viewing the structure against the backdrop of the surrounding mountains, it is easy to see how this symbol evolved as a stylized mirror of the topography.

Inside the small museum there is a display of Taíno cemis and artifacts, including a carved, pointed wooden tongue depressor used in ritual vomiting ceremonies. There is also a mural showing a series of petroglyphs that are also believed to have been created by the Taíno.

The museum is a 10-minute drive southeast of Jayuya. To see the Taíno petroglyphs in situ, go to La Piedra Escrita, a boulder in the Rio Saliente that is accesible from the shore. Ⓝ 18.209674 Ⓦ 66.561614

....

ST. KITTS AND NEVIS

COTTLE CHURCH

CHARLESTOWN, NEVIS

In 1824, Cottle Church became the first racially integrated place of worship in the Caribbean. John Cottle, a plantation owner and the former president of Nevis, established the church so that his family and slaves could worship together. At the time, black people were banned from attending Anglican services.

For his defiance of the law, Cottle is generally regarded as a kind and lenient figure. His lenience, however, did not extend to his construction methods: Black slaves built the church.

The Cottle Church ruins are hidden in the woods north of Charlestown. Look for the small sign on the main road just south of the Newcastle Airport and follow the dirt track. Ⓝ 17.196473 Ⓦ 62.596157

A house of worship for both slaves and their owners, Cottle Church was completed in 1824.

TRINIDAD AND TOBAGO

MYSTERY TOMBSTONE

PLYMOUTH, TOBAGO

The grave of Betty Stiven, who died in childbirth in 1783, aged 23, presents a riddle on its tombstone. "What was remarkable of her," the inscription reads, "she was a mother without knowing it and a wife without letting her husband know it, except by her kind indulgences to him."

Some local theories on this mystery point to a passionate and taboo interracial romance that would have required secrecy and the denial of matrimony. But what of the child who somehow came into the world without Stiven realizing? There's a theory for that, too—and it actually makes sense.

According to this version of events, Betty met Alex Stiven and they fell in love. He wouldn't marry her, so she got him drunk, then had a priest perform a sneaky marriage. This takes care of the "wife without letting her husband know" part.

As for the "mother without knowing it" detail, Betty allegedly got pregnant and then, before realizing she was with child, came down with meningitis. She spent the rest of her life in a coma, but not before delivering the baby, who died in the process

Within thefe Walls are Depofited the Bodies of Mrs BETTY STIVEN and her Child She was the beloved Wife of ALEX^B STIVEN to the end of his days will deplore her Death which happened upon the 25th day of Nov. 1783 in the 23rd Year of her Age what was remarkable of her She was a Mother without knowing it and a Wife without letting her Husband know it except by her kind indulgences to him

Betty Stiven's epitaph contains a puzzle.

and was buried beside her. (It is medically possible to give birth while comatose, so that part of the story, at least, could check out.)

The tombstone is enclosed by an orange fence and is well marked. Flights from Port of Spain in Trinidad arrive at A.N.R. Robinson International Airport near Canaan. From there, take the Claude Noel Highway toward Plymouth. Ⓝ 11.221079 Ⓦ 60.778723

PITCH LAKE

LA BREA, TRINIDAD

With the fragrance of a freshly paved road and a viscosity that varies with every step, Pitch Lake is not your average watering hole. The 250-foot-deep (76 m) asphalt lake is the size of around 75 football fields, making it the largest of the world's three naturally occurring pitch lakes—the others are located in Los Angeles and Venezuela.

In some spots, the surface of Pitch Lake is solid enough to walk on. In others, it is more like quicksand. To make matters more confusing, water collects in patches in the basin, resulting in a mix that is sometimes as thin as regular lake water, and other times as hard as rock.

English writer and explorer Sir Walter Raleigh encountered the lake in 1595 and used its pitch

The world's largest asphalt lake is thick enough to walk on.

to caulk his ship. More formalized mining began in 1867 and continues to this day. Asphalt from Pitch Lake has been used to pave roads in over 50 countries, including the runways at New York's JFK airport and the streets of Westminster Bridge in London.

Southern Main Road, La Brea. Pitch Lake is about a 90-minute drive from Trinidad's capital, Port of Spain. Ⓝ 10.232618 Ⓦ 61.628047

ST. VINCENT AND THE GRENADINES
MOONHOLE

BEQUIA, GRENADINES

Named for the stone arch through which you can see the moon set twice a year, Moonhole is a community of 19 homes made from stones and scavenged materials. The beachfront houses, built on the small island of Bequia during the 1960s, are open to the elements: There are no doors to lock and many of the walls have windowless archways.

Some of the homes are now available to rent and come complete with solar-powered refrigerators, hot water, and bars made from whale ribs.

Bequia is a 25-minute ferry ride from Kingstown, the capital of St. Vincent. Moonhole is a 20-minute taxi ride. Ⓝ 12.992146 Ⓦ 61.276701

Moonhole's residents live beneath a stone arch in homes made from salvaged materials.

Antarctica

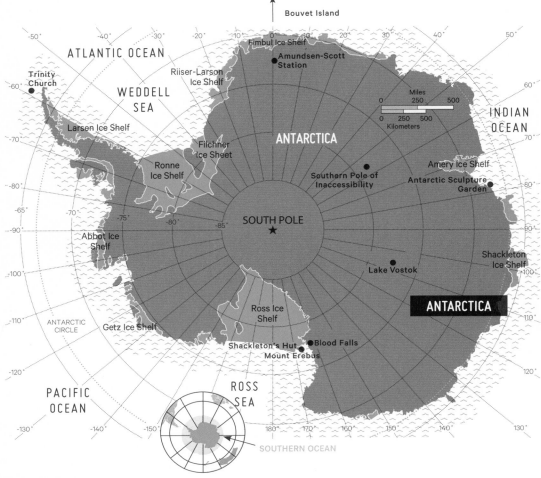

ATLANTIC OCEAN

Trinity Church

WEDDELL SEA

Riiser-Larson Ice Shelf

Fimbul Ice Shelf

Amundsen-Scott Station

ANTARCTICA

Larsen Ice Shelf

Filchner Ice Sheet

Ronne Ice Shelf

Southern Pole of Inaccessibility

Amery Ice Shelf

Antarctic Sculpture Garden

INDIAN OCEAN

Abbot Ice Shelf

SOUTH POLE

Lake Vostok

Shackleton Ice Shelf

ANTARCTICA

ANTARCTIC CIRCLE

Getz Ice Shelf

Ross Ice Shelf

PACIFIC OCEAN

Shackleton's Hut
Mount Erebus

Blood Falls

ROSS SEA

SOUTHERN OCEAN

Miles
0   250   500
Kilometers
0   250   500

BLOOD FALLS

MCMURDO DRY VALLEYS, VICTORIA LAND

Taylor Glacier is hemorrhaging blood—at least that's what it looks like. On the east end of the frozen mass, a five-story stream of rust-colored water flows into Lake Bonney, staining the ice.

This crimson water has been trapped beneath a glacier for 2 million years. Ocean water flooded East Antarctica 5 million years ago, creating a salty lake on the iron-rich bedrock. Taylor Glacier formed atop the lake, sealing it off from sunlight and oxygen and gradually burying it beneath 1,300 feet (396 m) of ice. Despite the absence of oxygen, the hidden reservoir of groundwater is rich with microbial life. At least 17 different microorganisms have been identified in Blood Falls's high-salt, high-iron water, which is now rising through fissures in the glacier. When it meets the air, the water oxidizes, resulting in the bloodlike hue.

Beyond its striking appearance, Blood Falls is interesting to scientists because the glacier's surviving microbes hint at the ecosystems that might be found on Mars and in other harsh, low-oxygen habitats.

The Dry Valleys are accessible only by helicopter from McMurdo Station, Scott Base, or via a cruise ship in the Ross Sea. Cruises depart from New Zealand.
Ⓢ 77.716686 Ⓔ 162.266765

Red water, trapped for 2 million years beneath a glacier.

✸ Antarctica: A note on how to get there

Though seven nations have claimed land in Antarctica, no country owns it. The continent operates according to the Antarctic Treaty of 1959, which defines the continent as a scientific preserve, regulates the research activities of each nation, and prohibits new territorial sovereignty claims.

Travel to Antarctica requires an abundance of two things: time and money. Many visitors arrive on a cruise ship from the Argentinian port of Ushuaia, located on the southern tip of South America. A few tourist ships also launch from Australia, New Zealand, Chile, and Uruguay. Trips take place during the austral summer (from November to March), and cost several thousand dollars depending on length and itinerary.

Commercial flights to the continent—most of which depart from Australia—do not land, but offer sightseeing from above. To fly into Antarctica, you generally need to nab a space on one of the military flights that shuttle in staff and supplies to the research stations. Supply flights also take place between November and March.

Regardless of how you get there, make sure you are healthy when venturing to Antarctica. There are no hospitals, and medical evacuation, when possible, is arduous and expensive.

In 1958, Soviet explorers staked their claim on the South Pole with a Lenin bust.

SOUTHERN POLE OF INACCESSIBILITY

The Southern Pole of Inaccessibility—as distinguished from the geographic South Pole, 550 miles (885 km) away—is the point in Antarctica farthest from the ocean. This inhospitable spot, on which few humans have trod, has an average yearly temperature of –72.8°F (–58.2°C). It is marked by a bust of Vladimir Lenin.

In 1958, a team of 18 determined Soviet explorers reached the Southern Pole of Inaccessibility for the first time. The team journeyed there from Mirny, a Soviet station established on the coast of the Davis Sea two years earlier. Aboard their tractors were components for the prefabricated wooden huts that would form a new four-person research station.

When they arrived, the team assembled the huts, raised the Soviet flag, and added one final touch to the top of the chimney on one of the buildings: a Lenin bust, positioned to face Moscow. The base station was used only for a few weeks to monitor weather before being abandoned to the elements. The communist revolutionary's sculpted head is often half-submerged in snow.

The exact point of the Southern Pole of Inaccessibility is the subject of debate—the movement and melting of ice sheets alter the coastline, affecting measurements. The bust is located at Ⓢ 82.099907 Ⓔ 54.967117

LAKE VOSTOK

VOSTOK STATION, PRINCESS ELIZABETH LAND

Buried in ice 2 miles (3.2 km) beneath Russia's Vostok research station lies a lake that's been sealed away from the world for an estimated 15 million years.

Measuring 160 miles long by 30 miles wide (258 km × 48 km), Lake Vostok is the largest of Antarctica's subglacial "ghost lakes." Its existence was confirmed in 1993, via radar altimeter data from a remote-sensing satellite.

The air temperature at Lake Vostok has plummeted as low as –128°F (–88.8°C), the lowest temperature ever recorded on Earth. The underground lake's average water temperature is comparatively balmy at 27°F (–2.7°C). Though that's below the usual freezing point, the lake stays in a liquid state due to the pressure exerted by the ice above, which also provides insulation. Geothermal heat from beneath the lake may also play a role.

The search for signs of life in the lake began during the late 1990s, when scientists began drilling into the ice core to retrieve samples for analysis. As their probes approached the water, concerns arose that the drilling fluids used, Freon and kerosene, might contaminate any samples they collected. With

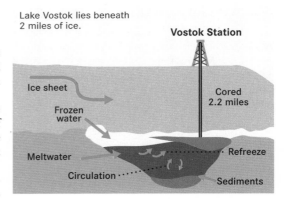

Lake Vostok lies beneath 2 miles of ice.

over 300 feet to go before they reached the lake, the team chose to stop drilling.

In 2012, using silicone oil as a drilling fluid, Russian scientists reached the surface of the lake. Analysis of the samples showed DNA from over 3,500 organisms. That so much life could exist in such an inhospitable environment gives hope to scientists that life may someday be found outside planet Earth.

Vostok Station, home of the big drill, is at the southern Pole of Cold, around 800 miles (1,287 km) southeast of the South Pole. Ⓢ 77.499996 Ⓔ 106.000028

ANTARCTIC SCULPTURE GARDEN

DAVIS STATION, PRINCESS ELIZABETH LAND

His proper name is *Man Sculpted by Antarctica*, but he's better known as "Fred the Head." He stands outside the meteorology building of Australia's Davis Station, and resembles Easter Island moai. Fred is the creation of a plumber named Hans, who carved the sculpture out of an old wooden pole during his winter stay in 1977.

Exposed to the winds and snow for decades, Fred has acquired a weathered complexion and become a totem for those enduring the long, dark Antarctic winters. In 2003, Davis artist-in-residence Stephen Eastaugh was so inspired by Fred that he created a sculpture garden using the wooden head as a centerpiece. Visitors are invited to add their own creations to Eastaugh's wood and metal sculptures, but they must first get permission from Davis's station leader and environmental adviser.

Flights to Davis via Casey, another Australian Antarctic station, leave from Hobart, Tasmania. Ⓢ 68.576206 Ⓔ 77.969449

There may not be any plants in this garden, but creativity flourishes.

SHRINES OF THE AMUNDSEN–SCOTT STATION ICE TUNNELS

SOUTH POLE

The web of ice tunnels beneath Amundsen-Scott Station serve as conduits for the research base's power lines, supply pipes, and sewage removal. But the tunnels aren't just utilitarian. Recesses carved into the frozen walls contain shrines to completed projects, departed scientists, and Antarctic in-jokes that defy explanation.

The tunnels were constructed over three summers as part of a station that replaced the original Amundsen-Scott base established in 1956. Completed by excavator, chainsaw, and pickax in 2002, the halls are a constant –60°F (–51.1°C) and measure 10 feet high by 6 feet (3 m × 1.8 m) wide. Or, at least they used to—much like a household

South Pole Station's winter 2003 crew left behind a pig's head shrine.

freezer, ice has been building up on the walls, reducing the size of the tunnels by several inches.

The extreme cold keeps each shrine beautifully preserved. Flowers, strings of popcorn, prayer candles, tinned caviar, and a hard hat are among the eclectic items on display.

Then there is the sturgeon, set in an alcove beside a document detailing its backstory. Back in the 1990s, Russian researchers staying in Antarctica gifted the fish to American scientists at McMurdo Station, located 850 miles (1,368 km) from the South Pole. The sturgeon sat unappreciated in a freezer for months, and was on its way to being thrown out when a researcher headed from McMurdo to Amundsen-Scott Station decided to bring it along as a gift.

The frozen fish, enshrined in the tunnels, is now the research base's only permanent resident. **Once you've checked out the tunnels, make sure to pose for a photo beside the ceremonial South Pole, marked by a red-and-white post surrounded by the flags of nations that have signed the Antarctic Treaty. ⑤ 72.294154 ⑥ 0.696294**

BOUVET ISLAND

BOUVET ISLAND

Located between Antarctica and South Africa, 1,404 miles (2,260 km) from the nearest human settlement, Bouvet is, by definition, the most remote island on Earth. Should you reach its shores, you will find it unwelcoming in the extreme. An ice-covered volcano lies at its core, and storms rage around the island for over 300 days per year.

Discovered in 1739, the island is so isolated that it was accidentally "lost" for nearly 70 years. No one actually set foot on Bouvet until 1927, when a group of Norwegians scrambled up its glacial cliffs. In 1964, an abandoned lifeboat with no national markings was discovered on the island. From the supplies spread about, it was clear someone had been there, yet ships did not normally come within 1,000 miles (1,609 km) of Bouvet, and the lifeboat was equipped only with oars and no mast. Despite a search, no bodies or tracks were found. As of today, fewer than a hundred people have ever set foot on the island—not counting whoever was in that lifeboat. **Assuming no budgetary restrictions, the best way to get on the island is to fly a helicopter from the deck of a ship and land delicately on Bouvet's icy surface. Bouvet is a dependent territory of Norway, so Norwegian laws apply. ⑤ 54.432711 ⑥ 3.407822**

ALSO IN ANTARCTICA

Wilson's Stone Igloo

Cape Crozier · See the remains of a shelter built during what Scott expedition survivor Apsley Cherry-Garrard described as "the worst journey in the world."

Discovery Hut

Hut Point · Built by the British in 1902, this wooden hut was used to store tinned meats, flour, and coffee for expeditions.

McMurdo Dry Valleys

McMurdo Sound · The snow-free peaks and troughs of this region comprise one of the world's most extreme deserts.

Chapel of the Snows

McMurdo · This is the third incarnation of the world's second-southernmost place of worship—the previous two burned down.

IceCube Research Station

South Pole · This particle detector uses a telescope buried a mile under the ice to look for neutrinos.

Buckminster Fuller's Dymaxion Map

All maps are, in one way or another, a lie. The problem isn't with the cartographer but with geography. When we flatten our globe to make a two-dimensional map, pulling and stretching it like taffy to make a rectangle, we distort the world and suddenly Greenland looks like it's the size of Africa. Greenland is not the size of Africa.

Inventor, architect, and designer Buckminster Fuller was aware of this map problem. "All flat surface maps are compromises with truth," he said. His solution, which premiered in *Life* magazine in 1943, was both beautiful and simple. Fuller turned the earth into an icosahedron, a three-dimensional shape made up of twenty equilateral triangles. These smaller triangular surfaces minimized the distortion problem. While it makes for a strange and jagged map, it creates a much truer picture of the world.

Unfolding the world this way did more than just fix the size of Greenland. It also created a new way to visualize the arrangement of Earth on a map. In this particular layout of Buckminster Fuller's map, the continents appear as one great land mass stretching from Australia to Antarctica.

It's for that reason that we chose the Dymaxion map for our world trip map. The exact trip you see may not be practical, or even entirely possible, but this map is a beautiful way to visualize an epic journey.

One great loop around the island of Earth.

We'll start in a cold place, the Davis Station Sculpture Garden in Antarctica, move through to the warmer climes of Australia, and then circle back again. It's everything you need to take the most incredible Atlas Obscura trip around the world possible . . . plane, boat, car, train, bicycle, horse, and camel not included.

R. Buckminster Fuller holding his folded Dymaxion map. Photographed by Andreas Feininger for Life *magazine, 1943.*

Amazon Bore Surfing,
p. 399

Snake Island
p. 402

5

6

3

4

dlândia,
p. 403

Glowing Termite Mounds,
p. 404

Trinity
Church
p. 458

Antarctic
Sculpture Garden,
p. 454

Start here

1

75

76

77

2

80

Keshwa Chaca
Bridge,
p. 416

Laguna del
Diamante,
p. 394

The Marble
Cathedral,
p. 407

Southern Pole
of Inaccessibility
p. 453

78

Mount Erebus,
p. 460

eball Festival,
p. 430

79

wallows
27

Blood Falls,
p. 452

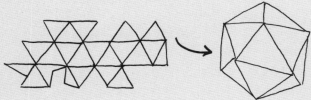

Buckminster Fuller's Dymaxion map represents his belief in equality.
When folded, it becomes a round Earth with one continuous landmass,
its face made up of 20 triangles all equal in size, with no definitive
center and no north/south or east/west.

Atlas Obscura's
Trip Around the World

Annual Festivals Around the World

JANUARY

Harbin International Ice and Snow Sculpture Festival

Harbin, China
For the world's competitive ice and snow sculptors, there is no bigger stage or more serious contest. The scale of these frozen structures is staggering: The festival grounds are split into an exhibition area for massive snow sculptures and an entire city constructed of ice blocks.
JANUARY 5

The Burning of the Clavie

Burghead, Scotland
In a ritual dating back to Roman times, villagers parade a "clavie"—a blazing wooden cask mounted on a pole—to the ruins of an ancient altar where it is built up into a great bonfire.
JANUARY 11
(New Year's Eve by the Julian Calendar)

FEBRUARY

Busójárás Monster Parade

Mohács, Hungary
Goat-horned monsters and mysterious women in lace masks parade through town, jangling bells and twirling noisemakers during this Carnival celebration.
SHROVE TUESDAY

Sa Sartiglia Festival

Oristano, Italy
Equestrian performers and competitors don blank-faced porcelain-doll masks and race elaborately decked-out horses through the streets. The riders attempt to spear star-shaped rings hung in front of the local chapel, which sometimes requires standing on top of their galloping horses.
LAST SUNDAY AND TUESDAY BEFORE LENT

Festival of the Exploding Hammers

San Juan de la Vega, Mexico
Commemorating Juan de la Vega's triumph, aided by San Juan Bautista, over local outlaws, the braver locals strap homemade firecrackers to the end of sledgehammers and slam them onto a sheet of metal, detonating the package to the delight of the crowd.
SHROVE TUESDAY
(Feast of San Juan Bautista)

Da Shuhua Molten Iron Throwing

Nuanquan, China
This annual tradition was dreamed up approximately 500 years ago by blacksmiths who wanted to participate in the annual Lunar New Year festivities but couldn't afford

Great Mosque of Djenné, p. 205

Richat Structure, p. 194

El Caminito del Rey, p. 71

Temples of Damanhur, p. 52

28

Maunsell Army Forts, p. 8

Fingal's Cave, p. 20

24

23

22

Borehole, p. 91

21

Svalbard Seed Vault, p. 104

20

North Pole

Diavik Diamond Mine, p. 269

19

Oymyakon, p. 93

63

Sourtoe Cocktail, p. 274

The Nekoma Pyramid, p. 324

Self-Mummifying Monks, p. 156

62

Cactus Dome, p. 253

The Fremont Troll, p. 303

Nutshell Dioramas, p. 362

Great Stalacpipe Organ, p. 358

Cryptozoology Museum, p. 375

Georgia Guidestones, p. 350

L'Anse aux Meadows, p. 276

15

14

12

11

13

House on the Rock, p. 340

The Devil's Kettle, p. 335

17

16

18

Bishop Castle, p. 306

69

68

70

71

72

67

65

66

Granite Mountain, p. 319

Neptune Reef, p. 348

9

Mississippi Basin Model, p. 354

10

74

73

Catatumbo Lightning Storm, p. 419

Boiling Lake, p. 440

7

8

Rain of Fish, p. 433

Cave of S...

Cave of the Crystals, p. 427

Lightning Field, p. 312

Integratron, p. 280

the luxury of traditional fireworks. Instead, the blacksmiths tossed cupfuls of molten iron against the city gate. The result was a spectacular shower of pyrotechnic beauty.
LUNAR NEW YEAR

MARCH
Frozen Dead Guy Days
Nederland, Colorado, USA
A rollicking festival honoring the town's one and only cryogenically frozen resident, Bredo Morstoel. Popular events include Ice Turkey Bowling, the Parade of Hearses, Tuff Shed Coffin Races, and the Salmon Toss.
SECOND WEEKEND IN MARCH

APRIL
The Chios Rocket War
Vrontados, Greece
Two churches shoot tens of thousands of small homemade rockets at one another. The goal, as much as there is one, is to hit the opposing church's bell tower.
ORTHODOX EASTER DAY

MAY
Festival of the Snake Catchers
Cocullo, Italy
The Snake Catchers procession, held in this small village in central Italy, honors St. Dominic, who cleared the area of snakes, and the pagan snake goddess, Agnizia. A statue of St. Dominic,

completely covered in snakes, is paraded away from the church, along with musicians and a coterie of snake catchers.
MAY 1

Rock-Throwing Battle
Santiago de Macha, Bolivia
Tinku, a Bolivian Aymara and Quechua tradition of ritualistic combat, represents the fight of indigenous peoples against colonial oppression. Tinku events take place all over the country in May, but the Macha festival is the most notoriously bloody. The festival begins with men and women dancing together, only for the women to back away from the men as things escalate into an all-out brawl.
EARLY MAY

JUNE
Kirkpinar Oil Wrestling Festival
Avarız Köyü, near Edirne, Turkey
In the world's oldest regular sporting event (besides the Olympics, that is), brawny wrestlers—naked but for their prized leather trousers—are doused in olive oil to grapple until one wrestler pins the other or lifts him above his head. The original wrestlers were said to have fought to the death. These days, competitors fight for the $100,000 cash prize.
LATE JUNE

JULY
L'Ardia di San Costantino Festival
Sedilo, Italy
One part horse race, one part religious pilgrimage, this yearly race is run by brave horsemen atop their faithful steeds, reenacting the victory at Rome of the emperor Constantine over his rival Maxentius in the year 312.
JULY 6–7

The Festival of Paucartambo
Paucartambo, Peru
Every year, *campesinos* from the Sacred Valley trek to a remote town to dance in the festival of the Virgen del Carmen. Each regional group performs in full regalia to reenact their village's origin story. They represent everything from wealthy Spanish-blooded landowners to tempting devils in Chinese dragon–style headdresses.
THIRD WEEK IN JULY

AUGUST

Dragon-Slaying Festival
Furth im Wald, Germany
The *Drachenstich*—one of the oldest folk plays in Germany—has been performed in this medieval "City of Dragons" since 1590. The festivities include 1,000 locals dressed in historical costume, at least 250 horses, a hero knight, and his wife who needs to be rescued from a 50-foot-long (15 m) fire-breathing dragon-robot.

SECOND WEEK IN AUGUST

Roggenburg Leiberfest
Roggenburg, Germany
On Assumption Day, this Bavarian village brings out their dead. Led by a large brass brand, the exquisitely adorned and bejeweled mummified bodies of four local saints are disinterred from their place of rest at the abbey and paraded through the village.

AUGUST 15

Drunken Dance Festival
Tokushima, Japan
Some 80,000 participants and 1.3 million spectators descend upon the city of Tokushima for Awa Odori, a three-day dance festival where colorfully dressed dance teams chant, sing, and perform different versions of a signature dance step. As the most famous Awa Odori song puts it, "The dancers are fools / the watchers are fools / both are fools / so why not dance?"

AUGUST 12–15

SEPTEMBER

Exeter UFO Festival
Exeter, New Hampshire, USA
UFO enthusiasts from around the country gather in this small town, the location of a 1965 UFO sighting, to listen to lectures and participate in alien fun for believers and skeptics alike.

LABOR DAY WEEKEND

OCTOBER

El Campello Moors and Christians Festival
El Campello, Spain
To commemorate the 13th-century struggle between the Christian kingdoms to the north of the Iberian Peninsula and the Moorish (Muslim) occupiers in the south for control of the country, the coastal town of El Campello puts on an intricately choreographed reenactment of the arrival of the Moorish fleet.

SECOND WEEK IN OCTOBER

Feast of St. Jude
San Hipólito Church, Mexico City
In honor of St. Jude, the patron saint of lost causes and desperate situations, nearly 100,000 of Mexico's working class, poor, and marginalized come to San Hipólito carrying red and white roses and hauling full-size wooden statues to sing, pray, and find strength.

OCTOBER 28

NOVEMBER

Bonfire Night
Ottery St. Mary, England
Each Guy Fawkes Day, Ottery's residents hoist burning barrels on their shoulders or heads and run through teeming crowds. Only established locals are allowed to carry the flaming barrels. And while the fire department is on hand, other health and safety regulations are encouraged to take the night off.

NOVEMBER 5

DECEMBER

Gävle Goat
Gävle, Sweden
Every year, the good people of Gävle build a huge straw Yule goat—an ancient Nordic Christmas (and probably pagan) celebration of the solstice—and display it in the town square. And every year (starting in 1966), vandals attempt to destroy the goat. It's been torched, stolen, smashed, run over by a car, and once nearly thrown in the river.

DECEMBER 1

Bonfires on the Levee for Papa Noël
St. James Parish, Louisiana, USA
Through the month of December, families of St. James Parish build pyres along the levee, some using fireworks as kindling. Each night the community lights one bonfire. On Christmas Eve, all the remaining pyres are lit, and the town comes out to walk around, taking in the displays.

DECEMBER 24

Harbin Ice Festival in the daytime, and at night, right.

Deadvlei, p. 221

Kolmanskop — Ghost Town, p. 220

Avenue of the Baobabs p. 226

Oklo Reactor, p. 202

Kane Kwei, p. 203

Capuchin Catacombs, p. 60

Easter Rocket War, p. 49

Abuna Yemata Guh, p. 212

Valley of Whales, p. 190

Socotra Island p. 122

Précontinent II, p. 199

Baalbek Trilithon, p. 118

Buzluda Monume p. 76

Kol

Chand Baori Stepwell, p. 130

Door to Hell, p. 139

Nek Chand's Garden, p. 127

Kyaiktiyo Pagoda p. 175

Skeleton Lake, p. 127

Root Bridges of Cherrapunji, p. 132

Dochula Pass Caves, p. 126

Tungus Event p. 90

Synchronized Fireflies p. 172

Hengshan Hanging Temple, p. 142

Guoliang Tunnel, p. 144

Third Tunnel of Aggression, p. 162

Jellyfish Lake p. 257

Skylab Remains, p. 238

Marree Man p. 240

Tana Toraja Funeral Rights, p. 166

Battleship Island, p. 160

G-Cans, p. 154

Waitomo Glowworm Caves, p. 249

Ball's Pyramid p. 245

Electrum, p. 246

SHACKLETON'S HUT

CAPE ROYDS

The shelves are lined with tins of curried rabbit, stewed kidneys, ox cheeks, and sheep tongues. Long johns hang off washing lines slung from the walls, and boots sit in rows underneath the makeshift beds built from wooden crates. Ernest Shackleton's hut looks just as it did when the explorer abandoned it in 1909.

Shackleton and his team of 14 men assembled the prefabricated timber hut in February 1908, having lugged it from London via New Zealand. The building was to serve as a base during the Nimrod Expedition, Shackleton's quest to be the first to reach the South Pole.

The four-man trip toward the pole had a bittersweet outcome: With their ponies dead and their food supply dwindling, the quartet turned back 112 miles (180 km) from the South Pole—the closest anyone had gotten at the time.

The contents of Shackleton's hut, naturally preserved in the frozen climate, offer a fascinating insight into the Heroic Age of Antarctic Exploration, the era beginning in the late 19th century in which pioneering explorers like Shackleton, Scott, and Amundsen led perilous, often fatal expeditions. Among the items on his packing list, intended to sustain 15 men for up to two years, were 1,600 pounds of the "finest York hams," 1,260 pounds of sardines, 1,470 pounds of tinned bacon, and 25 cases of whisky (that's 726 kg, 572 kg, and 667 kg, respectively).

Century-old curried rabbit and tins of ox cheek remain just as Shackleton left them in 1909.

The discovery of this last item was the cause of much excitement when, in 2010, conservators retrieved a crate of Mackinlay whisky from a stash of booze hidden beneath the hut. Three of the bottles underwent chemical analysis in Scotland, after which Mackinlay used the flavor profile to create a replica whisky. The original bottles were returned to the hut, where they sit in tribute to a long-departed gentleman explorer.

Visitors to the hut must be accompanied by a guide. A maximum of 8 people are allowed inside at one time, and all are required to sign the logbook. ⊗ 77.552922 ⊜ 166.168368

✏ Self-Surgery in the Antarctic

On April 29, 1961, 27-year-old Russian physician Dr. Leonid Rogozov experienced pain in his side and began to feel very unwell. Rogozov was one of a dozen men stationed at the Soviet Novolazarevskaya research base, and the only physician on staff. As the pain grew more intense, and a violent snowstorm made evacuation by plane impossible, Rogozov confronted a terrifying truth: He would have to perform his own appendectomy.

Lying on his back and using a mirror for guidance, Rogozov spent 2 hours digging around in his anesthetized abdomen as his colleagues looked on. Weakened and suffering from headspins, the determined doctor took 20-second rests every 5 minutes to gather his strength. Rogozov's journal entry on the operation recounts a close call: "At the worst moment of removing the appendix I flagged: My heart seized up and noticeably slowed; my hands felt like rubber. Well, I thought, it's going to end badly."

Despite the physical strain, Rogozov completed the operation and made a full recovery, going on to live for another 29 years. Some of the surgical instruments used during the operation are on display at the Museum of the Arctic and Antarctic in St. Petersburg.

Rogozov is not the only Antarctic patient who has resorted to self-administered surgery. In March 1999, Dr. Jerri Nielsen was stationed at the US Amundsen-Scott station at the South Pole when she found a lump in her breast. With transport off the continent impossible until October, Nielsen performed her own biopsy and sent images to oncologists in the United States for diagnosis.

When it was confirmed that she had breast cancer, Nielsen self-administered hormones and chemotherapy drugs using supplies air-dropped from military planes.

When the air temperature warmed to about –60°F (–51.1°C), Nielsen was flown to the US for treatment. She went into remission following a mastectomy, but the cancer returned and she died from the disease in 2009.

TRINITY CHURCH

BELLINGSHAUSEN STATION, KING GEORGE ISLAND

When it comes to unique weddings, it's hard to beat the union of Chilean Antarctic researcher Eduardo Aliaga Ilabaca and Russian scientist Angelina Zhuldybina. In January 2007, the couple donned their formal wear, walked through the snow, and tied the knot at Trinity Church, the southernmost Orthodox church in the world.

Standing 50 feet (15.2 m) tall and constructed from Siberian pine, Trinity was built in 2002 in Altai, a federal subject of Russia that borders Kazakhstan and Mongolia. From there, the church was transported 2,500 miles (4,023 km) to Kaliningrad (a Russian exclave on the Baltic Sea between Poland and Lithuania), dismantled, and brought to King George Island. A delegation of 20 religious leaders attended the official consecration, which took place in February 2004. Two priests now conduct services at the church, staying for a year before being swapped for new staff.

Trinity Church is within walking distance of Russia's Bellingshausen research station, making it a convenient Sunday morning trip for Russian Orthodox scientists. Weekly services, weddings, and baptisms are all on offer—with the Southern Ocean providing a natural, albeit alarmingly cold, source of water for the baptism ritual.

The church is on a hill near Bellingshausen Station, a Russian research base at Collins Harbor.
Ⓢ 62.196405 Ⓦ 58.972042

With its golden wall of saints, the southernmost Orthodox church in the world is the perfect spot for an adventurous elopement.

Sunlight illuminates the ceiling of an ice cave on the southernmost active volcano on Earth.

MOUNT EREBUS

ROSS ISLAND

When British polar explorer James Clark Ross came across this volcano in 1841, he named it after his ship. But Mount Erebus has since proven itself more like the original Erebus of Greek mythology, son of Chaos and god of darkness.

The mountain that bears Erebus's name is a 12,448-foot (3,974 m) active volcano brimming with boiling lava. It is an odd conflation of fire and ice: Heat and gas from the 1,700°F (926.6°C) lava lake filter up to the slopes, melting the packed snow and carving out ice caves.

Since 1979, an area on Mount Erebus's lower slopes has been regarded as a tomb. That year, an Air New Zealand sightseeing flight crashed into the mountain, killing all 257 passengers on board. The plane had been flying in a whiteout, using coordinates that differed from the approved route. Though an extensive recovery effort took place, wreckage from the crash is still on the mountain. A memorial cross stands nearby, along with a capsule containing messages from relatives of the victims.

Although the volcano is always active (sometimes hunks of molten rock shoot through the air), Mount Erebus is open to climbers in the Antarctic summer. Entry into the crash site area is prohibited unless you have a permit issued by the relevant New Zealand authorities. Ⓢ 77.527423 Ⓔ 167.156711

ACKNOWLEDGMENTS

In accordance with ancient Bolivian custom (see La Paz Witches' Market, page 395), our relationship with Workman Publishing began with the gift of a mummified llama fetus. When presented with the rank carcass, which is traditionally buried under the foundation of a new building as an offering to the fertility goddess Pachamama, Suzie Bolotin did not vomit. Instead she had it framed and mounted on her wall. We knew we had found the right publisher.

Indeed, we could not ask for better partners than the imaginative, exacting, and ever-patient team at Workman. Thank you, Suzie Bolotin, Maisie Tivnan, and Janet Vicario for making a bet on *Atlas Obscura*, and then for hanging with us through all the ups and downs of producing this insane book. We've had a great crew supporting us: Sun Robinson-Smith, Danny Cooper; Justin Krasner; our production editor Amanda Hong; our typesetters Barbara Peragine and Jaclyn Atkinson; Monica McCready; Doug Wolff; Carol White; Orlando Adiao; and our photo researchers Bobby Walsh, Melissa Lucier, Aaron Clendening, Sophia Rieth, and Angela Cherry. And, of course, Workman's wonderful publishing and marketing team: Selina Meere, Jessica Wiener, Rebecca Carlisle, and Thea James.

We're grateful to the incomparable Elyse Cheney and Alex Jacobs for shepherding this book through all its stages. To project manager Marc Haeringer for keeping us on track even when the train got really wobbly. To Meg Neal for her contributions to the revised edition. And to our entire team at AO HQ who have put so much of themselves into building *Atlas Obscura*. Thank you, David, David, Dan, Tyler, Megan, Sommer, Jordan, Alexa, Michael, Mike, Reyhan, Eric, Lex, Luke, Rachel, Sarah, Cara, Blake, Hana, Anika, Erik, Rose, Matt, Erin, Michelle, Rebecca, Ryan, Tao Tao, and Urvija. To the folks who were there with us in the early days, this book is yours too. Annetta, Seth, Allison, Nick, Adam, Aaron, and Rachel: Thank you.

JOSH:

It has been a tremendous pleasure working on this project with such extraordinary people. Ella, you are simply a wonder. I admire your diligence, good humor, and grace so much. Marc, my brain hurts imagining how we could have ever pulled this off without your organizational wizardry. Dylan, thank you for always being a wonderful partner, friend, and human being. How much longer can we keep Derweze waiting?

DYLAN:

Ella and Marc, that this book exists is the eighth wonder of the world. Thank you for five years of dedication and creativity. Maisie. Suzie. Janet. Dan. Megan. Allison. Nick. Annetta. Seth. Rachel. Eric. You all had a hand in *Atlas Obscura* becoming real. Thank you. Mom and Dad, as you know, you made me this way. Michelle Enemark for being there in every way for nearly half my life and to Phineas for coming into existence. Josh, thank you for everything. For trusting me, for guiding me. You are my friend and an inspiration. Derweze can't wait. I booked tickets.

ELLA:

Dylan and Josh, your philosophies on exploration and discovery have changed how I see the world. Thank you for trusting me with your most treasured wonders. Thank you, Marc Francois Haeringer, for your unwavering support, diplomacy, and artisanal charcuterie plates. Thank you, Opus and Jez, for making soothing noises as I lay on various floors and couches and moaned that this would never, ever be finished. Eric, thank you for all the Grund mornings. And thank you to Mum and Eclair for the many ways in which you tether my flailing self to the earth.

PHOTO CREDITS

INTRODUCTION, P. 6
Daniel Mihailescu/AFP/Getty Images.

GATEFOLD
Alamy Stock Photo: Tibor Bognar p. 2 (left); dpa picture alliance p. 2 (middle); Robert Harding p. 1 (middle); JASPERIMAGE p. 1 (left); Guido Paradisi p. 1 (right); Chris Willson p. 2 (right). **Shutterstock.com:** aphotostory pp. 1–2 (background), 2 (btm).

AFRICA
Adobe Stock: 3drenderings p. 227 (top); Marina Gorskaya p. 204 (top); Morphart p. 210 (top); Piccaya p. 205; R. Gino Santa Maria p. 186 (top); Siloto p. 219 (top). **Camille Moirenc/AGE Fotostock** p. 194 (btm). **Alamy Stock Photo:** age fotostock pp. 195 (btm), 201, 226 (btm); blickwinkel p. 228 (top); brianafrica p. 225 (btm); Rungtip Chatadee p. 188; Gilles Comlanvi p. 200 (top); Michael Runkel Egypt p. 186 (btm); Eddie Gerald p. 227 (btm); Oliver Gerhard p. 219 (btm right); Mike Goldwater p. 214 (top); Robert Harding p. 222 (btm); Blaine Harrington III p. 223 (top); Kim Haughton p. 199 (top); Hemis p. 197 (btm); Historic Collection p. 189; Seth Lazar p. 202 (top); Sulo Letta p. 196; Look Die Bildagentur der Fotografen GmbH p. 213 (inset); Henri Martin p. 208 (inset); Andrew Michael p. 223 (btm); National Geographic Image Collection p. 198 (top); B. O'Kane p. 185; Pictures Colour Library p. 187; Robert Estall photo agency p. 224 (btm); p. 203 (composite); Neil Setchfield p. 221 (btm); Mike P. Shepherd p. 192 (top); Kumar Sriskandan p. 225 (top); Fredrik Stenström p. 226 (top); Universal Images Group p. 194 (top); p. 224 (btm); Universal Images Group/DeAgostini p. 210 (btm); John Warburton-Lee Photography p. 228 (btm); Tim E White p. 207 (btm). **Matteo Bertolino/matteobertolino. com** p. 204 (btm). © **William Clowes** p. 211. **Getty Images:** Nigel Pavitt/AWL Images p. 215 (btm); Pascal Deloche/Corbis Documentary p. 209 (top); Anup Shah/Corbis Documentary p. 216 (top); Nik Wheeler/Corbis NX p. 208 (background); De Agostini/G. Dagli Orti p. 191; Marc Guitard/Editorial RF p. 213 (background); Leonid Andronov/iStock p. 199 (btm); pascalou95/iStock p. 193 (btm); DigitalGlobe/ScapeWare3d p. 206 (middle); Reinhard Dirscherl/WaterFrame p. 197 (top). **Ian Redmond/Nature Picture Library** p. 214 (btm).

Courtesy Photos: The following images are used under a Creative Commons Attribution 3.0 United States License (https://creativecommons.org/licenses/by/3.0/us) and belongs to the following Wikimedia Commons user: Ji-Elle p. 217 (btm left, middle & btm right). **Public Domain:** Dr. Steve Miller, from the Naval Research Laboratory/U.S. Navy p. 218 (top).

Atlas Obscura **Contributor:** Courtesy Megan E. O'Donnell p. 219 (btm).

ANTARCTICA
Aha-Soft/Adobe Stock p. 457 (btm). **Alamy Stock Photo:** B. O'Kane p. 458 (btm); Cavan Images p. 452 (btm); Robert Harding p. 458 (btm); Colin Harris/era-images p. 457 (top). **Courtesy of Stephen Eastnaugh/Australian Antarctic Division** p. 454 (btm). **Andreas Feininger/The LIFE Picture Collection/Getty Images**

p. 456. **Kristina Gusselin** p. 455 (top). **Carsten Peter/National Geographic Stock** p. 459. **Stein Tronstad** p. 453.

ASIA
Adobe Stock: Anthonycz p. 133 (top); bluebright p. 142 (btm); Rada Covalenco p. 144 (top right); evegenesis p. 145 (top); forcdan p. 119 (top); frog p. 136 (btm); kim1970 p. 144 (top left); R.M. Nunes p. 165 (top); SoulAD p.159 (top); Telly p. 145 (middle right). **AGE Fotostock:** Angelo Cavalli p. 175 (btm); Deddeda p. 177(top); Ivonne Peupelmann p. 166; Topic Photo Agency IN p. 164 (btm). **Alamy Stock Photo:** Aflo Co. Ltd p. 160; afrisson p. 133 (btm); age fotostock p. 134 (inset background); age fotostock p. 163; Agencja Fotograficzna Caro p. 146; Mark Andrews p. 150 (top right); Vladislav Ashikhmin p. 164 (top); Asia Images Group Pte Ltd p. 137; Don Bartell p. 150 (top left); Curtseyes p. 138 (top); Luis Dafos p. 136; DestinationImages p. 127 (btm); Paul Doyle p. 117 (btm); dpa picture alliance p. 141 (inset background); Kristaps Eberlins p. 118 (btm); Dominic Dudley p. 138 (btm); Michelle Gilders p. 162 (btm); Manfred Gottschalk p. 167; Simon Grosset p. 119 (top); hanohikirf p. 119 (btm); Marc F. Henning p. 145 (middle left); Imagebroker p. 122 (top) and p. 165 (btm); Ellen Isaacs p. 132 (btm left); LOOK Die Bildagentur der Fotografen GmbH p. 143; Don Mammoser p. 131; MJ Photography p. 130 (top); Will Moody p. 123; Nokuro p. 174; Novarc Images p. 132 (btm right); NPC Collection p. 157 (btm); NurPhoto. com p. 176; PhotoStock-Israel p. 117 (top); Paul Rushton p. 147; Olena Siedykh p. 126 (top); Jack Sullivan p. 118 (top); Keren Su/China Span p. 148 (btm); SuperStock p. 134 (inset); Jeremy Sutton-Hibbert p. 161; tonyoquias p. 178; Travel Asia p. 128 (btm); John Warburton-Lee Photography p. 121; Henry Westheim Photography p. 142 (top); Tim Whitby p. 139; Xinhua p. 152. **AP Photo:** David Guttenfelder p. 162 (top); Shizuo Kambayashi/STF p. 158. **Christian Caron** p. 130 (btm). **Getty Images:** AFP p. 124; Patrick AVENTURIER/Gamma-Rapho p. 180; Bloomberg p. 154 (btm); Amos Chapple/Lonely Planet Images p. 132 (top); Alireza Firouzi p. 113; gaiamoments p. 172 (top); Christian Kober/AWL Images p. 141 (inset); Eric Lafforgue p. 156; Quynh Anh Nguyen p. 129; Olive/Photodisc p. 128 (top); Brian J. Skerry/National Geographic Image Collection p. 155; George Steinmetz p. 168; Andrew Taylor/robertharding p. 173 (inset); YOSHIKAZU TSUNO/AFP p. 154 (top); VCG/Visual China Group p. 145 (top); Nik Wheeler p. 114; Fei Yang p. 144 (btm). **Chris Backe/worthygo.com** p. 179 (top). **Reuters:** Andrew Biraj p. 125; Amir Cohen p. 116 (btm); Thomas Peter p. 159 (btm). **Rehan Khan/Rex USA** p. 135 (btm). **Abedin Taherkonareh/EPA/Shutterstock** p. 112 (top). **UncorneredMarket.com** p. 135 (top).

Courtesy Photos Ehsan Abbasi p. 112 (btm); Ken Jeremiah p. 156.

Atlas Obscura **Contributors:** Chris Backe in South Korea p. 181 (top); Rachel Hallman p. 153; Nienna Mees p. 115; Sam Poucher p. 181 (btm); Jordan Samaniego p. 177 (btm); Anna Siri p. 169, 173 (inset background).

CANADA
Adobe Stock: PremiumGraphicDesign p. 269 (middle); Nadezda Razvodovska p. 276 (left). **Alamy Stock Photo:** 914

Collection p. 266 (top); All Canada Photos p. 268 (btm), p. 269 (top), p. 271 (middle right), p. 275, p. 276 (right), p. 281; Alt-6 p. 284 (top); blickwinkel p. 274; Yvette Cardozo p. 270 (btm), p. 282; Chronicle p. 277 (btm); Cosmo Condina p. 278 (btm); dpa picture alliance archive p. 272; iconim p. 265 (btm); INTERFOTO p. 273 (btm); Andre Jenny p. 267 (btm), p. 283 (btm); Lannen/Kelly Photo p. 284 (btm); Susan Montgomery p. 280 (btm); Radharc Images p. 273 (top). **Getty Images:** Bloomberg p. 281 (inset); DigitalGlobe/ScapeWare3d p. 265 (btm); Finn O'Hara/Photodisc p. 270 (middle); Carlos Osorio/Toronto Star p. 280(top); Chris Sheppard/500px p. 267 (top); Brian Summers/First Light p. 271(btm); xPACIFICA/National Geographic Image Collection p. 278 (top). **Rex USA:** Jon Freeman/Shutterstock p. 271 (top); REX Shutterstock p. 266 (btm). **LBNL/Science Source** p. 283 (top).

Courtesy Photos: Banff Indian Trading Post p. 264 (btm); Joshua Foer p. 271 (middle left); Keith Watson p. 279.

EUROPE
Adobe Stock: annexs2 p. 72 (top); Martina Berg p. 43 (btm); chicha1mk p. 84 (top); Rimas Jas p. 83; Igor Kisselev p. 101 (top); Jules Kitano p. 102 (btm); lenka p. 2; martialred p. 90 (btm); mino21 p. 95 (top); Nikokvfrmoto p. 51 (btm); Sved Oliver p. 97; Alexander Potapov 30 (top); sdp_creations p. 35 (top); skvoor p. 79 (top). **AGE Fotostock:** BEW Authors p. 85 (btm); DEA/A DAGLI ORTI p. 32; DOMELOUNKSEN p. 28 (btm); Patrick Forget p. 38 (top); Paula Mozdrzewska p. 85 (btm); Christine Noh p. 105; Werner Otto p. 44 (btm); Marco Scatagini p. 66 (btm). **Alamy Stock Photo:** Jon Arnold Images Ltd p. 50; ASK Images p. 41 (btm); Hans-Joachim Aubert p. 84 (btm); bilwissedition Ltd. & Co. KG p. 7 (btm); Michal Boubin p. 78 (btm); James Byard p. 17 (inset background); Christ Cathedral p. 17 (inset); Chronicle pp. 3, 15 (btm); CuboImages srl p. 57; deadlyphoto.com p. 104 (btm); Vincent Drago p. 86 (top); EmmePi Travel p. 96 (top); Julio Etchart p. 51 (top); Mark Eveleigh p. 69 (btm); Everett Collection Historical p. 14 (top); Everett Collection Inc p. 41 (top); Jerome Flynn p. 11; Peter Erik Forsberg p. 79 (btm); Leslie Garland Pictures p. 98 (btm); David Crossland/Germany Images p. 45; Michele and Tom Grimm p. 74 (btm); GL Archive p. 35 (btm); Robert Harding Picture Library Ltd p. 16, 100 (btm); HelloWorld Images p. 81; Hemis p. 63; Peter Horree p. 62; imageBROKER p. 33; ITAR-TASS News Agency p. 92; ITAR-TASS Photo Agency p. 90 (top); Ton Koene p. 65; Douglas Lander p. 107 (inset); Yannick Luthy p. 80; David Lyon p. 18; Paul Mayall Germany p. 43 (top); Jeff Morgan 05 p. 48 (btm); Eric Nathan p. 99; David Noton Photography p. 31 (top); OK-SANA p. 100 (btm); Pandarius p. 37; Sean Pavone p. 70 (top); Alan Payton pp. 9, 20 (left); Prisma by Dukas Presseagentur GmbH p. 72; Profimedia.CZ as p. 28 (top); QEDimages p. 5 (btm); Reciprocity Images Editorial p. 46; Bjarki Reyr MR p. 86 (btm); Peter Robinson p. 20 (right); Mauro Rodrigues p. 67; Denny Rowland p. 7 (btm); Adam Radosavljevic/Serbia Pictures p. 94 (btm); SPUTNIK p. 91 (top); Gerner Thomsen p. 93 (top); Urbanmyth p. 47; Ivan Vdovin p. 39; Guido Vermeulen-Perdaen p. 56 (btm); VPC Photo p. 58 (btm); Jasmine Wang p. 6; Sebastian Wasek p. 71; Rob Whitworth p. 69 (top); YAY Media AS p. 94 (top); Shau Hua Yi p. 82 (btm). **Scala/Art Resource, NY** p. 56

(top). **Natika/Can Stock Photo** p. 5 (top). **Christian Payne/Documentally** p. 10. **Getty Images:** William A. Allard/National Geographic pp. 60–61; Arctic-Images/Iconica p. 103; DEA/PUBBLI AER FOTO/DeAgostini p. 91 (btm); DEA/S. VANNINI p. 68; Geography Photos/Universal Images Group p. 76; Ton Honan/PA Images p. 17 (btm); Hulton Archive p. 12; ARIS MESSINIS/AFP p. 49 (btm); Daniel Mihailescu/AFP pp. viii, 88 (btm); nimu1956/E+ p. 59; Maryam Schindler/Picture Press p. 101 (btm); Science & Society Picture Library/SSPL p. 54 (top); Yulia-B/iStock p. 93 (btm). **Paul Léger:** p. 36. **Ricardo Ordonez/Reuters** p. 70 (btm). **ITV/Shutterstock/Rex USA** p. 21. **Science Source:** p. 108 (middle & btm). **Shutterstock.com:** Alexandr Makedonskiy p. 96 (btm); Gigi Peis p. 15 (top). **Temple dell'Umanita Association:** p. 52 (top). **Guido Alberto Rossi/Water Rights Images:** p. 53 (top).

Courtesy Photos: Stephen Birch p. 8; Jennifer Boyer p. 102 (top); © Hellbrunn Palace Administration p. 27; Paul Hyland p. 14 (btm); Collections Mundaneum p. 29; Jan Kempenaers/Courtesy of Little Breeze London p. 77 (all); Dawn Mueller p. 42; © Nick Palalino, 1999 p. 64 (btm); Kjartan Hauglid. © Emanuel Vigeland Museum/Bono p. 106. **Creative Commons:** The following image is used under a Creative Commons Attribution 3.0 United States License (https://creativecommons.org/licenses/by/3.0/us) and belongs to the following Wikimedia Commons user: Msemmett p. 7 (middle). The following image is used under a Creative Commons Attribution-ShareAlike 4.0 International License (https://creativecommons.org/licenses/by-sa/4.0) and belongs to the following Wikimedia Commons user: Romain Bréget p. 30 (btm). **Public Domain:** World Esperanto Association {{PD-old-70}} p. 26 (top).

Atlas Obscura **Contributors:** Scisetti Alfio p. 44 (top); Atlas Obscura p. 73; Michael Bukowski & Jeanne D'Angelo p. 87 (all); Christine Colby p. 13 (top); Peter Dispensa p. 40; Michelle Enemark p. 88 (top); Ophelia Holt p. 34; Courtesy of Nikolaus Lipburger, Kugelmugel p. 23; Michael Magdalena pp. 19, 49 (top); Roger Noguera p. 82 (top).

LATIN AMERICA
Adobe Stock: BuckeyeSailboat p. 436 (top); lacotearts p. 399 (middle); luciezr p. 412 (btm); martialred p. 439 (top); ocphoto p. 412 (top); Viktoria p. 446 (btm); VKA p. 433. **AGE Fotostock:** CSP_marconicouto p. 394 (btm); GUIZIOU Franck p. 430 (top). **Alamy Stock Photo:** age fotostock p. 431; Aurora Photos p. 423; Michele Burgess p. 405; Maria Grazia Casella p. 402 (btm); Cavan Images p. 443; dpa picture alliance archive p. 439 (btm); Exclusivepix Media p. 417 (inset); Julio Etchart p. 406 (top); Mark Eveleigh p. 435 (btm); Robert Fried p. 444 (top); Bernardo Galmarini p. 393; Mark Green p. 413; Robert Harding p. 418 (top); Hemis p. 442; Robert Adrian Hillman p. 429 (top); George H.H. Huey p. 440 (btm); imageBROKER pp. 394 (top), 396 (btm & top), 406 (btm), 407, 416; LatitudeStock p. 449 (btm); Eric Laudonien p. 438 (top); LOOK Die Bildagentur der Fotografen GmbH p. 447 (btm); Martin Norris Travel Photography p. 429 (btm); John Mitchell p. 441; Efrain Padro p. 448 (top); Dipak Pankhania p. 411 (btm); Wolfi Poelzer p. 438 (btm); MARIUSZ PRUSACZYK p. 432; Carrie Thompson p. 422; Tom Till p. 408 (top); travelstock44 p. 392

(top); Michael Ventura p. 448 (btm); Fabrice VEYRIER p. 420; Westend61 GmbH p. 446 (top); Xinhua p. 419 Zoonar GmbH p. 426. **Dave Bunnell/Caters News** p. 427 (btm). **Kiki Deere/kikideere.com** p. 403. **Getty Images:** AFP p. 430 (btm); cdwheatley/iStock p. 447 (top); DEA/G.SOSIO/De Agostini p. 437 (btm); diegograndi/iStock p. 434; Dan Herrick/Lonely Planet Images p. 435 (btm); Richard Maschmeyer/robertharding p. 414 (inset background); MSeis/iStock p. 410; Carsten Peter/Speleoresearch & Films/National Geographic p. 427 (top); rchphoto/iStock p. 415 (btm); Henryk Sadura p. 414 (inset); Frans Sellies p. 440 (top); Topical Press Agency/Hulton Archive p. 398 (btm); Rosanna U/Image Source p. 415 (top right); ullstein bild p. 399 (top); Uwe-Bergwitz/iStock p. 415 (top left); Edson Vandeira p. 404 (top); xeni4ka/iStockEditorial p. 417 (inset background). **Edgard Garrido/Reuters** p. 425 (btm). **Sofia Ruzo** p. 411 (top).

Creative Commons: The following images are used under a Creative Commons Attribution 3.0 United States License (https://creativecommons.org/licenses/by/3.0/us) and belong to the following Wikimedia Commons users: Shaddim p. 437 (top); Yurileveratto p. 398 (top).

Atlas Obscura **Contributors:** John Allen p. 421 (btm); cgracemo p. 397; Each Day I Dye p. 424 (btm); Mark Harrison p. 408 (btm); Allan Haverman pp. 391, 395; Courtesy of Kirk Horsted p. 450; Jason Decaires Taylor p. 436 (btm).

OCEANIA
Adobe Stock: Wolfgang Berroth p. 232; Tommaso Lizzul p. 236 (btm). **AGE Fotostock:** Jean-Marc La-Roque p. 243 (pineapple); Keven O'Hara p. 247 (btm); Joe Dovala/WaterFra p. 255. **Alamy Stock Photo:** Bill Bachman p. 244 (top); Robert Bird p. 248 (top); chris24 p. 242 (guitar); Steve Davey Photography p. 261; Philip Game p. 254; Iconsinternational.com p. 243 (Big Ned); National Geographic Image Collection p. 243 (koala); Martin Norris Travel Photography p. 238 (btm); Christine Osborne Pictures p. 242 (banana); Aloysius Patrimonio p. 238 (top); Stefano Ravera p. 242 (mango); Andrew Sole p. 248 (btm); Jack Sullivan p. 243 (prawn); John White Photos p. 243; Big Galah; Wiskerke p. 242 (merino); Ian Woolcock p. 239; Zoonar GmbH p. 242 (crocodile). **Getty Images:** The Asahi Shimbun p. 253; Ben Bohane/AFP p. 262; Don Kelsen/Los Angeles Times p. 247 (top); Desmond Morris Collection/UIG p. 251 (btm); Mitch Reardon/Lonely Planet Images p. 237 (top); Oliver Strewe/Lonely Planet Images p. 258; Michele Westmorland/The Image Bank p. 257; Whitworth Images p. 245 (background).

Steven David Miller/Nature Picture Library p. 237 (btm). **Clive Hyde/Newspix** p. 233. **Martin Rietze/MRIETZE.COM** p. 249; **Patrick Horton/Shutterstock.com** p. 252.

Courtesy Photos: Patrick J. Gallagher p. 241 (btm); David Hartley-Mitchell p. 246 (top); Malcolm Rees p. 250 (btm); **Creative Commons:** The following image is used under a Creative Commons Attribution 3.0 United States License (https://creativecommons.org/licenses/by/3.0/us) and belongs to the following Wikimedia Commons user: Peter Campbell p. 240. The

following image is used under a Creative Commons Attribution-ShareAlike 2.5 Generic License (https://creativecommons.org/licenses/by-sa/2.5) and belongs to the following Wikimedia Commons user: Peter Halasz p. 245 (inset). **Public Domain:** NASA p. 251 (top).

Atlas Obscura **Contributors:** Céline Meyer p. 246 (btm); Amanda Olliek p. 241 (top).

USA
Adobe Stock: airindizain p. 381 (btm); davidevison p. 369; Diverser p. 359 (btm); Dominic p. 310 (top); Imagewriter p. 326 (top); kovalto1 p. 340 (btm); opin47 p. 300 (top); Sean Pavone Photo p. 350 (top); Alexander Potapov p. 321 (top); sljubisa p. 323 (top); tanais p. 354 (btm); valdezrl p. 312 (top); Vector Tradition p. 334 (btm). **Alamy Stock Photo:** Irene Abdou p. 371 (top); Nathan Allred pp. 314–315 (background); Alan Bozac p. 329 (btm); Pat Canova p. 346 (top); Chronicle p. 382 (top); Cultura RM p. 372 (inset background); D Guest Smith p. 329 (top); Danita Delimont p. 322 (btm), p. 367 (top); Design Pics Inc p. 388; Don Despain p. 317 (btm); dpa picture alliance p. 315 (top); Randy Duchaine p. 375 (btm); Education & Exploration 3 p. 348 (top); Richard Ellis p. 355 (btm); Everett Collection Historical p. 364 (btm); Franck Fotos p. 332 (btm); Zachary Frank p. 301 (top); Oliver Gerhard p. 373; Joseph S. Giacalone p. 298 (inset); Michelle Gilders p. 320 (top); Jay Goebel p. 317 (top); Robert Harding p. 311; Blaine Harrington III p. 307 (top right); Clarence Holmes Photography p. 365; Janet Horton p. 302 (top); Independent Picture Service p. 340 (top); Inge Johnsson p. 308; Richard Levine p. 370 (top); Ilene MacDonald p. 307 (top left); Mary Evans Picture Library p. 350 (btm); National Geographic Image Collection p. 310 (btm); Luc Novovitch p. 305 (top); B. O'Kane p. 290 (top); Edwin Remsberg p. 362 (top); RMF stock p. 287 (background); RosalreneBetancourt 7 p. 339 (top); Philip Scalia p. 360; SCPhotos p. 387; Jim West p. 323 (btm); Zanna Pesnina p. 294; ZUMA Press, Inc. pp. 287 (inset), 288, p. 316 (top), 347, 361, 379 (btm); **Brad Andersohn** p. 295 (inset). **AP Photo:** Wilfredo Lee 348 (btm); Douglas C. Pizac p. 319 (top). **Boston Athenæum** p. 379 (top). **Brendan Donnelly** p. 291 (middle left, center & right). **Zach Fein** p. 338. **Getty Images:** Allentown Morning Call/MCT p. 370 (btm); John Ashmore p. 336; Raymond Boyd p. 351; John B. Carnett/Bonnier Corporation p. 357 (middle); Bryan Chan/Los Angeles Times p. 293 (btm); Scott Gries p. 364 (btm); Paul Hawthorne p. 333 (top); Kevin Horan/The LIFE Images Collection p. 333 (btm); Karen Kasmauski p. 378; MagicDreamer/iStock p. 318 (btm); Keith Philpott/The LIFE Images Collection p. 337 (btm); Rischgitz/Stringer p. 368; Joel Sartore/National Geographic p. 358 (top); Dieter Schaefer p. 309; Scott T. Smith/Corbis Documentary p. 314 (inset); Universal History Archive p. 299.

Ahenredon/Shutterstock.com p. 383. **Skeletons: A Museum of Osteology/Skulls Unlimited International, Inc.** p. 325 (all).

Courtesy Photos: Richard Reames-Arborsmith Studios p. 300 (btm); Clark R. Arrington p. 356; Allison Meier/flickr.com p. 326 (btm); Ryan Cheung/flickr.com p. 353 (btm); © 2016 Google, Imagery © 2016 DigitalGlobe, Landsat, U.S. Geological Survey p. 292 (btm); Paul Hall p. 327 (top); Megan

INDEX

Atlas Obscura, publisher of the definitive guide to the world's hidden wonders, is a travel company defined by storytelling. It was founded in 2009 by Dylan Thuras and Joshua Foer with a mission to inspire wonder and curiosity about the world. Our editorial team reports on hidden places, incredible histories, scientific marvels, and gastronomic wonders. Atlas Obscura creates hundreds of unique global trips and local experiences every year, bringing our community with us to visit the world's most unusual places and try the world's most extraordinary foods.